D1272880

Reliability Engineering

Contributors

M. B. Adelson	A. Drummond	E. J. Rappaport
D. O. Baechler	G. T. Harrison	J. A. Scanga
H. S. Balaban	R. C. Horne	N. J. Scarlett
L. W. Ball	H. R. Jeffers	D. W. Sharp
G. J. Blakemore	H. R. Leuba	D. E. Van Tijn
F. D. Callison	A. E. Martin	W. H. von Alven
H. Dagen	P. R. Oyerly	E. L. Welker
A. P. Dinsmore	M. M. Ragsdale	

Reliability Engineering

Prepared by the Engineering and Statistical Staff
of ARINC Research Corporation

Edited by
WILLIAM H. VON ALVEN
Staff Consultant

ARINC Research Corporation

A subsidiary of Aeronautical Radio, Inc.

Prentice-Hall, Inc. Englewood Cliffs, New Jersey

PRENTICE-HALL INTERNATIONAL, INC. London
PRENTICE-HALL OF AUSTRALIA, PTY., LTD. Sydney
PRENTICE-HALL OF CANADA, LTD. Toronto
PRENTICE-HALL OF INDIA (PRIVATE) LTD. New Delhi
PRENTICE-HALL OF JAPAN, INC. Tokyo

Current printing (last digit):

12   11   10   9   8   7   6   5   4   3

© 1964 by ARINC Research Corporation.

Library of Congress Catalog Card Number 64-12217
Printed in the United States of America
C77312

Preface

"Reliability," only a few years ago, was a qualitative and somewhat equivocal term, and the subject was accorded scant attention either in technical education or in the design, manufacture, and use of systems. Now, reliability and the concepts of maintainability, system effectiveness, and system worth become inextricably involved in all major phases of the genesis and use of commercial, military, and space systems, including design, development, procurement, production, operation, and maintenance. Where emphasis has been placed on these concepts, dramatic improvements often have been obtained, in the form of reduction in costs, superior workmanship, a higher probability of satisfactory operation, and better customer-supplier relationships.

The trend toward inclusion of increasingly rigorous reliability and maintainability requirements in contracts intensifies the importance of specialized training for contractors and users. The contractor must demonstrate that his product meets specific performance, reliability, and maintainability requirements, while the customer must be able to specify and monitor appropriate test programs and to interpret test reports. Maintenance organizations are inescapably concerned with reliability and maintainability concepts, since they affect the estimation of spares requirements, the adequacy of repairs, and the feedback of repair information for use in system modification or improvement.

With the recognition of reliability and maintainability as vital factors in the development, production, operation, and maintenance of today's complex systems, greater emphasis must be placed on the training of management, engineering, procurement, and administrative personnel in the application of these concepts. The material in this volume has been prepared as a contribution to this important educational work.

The first edition of this text was prepared in 1960 for a specialized training program in reliability engineering which was presented to personnel of the Weapons Guidance Laboratory at Wright-Patterson Air Force Base, Ohio. Rapid growth in the development of techniques associated with reliability, maintainability, and system effectiveness has necessitated several revisions to meet the needs of hundreds of engineers who have participated in subsequent training programs conducted by ARINC Research Corporation for many industrial organizations and government agencies.

Reliability Engineering begins with a discussion of system effectiveness concepts, relating operational readiness, mission reliability, and design adequacy. Chapters 2 through 5 present basic mathematical concepts and data analysis techniques which are used in quantifying reliability, maintainability, and system effectiveness. Chapters 6 through 11 describe and illustrate several useful system analysis techniques, including feasibility, reliability functional analyses (such as allocation, reliability models for serial, redundant, and multimodal systems, and reliability prediction for electronic and propulsion systems); methods for maintainability analysis; and methods for quantification of system effectiveness. Chapter 12 presents engineering concepts and techniques of established value in designing for reliability and maintainability. Chapter 13 lays the groundwork for understanding of reliability and quality guarantees in part and equipment specifications employing sampling procedures, while Chapter 14 outlines procedures for reliability demonstration tests. Chapter 15 discusses the relationships between cost and effectiveness, and presents an elementary discussion on costing of a support operation. (The concepts discussed in Chapter 11 and Chapter 15 form the basis for cost and effectiveness analyses which are useful in comparing new design alternatives, optimizing a given design, evaluating field operations of a given system, or evaluating proposed field modifications before they are installed on a fleet-wide basis.) Chapter 16 concludes with a discussion of well-proven reliability management concepts.

Appendix A lists government documents establishing reliability and maintainability requirements. The listing is reprinted through the courtesy of the Martin Marietta Corporation.

This text has been addressed to senior-level engineering students and to practicing engineers. Adequate material is presented for a lecture program for 30 classroom hours; and for as much as 45 classroom hours if problem sets are worked and discussed. Answers to all problems requiring numerical results appear in Appendix B.

Grateful acknowledgment is made to each member of the technical staff of ARINC Research Corporation who contributed to this text; to our students and course sponsors whose many suggestions were incorporated in this edition; to Dr. L. W. Ball, of The Boeing Company, whose contributions and philosophy are reflected in Chapter 16, *Reliability Engineering, Programming, and Management;* to Dr. E. L. Welker for his contributions and guidance in the preparation of this text; and to C. R. Knight, Vice President of ARINC Research Corporation, for his interest in, and support of, this project. Thanks are due, also, for the invaluable assistance given by C. J. Hedetniemi, Manager of the ARINC Research Reports Department, and the editorial, drafting, and typing personnel under his direction.

William H. von Alven
Editor

Table of Contents

System Effectiveness*

1.1 Introduction

The ultimate output of any system is the performance of some intended function. This function is frequently called the *mission* in the case of a weapons system. For other types of systems, it may be described by some system-output characteristic, such as satisfactory message transmission in a communication system or weather identification in an airborne weather radar. The term often used to describe the over-all capability of a system to accomplish its mission is *system effectiveness*. A more precise definition of this expression will be given later, but for the present it is sufficient to observe that system effectiveness relates to that property of system output which was the real reason for buying the system—namely, the carrying out of some intended function. If the system is effective, it carries out this function well. If it is not effective, attention must be directed to those system attributes which are deficient.

Of the major attributes determining system effectiveness, the one that has received the most thorough and systematic study in recent years is reliability. One reason for this is that reliability problems, particularly in complex

---

* The material in this chapter is adapted from *Concepts Associated with System Effectiveness* by E. L. Welker and R. C. Horne, Jr., ARINC Research Monograph No. 9 (July 15, 1960).

weapons systems, became so acute in the period after World War II that there was no real alternative to giving them immediate, concentrated attention. Another reason is that, among the more important and interrelated system attributes, reliability is the most susceptible to investigation out of context. Even in reliability studies, however, there comes a time when the investigation can no longer be pursued unless it is extended to include related characteristics that influence system effectiveness.

1.2 Factors Influencing System Effectiveness

Equipments of New Design

A typical history of the development of a new equipment would reveal a number of interesting steps in the progression from original concept to acceptable production model. These steps are particularly marked if the equipment represents a technical innovation—i.e., if it pushes the state of the art by introducing entirely new functions or by performing established functions in an entirely new way. Starting with a well-defined operational need, the research scientist, designer, reliability engineer, statistician, and production engineer all combine their talents to execute a multitude of operations leading to one ultimate objective: the production of an equipment that will perform as intended, with minimum breakdowns and maximum speed of repair. All this must be done at minimum cost, and usually within an accelerated time schedule.

These program requirements are severe, to say the least. In order to meet them, many compromises are required. One of the first of these compromises is often a sharp curtailment in the basic research time allotted to the job of proving the feasibility of the new design. After only brief preliminary study, a pilot model of the equipment is built. With luck, it will work; but it is likely to be somewhat crude in appearance, too big and too heavy, not well-designed for mass production, subject to frequent failure, and difficult to repair. Indeed, at this early stage in the program, it is quite possible that the first model might be incapable of working if it were taken out of the laboratory and subjected to the more severe stresses of field operation, whether this be military or civilian. By the time this situation is corrected, the development program will have included many design changes, part substitutions, reliability tests, and field trials, eventually culminating in a successful operational acceptance test.

Usually, it is not until the equipment appears to have some chance of reaching this ultimate goal of acceptance that attention is focused on reduction of the frequency of failure, thus providing the impetus for a serious reliability effort. Experience has shown that this is unfortunate. Ideally, such an effort should begin immediately after the feasibility study, because some problems can be eliminated before they arise, and others can be solved at an early development stage, when design modifications can be effected most easily and economically. Even with this early start, reliability will continue to be a primary problem in new equipment, especially when it is of novel design. Early neglect of reliability must be compensated by extraordinary efforts at a later period, because an equipment simply is not usable if it fails too frequently to permit suitable reliance on the likelihood of its operating when needed. Since such early neglect has been common in the past, reliability has received strong emphasis in the research designed to bring equipment performance characteristics up to satisfactory levels.

The description just given is generally applicable to the development of radically new equipment. However, when attention is directed to equipment in everyday use or to new equipment built predominantly on standard design principles and from well-tested parts, it becomes evident that effectiveness is dependent not only on performance capabilities and reliability but also on a number of other factors, including operational readiness, availability, maintainability, and repairability. Definitions for these concepts will be given later; at present, it suffices to observe that they are all so interrelated that they must be viewed together and discussed, not as separate concepts, but within the framework of the over-all system to which they contribute.

Interrelationships Among Various System Properties

The discussion above implies that it is probably not practicable to maximize all of the desirable properties of a system simultaneously. Clearly, there are "trade-off" relationships between reliability and system cost, between maintainability and system cost, between reliability and maintainability, and between many other properties. It would be most helpful to have a numerical scale of values for each of the several properties and a multidimensional plot or chart showing the interrelationship among those values. Before such relationships can be obtained, it is first necessary to define in a precise and quantitative manner the properties with which we are concerned. The following outline is intended to show some of the factors which must be considered:

A. *System Performance* (*Design Adequacy*)

   1. Technical Capabilities

     (a) Accuracy
     (b) Range
     (c) Invulnerability to countermeasures
     (d) Operational simplicity

2. Possible Limitations on Performance

   (a) Space and weight requirements

   (b) Input power requirements

   (c) Input information requirements

   (d) Requirements for special protection against shock, vibration, low pressure, and other environmental influences

B. *Operational Readiness*

  1. Reliability

    (a) Failure-free operation

    (b) Redundancy or provision for alternative modes of operation

  2. Maintainability

    (a) Time to restore failed system to satisfactory operating status

    (b) Technical manpower requirements for maintenance

    (c) Effects of use cycle on maintenance (Can some maintenance be performed when operational use of the system is not required?)

  3. Logistic supportability

C. *System Cost*

  1. Development cost, and particularly development time, from inception to operational capability

  2. Production cost

  3. Operating and operational support costs

The optimization of *system effectiveness* by judicious choice or balancing of the conflicting requirements suggested in the list above is an extremely complex problem. There is a high degree of interaction among the factors which enter into the consideration of this problem. One of the first areas which must be attacked is the defining of terms for precise communication among individuals or groups working in the field. A survey of the literature reveals astonishing semantic problems. The word *maintainability*, as used by some, is synonymous with *availability*, as used by others. For some groups, *reliability* means what is called *system effectiveness* here. There are many other conflicting interpretations, as even a casual survey of maintainability literature will quickly show.

The interdependence of the concepts influencing system effectiveness has naturally contributed to a certain amount of confusion in the applicable terminology. Even the word *reliability* is used in different ways by different people, although it is probably subject to less variation in meaning than any of the other concept names to be discussed. The word *maintainability*, for example, is used in an extremely wide range of meanings. Some restrict its use to the distribution of active repair times; others interpret it as including practically everything, in which case it becomes more like the concepts of *effectiveness* or *operational readiness* as they are defined later in this chapter.

Completely uniform usage is probably an unattainable goal, but the necessity for some uniformity is obvious. Precise communication is impossible if a single statement has a different meaning to each individual. In the following section, definitions of the concepts influencing system effectiveness are presented. An effort has been made to choose terms that (1) are in accordance with the usage already accepted by a considerable number of the people working in this field, and (2) permit clear discrimination between concepts and thus facilitate study of their interrelationships.

1.3 Definitions of Concepts

System Effectiveness

In the introductory paragraphs, mention was made of the inclusiveness of the concept of system effectiveness. It was described rather loosely as the ability of the system to do the job for which it was purchased. A more formal definition is as follows:

1. *System effectiveness is the probability that the system can successfully meet an operational demand within a given time when operated under specified conditions.*

Effectiveness is obviously influenced by the way the equipment is designed and built. However, just as critical are the ways in which the equipment is used and maintained. In other words, system effectiveness can be materially influenced by the design engineer, the production engineer, the operator, and the maintenance man. It can also be influenced by the logistic system that supports the operation, and by the administration through personnel policy, rules governing equipment use, fiscal control, and many other administrative policy decisions.

To apply the general definition to a one-shot device such as a missile, it need only be modified to the following:

2. *System effectiveness is the probability that the system (missile) will operate successfully (kill the target) when called upon to do so under specified conditions.*

The major difference between these two definitions lies in the fact that, in Definition 2 (for a one-shot device), time is relatively unimportant. In the first, more general definition, operating time is a critical element and effectiveness is a function of time. Another difference is that the first definition provides for the repair of failures, both at the beginning of the time interval (if the equipment is inoperable then) and also during the operating interval (if a failure occurs after a successful start); the second definition assumes no repair.

Both definitions imply that the system fails (1) if it is in an inoperable condition when needed, or (2) if it is operable when needed but fails to complete the assigned mission successfully. The expression "specified conditions" implies that system effectiveness must be stated in terms of the requirements placed upon the system, indicating that failure and use conditions are related. As the operational stresses increase, failure frequency may also be expected to increase.

If continuous operation is required, any cessation due to failure or scheduled maintenance reduces system effectiveness. If the demands on the equipment are such that an on-off use cycle provides significant free time for maintenance, system effectiveness is enhanced. Maintenance of a state of readiness on a continuous basis may (or may not) increase the percentage of equipments which reach an inoperable condition prior to demand for use. If it does, removal from the readiness state for a portion of time each day might increase effectiveness.

It should also be mentioned that operational requirements sometimes exceed design objectives. A decrease in target vulnerability can result in a decrease in system effectiveness. Surface-to-air missiles designed to be effective against subsonic aircraft can well have almost no system effectiveness when called upon to engage supersonic targets.

Reliability

The most commonly accepted definition of reliability is given below:

Reliability is the probability that a system will perform satisfactorily for at least a given period of time when used under stated conditions.

A *reliability function* is this same probability expressed as a function of the time period. Thus, reliability relates to the frequency with which failures occur. Here "failure" means "unsatisfactory performance," usually representing a judgment of an operator or a maintenance man. This does not preclude the possibility of clear-cut failure, such as complete inoperability, in which case judgment does not enter at all.

Mission Reliability

Mission reliability is defined as the probability that a system will operate in the mode for which it was designed for the duration of a mission, given that it was operating in this mode at the beginning of the mission.

Mission reliability thus defines the probability of nonfailure of the system for the period of time required to complete a mission. The probability is a point on the reliability function corresponding to a time equal to the mission length. All possible redundant modes of operation must be considered in describing reliability, mission reliability, and system effectiveness.

Operational Readiness and Availability

The capability of a system to perform its intended function when called upon to do so is often referred to by either of two terms: operational readiness and availability. It is the emphasis on the phrase "when called upon" that differentiates this concept from the more general one of system effectiveness. This emphasis restricts attention to probability at a point in time rather than over an interval of time, the latter being descriptive of system effectiveness. It should be noted that sometimes this interval can be extremely long. There is an additional major difference. System effectiveness includes the built-in capability of the system—its accuracy, power, etc. Operational readiness excludes these native system characteristics; i.e., it excludes the ability of the system to do the intended job and includes only its readiness to do it at a particular time.

In order to differentiate between two separate and useful concepts, it is well to formalize a distinction between the terms—operational readiness and availability. It has been apparent in past discussions of system effectiveness that the terms are used by some to represent different concepts but are used almost synonymously by others. Both concepts relate the operating time between failures to some longer time period; they differ in what is to be included in this longer time period. Availability is defined in terms of operating time and down time, where down time includes active repair time, administrative time, and logistic time. On the other hand, operational readiness is defined in terms of all of these times, and, in addition, includes both free time and storage time—i.e., all calendar time. Availability and operational readiness are defined as follows:

> *The availability of a system or equipment is the probability that it is operating satisfactorily at any point in time when used under stated conditions, where the total time considered includes operating time, active repair time, administrative time, and logistic time.*

> *The operational readiness of a system or equipment is the probability that at any point in time it is either operating satisfactorily or is ready to be placed in operation on demand when used under stated conditions, including stated allowable warning time. Thus, total calendar time is the basis for computation of operational readiness.*

Design Adequacy

An additional comment with respect to operational readiness and availability is required to emphasize a restriction mentioned above. These two concepts exclude from consideration the built-in capability of a system to do the job for which it is being used. Thus, misapplication of a system is entirely excluded from measurements of either availability or operational readiness but must be reflected in measurement of system effectiveness.

As an extreme example, machine guns have often been used against attacking aircraft. Obviously this is not the designer's intent: the machine gun was not designed to provide antiaircraft fire power. In such an emergency application, system effectiveness is certainly low, but this circumstance has nothing to do with an evaluation of the availability or operational readiness of the machine gun in field use. As another example, the differences between 75mm and 90mm tank guns in range, accuracy, and penetrating power against enemy tanks have a significant bearing on measurements of system effectiveness but are irrelevant to evaluations of their operational readiness and availability.

The characteristic discussed in the preceding paragraph can be identified by the term, system design adequacy.

System design adequacy is the probability that a system will successfully accomplish its mission, given that the system is operating within design specifications.

The design may include alternative modes of operation, which are equivalent to built-in automatic repair, usually with allowable degradation in performance. These alternative modes of operation are, of course, included in the definition of system design adequacy. The probability itself is a function of such variables as system accuracy under the conditions of use, the mission to be accomplished, the design limits, system inputs, and the influence of the operator.

Repairability

Repairability is defined as the probability that a failed system will be restored to operable condition in a specified active repair time.

It is useful to express this probability in two forms, the density function and the cumulative function. These are called the *active repair time density function* and the *repairability function*, respectively. The repairability function expresses the probability that the active repair time does not exceed any given total time.

Maintainability

Maintainability is defined as the probability that a failed system is restored to operable condition in a specified down time.

This is directly analogous to repairability. The difference is merely that maintainability is based on total down time (which includes active repair time, logistic time, and administrative time), while repairability is restricted solely to active repair time.

The analogy holds with respect to the associated functions as well. The *maintainability function* is the cumulative probability that the failed system is restored to operable condition in not more than a specified down time, expressed as a function of this down time. The corresponding density function is called the *maintenance time density function*.

Serviceability

Intuitively it would seem that some term should be used to represent the degree of ease or difficulty with which an equipment can be repaired. The term *serviceability* has been selected for this concept. Serviceability has a strong influence on repairability, but the two are essentially different concepts. Serviceability is an equipment design characteristic while repairability is a probability involving certain categories of time.

Although the definition of serviceability is stated in a manner that suggests a quantitative concept, it is often necessary to accept a qualitative evaluation of the serviceability of an equipment. The definition as given does accentuate the idea that comparison of equipments can yield a conclusion that Equipment A is more serviceable than Equipment B. Actually, this kind of conclusion may be entirely satisfactory, for the numerical evaluation can be made when repairability is measured. That is to say, the better the serviceability, the shorter the active repair time. Hence, repairability is a reflection of serviceability even though the two concepts are quite distinct.

Serviceability is dependent on many hardware characteristics, such as engineering design, complexity, and the number and accessibility of test points. These characteristics are under engineering control, and poor serviceability traceable to such items is the responsibility of design engineers. However, many other characteristics which can cause poor serviceability are not directly under the control of the design engineers. These include lack of proper tools and testing facilities, shortage of work space in the maintenance shop, poorly trained maintenance personnel, shortage of repair parts, and other factors that can increase the difficulties of maintenance.

Intrinsic Availability

It is also useful to define another term, intrinsic availability.

The intrinsic availability of a system or equipment is the probability that it is operating satisfactorily at any point in time when used under stated conditions, where the time considered is operating time and active repair time.

Thus, intrinsic availability excludes from consideration all free time, storage time, administrative time, and logistic time. As the name indicates, intrinsic availability refers primarily to the built-in capability of the system or equipment to operate satisfactorily under stated conditions.

The effect of these definitions is essentially to allow realistic assignment of responsibility in case an unsatisfactory situation exists. If an improvement in intrinsic availability is required, responsibility can properly be assigned to the design and production engineers—assuming, of course, that the operating conditions are compatible with design specifications. On the other hand, if availability is unsatisfactory and improvement in intrinsic availability is not indicated, the responsibility is properly placed on the commander or civilian administrator to effect the required improvement by reducing administrative and logistic delay. If neither of these steps is indicated, and operational readiness is not satisfactory, improvement depends on changes in free time and storage time, implying more efficient use of the system equipment.

1.4 Definitions of Time Divisions

Time is of fundamental importance in the quantification of the concepts which were defined in the previous section, for it is this factor which permits the attributes to be measured rather than described in merely qualitative terms. The usual measures of time—the year, month, day, and hour—form the basis for the computation of reliability. However, there are so many ways of delineating these intervals that the method adopted in each investigation must be carefully developed in order to provide the desired results.

In general, the interval of interest is the total calendar time during which the system is in use. As shown in Figure 1.1, this interval may be divided into available time and unavailable time. *Available time* is that time during which the system is available for use by the intended user; *unavailable time* is that time during which the system is being supplied, repaired, restored, or kept in condition for its intended use. Available time may be further broken down into *usage time* (during which the carrier of the system is employed for its intended tactical purpose) and *ready time* (during which the carrier is available for use but is not in tactical service). Usage time may be further sub-divided into operate time and standby time. *Operate time* is that time during which any portion of the system is fully energized. During *standby time,* any portion or all of the system is partially energized, but no portion of the system is fully energized.

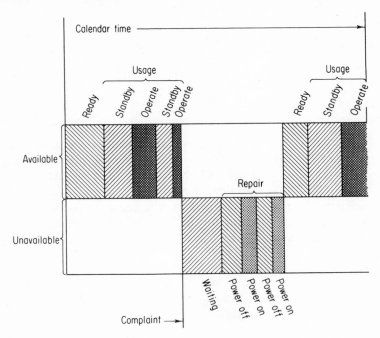

Fig. 1.1 Principal divisions of calendar time.

Insofar as unavailable time is concerned, the interval of primary interest is *repair time*—that time in which maintenance is being done or manpower is otherwise being expended on the system. The other component of unavailable time is *waiting time.* This term applies primarily to time lost for administrative and logistic reasons, such as time until maintenance personnel have an opportunity to start repair work on an item. A period of waiting time can occur during an interval of repair time—for example, when parts required for repair are not available. The foregoing general categories describe the manner in which the over-all period of observation is divided.

The primary element in this discussion has been a time interval, with differentiation between time intervals being based on the state of the system: whether or not it is operating; and if it is not, what, if anything, is being done to it during the interval in question. Thus, there are really two time-division criteria, one being the equipment state of operability and the other being the demand for its use. These can be written as follows:

Equipment State of Operability

1. Operable
2. Inoperable

   (a) Administrative time
   (b) Logistic time
   (c) Active repair time

Use Demand

1. Use is required
2. Use is not required

   (a) Time in storage, as a spare
   (b) Free time, as off hours when no operation is in process.

This unavoidable dual time breakdown is responsible for many of the problems involved in defining concepts and determining interrelationships. If continuous operation were required, the dual breakdown would disappear because there would be no free time or storage time.

In order to use the probability definitions discussed in the previous section, the following definitions are given for the time elements which must be considered in an evaluation of system effectiveness.

Operating Time

Operating time is the time during which the system is operating in a manner acceptable to the operator.

This includes the time when the operator may be somewhat dissatisfied with the manner of operation, but is not sufficiently dissatisfied to shut the system down and request repair action.

Down Time

Down time is the total time during which the system is not in acceptable operating condition.

Down time can, in turn, be subdivided into a number of categories:

Active repair time is that portion of down time during which one or more technicians are working on the system to effect a repair. This time includes preparation time, fault-location time, fault-correction time, and final checkout time for the system.

Logistic time is that portion of down time during which repair is delayed solely because it is necessary to wait for a replacement part or other subdivision of the system.

Administrative time is that portion of down time not included under active repair time and logistic time. This includes both necessary administrative activities and unnecessarily wasted time. These two times are grouped for convenience only, and not with any implication that administrative time is wasted.

Free Time

Free time is time during which operational use of the system is not required. This time may or may not be down time, depending on whether or not the system is in operable condition.

Storage Time

Storage time is time during which the system is presumed to be in operable condition, but is being held for emergency—i.e., as a spare.

1.5 Relationships Among Time Elements

The previous section described in a general way the framework within which a variety of concepts must be placed, and also proposed reasonably concise definitions of these concepts and the time elements associated with them. Additional insight into the nature of these equipment properties can be achieved only by more careful presentation of their interrelationships. As an aid in doing this, Figure 1.2 has been prepared as a chart relating the terms that have been defined and the time divisions that have been described. This approach will also assist in clarifying some of the mathematical relationships which must be developed in order to provide quantitative measurement. For convenience and quick reference, the concept and time-element definitions previously given are repeated in summary form adjacent to the chart in Figure 1.2. In the interest of simple terminology, the chart has been phrased for a system. It must be remembered that the word "equipment" could have been used in place of the word "system." However, it is always possible to define the equipment as a system. Hence, the terminology is entirely general and the treatment to follow will be made as general as possible.

Relationships Among Time Intervals

The purpose of Figure 1.2 is to indicate graphically how the various time intervals are combined to produce the different measures of system characteristics, which in turn are combined and included under the general heading of

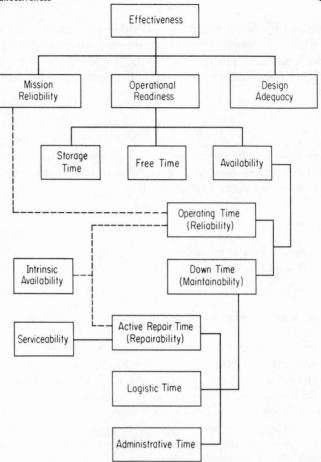

Fig. 1.2 Concepts associated with system effectiveness.

Definitions of Concepts

System effectiveness is the probability that the system can successfully meet an operational demand within a given time when operated under specified conditions.

System effectiveness (for a one-shot device such as a missile) is the probability that the system (missile) will operate successfully (kill the target) when called upon to do so under specified conditions.

Reliability is the probability that the system will perform satisfactorily for at least a given period of time when used under stated conditions.

Mission reliability is the probability that, under stated conditions, the system will operate in the mode for which it was designed (i.e., with no malfunctions) for the duration of a mission, given that it was operating in this mode at the beginning of the mission.

Operational readiness is the probability that, at any point in time, the system is either operating satisfactorily or ready to be placed in operation on demand when used under stated conditions, including stated allowable warning time. Thus, total calendar time is the basis for computation of operational readiness.

Availability is the probability that the system is operating satisfactorily at any point in time when used under stated conditions, where the total time considered includes operating time, active repair time, administrative time, and logistic time.

Intrinsic availability is the probability that the system is operating satisfactorily at any point in time when used under stated conditions, where the time considered is operating time and active repair time.

Design adequacy is the probability that the system will successfully accomplish its mission, given that the system is operating within design specifications.

Maintainability is the probability that, when maintenance action is initiated under stated conditions, a failed system will be restored to operable condition within a specified total down time.

Repairability is the probability that a failed system will be restored to operable condition within a specified active repair time.

Serviceability is the degree of ease or difficulty with which a system can be repaired.

Definitions of Time Categories

Operating time is the time during which the system is operating in a manner acceptable to the operator, although unsatisfactory operation (or failure) is sometimes the result of the judgment of the maintenance man.

Down time is the total time during which the system is not in acceptable operating condition. Down time can, in turn, be subdivided into a number of categories such as active repair time, logistic time, and administrative time.

Active repair time is that portion of down time during which one or more technicians are working on the system to effect a repair. This time includes preparation time, fault-location time, fault-correction time, and final check-out time for the system, and perhaps other subdivisions as required in special cases.

Logistic time is that portion of down time during which repair is delayed solely because of the necessity for waiting for a replacement part or other subdivision of the system.

Administrative time is that portion of down time not included under active repair time and logistic time.

Free time is time during which operational use of the system is not required. This time may or may not be down time, depending on whether or not the system is in operable condition.

Storage time is time during which the system is presumed to be in operable condition, but is being held for emergency—i.e., as a spare.

effectiveness. Details of the numerical measurements will be discussed in later chapters. It is sufficient to note here that, for most concepts, it is not enough to think in terms of averages. Instead, it is necessary to describe the item in question by one or more functions such as density and cumulative distribution. This parallels the reliability approach, which properly includes many measures such as time-to-failure density functions, reliability and unreliability functions, variances, coefficients of variation, and failure rates, in addition to mean time to failure or mean life.

In order to emphasize the basic nature of the time breakdown, each box in Figure 1.2 is labeled by the time-interval designation where possible, with the associated concept name shown in parentheses. A time designation is not applicable where operating time is combined with one or more categories of nonoperating time, because such combinations result in some concept which is, by definition, not a simple time measurement. Availability, for example, is not a time element but a probability involving a combination of time elements. The same statement applies also to operational readiness and intrinsic availability. A different situation exists with respect to effectiveness and design adequacy. Effectiveness is dependent upon all of the time elements while design adequacy does not involve any time element.

Usual relationships are shown by solid lines, and special types of relationships are shown by dotted lines. The combination of active repair time and operating time to give intrinsic availability is shown by a dotted line for two reasons. First, improvement in this characteristic must be achieved by the manufacturer, since it is primarily concerned with built-in equipment properties—assuming, of course, that the user is staying within design limits. In the second place, it does not fit easily in the chain with maintainability, since active repair time is involved in both maintainability and intrinsic availability, while operating time is included in intrinsic availability but is not involved in maintainability.

For simplicity, the chart omits one relationship which could have been shown by a dotted line from free time to down time. This would have indicated the problem arising from noncontinuous equipment use. Since free time does not exist for equipment in continuous use, a simple chaining of time categories is possible for this case. However, under conditions of intermittent use, there is a chance that down time may overlap free time, thus reducing the degrading effect of down time on operational readiness and hence on effectiveness.

Assignment of Responsibility

Even before discussing quantitative measurement for the concepts included in Figure 1.2, it is possible to demonstrate how such measures can be helpful in locating trouble areas and assigning responsibility for remedial action to improve effectiveness. When properly expressed in numerical form, these concepts will provide bases for comparative evaluation of competing equipments or systems and for determining the particular characteristics which are responsible for the differences.

The property of the time breakdown that leads to these results is the relationship between the lengths of various time intervals and the responsibilities of various personnel groups. First of all, it is apparent that administrative personnel are responsible for controlling free time, storage time, administrative time, and logistic time, while production and design engineers are responsible to a large degree for operating time (failure frequency) and active repair time. Of course, maintenance and design engineers share the responsibility for active repair time, but the maintenance engineers are limited by the serviceability characteristics that have been built into the equipment.

To achieve utmost effectiveness, it is necessary to maximize operating time and minimize down time. The role of nonuse time (free time and storage time) is that of a safety valve or safety factor. This is because maximum free time means minimum pressure for system use, while storage time results from the existence of spares to carry the operational load in case of emergency. Since the deterioration rate during storage may be different from that during use, and since by error some inoperable equipments may be placed in storage, this time element must be considered in determining operational readiness. The amount of equipment redundancy is determined by administrative decision, but the need for such redundancy is due to the relative magnitudes of down time and operating time. Thus storage time exists because of deficiencies in reliability and maintainability.

Large amounts of free time result from a requirement for short operating time. This happens when an equipment is needed relatively infrequently and there is firm scheduling of the need. It can also happen if working hours are restricted instead of continuous. Examples are numerous. Banks need time locks on safes at night but not during the day. Some communication equipment regularly has free time. Automatic answering services are needed only when the operator is absent. Television stations have regular hours during which there is no telecast. It is clear that operational readiness can be enhanced by using free time for maintenance, and free time can thus compensate to some extent for poor maintainability and poor reliability.

The important point with respect to free time and storage time is that they permit administration to somewhat alleviate the effects of equipment inadequacies and thus to gain operational readiness. However, it is important to note that free time and storage time have no connection with *improving* poor equipment. They provide an inferior but sometimes necessary alternative to the preferred solution of obtaining better equipment. They are a substitute for quality, but not a way of achieving quality.

It follows from the foregoing discussion that the more significant indicators of equipment characteristics are to be found in times other than free time and storage time. Figure 1.2 shows that these other types of time are all involved in the concept of availability which combines operating time with total down time, including the three subcategories of down time—i.e., administrative time, logistic time, and active repair time. These subcategories involve both administrative and engineering responsibilities.

Administrative Time

The administrative time category is almost entirely determined by administrative decision concerning the processing of records and the personnel policies governing maintenance engineers, technicians, and those engaged in associated clerical activities. Establishing efficient methods of monitoring, processing, and analyzing repair activities is the responsibility of administration.

In addition, administrative time has been defined to include wasted time because such time is the responsibility of administration. It is independent of engineering as such, and is not the responsibility of the equipment manufacturer.

Logistic Time

Logistic time is the time consumed by delays in repair due to the unavailability of replacement parts. This is a matter largely under the control of administration, although the requirements for replacements are determined by operating conditions and the built-in ability of the equipment to withstand operating stress levels. Policies determined by procurement personnel can, if properly developed, minimize logistic time. Therefore, the responsible administrative officials in this area are likely to be different from those who most directly influence the other time categories. This is the reason for separate consideration of logistic time.

Active Repair Time and Operating Time

Active repair time and operating time are both determined principally by the built-in characteristics of the equipment, and hence are primarily the responsibility of the equipment manufacturer. Improvement in this area requires action to reduce the frequency of failure or to increase ease of repair, or both. Operating time and active repair time are associated, respectively, with the concepts of reliability and repairability, which are related through the concept of intrinsic availability.

Administration can do little to reduce active repair time and increase operating time. Administrators can influence these time elements to a limited extent by seeing that operating stress levels are within design specifications and that the maintenance shop is supplied with proper tools and adequately trained personnel.

1.6 System Worth

Every system we build must be one that will perform its intended function at the lowest total cost. Cost studies show that the total cost of ownership (including initial and operating costs for the service life of the equipment) can be materially reduced if proper attention is given to reliability and maintainability early in the design of the system. These considerations lead to the concept of system worth, which is illustrated in Figure 1.3, and relate system effectiveness to total cost, scheduling, and personnel requirements.

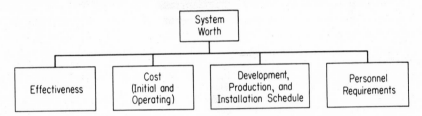

Fig. 1.3 Concepts associated with system worth.

To optimize system worth, program managers face the difficult task of striking balances to maximize system effectiveness while minimizing total cost,* development time, and personnel requirements. The Department of Defense has taken an important step in furthering this practice through its recent instruction, *Reporting of RD&E Program Information* (Instruction 3200.6, 7 June 1962). In practice, the result is a selection from several alternatives of the most promising system or component for which development effort is required. A Technical Development Plan (TDP) is outlined in Figure 1.4.

At this point it should be noted that the system effectiveness concept, as described earlier in this chapter, applies to the operation of a system in its use environment and is capable of being measured (see Chapter 11). However, as the actual use environment is often unknown or beyond the control of the system manufacturer, only certain elements of the system effectiveness concept can be specified for contractual purposes. From a practical point of view, a mission analysis must be conducted to determine the required level

* See Chapter 15.

of *intrinsic availability* as well as the needed performance characteristics (*design adequacy*). The problem of specifying system requirements becomes increasingly complex if redundant or multimodal operation is employed.

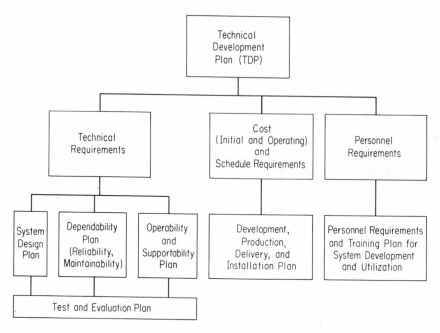

Fig. 1.4 Outline of requirements for a technical development plan.

For example, to achieve the required level of availability at the optimum cost level the system designer may have to consider several alternative approaches. Should he recommend the use of several redundant systems so that one or more spare systems will always be available and the failed system can be repaired under less pressing circumstances, or should he recommend building one highly reliable system which can be repaired quickly? In many cases more effort spent in improving reliability and repairability by the use of higher-quality parts, redundant circuitry, plug-in assemblies, and simpli-fied or semiautomatic trouble-location devices can do much to improve availability and reduce total costs to a fraction of what they might be if these two factors were not considered. Trade-off analyses, therefore, are often essential to determine minimum requirements for achieving the needed level of availability at the optimum expenditure of money, personnel, and time. The word "optimum" is used here because trade-offs will also be required between these factors of system worth.

1.7 Summary

An analysis of over-all calendar time in terms of the items indicated in Figure 1.2 is useful in determining the worth of competing systems. This analysis based on calendar time is essentially an operational-readiness analysis, with emphasis on the relative contributions of component times.

If the use cycle of an equipment is such that free time is a large percentage of total calendar time, then availability and all items in lower classifications could conceivably have only a negligible effect on operational readiness. On the other hand, for equipment for which continuous operation is required, even active repair time may have a significant effect on availability and operational readiness.

The consequences of failure are critical in most military applications. Operational readiness can be improved by establishing a use cycle that permits relatively large amounts of free time and by providing redundancy (storage time) in the form of spare system components. However, a false sense of security can be built up by merely increasing free time and storage time without considering all the other time breakdowns.

Active repair time and operating time combine to determine intrinsic availability. In a similar manner, down time and operating time combine to determine availability. The effects of free time and storage time operate to modify availability and thus determine operational readiness.

The three attributes, operational readiness, mission reliability, and design adequacy, are combined to determine system effectiveness—the ultimate measure of how well the system performs its intended function. System worth relates system effectiveness, total costs, schedules, and personnel requirements.

The attributes shown on Figure 1.2 will be treated in considerably more detail, individually and collectively, in later chapters.

PROBLEMS

1. What factors determine design adequacy for (a) an aircraft tracking radar system, (b) an airborne communication system, and (c) a surface-to-air missile system? How would you measure design adequacy of each of these systems?

2. What three factors constitute the ultimate measure of how well a system performs its intended function? Which of these factors does not involve any time elements?

3. Which of the concepts associated with system effectiveness might be included in a contract for the development of a new radar system?

4. What is the relationship between maintainability, repairability, and serviceability? How might these characteristics be measured?

5. In evaluating the worth of competitive systems, what elements of cost must be considered?

Mathematical Concepts: Probability

2.1 Introduction

The preceding chapter implied that the subjects of probability and statistics are essential tools of reliability analysis. In the present treatment of mathematical concepts, probability is discussed at some length in order to provide the necessary background for an understanding of later chapters. In the chapter on statistics, emphasis is placed on the mathematical relationships encountered in reliability work.

Basic probability theory and statistical methods are two distinct yet related fields of mathematics. We naturally think of probability and gambling together, since some of the early developments in probability were fostered by an interest in games of chance. Early statistical research was encouraged by a variety of interests, such as anthropometry, genetics, actuarial science, and demography. Some of the pioneers in statistics were engineers, but persons engaged in agricultural research made major contributions to the development of statistical theory

Those readers who have had a formal course in probability will realize that this chapter uses an uncommon order of discussion. Usually, we find the subject of probability introduced by a treatment of permutations and combinations, terms to be defined later. However, we have adopted a different approach for a number of reasons. In the first place, the initial emphasis on probability may make the discussion more interesting to those for whom this chapter is a review of previous courses. In the second place, most of the statements in reliability theory are phrased in terms of probability rather than permutations and combinations. In other words, probability is the most important for our purposes. Finally, it seems reasonable to say that probability can be viewed as a logical first concept, capable of direct treament in simple terms which can be illustrated by engineering examples. This latter statement refers specifically to Section 2.2, "A General Method for Solving Probability Problems," which depends entirely on the probability concept definition. However, no harm will be done if the reader wishes to turn first to the sections on permutation and combinations.

Definitions of Probability

There are many definitions of *probability*, and it is a popular pastime of mathematicians to debate the relative merits of each. No attempt will be made here to participate in the debate, or even to reconcile the different interpretations, but is necessary to mention some of them.

The probability that an event will occur can be thought of as the proportion of the time we would expect it to occur in a very large number of trials. If there were no sampling variations, we would expect 120 throws of a single unbiased die to yield 20 ones, 20 twos, and so on up to 20 sixes. The word *unbiased* means that each face would appear as often as each other face. Since there are six faces, the probability of each appearing is one sixth.

The foregoing is a rough statement of the relative-frequency definitions of probability. This definition says that if an event occurs in s out of a total of n trials, and does not occur in the remaining $(n-s)$ trials, the probability of its occurrence is the limit of the ratio of s to n as n goes to infinity. Some mathematicians argue that this definition is not logically consistent, but they all agree that, in a proper experiment, s/n is a fairly good approximation of the probability if n is large.

Another definition is based on a count of the number of possible results of a trial. The dice example applies here. Since there are six possible outcomes of a throw of a single die, the probability of any one number occurring is one sixth, assuming no bias. In general, if an event can occur in m ways and can fail in n ways, the probability of its occurrence is $m/(m + n)$, provided the ways are exhaustive, equally likely, and mutually exclusive.

The concept *equally likely* is basic; it can play the role of the undefined element. Indeed, it is quite difficult to define this term without using the word "probability"—which would be a circular logical inconsistency. However, the term can be described as being the lack of any bias favoring one way over another in the random trial.

The ways are *mutually exclusive* if, when one is known to occur, the other is known not to occur. For instance, if the event is the drawing of a single card from a deck and obtaining an ace, there are four mutually exclusive ways of doing it: by drawing an ace of spades, hearts, diamonds, or clubs. If a single card is drawn and it is the ace of spades, it cannot be the ace of another suit. Here the ways are equally likely if the drawing procedure is not biased in favor of any one card. The probability of drawing an ace is 4/52 under these conditions.

Finally, there is a so-called *point set theory* or *measure theory* definition of probability. It merely defines the probability of an event as a positive proper fraction associated with the event. It further states certain conditions about this number. Essentially these conditions require that the probability be unity if there is no alternative event, and certain combinatorial properties of probabilities are stated. The measure theory approach is not used in the present discussion, but there is much to recommend it as a sound basis for a rigorous mathematical treatment of probability.

Definitions of Related Terms

Some of the terms which were used in the definitions of probability and some other terms which will be needed later are defined as follows:

1. *Exhaustive.* As used in the definitions of probability, this term means that all possible ways for an event to happen are included. The reasons for this restriction in the definition should be obvious.

2. *Trial.* A trial is an attempt under a certain set of rules to produce an event A where the outcome of the event A is uncertain. Thus, each repeated throw of dice in a game of craps is an attempt to make one's point. One usually speaks of a random trial. The term "random" implies without bias.

3. *Independent Trials.* If the outcome of one trial does not influence the outcome of a subsequent trial, the two trials are said to be independent. Each throw of dice in a crap game meets this criterion. However, the drawing of cards from a deck without replacement does not meet it, since the number of ways an event (drawing a specified card, for example) can happen changes with each drawing.

4. *Conditional and Unconditional Probabilities.* The conditional probability of an event is encountered when information about the occurrence of some other event is available. If one is informed that a certain event has occurred, does this tell him anything about the probability of the occurrence of another event? Knowing that B has occurred, what is the probability of A occurring? If the events are dependent, the knowledge that one has occurred does modify the probability of the other, and this probability is conditional. If no information is available as to the occurrence of an event on a previous trial, the probability is unconditional.

Some of the more common instances of conditional probabilities in reliability are worth noting. The probability of the failure of a set of redundant elements gives rise to conditional probabilities. Thus, it is often necessary to consider the conditional probability of failure of the redundant system, given that a certain number of the parallel elements are still operating. The probability of failure-free operation for a specified time, given that the equipment in question has been previously operated for a specified time, is also conditional. In actuarial theory, the analogue is the probability that death does not occur in the next year given the attained age at the beginning of the year. The probability of failure-free operation for a given time interval after an equipment leaves maintenance is dependent on performance levels specified in maintenance procedures. Thus, there must be conditional probabilities for various performance levels.

Discrete and Continuous Probabilities

The foregoing discussion of probability was phrased in terms of the discrete case, a set of distinct events, each of which is present as a go-no-go dichotomy. If one associates numbers with the separate events, they become a set of isolated values like the integers 1, 2, 3, . . ., and so on up to a finite limit, or they might be without limit—a situation called the infinite case.

We might suspect at first glance that the infinite case is unusual, but this is not true. It is easy to find examples of the following type. Suppose that two men, Joe and Sam, decide to play a game of coin-matching until one of the two "goes broke." In order to determine the probability that Joe loses all his money to Sam, it is necessary to consider an infinite number of matching attempts, for no matter how many times they match coins, there is still a chance that the game is not finished.

The concept of continuous probabilities is included quite easily under the definitions of probability given above, especially in the measure theory approach. It is assumed there that *continuous* is understood to mean "not discrete." More explicitly, the continuous case is one in which the numerical values assigned to the events consist of any number between two limits which may on occasion be infinite. There may, of course, be gaps in which the numbers do not correspond to events. Examples of continuous cases are: the time between equipment failures, the height of a man, the position of a satellite in orbit, etc.

A simple description of the continuous case is as follows: Think of the continuous range as being subdivided into a set of intervals, and number these intervals by the positive integers. The number of intervals may be either finite or infinite. Consider all events in a given interval as being the same—that is, forget the continuous variation within the interval. Now the true continuous case is approximated more accurately if the intervals are made smaller, and the true situation in the continuous case can be considered as a limit approached by decreasing each and every interval length toward zero. If this process is carried out, by means of precise mathematical methods, we obtain continuous probability.

It is interesting to point out that in the truly continuous case we obtain a zero probability for a single point, contrary to the discrete case. This is illustrated as follows: If we chose to place a mark at random on a line six inches long, is it not reasonable to believe that we would never mark the line *exactly* three inches from one end, the midpoint? Using the discrete analogy, there is only one way to win but an infinite number of ways to lose. Hence, the answer is zero. There are combinations of discrete and continuous probabilities in which special points do have non-zero probabilities. An example will be given later. It is apparent now that we must speak of continuous probabilities in terms of observing a value within an interval. The event here is within an interval and not at a point or assuming a given value as in the discrete case.

2.2 A General Method for Solving Probability Problems

The relative-frequency definition of probability is by itself a very powerful tool in solving probability problems. It does two things: (1) it permits the use of observed data on the proportion of successes as an estimate of the probability of success, and (2) it permits a reversal of this procedure to use probability (or an estimate thereof) in predicting the proportion of future successes. In other words, the first item is the input data needed in solving the problem; the second item is the method of solution. However, the latter point is not obvious without further comment. It can be clarified by applying the stated method of solution to a specific reliability problem.

The technique to be used can be described somewhat as follows: Suppose that we were to try an experiment over and over again, say N times. The number of times we would expect the experiment to have a specified outcome would be proportional to the probability of the occurrence of this outcome. In algebraic form, let p be the probability of the specified outcome and n be the expected number of occurrences of the outcome. Then,

$$n = pN.$$

This says that we can solve a probability problem by assuming that events will turn out exactly according to their probabilities. This statement is sometimes called *the law of averages*, a rather lucid phrase which can sometimes lead to confusion in thinking, but which does clarify the general method that is used in the following example. Of course, we shall use the concept above in reverse. Thus, we will have occasion to count the expected number of occurrences of events and take the ratio of this number to the total number of experiments in order to compute the probability in the form $p = n/N$. These are the only two relationships needed in the example.

Consider an airborne equipment consisting of three black boxes denoted by A, B, and C. Assume that in a particular use of this equipment, the failure of one of the boxes does not influence the failure of either of the others. (It will be noted later that this assumption is not really a restriction of any significance.) Denote by a, b, and c the probability of the successful operation of boxes A, B, C, respectively; and denote by $\bar{a}$, $\bar{b}$, and $\bar{c}$ the probability of the failure of A, B, and C, respectively. Now suppose that success and failure have occurred in the following proportions in previous flights:

Box	Proportions	
	Success	*Failure*
A	$a = 1/2$	$\bar{a} = 1/2$
B	$b = 2/3$	$\bar{b} = 1/3$
C	$c = 4/5$	$\bar{c} = 1/5$

For simplicity, let a trial be a mission involving a flight of fixed duration. The table expresses the equality of the observed relative frequencies to the corresponding probabilities. Although this is not exactly correct in the sense that the data are not expected to reflect the true value precisely, for present purposes it is satisfactory to accept the stated equalities as being true. A discussion of this point will be deferred until the study of sampling problems in Section 3.4.

From the probabilities above, we can compute the number of failures expected in any number of future flights. Each of the three boxes will or will not fail in all possible combinations. All we need to do to obtain probabilities for the separate combinations is to consider each black box in turn, as follows:

Consider 60 future flights. In $60a = 30$ of these, A will operate properly;* in $60\bar{a} = 30$ of them, A will fail. Of the 30 in which A operates properly, B will operate in $30b = 20$ and B will fail in $30\bar{b} = 10$. Similarly, of the 30 flights in which A will fail, B will operate in $30b = 20$ flights and B will fail

* For brevity, the phrase "will operate" is used in place of the more accurate phrase "will operate on the average."

in $30\bar{b} = 10$ flights. Each of the last four numbers must be split, according to whether or not C operates properly or fails, by multiplying c and $\bar{c}$.

At this point, it is apparent that the process becomes difficult to follow unless the computations are expressed in some systematic form, such as that given below:

60			
$60a = 30$		$60\bar{a} = 30$	
$30b = 20$ $\quad$ $30\bar{b} = 10$		$30b = 20$ $\quad$ $30\bar{b} = 10$	
$20c = 16$ $\;$ $20\bar{c} = 4$ $\;$ $10c = 8$ $\;$ $10\bar{c} = 2$		$20c = 16$ $\;$ $20\bar{c} = 4$ $\;$ $10c = 8$ $\;$ $10\bar{c} = 2$	

The results can be summarized in two forms, each of which identifies the failure combinations in a systematic way.

abc 16

$ab\bar{c}$ 4

$a\bar{b}c$ 8

$a\bar{b}\bar{c}$ 2

$\bar{a}bc$ 16

$\bar{a}b\bar{c}$ 4

$\bar{a}\bar{b}c$ 8

$\bar{a}\bar{b}\bar{c}$ 2

Total 60

$60 \begin{cases} a(30) \begin{cases} b(20) \begin{cases} c & 16 \\ \bar{c} & 4 \end{cases} \\ \bar{b}(10) \begin{cases} c & 8 \\ \bar{c} & 2 \end{cases} \end{cases} \\ \bar{a}(30) \begin{cases} b(20) \begin{cases} c & 16 \\ \bar{c} & 4 \end{cases} \\ \bar{b}(10) \begin{cases} c & 8 \\ \bar{c} & 2 \end{cases} \end{cases} \end{cases}$

Probabilities for any combination can now be computed as the ratio of the number of missions with the particular failure combination to the total number of missions flown, 60 in this example. Thus, the probability that A and B fail while C does not is $8/60 = 2/15$, the 8 being indicated above by $\bar{a}\bar{b}c$. Note that taking the ratio eliminates the effect of the choice, 60. Any number would have yielded the same ratios.

There is a special significance in the method of identification: the indicated product is the probability. For example, the $\bar{a}\bar{b}c$ product is (1/2) (1/3)

(4/5) = 2/15 as computed above. Furthermore the 8 missions noted for this case can be computed by the formula

$$60\bar{a}\bar{b}c$$

since $\bar{a}$, $\bar{b}$, and c are used to denote probabilities. The product $\bar{a}\bar{b}c$ constitutes an illustration of the "both-and" theorem in probability to be discussed later. The multiplication by 60 merely reflects the use of the definition of probability as the expected number of occurrences of an event in a given number of trials, 60 in this case.

It is also possible to illustrate the "either-or" theorem of probability in which probabilities are added. Thus, the probability that A and B both fail, regardless of C, is shown as 10/60 = 1/6, the numerator 10 being obtained by the computations listed above. It could be denoted by $\bar{a}\bar{b}$, the formula for this probability. It can also be obtained as $\bar{a}\bar{b}c + \bar{a}\bar{b}\bar{c} = \bar{a}\bar{b}(c + \bar{c}) = \bar{a}\bar{b}$. In terms of the number of missions, this is 8 for $\bar{a}\bar{b}c$ and 2 for $\bar{a}\bar{b}\bar{c}$.

To talk about the probability of equipment failure, we must consider the way in which the boxes are connected. The failure of a black box may or may not mean equipment failure, depending on the series or parallel arrangement of the boxes. Hence it is necessary to look at a number of possible arrangements of boxes A, B, and C in the present example. Four possible cases are diagramed below.

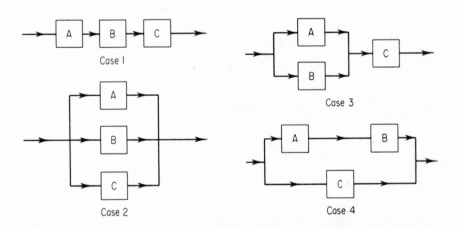

Case 1

Case 3

Case 2

Case 4

Failure-free operation of the equipment in each of the four cases requires the following conditions:

Case 1. None of the three boxes can fail.

Case 2. Not more than two of the boxes can fail.

Case 3. Box C must not fail and at least one of the other two boxes must also operate without failure.

Case 4. Boxes A and B must both operate without failure if box C fails, but the equipment will operate if box C does not fail regardless of what happens to boxes A and B.

Using these conditions, it is now possible to tabulate the number of successful missions for each of the four cases. The failure combinations are identified by the associated probabilities in formula form in the first column.

Successful Missions

Failure combination	Case 1	Case 2	Case 3	Case 4
abc	16	16	16	16
$ab\bar{c}$		4		4
$a\bar{b}c$		8	8	8
$a\bar{b}\bar{c}$		2		
$\bar{a}bc$		16	16	16
$\bar{a}b\bar{c}$		4		
$\bar{a}\bar{b}c$		8		8
$\bar{a}\bar{b}\bar{c}$				
Total	16	58	40	52
Probability	16/60	58/60	40/60	52/60

Since the identification of the failure combination is the formula for the probability of the combination, the formula for the probability of successful equipment operation in each of the four cases can be written as the sum of the identifications for the recorded entities. These are as follows:

Case	Probability of Success
1	abc
2	$abc + ab\bar{c} + a\bar{b}c + a\bar{b}\bar{c} + \bar{a}bc + \bar{a}b\bar{c} + \bar{a}\bar{b}c$
3	$abc + a\bar{b}c + \bar{a}bc$
4	$abc + ab\bar{c} + a\bar{b}c + \bar{a}bc + \bar{a}\bar{b}c$

These formulas can be simplified for Cases 2, 3, and 4. Thus it can be shown that for Case 2 the formula is

$$1 - \bar{a}\bar{b}\bar{c} \text{ or } a + b + c - ab - ac - bc + abc.$$

For Case 3, the formula is

$$ac + \bar{a}bc \text{ or } c(a + \bar{a}b) \text{ or } c(a + b - ab).$$

For Case 4, the formula is

$$c + ab\bar{c} \text{ or } c + ab - abc.$$

In the foregoing example, the system was defined as an airborne equipment, and success was defined as failure-free performance for the duration of a mission of a fixed length. If the equipment were used for a longer time on a longer mission, we would expect lower probabilities of success. This can be represented in general by replacing the numerical probabilities by functions of time. Let $R_a(t)$, $R_b(t)$, and $R_c(t)$ represent the probabilities of failure-free operation for a mission of length t for boxes A, B, and C, respectively. Failure probabilities will be $1 - R_a(t)$, $1 - R_b(t)$, and $1 - R_c(t)$.

As an example, let

$$R_a(t) = e^{-\alpha t}$$

$$R_b(t) = e^{-\beta t}$$

$$R_c(t) = e^{-\gamma t}.$$

Then the probabilities of equipment success for the same four cases are listed below.

Case	Probability of Success
1	$e^{-(\alpha+\beta+\gamma)t}$
2	$e^{-\alpha t} + e^{-\beta t} + e^{-\gamma t} - e^{-(\alpha+\beta)t} - e^{-(\alpha+\gamma)t} - e^{-(\beta+\gamma)t} + e^{-(\alpha+\beta+\gamma)t}$
3	$e^{-\gamma t}(e^{-\alpha t} + e^{-\beta t} - e^{-(\alpha+\beta)t})$
4	$e^{-\gamma t} + e^{-(\alpha+\beta)t} - e^{-(\alpha+\beta+\gamma)t}$

It is interesting to note that the function for Case 1 is of the same form as the functions for the black boxes—namely, an exponential. In all other cases the equipment probability of success is not exponential, but instead is quite complex. For other black box reliability functions, all four cases are usually complex. This important point is discussed further in the chapters on redundancy (Chapters 7 and 8).

2.3   The Elements of Probability Theory

The definitions of probability and the use of these definitions in solving a particular probability problem of the type encountered in reliability work are sufficient evidence to show the significance of the theory of probability in the material of this text. Obviously, however, we cannot hope to reproduce all of this theory. Rather, we must restrict our treatment to a brief summary which indicates basic principles and fundamental concepts. We must omit some desirable mathematical rigor in the interest of brevity, but we will try to give an indication of the mathematical techniques by a few examples. Readers not concerned with this aspect may neglect the proofs and merely note the results. It is hoped that most of the readers will become sufficiently intrigued by the presentation to pursue the subject by further study.

Notation

Even though problems in probability require the application of all of the usual operations of mathematics, it is necessary to devise certain symbolism peculiar to probability theory. We shall use the following notation:

1. An event will be identified by a capital letter, as A, B, C, etc.

2. The probability that a specified event occurs, say event A, will be denoted by either (A) or $p(A)$.

3. We shall speak of the simultaneous occurrence of two or more events. For such combinations of events we use (AB) or $p(AB)$ as the probability of the simultaneous occurrence of A and B, and similarly (ABC) or $p(ABC)$ for the three events A, B, and C, etc.

4. We use the symbol $(A|B)$ or $p(A|B)$ to denote the probability that A occurs if it is known that B has occurred. Of course compound events could be used in the same symbols. Thus $(AB|CDE)$ is the probability that both A and B will occur if it is known that the events C, D, and E have all occurred. Such probabilities are called *conditional*, meaning probability conditioned on having certain special information in the form of the known occurrence of some other events.

While the notation above is not universally accepted, it is widely used. Unfortunately, no serious attempt at standardization has been made. We have adopted what we believe to be a symbolism suitable to most of the problems which arise in reliability work. We will first develop the elementary rules for combining these probabilities using the notation given above.

The " Either-Or " Rule, or Additive Rule

The simplest theorem in probability can be stated in either of the following two forms:

1. If A and B are mutually exclusive events, then the probability of *either A or B* in a single random trial is the sum of the probabilities of events A and B, or, in formula form,

$$(A) + (B).$$

2. The probability of an event is the sum of the probabilities of its mutually exclusive forms:

In the second statement, the theorem covers more than two events. Of course, the first form can be easily extended to apply to the more general case as suggested by the following example:

Suppose a bag contains 100 marbles, 10 of which are white, 15 are red, 20 are blue, 25 are green, and 30 are black. If one marble is drawn at random (without looking), the probabilities associated with the various colors are:

Color	White	Red	Blue	Green	Black
Probability	0.10	0.15	0.20	0.25	0.30

The probability of *either* white *or* green is $0.10 + 0.25 = 0.35$ according to the rule stated above—the *either-or rule*. This is obviously correct since there are $10 + 25 = 35$ white and green marbles out of the 100 in the bag.

It is important to stress the mutually exclusive restriction in the statements of the either-or rule. Addition of probabilities is permissible only when the events are nonoverlapping, the real meaning of mutually exclusive. This can be illustrated by considering a slightly more complicated problem of drawing marbles. Suppose we draw a marble at random as before, note the color, and then return it to the bag. Mix the marbles up again and draw a second marble. Then ask the question, what is the probability that at least one of the two is not black?

First consider an incorrect solution—one which results from a misapplication of the either-or rule. The probability of other than black in the first (or second) draw is 0.70. Hence, by adding probabilities of 0.70 for each draw, we obtain $0.70 + 0.70 = 1.40$. This answer is impossible since a probability cannot be larger than unity. The error occurs because the two events are not mutually exclusive—the occurrence of one does not preclude the other. "Not black" on the first draw does not preclude "not black" on the second draw.

To obtain the correct solution, we must modify the analysis. It will be necessary to use the general method of Section 2.2. This will not only yield the correct answer, but it will also introduce the both-and rule to be discussed below. Suppose we were to try this experiment 1000 times. We would expect them to turn out as follows:

Total	First Draw		Second Draw	
			Black	90
	Black	300		
			Not Black	210
1000				
			Black	210
	Not Black	700		
			Not Black	490

There are three categories which satisfy the required conditions. They are:

Category	First Draw	Second Draw	Number	Probability
1	Black	Not Black	210	0.21
2	Not Black	Black	210	0.21
3	Not Black	Not Black	490	0.49
	Sum		910	0.91

Notice that these three categories are mutually exclusive. If one of the three occurs, then neither of the others occurs.

We can easily satisfy ourselves of the correctness of this answer by observing that the other category, Black-Black, is the only loser. Its probability is 0.09, the complement of 0.91 computed previously. It can also be observed that the answer can be obtained by the following argument. Denote "not black on the first draw" by $(\bar{B}_1)$, "not black on the second draw" by $(\bar{B}_2)$, and "not black on either draw" by $(\bar{B}_1 \bar{B}_2)$. The probability can be expressed as:

$$(\bar{B}_1) + (\bar{B}_2) - (\bar{B}_1\bar{B}_2) = 0.70 + 0.70 - 0.49 = 0.91.$$

Event $\bar{B}_1$ includes two of the categories above: Not Black-Black and Not Black-Not Black. Similarly, $\bar{B}_2$ includes Black-Not Black and Not Black-Not Black. Notice that both $\bar{B}_1$ and $\bar{B}_2$ include Not Black-Not Black. Hence it must be subtracted out since it certainly should not be counted twice.

The " Both-And " or Multiplicative Rule

The *both-and rule* expresses the probability of the occurrence of both of two (or more) independent events, the combination being called a *compound event*. It is stated in the following way:

If A and B are independent events, then the probability of obtaining *both A and B* is the probability of obtaining A times the probability of obtaining B. In the formula form,

$$(AB) = (A)(B).$$

The extension to more than two events is simple.

This rule can also be illustrated by an example of drawing marbles from a bag. However, since we wish to consider two independent events, assume we first draw a marble at random from a bag containing 3 white and 7 black marbles, and then we draw one marble from another bag containing 4 white and 6 black marbles. All possible outcomes and their respective probabilities for the two separate draws are:

	First Draw	Second Draw
White	0.3	0.4
Black	0.7	0.6

For the combination of the two draws, the both-and rule gives the following results:

First Draw	Second Draw	Probability
White	White	(0.3) (0.4) = 0.12
White	Black	(0.3) (0.6) = 0.18
Black	White	(0.7) (0.4) = 0.28
Black	Black	(0.7) (0.6) = 0.42

These answers can be easily checked and explained by solving this problem by the general method of Section 2.2.

Notice that the probability that the two draws yield different colored marbles is obtained as the sum $0.18 + 0.28 = 0.46$, the probabilities for the second and third categories. Addition of probabilities is justified by the either-or rule. The probability of at least one white marble is $0.12 + 0.18 + 0.28 = 0.58$, the sum of the probabilities of the first three categories.

The " Both-And " Rule for Events Which May Not Be Independent

The both-and rule or theorem covered only independent events. It is very useful to consider a corresponding theorem for the case where the events are not necessarily independent. That is, the theorem is to be stated in the more general form to cover both the dependent and the independent cases. We shall first state the theorem and then explain its meaning by an example. Proof would depend on an argument like the general method used so many times before. The theorem is as follows:

The probability of the joint occurrence of both events A and B is given by the product of the unconditional probability of event A and the conditional probability of B, given that A has occurred. In formula form

$$(AB) = (A) (B|A).$$

Obviously the roles of A and B could be reversed, giving an alternative formula

$$(AB) = (B) (A|B).$$

Consider the following example. A bag contains 4 red and 6 black marbles. One marble is drawn and, without replacing it, a second marble is drawn. What is the probability that both marbles are red?

To use the theorem, let A stand for the event "a red marble on the first draw" and let B stand for the event "a red marble on the second draw." We shall use the formula

$$(AB) = (A)(B|A).$$

Now $(A) = 4/10$ and $(B|A) = 3/9$. The value of $(B|A)$ is obtained as follows: If A has indeed occurred, the bag contains 3 red and 6 black marbles after the first draw, and $(B|A)$ is the probability of drawing a red marble from this bag. Hence

$$(AB) = \left(\frac{4}{10}\right)\left(\frac{3}{9}\right) = \frac{2}{15}.$$

In this particular example, it is difficult to use the other form of the theorem,

$$(AB = (B)(A|B).$$

Actually, $(B) = 4/10$ and $(A|B) = 3/9$, values which are reasonable by symmetry but not as easily explained as the probabilities in the first solution. As usual, they can be explained by using the general method. Suppose the experiment is performed 90 times. The following frequency of outcomes would result:

First Draw	Second Draw
	Red $\frac{3}{9}(36) = 12$
Red $\frac{4}{10}(90) = 36$	
	Black $\frac{6}{9}(36) = 24$
	Red $\frac{4}{9}(54) = 24$
Black $\frac{6}{10}(90) = 54$	
	Black $\frac{5}{9}(54) = 30$

The probability of red on the second draw is $\frac{1}{90}(12 + 24) = 0.4 = (B)$. Of the $12 + 24 = 36$ results with red on the second draw, only the 12 had red on the first. Hence,

$$(A|B) = \frac{12}{12 + 24} = \frac{1}{3}.$$

Hence,

$$(AB) = (B)(A|B) = (0.4)(1/3) = 2/15,$$

checking the previous solution.

The " Either-Or " Theorem for Events Which May Not Be Mutually Exclusive

Previously we gave an example of an either-or probability in which the events were not mutually exclusive. By using all of the discussed theorems, we can now develop a formula for this more general case. It will express the either-or formula for events if they are mutually exclusive or if they are not. Actually this general case does not give rise to new rules, but it depends on a modification of the events in question to a new set of events which satisfy the conditions of the theorems as previously stated. The formula will be explained in terms of an example of two events which are not mutually exclusive. Then the formula will be extended to more than two events.

Consider the failure of a particular electronic equipment. There may be failures arising from tubes and failures arising from resistors—to mention only two possibilities. These failures are not mutually exclusive since a failure from a tube does not exclude failure from resistors. If these events are separated into mutually exclusive subclasses, the following events and probabilities (let A be failure from tubes and B failure from resistors) are apparent:

1. Failure from tubes alone, assuming no simultaneous failures from resistors, with probability

$$(A) - (AB).$$

2. Failure from resistors alone (no simultaneous failures from tubes), with probability

$$(B) - (AB).$$

3. Failure from both tubes and resistors simultaneously (assuming that they are independent), with probability

$$(AB).$$

The probability of failure from either tubes or resistors is obtained by applying the additive rule—a valid procedure since, as written, the three events are mutually exclusive. This gives

$$[(A) - (AB)] + [(B) - (AB)] + (AB) = (A) + (B) - (AB).$$

The same procedure is extendible to more than two cases.

The method of proof depends on an argument of the following type: The probability of the occurrence of either A or B can be obtained as the probability of A plus the probability of "not A" and B. This is

$$(A) + [1 - (A)] (B) = (A) + (B) - (A) (B)$$
$$= (A) + (B) - (AB).$$

Similarly, we can write the probability of the occurrence of either A, B, or C as the sum of the two probabilities:

1. the probability of A, and
2. the probability of "not A" but either B or C.

These are

$$(A)$$

and

$$[1 - (A)] [(B) + (C) - (BC)],$$

respectively. The sum is

$$(A) + (B) + (C) - (A) (B) - (A) (C) - (BC) + (A) (BC)$$
$$= (A) + (B) + (C) - (AB) - (AC) - (BC) + (ABC).$$

The generalization for either $A_1, A_2, \ldots, A_n$ is

$$\sum_i (A_i) - \sum_{i \neq j} (A_i A_j) + \sum_{i \neq j \neq k} (A_i A_j A_k) + \ldots + (-1)^{n+1} (A_1 A_2 \ldots A_n).$$

The formula can be easily proved by induction, as follows:

It is obvious that the event either A_1 or A_2 or A_3 .... or A_n can be broken up into a series of mutually exclusive events by a simple scheme. Let $\bar{A}_i$ denote the events "not A_i"; then the series of mutually exclusive events is

$$A_1, \bar{A}_1 A_2, \bar{A}_1 \bar{A}_2 A_3, \bar{A}_1 \bar{A}_2 \bar{A}_3 A_4 \ldots \bar{A}_1 \bar{A}_2 \ldots \bar{A}_{n-2} A_{n-1}, \bar{A}_1 \bar{A}_2 \ldots \bar{A}_{n-1} A_n.$$

Hence, the probability of at least one of the events $A_1, \ldots, A_n$ is the sum of the probabilities of the series of mutually exclusive events above.

These events can be seen graphically in Figure 2.1. Let A_i be represented by a circle for each i. All events will then appear as overlapping circles. The first event in the series is the A_1 circle. The second event is the part of the A_2 circle that is not inside the A_1 circle. The sum of the first two events is the sum of these two areas. The third event is that part of the area in the A_3 circle which is not in the A_1 and A_2 circles, or, stated another way, it is the portion of the A_3 circle not included in the area represented by the first two events in the series, etc.

The probability for at least one of $(n - 1)$ events is assumed to be given by the formula above. It has been tested for $n - 1 = 2$ and 3. Now the probability for the n events can be written in the forms

$$(A_1) + (\bar{A}_1 A_2) + (\bar{A}_1 \bar{A}_2 A_3) + \ldots + (\bar{A}_1 \bar{A}_2) \ldots (\bar{A}_{n-1} A_n)$$
$$= (A_1) + (\bar{A}_1)\,[(A_2) + (\bar{A}_2 A_3) + \ldots + (\bar{A}_2 \ldots \bar{A}_{n-1} A_n)]$$
$$= (A_1) + (\bar{A}_1)\,(B)$$

where the event B is at least one of the events $A_2, A_3, \ldots, A_n$ (the probability of event B being the expression in the square brackets). Since this event B consists of at least one of $n - 1$ events, the general formula holds by the assumption above. Hence, the probability of at least one of the n events $A_1, A_2, \ldots, A_n$ is

$$(A_1) + (\bar{A}_1)\sum_{2}^{n} (A_i) - \sum_{2}^{n} (A_i A_j) + \ldots + (-1)^n (A_2 \ldots A_n).$$

(Summation indices and symbols are abbreviated.) Replace $(\bar{A}_1)$ by its equivalent, $1 - (A_1)$. Multiplying and collecting terms gives

$$\sum_{1}^{n} (A_i) - \sum_{1}^{n} (A_i A_j) + \ldots + (-1)^{n+1} (A_1 \ldots A_n),$$

which was to be proved.

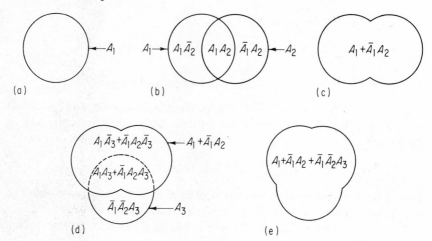

Fig. 2.1 Probabilities of mutually exclusive events.

2.4 Permutations and Combinations

One of the definitions of probability was based on a count of the number of ways an event can occur and the number of ways it can fail in a single trial. For example, if a card is drawn at random from a deck, we can get a heart in

13 ways (there are 13 hearts) and a non-heart in 39 ways (there are 39 cards which are not hearts). Hence, we can say that the probability of getting a heart is

$$\frac{13}{13 + 39} = 0.25.$$

Now there are many problems in which the number of ways is not so easily obtained. In many such problems we are concerned with counting the number of ways some compound event will occur, as, for instance, the number of ways in which there can be x occurrences of an event in n repeated trials. We can simplify counting the number of ways by developing certain theorems analogous to those on probabilities which were discussed above. These theorems constitute the simpler portions of combinatorial analysis, dealing with the concepts of permutation and combination, which we will now define.

A *permutation* of a set of objects is an arrangement of these objects. In many cases we are interested in selecting only a subset and arranging them, so we can speak of a permutation of x objects selected out of a group of n objects. Thus, permutations can differ either because they are different orderings or arrangements of a particular subset or they can differ because they do not contain identical subsets. To illustrate, consider the set of $n = 5$ to be the first five letters of the alphabet, a, b, c, d, and e. Consider permutations of three of the set. Examples of different permutations are

1. *a   b   c,*
2. *a   c   b,*
3. *a   c   e.*

A *combination* is a subset without regard for order. Thus, while a, b, c and a, c, b were different permutations in the example above, they are not different combinations. However, a, b, c and a, c, e are different combinations since the elements in the two are not exactly the same.

In the definitions and examples above, we assumed that no repetition was permitted. There are numerous cases, however, in which we wish to allow an object to enter more than once—even after it has been selected, we wish to permit it to be selected again. For example, how many numbers can be written with the three digits 1, 2, and 5, allowing repetition of digits? In answering this question, we would include as acceptable numbers 125, 152, 221, 555, etc. We usually exclude such repetitions when we discuss permutations and combinations by name, but we are required to develop special theorems and formulas to handle them. In other words, if we speak of permutations and combinations, we assume no repetition unless it is specifically stated such can occur.

We must now write some of the basic theorems and formulas for combinatorial analysis. The theorems will be used to justify formulas for the number of permutations and combinations of n things taken x at a time. The treatment will of necessity be abbreviated. Interested readers are referred to one of the many elementary probability and college algebra texts for further details.

The two most important theorems are as follows:

Theorem. If events A and B are mutually exclusive and if event A can occur in a ways and event B can occur in b ways, then either A or B can occur in $a + b$ ways.

Theorem. If events A and B are independent and if event A can occur in a ways and event B can occur in b ways, then both A and B can occur in ab ways.

Each of these theorems can be extended to more than two events by very simple arguments. It is also of interest to extend them to more general conditions. Thus we shall extend the first theorem to cover cases in which the two events are not necessarily mutually exclusive and the second to cover the case in which they are not necessarily independent.

Theorem. If event A can occur in a ways and event B can occur in b ways, then either event A or event B can occur in $(a + b - c)$ ways where c is the number of ways both events can occur. (If A and B are mutually exclusive, $c = 0$.)

Theorem. The number of ways events A and B can both occur is given by the product of the number of ways A can occur and the number of ways B can occur given that A has occurred. Obviously the A and B can be interchanged. (If A and B are independent, the number of ways B can occur given that A has occurred is the same as the number of ways B can occur without knowledge about A.)

All of these theorems can be made quite plausible by simple examples. Consider drawing a card from a standard deck. We shall speak of the number of ways we can obtain a particular suit or a face card under a variety of conditions to illustrate each of the theorems above in order.

There are 13 ways in which we can obtain a spade and 3 ways in which we can obtain a heart face card. Since they are mutually exclusive we can add 13 and 3 to give the number of ways to obtain either a spade or a heart face card.

Consider two decks of cards. The number of ways to obtain a heart from a particular one of the two decks is 13, and the number of ways we can obtain a face card of any suit from the other deck is 12. Since these events are independent, the number of ways both events can happen is the product of 13 and 12, giving 156 ways in all.

Consider the same events as in the previous example, but assume we have only one deck. We wish to find the number of ways we can obtain either a heart or a face card. Now these events are not mutually exclusive since there are 3 face cards which are hearts. Hence, we can apply the third theorem to obtain $13 + 12 - 3 = 22$ ways to obtain either a heart or a face card. We could get the same answer by defining two events which are equivalent in their effect: (1) obtaining a heart (which includes the heart face cards), and (2) obtaining a face card which is not a heart. These are 13 and 9, respectively, and the sum is 22, checking the previous solution.

To illustrate the last of the four theorems, consider the number of ways we can draw a club from the deck, 13, and then without replacing this card, the number of ways we can draw another club, 12. To compute the number of ways both events can occur we take the product of 13 and 12, or 156. Notice that the number of ways the second event can occur was computed on the knowledge that the first one had occurred as required by the theorem which we are illustrating.

The preceding example illustrates the so-called nonreplacement case—the first card drawn is not replaced prior to drawing the second card. In the replacement case the card would have been returned prior to the second draw, and the number of ways of drawing two clubs would have been $13^2 = 169$. This is a repetition of the same experiment. In general, if an event can occur in m ways and if it be repeated x times as in the replacement case, then the number of ways in which it can occur in the x trials is m^x. Repetitions in the nonreplacement case lead to the permutation formula which we shall now consider.

Permutation Formula

The number of permutations of n things taken x at a time is given by the formula

$$P(n, x) = n(n - 1)(n - 2) \ldots (n - x + 1).$$

The justification of this formula merely requires noting that it is repetition in the nonreplacement case. Thus, there are

n ways for the first object,
$(n - 1)$ ways for the second object,
$(n - 2)$ ways for the third object,
$(n - x + 1)$ ways for the xth object,

and we merely need to form the product according to the theorems quoted above.

The formula can be simplified by the use of factorial notation. Factorial n, denoted by $n!$, means

$$n! = n(n - 1)(n - 2) \ldots 3 \cdot 2 \cdot 1,$$

the product of integers from n down to 1. Thus

$$1! = 1,$$
$$2! = 2 \cdot 1 = 2,$$
$$3! = 3 \cdot 2 \cdot 1 = 6, \text{etc.}$$

It is easy to see that

$$P(n, x) = n(n - 1)(n - 2) \ldots (n - x + 1)$$

$$= \frac{n(n - 1)(n - 2) \ldots (n - x + 1)[(n - x)(n - x - 1) \ldots 3 \cdot 2 \cdot 1]}{(n - x)(n - x - 1) \ldots 3 \cdot 2 \cdot 1}$$

$$= \frac{n!}{(n - x)!}.$$

Notice that, if $n = x$, the formula gives

$$P(n, n) = \frac{n!}{(n - n)!} = \frac{n!}{0!}.$$

Now the symbol $0!$ is not covered by the definition of the factorial notation given above. We choose to let

$$0! = 1,$$

and we find that the formula above reduces to

$$P(n, n) = n!,$$

the correct answer for the number of permutations on n things taken n at a time. There are of course other reasons for defining $0!$ to be equal to one, but it is not possible to cover this subject completely here.

Combination Formula

The number of combinations of n things taken x at a time was defined previously in a general way. It is now necessary to make the definition more precise. We can define the number of combinations of n things taken x at a time as the number of ways in which the n things can be divided into two groups such that one group contains x of the items and the other contains the remaining $(n - x)$. It is denoted by a variety of equivalent symbols, such as

$$\binom{n}{x}, {}_nC_x, C(n, x),$$

and the formula is

$$C(n, x) = \frac{n!}{x!(n - x)!}.$$

(We shall use only the first and third symbols in this treatment.)

The easiest way to establish this formula is to consider the relationship between permutations and combinations. The argument will be simplified if we speak in terms of a special example. Consider five letters, a, b, c, d, and e. Let us write a particular combination of three of the five letters, say

$$b \quad d \quad e.$$

Now there are $3! = 6$ permutations of these three letters. They are bde, bed, dbe, deb, ebd, edb. While these 6 are different permutations, they are not different combinations. That is, for each combination, we can form $3!$ permutations. Thus,

$$3! \, C(5, 3) = P(5, 3),$$

or
$$C(5, 3) = \frac{P(5, 3)}{3!}$$

$$= \left(\frac{5!}{2!}\right) \frac{1}{3!}$$

$$= \frac{5!}{3! \, 2!}.$$

The same argument holds in general.

$$x! \, C(n, x) = P(n, x)$$

$$= \frac{n!}{(n - x)!},$$

giving
$$C(n, x) = \frac{n!}{x! \, (n - x)!}.$$

These ideas can be clarified by solving a number of problems involving the formulas and theorems which have been stated. Many other examples can be found by reference to almost any elementary text in college algebra.

Selecting a Committee

In how many ways can a committee of 3 people be selected from a group of 6? Obviously this is a simple combination illustration since we are interpreting the statement to mean that two committees are different if not more than two members are on both—there must be at least one member different. Hence, the answer is

$$C(6, 3) = \frac{6!}{3! \, 3!} = 20.$$

Probabilities and Selecting a Committee

We can illustrate the use of combinations in the solution of probability problems by the following examples in selecting a committee.

Assume a group of 9, consisting of 4 males and 5 females. What is the probability that a committee of 3 selected at random will be all male? As before, the number of ways of selecting the committee is

$$\binom{9}{3} = \frac{9!}{3!\,6!} = 84.$$

The number of ways in which 4 males can be divided into two groups with 3 in one group and 1 in the other is

$$\binom{4}{3} = \frac{4!}{3!\,1!} = 4.$$

These are the favorable events which are included in the total of 84. Thus,

$$\frac{4}{84} = \frac{1}{21}$$

is the probability in question.

Consider the question: What is the probability that the committee is composed of 2 males and 1 female? In this example, the denominator remains the same. The numerator, however, is $\binom{4}{2}\binom{5}{1}$, since there are $\binom{4}{2}$ ways of selecting the males, and there are $\binom{5}{1}$ ways of selecting the females. This gives a probability of 30/84.

2.5   Permutations and Combinations in Probability

We can often simplify the solution of probability problems by using permutations and combinations. Frequently we can use either one depending on how we view the problem. Actually there is no rule governing the choice except that we should use whichever method yields the simplest development, other things being equal. It turns out in practice that the choice is usually one of preference.

A Card Problem

For a simple example of the two approaches, consider the drawing of three cards from an ordinary deck. What is the probability that exactly one card will be a spade? By the use of combinations, as before, this can be computed as

$$\frac{\binom{39}{2}\binom{13}{1}}{\binom{52}{3}} = \frac{39!\,13!\,3!\,49!}{2!\,37!\,12!\,52!} = \frac{39 \cdot 38 \cdot 13 \cdot 3}{52 \cdot 51 \cdot 50} = \frac{741}{1700}.$$

On the other hand, by use of permutations, the problem may be considered in the following way: There are 52 ways to draw the first card, 51 ways to draw the second card, and 50 ways to draw the third. Thus, the number of three-card arrangements which can be made from a group of 52 cards is

$$P(52, 3) = \frac{52!}{49!} = 52 \cdot 51 \cdot 50.$$

If the last card drawn is a spade, the probabilities are

$$\frac{39 \cdot 38 \cdot 13}{52 \cdot 51 \cdot 50}.$$

There are 39 ways to draw a non-spade on the first draw out of a total of 52. There remain 38 ways out of a total of 51 to draw a second non-spade, and 13 ways out of the remaining 50 to select a spade on the third drawing. However, the spade can occur in the first or second drawing as well as in the third. Thus, there are 3 ways in which numerators can be arranged. (This amounts to $P(3, 1)$ since we are only arranging the order of the spade drawing —i.e., the "39" will always precede the "38.") Therefore, the desired probability is

$$\frac{39 \cdot 38 \cdot 13 \cdot 3}{52 \cdot 51 \cdot 50} = \frac{741}{1700}$$

as before.

A few additional examples may further clarify the application of these rules. It is difficult to classify problems into types which automatically dictate the application of any given rule. The major problem is that of determining all ways in which an event can happen. If this is done without error, the selection of the applicable rule is a minor problem.

A Dice Problem in Probability—Different Approaches

To begin with a simple problem, consider the probability of obtaining at least *one* 6 in two rolls of one die. It is known that the rolls (events) are independent but are not mutually exclusive—i.e., a 6 can occur on *both* rolls. The probability of obtaining a 6 on one roll is 1/6, but the probability of obtaining it on either the first or the second roll is not $1/6 + 1/6$. The probability of obtaining a 6 on each of the rolls of the die is computed by using the product

$$(1/6)(1/6) = 1/36.$$

Thus the probability of obtaining a 6 on the *first* roll only is

$$1/6 - 1/36;$$

i.e., the probability of obtaining a 6 on the first roll minus the probability of obtaining a 6 on both the first and second rolls.

The probability of obtaining a 6 on the second roll only is also

$$1/6 - 1/36,$$

and the probability that a 6 will occur on both rolls is already computed to be 1/36. Thus, if a 6 on the first roll is called event A, a 6 on the second roll is called event B, and a 6 on both rolls is called event AB, then there are the three mutually exclusive compound events with probabilities as follows:

Compound Event	Probability
6 on the first roll only	$(A) - (AB) = 1/6 - 1/36 = 5/36$
6 on the second roll only	$(B) - (AB) = 1/6 - 1/36 = 5/36$
6 on each roll	$(AB) = 1/36 = 1/36.$

Addition of these probabilities—a valid step, since they are mutually exclusive—gives

$$(A) + (B) - (AB) = 1/6 + 1/6 - 1/36 = 11/36$$

as the probability of at least one 6 in two rolls.

Another way to perform the computation above is to make separate estimates of the numerator and the denominator—i.e., the number of ways a 6 can occur and the number of ways the die can come up. There are 6 ways in which each throw can come up. Thus, by multiplying, we find that there are 36 ways in which the event (two throws) can occur. For event A—a 6 on the first throw and a non-6 on the second—there are 1×5 ways. For a non-6 on the first throw and a 6 on the second, there are 5×1 ways for the event to occur. For a 6 on each throw, there is only $1 \times 1 = 1$ way for the event to occur. These are mutually exclusive. Thus the total number of ways is $5 + 5 + 1$, or a total of 11, giving a probability of 11/36 that at least one 6 will occur in two rolls.

A Dichotomy—The Binomial

Many problems are concerned with the events where there are but two outcomes of each trial. This situation exists in the example above. Thus, on one roll of a die, the probability is 1/6 that a 6 will occur and 5/6 that it will not occur. These are two mutually exclusive events. The rules of addition and multiplication can be applied routinely by use of the binomial expression

$$\left(\frac{1}{6}X+\frac{5}{6}\right).$$

If two dice are thrown, the probabilities of the various numbers of times that a 6 will occur can be computed by writing

$$\left(\frac{1}{6}X+\frac{5}{6}\right)^2 = \frac{1}{36}X^2+\frac{10}{36}X+\frac{25}{36}.$$

The probability of two 6's is 1/36, the coefficient of X^2. The probability of exactly one 6 is 10/36, the coefficient of X. The probability of no 6's is 25/36, the constant term. Note that the number of successful trials—i.e., the number of 6's—is indicated by the exponent of X. The probability of at least one 6 in the two throws is the sum of the coefficients of the X^2 and X, $1/36 + 10/36 = 11/36$, as computed before. Note that the second term includes two cases: (1) a 6 on the first throw and other than a 6 on the second, and (2) other than a 6 on the first throw and a 6 on the second.

For three throws, the probabilities are given by the binomial

$$\left(\frac{1}{6}X+\frac{5}{6}\right)^3 = \frac{1}{216}X^3+\frac{15}{216}X^2+\frac{75}{216}X+\frac{125}{216}.$$

Again, the number of successes is given by the exponent of X and the probability by the numerical coefficient. It is this property of the exponent of X—identification of number of successes for which the coefficient is the probability—that is of interest. This property is general. For any number of tosses of the die—say k—the probabilities are given by the coefficients of powers of X in the expansion of the binomial,

$$\left(\frac{1}{6}X+\frac{5}{6}\right)^k.$$

This is valid only if the tosses are independent.

The probability of at least one success in k trials can be computed by adding all coefficients of powers of X from 1 to k. However, it is easier to compute the complementary probability—i.e., the probability of no successes—and subtract it from unity, especially if the number of tosses is large. For $k = 3$, the two methods are

$$\frac{1}{216}+\frac{15}{216}+\frac{75}{216}=\frac{91}{216}$$

and

$$1-\frac{125}{216}=\frac{91}{216}.$$

The general formula for at least one success in k trials is

$$1 - q^k = 1 - (1 - p)^k,$$

where p is the probability of success in a single trial and $q = 1 - p$ is the probability of failure in a single trial.

The validity of the binomial expansion method of computing probabilities can be established by showing that the coefficient of X_i in the expansion of $(pX + q)^k$ is in fact the probability of exactly i successes in k independent trials. This coefficient is

$$\binom{k}{i} p^i q^{k-i}.$$

The expression $p^i q^{k-i}$ is the probability of obtaining i successes and $k - i$ failures in a particular order. The combinatorial coefficient, $\binom{k}{i}$, is the number of such orders. Hence, the probability of exactly i successes, independent of order, is the sum of $\binom{k}{i}$ terms each equal to $p^i q^{k-i}$, which is the binomial coefficient itself, giving the answer

$$\binom{k}{i} p^i q^{k-i}.$$

Drawing Balls from an Urn

Suppose that 2 balls are drawn from an urn containing 5 balls—3 white and 2 black. What is the probability that the 2 balls will both be white?

If the first ball is replaced before the drawing of the second, the probability can be determined as in the previous example by expanding the binomial,

$$\left(\frac{3}{5} w + \frac{2}{5}\right)^2.$$

However, if the balls are considered as being drawn without replacement, or if 2 are drawn simultaneously, the answer is different. In this case, there are 5 ways to draw the first ball and 4 (remaining) ways to draw the second. According to the multiplicative rule, the number of ways for the events to happen is

$$(5)(4) = 20.$$

There are 3 white balls. Thus, there are 3 ways in which the first white ball can be drawn and there are 2 ways in which a second white ball can be drawn: $3 \cdot 2 = 6$. Then the probability is: $6/20 = 3/10$. The binomial does not apply in this case because the trials are not independent.

If 3 balls are drawn, the probability that at least 2 of them will be white is obtained as follows: Reasoning as before, if there is no replacement, there are 5 ways to draw the first ball, 4 ways to draw the second, and 3 ways to draw the third, giving $5 \times 4 \times 3 = 60$ ways of drawing the 3 balls. This is the denominator. A success could occur if the first ball were white, if the second ball were white, and if the third ball were black. This event could occur in $3 \times 2 \times 2$ ways. However, another success could occur if the first ball were black and the next 2 were white, or, lastly, if the middle drawing were black and the other 2 were white—a total of 3 different ways. This is equivalent to dividing 3 things into two groups, with 1 black ball in one group and 2 white balls in the other,

$$\binom{3}{1} = 3.$$

Each of these 3 events could occur in $3 \times 2 \times 2 = 12$ ways, giving $3 \times 12 = 36$ ways in which 1 black and 2 white balls could result from the draw of 3. The number of ways in which 3 white balls could be drawn is $3 \times 2 \times 1 = 6$. Hence, the probability in question is

$$\frac{36 + 6}{60} = 0.7.$$

The solution could also be expressed as the ratio

$$\frac{\binom{3}{2}\binom{2}{1} + \binom{3}{3}}{\binom{5}{3}} = \frac{6 + 1}{10} = 0.7.$$

This illustrates the use of combinations (excluding discussion of order). The first term in the numerator is the number of ways to select 2 white balls and 1 black ball; the second numerator term is the number of ways to select 3 white balls; the denominator is the number of ways to select 3 balls regardless of color. The experiment is equivalent to assuming that 3 balls are drawn all at once. A little reflection will show that drawing 3 at once is equivalent to drawing one after another without replacement.

Probability of Matching Birthday Anniversaries

A rather unusual probability problem is concerned with coincidence of birthdays. Suppose 25 people are gathered in a room, and one asks for the probability that at least 2 have the same birthday anniversary date—i.e., the same day of the year, but not the same year. The solution is most easily obtained by first computing the complementary probability—that there are no birthday matches. To compute the number of ways in which the birthdays might be distributed, suppose the people are arranged in some order and numbered from 1 to 25. Each one has 365 choices for a birthday, giving a total of 365^{25} ways in which the birthdays might occur. If there are no matches, we could say that there are 365 choices for the first person; 364 for the second person; 363 for the third person; and so on down to 341 for the twenty-fifth person. The number of ways decreases because it is a no-replacement type of experiment. Once a date is selected, it is not available for the other people. Hence, the number of ways the 25 selections can be made is the product

$$365 \cdot 364 \cdot 363 \cdots 341.$$

Hence the probability of no matches is

$$\frac{365 \cdot 364 \cdot 363 \cdots 341}{365^{25}}.$$

This can be simplified to

$$\frac{365!}{340! \; 365^{25}} = \frac{P(365 \cdot 25)}{365^{25}}$$

which can be computed to be 0.4314. Then the probability of at least 1 match is $1 - 0.4314 = + 0.5686$, or approximately 1 chance in 2. Perhaps this is higher than one might guess.

It is instructive to look at this problem as a product of conditional probabilities. Suppose that all of the 25 people, in numerical order, name their birthdays; consider also the probability that, as each names his birthday, it does not match any of the birthdays previously named. The product of these probabilities will give the probability of no matches. Now the first person to name his birthday is certain not to match any previously named, since there are none. Therefore, the first probability is one, which, to agree with the previous solution, will be written as 365/365. The probability for the second person is 364/365, since one date has been used. The third is 363/365, the fourth is 362/365, and so on to 341/365 for the twenty-fifth person. The product is the same as the previously computed probability.

Conditional probabilities have been involved in a number of the previous examples. Additional insight into these conditional probabilities can be obtained from a slightly more complex problem.

When discussing conditional probabilities, we are speaking of a compound event, an event which consists of a combination of two or more other events. If two events are involved, say A and B, the compound event is denoted by AB. The probability of A and B, (AB), has previously been written in the forms $(AB) = (A)(B|A) = (B)(A|B)$. These were explained as follows: We can compute the probability (AB) as the product of two probabilities—the probability of A times the probability that B occurs, given that A has occurred; this is $(AB) = (A)(B|A)$. A similar statement with A and B interchanged gives

$$(AB) = (B)(A/B).$$

Now consider the equation

$$(B)(A|B) = (A)(B|A).$$

Solving for $(A|B)$ gives

$$(A|B) = \frac{(A)(B|A)}{(B)},$$

which is equivalent to

$$(A|B) = \frac{(AB)}{(B)}.$$

These formulas—comprising Bayes' Theorem—state a very curious relationship if we think of A and B as being related in time in such manner that A is the event resulting from a trial which occurred before the trial resulting in B. The probabilities can then be described as follows:

(A) is the probability that the first trial yields result A with no knowledge of the second trial.

(B) is the probability that the second trial yields B with no knowledge of the first trial.

(AB) is the probability that B follows A.

$(B|A)$ is the probability that the second trial results in B if it is known that A has occurred.

$(A|B)$ is the probability that the first trial resulted in A if it is known that the second trial resulted in B. Thus, it is an "after-the-fact" or an *a posteriori* probability that something did in fact occur on a trial which has taken place, but the outcome was not observed.

Thus, the formula called *Bayes' Theorem* is a way to interpret new knowledge as an influence upon probability of a trial which may actually have preceded the acquisition of this new knowledge.

Bertrand's " Box Paradox "

The most common example of the use of Bayes' Theorem is the famous *Bertrand's Box Paradox*. The word "paradox" is actually a misnomer, since the only paradoxical aspect of the problem is the tendency to make various errors in logic and get various incorrect answers to a simple probability problem.

Three boxes, identical in external appearance, each contain two coins. In Box A, both coins are gold; in Box B, one coin is gold and one is silver; and in Box C, both coins are silver. A box is chosen at random and a coin is removed at random from the box. The coin turns out to be gold. What is the probability that the other coin in the box is gold?

The question is equivalent to asking for the probability that Box A was selected, since only in this case will the other coin be gold. There are two easy ways to get the wrong answer. One way is to say that since the choice of boxes was made at random, the probability of having chosen Box A is one third. This is wrong because it disregards the information that a gold coin was drawn from the selected box. From this information it is obvious that Box C was not chosen since Box C contains only silver coins. This knowledge in turn tempts one to make another error and to give another false answer. Since it is known that either Box A or Box B must have been selected, there is a probability of one-half that it was Box A. Again the answer is false, because it overlooks the difference between the probabilities of selecting a gold coin from Boxes A and B.

To explain this more subtle difference, it is convenient to apply the general method previously discussed. Suppose that the box-coin selection problem were tried a number of times and the results of the trials turned out exactly as the probabilities say they should. A count of the times that different results occur can be used just as a count of the number of ways an event can occur, as illustrated in a number of previous examples. In this case, suppose 300 experiments were performed. We would expect Box A to be chosen 100 times; Box B, 100 times; and Box C, 100 times. Out of the 100 times Box A was chosen, there would be 100 cases in which a gold coin was drawn and zero cases in which a silver coin was drawn. For the 100 times Box B was chosen, 50 would result in a gold coin being drawn and 50 would result in a silver being drawn. For the 100 times Box C was chosen, there would be 100 cases in which a silver coin was drawn and zero cases in which a gold one was drawn. These numbers can be conveniently written in the following two-way table:

Coin	*Box*			*Total*
	A	B	C	
Gold	100	50	0	150
Silver	0	50	100	150
Total	100	100	100	300

Since a gold coin did appear, the event in question must be in the first row of the table, thus restricting it to 150 of the plays of the game. Of these 150 plays, 100 are Box A and 50 are Box B. Hence, the probability that Box A was in fact chosen (the other coin is gold) is equal to $\dfrac{100}{150} = 2/3$.

This answer is obtained also by a direct application of Bayes' Theorem. Use the form

$$(A|G) = \frac{(AG)}{(G)}$$

where G refers to drawing a gold coin and A refers to drawing Box A. It is also necessary to speak of Boxes B and C. The formulas for (AG) and (G) are

$$(AG) = \binom{1}{3}(1) = 1/3$$
$$(G) = (A)(G|A) + (B)(G|B) + (C)(G|C)$$
$$= 1/3(1) + 1/3\,(1/2) + 1/3\,(0)$$
$$= 1/2.$$

Therefore,

$$(A|G) = \frac{1/3}{1/2} = 2/3, \text{ as before.}$$

It is interesting to observe that this is one of the first times we encounter the problem of estimating from experimental data what the underlying probabilities were prior to data collection. The example serves to introduce the sample-population relationship discussed in the next chapter.

PROBLEMS

1. From a box of 10 items, a random sample of 3 is to be selected. How many possible ordered samples can be drawn if (a) sampling is with replacement? (b) sampling is without replacement?

2. With reference to Problem 1, if there are 8 good items and 2 defectives in the box, how many ways can ordered random samples of 3 be drawn that will contain 0, 1, or 2 defectives if (a) sampling is with replacement? (b) sampling is without replacement?

3. From Problems 1 and 2 compute the probabilities of 0, 1, and 2 defectives for sampling with and without replacement.

4. A bartender is to put 3 olives and 2 cherries into 3 martinis and 2 manhattans. If he is drunk and puts them in at random, what is the probability that the olives go in the martinis and the cherries in the manhattans? (Assume that each drink has either an olive or a cherry.)

5. A system designer has the choice of the 3 configurations shown below, with element reliabilities as indicated. Assuming independence, obtain the reliability of each configuration.

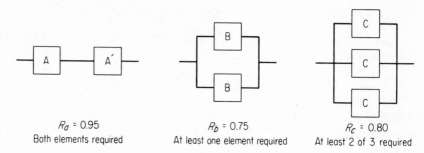

R_a = 0.95	R_b = 0.75	R_c = 0.80
Both elements required	At least one element required	At least 2 of 3 required

6. Assume a designer has the freedom to use as many elements in parallel as he wishes. If an element has a reliability of 0.6 over a fixed time interval, determine the minimum number of parallel elements he must use to achieve a unit reliability of at least 0.95 for the following 2 cases: (a) successful unit operation will result if at least 1 element operates; (b) at least 2 parallel elements are required.

7. A missile booster consists of a complex of 8 rocket engines—4 outboard and 4 inboard. If the probability is 0.95 that each engine will operate successfully over a fixed time interval, give the formulas for the probability of successful rocket operation if the design is the following (assume independence): (a) All 8 engines are required; (b) at least 6 of the 8 engines are required; (c) at least 3 inboard and 3 outboard engines are required.

Mathematical Concepts: Statistics

3.1 Introduction

This chapter presents a discussion of some of the fundamentals of mathematical statistics in terms of their application in reliability theory, including such basic concepts as random variables, density functions, and distribution functions. Special attention is paid to several specific functions commonly encountered in reliability work. This is followed by a brief introduction to the theory of sampling—the use of observed data in estimating characteristics of the unknown population from which the observations were drawn.

3.2 Basic Concepts

Discrete vs. Continuous Random Variables in Reliability

Reliability studies are concerned with both discrete and continuous random variables. The number of failures in a given interval of time and the number of defective parts in a sample are discrete variables, while the time from part installation to failure and the time between successive equipment failures are continuous. These examples suggest that test data can sometimes be collected in either form, as discrete or as continuous random-variable observations, at the discretion of the designer of the data collection system. Thus, in a life test, exact failure times can be observed if continuous monitoring is possible. On the other hand, if periodic monitoring is used—say, an hourly or a daily count of failures—then the data are recorded as observations of a discrete random variable.

The life-test illustrations point out the important fact that we can choose to view a discrete random variable as merely a "grouped data" equivalent of a continuous random variable. With continuous monitoring, actual failure times are observed accurately; but with periodic monitoring, failure times are known only within an interval of time—i.e., they consist of a grouping of the data in units equal to the length of this interval. In the solution of problems of a theoretical nature, it is often helpful to keep this relationship in mind, for it permits us to approximate a discrete random variable by a continuous one, and vice versa.

Density Functions

The *density function* of a discrete random variable is the probability that a certain value of a discrete variable will occur, expressed as a function of that value. In reliability studies, we usually think in terms of a time variable, t, and we shall use the notation $f(t)$ to represent its density function. Other representations are found in reliability literature—for example, x for the variable and $f(x)$ for the density—but the time notation will be used throughout most of the discussion in this chapter.

For a discrete random variable, t, the k possible values of the variable are denoted by $t_1, t_2, t_3, \ldots, t_k$, where k can be either finite or infinite. For a selected value of t, for example, t_3, the probability that a random trial will yield the value of t_3 is then denoted by $f(t_3)$. Since there are k possible values of t, it is necessary that

$$\sum_{i=1}^{i=k} f(t_k) = 1. \tag{3.1}$$

The statements above must be modified when we discuss a continuous random variable. The modification is required by the fact that, for most cases, the continuous random variable must be described as falling within an interval rather than exactly at a point. Indeed, it is customary to say that the probability of an event occurring at a point in time is zero (actually it is more than customary, but some of the niceties of mathematical rigor must be foregone in this discussion). However, we can logically speak of a non-zero value of the probability in an interval of time. For example, it is unreasonable to expect an electron tube to fail at the conclusion of exactly 100 hours of use, but it is reasonable to speak of its failure after 99 hours and 45 minutes but before 100 hours and 15 minutes, or in any other desired interval of time.

There are certain important exceptions to the general statement about zero probability at a point for a continuous variable. For example, perhaps a portion of new electron tubes of a certain type are defective, and they never work. These cause installation failures, and the probability associated with $t = 0$ is the fraction of defectives which are inoperable when installed. These tubes could obviously be handled separately from other failures, but the mathematical concept of integration has been extended to make this unnecessary. No difficulty will be encountered if we disregard this rare case in the discussion of continuous random variables to follow.

We can define the density function of a continuous random variable, $f(t)$, in a manner entirely consistent with that of a discrete random variable. The probability that a discrete variable will assume any of a set of values is the sum of the density function evaluated for the separate values. For the continuous case, this corresponds to the probability that the continuous variable falls in an interval, an infinite set of values, requiring that we replace the summation operation by an integration. Thus, if the probability that a random trial yields a value of t within the interval from t_1 to t_2 is

$$\int_{t_1}^{t_2} f(t)dt,$$

then $f(t)$ is the density function for the continuous random variable, t. If t can never be less than a nor greater than b then

$$\int_{a}^{b} f(t)dt = 1, \tag{3.2}$$

and, in the definition, $a \leq t_1 \leq t_2 \leq b$. Analogous to the discrete case, the interval a to b can be $-\infty$ to $+\infty$.

If the interval t_1 to t_2 is short, then $t_2 - t_1$ is small, and the integral can be reasonably approximated by

$$f(t_1)(t_2 - t_1).$$

A more usual statement of this approximation is as follows: The probability that t falls in the interval t_1 to $(t_1 + dt_1)$ is $f(t_1)dt_1$. In most cases the subscript is dropped and the probability is written as $f(t)dt$. Indeed, some authors use this as the basic expression in the definition of the density function.

Note that the definition of a density function always requires knowledge or explicit statement of the interval a to b over which the function applies. For time as a variable, the usual range is 0 to ∞, but there are cases in which it is $-\infty$ to $+\infty$, as shown earlier. Even when a and b are finite, it is convenient to use the convention that limits will be taken as $-\infty$ to $+\infty$ with $f(t) = 0$ for $-\infty < t < a$ and $b < t < \infty$, and this convention will be used herein.

Distribution Functions

The term *distribution function* has been used in many senses in literature. Recently, authors in probability and statistics have tended toward uniformity of usage, as follows: The distribution function, $F(t)$, is the probability that in a random trial, the random variable is not greater than t. Thus,

$$F(t) = \int_{-\infty}^{t} f(t)dt, \tag{3.3}$$

and this is recognized as the *unreliability function* when speaking of failures.* The name *cumulative distribution function* is also used, but the modifier "cumulative" is really not necessary. If the random variable is discrete, the integral is replaced by a summation.

The reliability function, $R(t)$, is given by

$$R(t) = 1 - F(t). \tag{3.4}$$

It expresses the probability that the variable is at least as large as t. In integral form,

$$R(t) = \int_{t}^{\infty} f(t)dt. \tag{3.5}$$

Differentiating Equation (3.5),

$$\frac{dR(t)}{dt} = -\frac{dF(t)}{dt} = -f(t). \tag{3.6}$$

Failure Rate

The probability of failure in a given time interval, t_1 to t_2, can be expressed in terms of either unreliability functions or reliability functions. Starting with the integral form previously used, the relationships are as follows:

$$\int_{t_1}^{t_2} f(t)dt = \int_{-\infty}^{t_2} f(t)dt - \int_{-\infty}^{t_1} f(t)dt = F(t_2) - F(t_1). \tag{3.7}$$

$$\int_{t_1}^{t_2} f(t)dt = \int_{t_1}^{\infty} f(t)dt - \int_{t_2}^{\infty} f(t)dt = R(t_1) - R(t_2). \tag{3.8}$$

The rate at which failures occur in the interval t_1 to t_2, naturally called the *failure rate*, will be defined as the ratio of the probability that failure occurs

---

*Some people object to the use of the same letter in the integrand and also in the limits of the integral. This is done here in spite of the objection in order to simplify the reference to time as the variable in each of the functions $f(t)$ and $F(t)$.

in the interval, given that it has not occurred prior to t_1, the start of the interval (cf. Section 2.3.) to the interval length. Thus, the expression for the failure rate is

$$\frac{R(t_1) - R(t_2)}{(t_2 - t_1) R(t_1)}. \tag{3.9}$$

Another familiar form of the failure rate expression is obtained by the substitutions

$$t_1 = t, \quad t_2 = (t + h).$$

Then the expression is

$$\frac{R(t) - R(t + h)}{h R(t)}. \tag{3.10}$$

Reliability literature shows some variations in the definition of failure rate. The expression is sometimes written without the interval length in the denominator, as

$$\frac{R(t) - R(t + h)}{R(t)}. \tag{3.11}$$

This expression is, of course, the probability that, given survival up to time t, failure occurs in the interval t to $(t + h)$. It is equivalent to the definition of failure rate as first given if $h = 1$. The word "rate" usually is interpreted as something per unit time—such as miles per hour or revolutions per minute. Hence, if h is not unity, it seems only natural to divide by h and thereby reduce failure rate to a per-unit-time basis. Therefore, the first definition will be adopted, and there will be no further mention of the second.

Hazard Rate

The *hazard rate* or *instantaneous failure rate*, denoted by $z(t)$, is defined as the limit of the failure rate as the interval length approaches zero. Then the hazard rate is

$$z(t) = \lim_{h \to 0} \left(\frac{R(t) - R(t + h)}{h R(t)} \right) = - \frac{1}{R(t)} \frac{dR(t)}{dt} = \frac{f(t)}{R(t)}, \tag{3.12}$$

which can also be written as

$$z(t) = - \frac{d \ln R(t)}{dt}. \tag{3.13}$$

Perhaps the simplest explanation of hazard rate and failure rate is made by analogy. Suppose a family takes an automobile trip of 120 miles and completes the trip in 3 hr. Their average rate was 40 mph, although they drove faster at some times and slower at other times. The rate at any given instant could have been determined by reading the speed indicated on the speedo-meter at that instant. The 40 mph is analogous to the failure rate, and the speed at any point is analogous to the hazard rate. Actuarial theory also provides an analogy, with death rate analogous to failure rate and force of mortality analogous to hazard rate.

Moments

Even though all of the properties of a density function are determined when the specific function is given, we are usually unable to visualize these pro-perties without the aid of certain additional relationships and pictorial representations. We shall not attempt to cover the pictorial aspects in the present treatment, but certain principles of graphical presentation will be illustrated. It is important, however, to examine some of the numerical relationships which are widely used in describing and comparing density functions.

Reliability studies make frequent reference to mean life and mean time between failures. It is also common to encounter the terms—variance and standard deviation. The *mean* and *variance* are names applied to two of the so-called moments of a density function. We shall find it useful in the present context to devote special attention to the concept of moments and to describe methods of calculation and interpretation of moments. These objectives will be accomplished by, first, giving the definition; second, illustrating the com-putation; and, finally, discussing the interpretation of moments.

For a continuous density function $f(t)$, $-\infty < t < \infty$, the moment of order k about $t = a$, denoted by v_k, is defined by the relationship:

$$v_k = \int_{-\infty}^{\infty} (t - a)^k f(t)dt. \tag{3.14}$$

If $a = 0$, this formula becomes

$$v_k = \int_{-\infty}^{\infty} t^k f(t)dt. \tag{3.15}$$

We are especially interested in moments about the value $a = v_1$, denoted by μ_k and given by the formula

$$\mu_k = \int_{-\infty}^{\infty} (t - v_1)^k f(t)dt. \tag{3.16}$$

For the discrete case we merely need to replace the integral by a summation. We shall illustrate the computational procedures for both types of density functions.

The ν_k and μ_k moments are dependent on the units used for measuring the variable t. Furthermore, the ν_k moments are dependent on the zero point of t although the μ_k moments are not. It is convenient to define another moment, α_k, which is independent of both unit and zero point. This is accomplished by the formula

$$\alpha_k = \frac{\mu_k}{\mu_2^{k/2}}. \tag{3.17}$$

We shall have little need for moments of higher order (large values of k) in the present treatment. Indeed it will be possible to restrict our attention primarily to the values $k = 1$ and $k = 2$. Previous experience probably leads to the recognition of ν_1 as the mean and of μ_2 as the variance, commonly denoted by σ_2 where σ, the positive square root of μ_2, is called the *standard deviation*. The third and fourth moments, α_3 and α_4, are used to describe *skewness* (departure from symmetry, if any) and *kurtosis* (relative peakedness around the mean), respectively. Later we shall give examples to clarify some of these concepts. The meanings of these four moments are noted here only to emphasize the usefulness of moments as descriptions of properties of the density function.

Other types of measures could be and often are used in place of the moments defined above. However, even the simpler of these other measures frequently have complex definitions and in many instances are more difficult to compute than are the moments. For example, the median, denoted by M, is defined by

$$\int_{-\infty}^{M} f(t)\, dt = \int_{M}^{\infty} f(t)dt = 0.5. \tag{3.18}$$

Since M occurs as a limit on the integral sign, its computation in formula form is frequently very difficult. The mode also can be used to illustrate this point. It is defined as the value or values of t which satisfy

$$\frac{df(t)}{dt} = 0, \tag{3.19}$$

provided that the value or values of t are at relative maxima of $f(t)$. While these two examples are stated for the continuous case, it is obvious that corresponding difficulties occur for the discrete case. Moments are no doubt popular because they can be so easily defined and so readily computed from quite simple formulas.

Numerical values of the moments can be obtained by a variety of arithmetical procedures. If the density function is given by a formula, the definitions can be used directly to obtain formulas for the moments. Numerous other devices are also available and illustrations will be given later. The more interesting case arises when the density function itself is expressed by a table of numerical values. In this situation, many alternative methods are available.

To illustrate these, it is necessary to develop a relationship between ν_k and μ_k types of moments, as follows. Consider the discrete case for this purpose. This is not a restriction, since the same formulas can be derived for the continuous density by using integrals in place of summations.

Suppose there are n values of t, say $t_1, t_2, \ldots, t_n$ with values of the density $f(t_1), f(t_2), \ldots, f(t_n)$, respectively. By definition,

$$\mu_k = \sum_{i=1}^{i=n} (t_i - \nu_1)^k f(t_i), \tag{3.20}$$

and

$$\nu_1 = \sum_{i=1}^{i=n} t_i f(t_i). \tag{3.21}$$

Consider successive values of k. For $k = 1$,

$$\mu_1 = \sum_{i=1}^{i=n} (t_i - \nu_1) f(t_i)$$

$$= \sum_{i=1}^{i=n} [t_i f(t_i) - \nu_1 f(t_i)]$$

$$= \sum_{i=1}^{i=n} t_i f(t_i) - \nu_1 \sum_{i=1}^{i=n} f(t_i)$$

$$= \nu_1 - \nu_1(1)$$

$$= 0. \tag{3.22}$$

[Note that $\sum_{i=1}^{i=n} f(t_i) = 1$ since $f(t)$ is a density function.]

The result above indicates that $\mu_1 = 0$ for all density functions and hence is of no value as a descriptive number. For $k = 2$,

$$\mu_2 = \sum_{i=1}^{i=n} (t_i - \nu_1)^2 f(t_1)$$

$$= \sum_{i=1}^{i=n} (t_i^2 - 2\nu_1 t_i + \nu_1^2) f(t_i)$$

$$= \sum_{i=1}^{i=n} t_i^2 f(t_i) - 2\nu_1 \sum_{i=1}^{i=n} t_i f(t_i) + \nu_1^2$$

$$= \nu_2 - 2\nu_1^2 + \nu_1^2$$

$$= \nu_2 - \nu_1^2. \tag{3.23}$$

The algebraic steps depend on the common rules of addition and the definition of ν_k moments. For $k = 3$,

$$\mu_3 = \sum_{i=1}^{i=n} (t_i - \nu_1)^3 f(t_i)$$

$$= \sum_{i=1}^{i=n} (t_i^3 - 3\nu_1 t_i^2 + 3\nu_1^2 t_i - \nu_1^3) f(t_i)$$

$$= \nu_3 - 3\nu_1\nu_2 + 3\nu_1^3 - \nu_1^3$$

$$= \nu_3 - 3\nu_1\nu_2 + 2\nu_1^3. \tag{3.24}$$

The algebraic steps are dependent on the same rules as noted above.

It is easy to see that for any k we merely need to expand a binomial. Thus

$$(t_i - \nu_1)^k = t_i^k - k\nu_1 t_i^{k-1} + \frac{k(k-1)}{2} \nu_1^2 t_i^{k-2} + \cdots$$

$$+ k(-\nu_1)^{k-1} t_i + (-\nu_1)^k. \tag{3.25}$$

In Equation 3.20 for μ_k, these terms of the binomial expansion are multiplied by $f(t_i)$ and then summed, yielding a series of terms involving ν_k-type moments and binomial coefficients. Notice that the last two terms of the resulting series can be combined as follows:

$$k(-1)^{k-1} \nu_1^k + (-1)^k \nu_1^k = (-1)^{k-1} \nu_1^k (k-1). \tag{3.26}$$

This description permits us to write formulas for any μ_k as follows:

$$\mu_2 = \nu_2 - \nu_1^2,$$

$$\mu_3 = \nu_3 - 3\nu_1\nu_2 + 2\nu_1^3,$$

$$\mu_4 = \nu_4 - 4\nu_1\nu_3 + 6\nu_1^2\nu_2 - 3\nu_1^4, \tag{3.27}$$

$$\mu_5 = \nu_5 - 5\nu_1\nu_4 + 10\nu_1^2\nu_3 - 10\nu_1^3\nu_2 + 4\nu_1^5,$$

$$\mu_6 = \nu_6 - 6\nu_1\nu_5 + 15\nu_1^2\nu_4 - 20\nu_1^3\nu_3 + 15\nu_1^4\nu_2 - 5\nu_1^6, \text{ etc.}$$

Therefore, we can compute ν_k for all values of k in which we are interested and then use the formulas above to compute the μ_k. We could also compute ν_1 and then use Equation 3.20:

$$\mu_k = \sum_{i=1}^{i=n} (t_i - \nu_1)^k f(t_i)$$

by subtracting as indicated, $t_i - \nu_1$, for each i, raising the difference to the kth power without expanding the binomial, multiplying each such power by the appropriate $f(t_i)$, and summing these products. We would obtain α_k from the formula

$$\alpha_k = \frac{\mu_k}{\mu_2^{k/2}} = \frac{\mu_k}{\sigma^k}. \tag{3.28}$$

The density function may be a tabulation which represents a theoretical distribution, such as a population density, or it may be obtained from observed data with or without smoothing by some technique such as curve fitting. We shall use as an illustration a set of data which might have been generated by observations of times between failures of certain equipments in field use. The development of the observed density function from this data is shown in Table 3.1.

Table 3.1 Development of Observed Density Function

| Hours Between Failures | | Observed | |
Interval	Midpoint t_i	number f_i	Density $f(t_i)$
0 – 10	5	4	0.08
10 – 20	15	6	0.12
20 – 30	25	10	0.20
30 – 40	35	13	0.26
40 – 50	45	7	0.14
50 – 60	55	5	0.10
60 – 70	65	3	0.06
70 – 80	75	2	0.04
	Sum	50	1.00

Time between failures in hours is shown in the first two columns, with the first column showing the time intervals into which observations were grouped and the second column showing interval midpoints. When data is grouped in this way, it is common practice to treat the interval midpoint as the failure time or variable value in the general case, for all occurrences in the interval. The number of observations is shown in the third column. Thus, 4 failures were observed between 0 and 10 hr of equipment operation, and these 4 will be treated as though each of them failed after exactly 5 operating hours. In this example, a total of 50 failures were observed. If the frequencies in column three are each divided by 50, the observed density function values are obtained as shown in column four. These are often called *relative frequencies*, a term more descriptive than *observed density function* but less suitable for present purposes.

In practice it is common to use the observed frequencies rather than the density function to compute moments. This procedure usually reduces the arithmetical work load, since observed frequencies are integers and density function values are fractions. Of course, the difference between the two methods is merely that in one method we sum first and then divide, while in the other we divide initially. For example, the first moment, v_1, is given by either of the two arithmetical procedures set forth below.

1. Observed frequency method

$$\nu_1 = \frac{1}{50}\,[4(5) + 6(15) + 10(25) + 13(35) + 7(45) + 5(55) + 3(65) + 2(75)]$$
$$= 0.02\,[20 + 90 + 250 + 455 + 315 + 275 + 195 + 150]$$
$$= 0.02\,(1750)$$
$$= 35 \text{ hr.}$$

2. Density function method

$$\nu_1 = (0.08)\,(5) + (0.12)\,(15) + (0.20)\,(25) + (0.26)\,(35) + (0.14)\,(45)$$
$$+ (0.10)\,(55) + (0.06)\,(65) + (0.04)\,(75)$$
$$= 0.40 + 1.80 + 5.00 + 9.10 + 6.30 + 5.50 + 3.90 + 3.00$$
$$= 35 \text{ hr.}$$

The first of the two methods will be used in most of the numerical illustrations to follow.

Worksheets for moment computations are reproduced in the tables below. Each table is intended to illustrate one of the many available numerical schemes, all being based on the moment definition formulas. Table 3.2 illustrates the computation of ν_1, ν_2, and ν_3 from the data in its original form. Resulting values of these moments are then used to compute μ_2 and μ_3.

Table 3.2 COMPUTATION OF MOMENTS FROM ORIGINAL DATA

t	f	ft	ft^2	ft^3
5 ✕	4	20	100	500
15	6	90	1,350	20,250
25	10	250	6,250	156,250
35	13	455	15,925	557,375
45	7	315	14,175	637,875
55	5	275	15,125	831,875
65	3	195	12,675	823,875
75	2	150	11,250	843,750
Σ	50	1,750	76,850	3,871,750

Necessary moment computations are as follows:

$$\nu_1 = \frac{1,750}{50} = 35,$$

$$\nu_2 = \frac{76,850}{50} = 1,537,$$

$$\nu_3 = \frac{3,871,750}{50} = 77,435,$$

$$\mu_2 = \nu_2 - \nu_1^2 = 1{,}537 - 35^2 = 1{,}537 - 1{,}225 = 312,$$

$$\mu_3 = \nu_3 - 3\nu_1\nu_2 + 2\nu_1^3 = 77{,}435 - 3(35)\,(1{,}537) + 2(35^3)$$

$$= 77{,}435 - 161{,}385 + 85{,}750$$

$$= 1{,}800.$$

Table 3.3 shows the worksheet for computing μ_2 and μ_3 by an alternate method if it is known from previous computation, as above, that $\nu_1 = 35$. This worksheet also shows a check of the fact that $\mu_1 = 0$. To introduce a common notation, the table headings use $\bar{t}$ as a symbol for the mean, ν_1. The Table 3.3 computation gives the values of μ_2 and μ_3 as follows:

$$\mu_2 = \frac{15{,}600}{50} = 312$$

$$\mu_3 = \frac{90{,}000}{50} = 1{,}800,$$

thereby confirming the values obtained from Table 3.2. It is possible to find the standard deviation, σ, from the formula

$$\sigma = \sqrt{\mu_2} = \sqrt{312} = 17.66.$$

Table 3.3 WORKSHEET FOR ALTERNATE COMPUTATION METHOD

$(t-\bar{t})$	f	$f(t-\bar{t})$	$f(t-\bar{t})^2$	$f(t-\bar{t})^3$
−30	4	−120	3,600	−108,000
−20	6	−120	2,400	−48,000
−10	10	−100	1,000	−10,000
0	13	0	0	0
10	7	70	700	7,000
20	5	100	2,000	40,000
30	3	90	2,700	81,000
40	2	80	3,200	128,000
Σ	50	0	15,600	90,000

A third computation is developed in Table 3.4. In this case the arithmetical problems are eased by a modification of the time scale which, in effect, considers a 10-hr time unit and measures time from the midpoint of the first interval as a zero point. This time transformation is accomplished by computing moments for a new time variable, x, related to time variable t by the equation

$$x = \frac{t-5}{10}.$$

The method for computing variable t moments from those for variable x will be illustrated numerically.

Table 3.4 COMPUTATION WITH MODIFIED TIME SCALE

t	$x = \dfrac{t-5}{10}$	f	fx	fx^2	fx^3
5	0	4	0	0	0
15	1	6	6	6	6
25	2	10	20	40	80
35	3	13	39	117	351
45	4	7	28	112	448
55	5	5	25	125	625
65	6	3	18	108	648
75	7	2	14	98	686
Σ		50	150	606	2,844

Since moments in terms of two variables are involved, the moment notation will give variable identification by a second subscript. Using the numbers in Table 3.4 we obtain

$$\nu_{1x} = \frac{150}{50} = 3$$

$$\nu_{2x} = \frac{606}{50} = 12.12$$

$$\nu_{3x} = \frac{2,844}{50} = 56.88$$

$$\mu_{2x} = \nu_{2x} - \nu_{1x}^2 = 12.12 - 3^2 = 3.12$$

$$\mu_{3x} = \nu_{3x} - 3\nu_{1x}\nu_{2x} + 2\nu_{1x}^3$$

$$= 56.88 - 3(3)(12.12) + 2(3^3)$$

$$= 56.88 - 109.08 + 54$$

$$= 1.80.$$

It is now possible to compute t moments as follows:

$$\nu_{1t} = 10\nu_{1x} + 5 = 35$$

$$\mu_{2t} = 10^2\mu_{2x} = 312$$

$$\mu_{3t} = 10^3\mu_{3x} = 1,800.$$

The Population and the Sample

Before any specific functions are discussed, it is important to state the usual nature of the statistical problem in order to explain how we should view the formulas and curves to be presented. Collected data received from any testing or surveillance program should be viewed as information to be used in guessing what will happen in the future—i.e., in prediction. Of course, there is an element of historical interest in knowing what did happen just for the sake of knowledge itself. The real payoff, however, lies in using the data as a sample from some population, and then describing this population from some of the characteristics of the sample data. For example, we would like to be able to take field data on failures in a specific aerospace or weapons system and write the formula for the failure density of all future failure experience which will be met with this same system and even with improved versions of the system, using adjustment for the system modifications.

The moral of this view for the reliability analyst is that field data constitute a sample, that random sampling peculiarities must be smoothed out, that population density parameters must be estimated, that the estimation errors must themselves be estimated, and—what is even more difficult—that the very nature of the population density must be estimated. To achieve these ends, it is necessary to learn as much as possible about the population density functions, and especially what kind of results we can expect when samples are drawn, the data are studied, and we attempt to go from data backward to the population itself. It is also important to know what types of population densities are produced from any given set of engineering conditions. This implies the necessity for developing probability models, or going from a set of assumed engineering characteristics to a population density. Thus, the following discussion of densities relates to the population formulas when idealized formulas are used, and then to sampling problems derived from assumptions as to which formulas do actually apply.

3.3   Specific Functions

It has been traditional to restrict the number of functions used in reliability work. This has been done for two reasons. First, it has been found that a relatively small number of functions satisfy most of the needs. Second, it is an unfortunate fact that statistical theory is not very well developed for many functions. We may cite the binomial, Poisson, and uniform densities for discrete variables, and the exponential, normal (or Gaussian), gamma, Weibull, and rectangular densities for continuous variables. However, statistical theory is very incomplete even for these functions. In this presentation, only the major aspects of each function will be touched upon.

Figure 3.1 gives formulas and curves for some of the functions associated with the selected densities. Perhaps the binomial, exponential, and Gaussian can be singled out as being of prime value in reliability work and, hence, deserving of our special attention; however, it would be difficult to please everyone in any selection. Nevertheless, we shall give more consideration to these functions—recognizing that others might make a different choice.

Difference in failure rates and hazard rates are significant elements in the comparison of densities. The exponential has a constant hazard rate, independent of time. By contrast, the Gaussian density hazard rate increases with time. For the Weibull density, the hazard rate decreases or increases with time, depending on the parameters of the density function. A constant hazard rate means that the probability of failure is independent of age—i.e., an old equipment which is still operating is just as good as a new one. This is equivalent to saying that deterioration is not involved in failure. On the other hand, an equipment which deteriorates, and has a tendency to failure that increases with age, will be described by a density with an increasing hazard rate. For the equipment failures to be described by a density with a *decreasing* hazard rate, the equipment must improve with age.

Since it is quite difficult to visualize a part or an equipment which does not eventually deteriorate, it is natural to ask why we should consider any case other than the one with increasing hazard rate. There appear to be a few reasons for considering decreasing hazard rates, but there are many reasons for looking at the constant hazard rate. Some of these reasons are examined below.

At the outset, it is important to observe that constant hazard rate means exponential density. This can be proved as follows: Let the constant rate be denoted by λ. Then, using Equation 3.13:

$$z(t) = -\frac{d ln R(t)}{dt} = \lambda.$$

Solving this differential equation gives

$$ln\ R(t) = -\lambda t + c$$

where c is the constant of integration. By noting that $R(0) = 1$, it follows that $c = 0$. Then

$$ln\ R(t) = -\lambda t \quad \text{or} \quad R(t) = e^{-\lambda t}, \tag{3.29}$$

and

$$f(t) = -\frac{dR(t)}{dt} = \lambda e^{-\lambda t}. \tag{3.30}$$

Hence, we can speak of exponential density as equivalent to constant hazard rate.

Name	Density Function $f(t)$	Reliability Function $R(t)$	Hazard Rate $z(t) = \dfrac{f(t)}{R(t)}$
Normal or Gaussian	$f(t) = \dfrac{1}{\sigma\sqrt{2\pi}}\, e^{-\frac{(t-\theta)^2}{2\sigma^2}}$	$R(t) = \int_t^\infty \dfrac{1}{\sigma\sqrt{2\pi}}\, e^{-\frac{(t-\theta)^2}{2\sigma^2}}\, dt$	$z(t) = \dfrac{f(t)}{R(t)}$
Exponential	$f(t) = \dfrac{1}{\theta}\, e^{-t/\theta}$	$R(t) = e^{-t/\theta}$	$z(t) = \dfrac{1}{\theta}$
Gamma	$f(t) = \dfrac{1}{a!\,\beta^{a+1}}\, t^a e^{-t/\beta}$ $\beta = 1,\ a = 0, 1, 2, 3, 4,$	$R(t) = \int_t^\infty \dfrac{1}{a!\,\beta^{a+1}}\, t^a e^{-t/\beta}\, dt$ $\beta = 1,\ a = 0, 1, 2, 3, 4$	$z(t) = \dfrac{f(t)}{R(t)}$ $\beta = 1,\ a = 0, 1, 2, 3, 4$
Weibull	$f(t) = \dfrac{\beta t^{\beta-1}}{a}\, e^{-\frac{t^\beta}{a}}$ $a = 1,\ \beta = 1, 2, 3, 4$	$R(t) = e^{-\frac{t^\beta}{a}}$ $a = 1,\ \beta = 1, 2, 3, 4$	$z(t) = \dfrac{\beta t^{\beta-1}}{a}$ $a = 1,\ \beta = 1, 2, 3, 4$

Fig. 3.1(a) Density and reliability functions and hazard rates of the normal, exponential, gamma, and Weibull distributions.

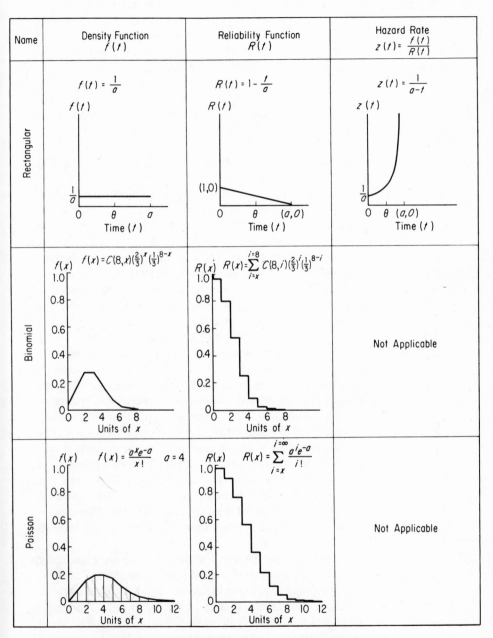

Fig. 3.1(b) Density and reliability functions of the rectangular, binomial, and Poisson distributions; hazard rate of the rectangular distribution.

In Figure 3.1(a) the exponential reliability function is expressed as:

$$R(t) = e^{-t/\theta}, \tag{3.31}$$

where θ is called the *mean life* or *mean time to failure*. It must now be indicated that for the exponential distribution function:

$$\theta = 1/\lambda. \tag{3.32}$$

This is not a matter of arbitrary definition. Rather, it is a basic property of the exponential distribution function which may be evaluated from:

$$\theta = \int_0^\infty R(t)dt. \tag{3.33}$$

There is another situation for encountering failures exhibiting this constant characteristic, but not properly a constant hazard rate. This is the situation in which failures are repaired as they occur. The reliability function can be thought of in terms of an experiment in which a set of new equipments are all started in operation at the same time. The reliability function is the proportion of these equipments which we theoretically expect to be still operating at any time t, provided no repairs are made. In this case the number of equipments at risk decreases as failures occur. If repairs are made, the number at risk would not decrease. However, even if the new equipment density has an increasing hazard rate, it can be shown that a constant failure pattern is approached. This occurs as follows: As the equipments age and begin to fail, the failure frequency rises to a peak. With repair, usually thought of as replacement with new equipments, the average age of equipments tends downward, and so does the failure frequency. This cycling of age and failure frequency continues, but it diminishes (quite rapidly in many cases) and approaches a stable state with constant failure frequency. The latter result can be illustrated by a simple example. If a man starts a new taxicab company with 50 new cabs, he can expect his maintenance shop load to approach a level in a fairly short time. This illustration indicates why this constant hazard rate is really not a constant hazard rate. The individual equipments do not have an exponential time-to-failure density. However, it is not difficult to think of the constant hazard rate as being a proper term to apply to the *system* of equipments.

This discussion of stable state in the repair situation leads quite naturally to another example in which a constant hazard rate is observed. This case will not be justified mathematically, but it is easy to visualize that a complex equipment might well act like the stable state in the repair case. An equipment with many parts which have many different failure patterns will indeed have a time-to-failure density closely approximated by an exponential. Furthermore, if such an equipment is repaired and returned to operation, the age and failure cycles damp out very rapidly.

Another use for which the exponential is particularly suited is as an approximation of some other density for a relatively short interval of time. Short interval of time cannot be defined in exact terms—however, it depends on the seriousness of error resulting from use of the approximation, and, in general, it refers to an interval in which the true density hazard rate does not change very much. Of course this change is usually unknown and must be estimated from field or test data. Such a procedure is effective, but one warning must be stressed: If mean life or mean time to failure is estimated by taking test data over such a short interval of time, on the assumption of an exponential density over an extended time range, very severe bias can result. Such a procedure can produce a very large overestimation of mean life. The difficulty is that the data are collected before deterioration has a chance to increase failure rates. Under these circumstances, to assume the exponential is the same as to assume that this increase never occurs. Statistics can lie if they are used improperly, and the warning above applies directly to a case in which they do lie.

Density and Reliability Functions of a Complex Equipment in Terms of Functions of Parts

Consider an equipment made up of k parts in series, which means that the failure of any one part causes failure of the equipment. Assume that the failures of parts themselves are independent events. Denote the part reliability functions by $R_i(t)$, $i = 1, 2, \ldots, k$ and the equipment reliability function by $R(t)$, and denote the density functions by $f_i(t)$ and $f(t)$, respectively. For the equipment to be failure-free for time t, every part must operate without failure for that time. Thus, the product rule for probabilities gives the equation

$$R(t) = R_1(t) \cdot R_2(t) \cdots R_k(t) \qquad (3.34)$$

for the reliability function of the equipment. The density function is given by

$$f(t) = \frac{-dR(t)}{dt}.$$

We can perform the differentiation process most easily by considering the hazard rate from Equations 3.12 and 3.13.

$$z(t) = \frac{f(t)}{R(t)} = -\frac{dlnR(t)}{dt}.$$

Now
$$lnR(t) = \sum_{i=1}^{k} lnR_i(t), \qquad (3.35)$$

and
$$-\frac{dlnR(t)}{dt} = \frac{f(t)}{R(t)} = \sum_{i=1}^{k} \frac{f_i(t)}{R_i(t)} \qquad (3.36)$$

or

$$f(t) = R(t) \sum_{i=1}^{k} \frac{f_i(t)}{R_i(t)}.$$ (3.37)

Note that the proof above includes the statement that the equipment hazard rate is the sum of the part hazard rates under the assumption of independence of part failures and that this is true regardless of the forms of the part density functions.

An especially important case arises when the parts each have an exponential time-to-failure density. Let

$$f_i(t) = \lambda_i e^{-\lambda_i t}, \; R_i(t) = e^{-\lambda_i t}, \; i = 1, 2, \ldots, k.$$ (3.38)

Then

$$R(t) = e^{-\lambda_1 t}, e^{-\lambda_2 t}, \ldots, e^{-\lambda_k t}$$

$$= e^{-(\lambda_1 + \lambda_2 + \ldots \lambda_k)t},$$ (3.39)

or

$$R(t) = e^{-\lambda t}$$ (3.40)

where

$$\lambda = \lambda_1 + \lambda_2 + \ldots + \lambda_k \quad \text{and} \quad f(t) = \lambda e^{-\lambda t}.$$ (3.41)

In this case the equipment also has an exponential time-to-failure density, and the hazard rate is the sum of the part hazard rates.

Such a result is very important in the prediction of equipment reliability. Indeed, since equipment hazard rate is the sum of part hazard rates for any forms of part reliability functions, it is logical to add hazard rates as a distribution-free approximation for a moderate time interval prediction. This result depends on an assumption of independence of part failures. Sometimes such an assumption is intentionally overlooked on the basis of the argument that data used to estimate hazard rates are collected under the influence of part interactions. Thus, by the conditioning influence of part interactions, the independence among failure rates is simulated—or approximated. Again, it can be argued that this may be a reasonable alibi for short-range prediction. A more complete treatment of this problem will be given in the chapters on prediction methods.

The Use of Density Functions in Developing a Failure Pattern Model

As indicated earlier, the reliability analyst is frequently faced with the task of developing the mathematical representation of a failure pattern from a set of conditions expressed in engineering terms. A very simple case of this was shown in the derivation of the exponential from the condition that hazard rate was independent of age. It is instructive to consider a more complex example which has many applications in engineering design. We will phrase the discussion in terms of a part, although it could be applied directly to a component or even an equipment. The problem to be solved is the selection of a part of sufficient strength to yield an acceptable design safety factor, assuming that we know the distributions of the part strength and of the loads to be experienced. It is further assumed that safety-factor design criteria are given.

As a first step, consider merely the problem of developing a probability model which expresses the probability that the load will exceed the strength. The design safety factor will be considered later. Use the following notation:

$$x = \text{the load with density function } f(x);$$

$$y = \text{the strength with density function } f(y);$$

$$z = x - y = \text{the load minus the strength; and failure occurs if}$$
$$z > 0, \text{ i.e., if load exceeds strength.}$$

Of course, x, y, and z are all expressed in the same units. They are the same type of variable, such as pounds per square inch. The problem is first to find the probability density function of z, say $g(z)$.

To visualize the problem, consider the densities $f(x)$ and $f(y)$ shown in Figure 3.2. The two densities are drawn on the same pair of axes, and x and y were observed to be the same type of variable. Thus, the horizontal axis is x or y and the vertical axis is $f(x)$ or $f(y)$, depending on whether we are speaking of load or strength, respectively. Other symbols on the figure will be explained later. The point of interest is the probability that a randomly selected x exceeds an independently and randomly chosen y.

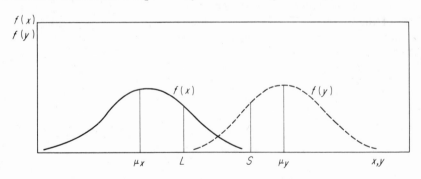

Fig. 3.2 Density functions, $f(x)$ and $f(y)$.

By virtue of independence, the joint density of x and y is, in differential form,

$$f(x)\ f(y)dxdy.$$

Since we are primarily interested in $z = (x - y)$, this joint density will be transformed into

$$f(x)\ f(x - z)dxdz.$$

Then, the density of z, denoted by $g(z)$, is, in differential form,

$$g(z)dz = \int_b^a f(x) \, f(x - z)dxdz, \tag{3.42}$$

where a and b are the extreme values of x, the smallest and the largest possible loads, respectively.

Now one design procedure is as follows: A load level L is specified at some number, n_x, of standard deviations of $f(x)$ above the mean load, θ_x; and a strength level, S, is specified at some number, n_y, of standard deviations of $f(y)$ below the mean strength, θ_y. Denote the standard deviations by σ_x and σ_y, respectively. Then

$$L = \theta_x + n_x\sigma_x, \tag{3.43}$$

and

$$S = \theta_y - n_y\sigma_y. \tag{3.44}$$

Now define a safety factor, m, by the formula

$$m = \frac{S}{L} = \frac{\theta_y - n_y\sigma_y}{\theta_x + n_x\sigma_x}. \tag{3.45}$$

The design criterion is to find a part with strength properties such that a preselected m can be achieved for given $f(x)$ (and hence, θ_x and θ_x), n_x, and n_y. To be more exact, the engineer should then check to determine the probability that load exceeds strength; i.e., $z > 0$. Of course, this probability may be determined from the conditions above and knowledge of the form of the densities $f(x)$ and $f(y)$. The whole process will be illustrated for the case in which these two densities are normal.

In this case

$$f(x) = \frac{1}{x\sqrt{2\pi\sigma}} \exp \frac{-(x - \theta_x)^2}{2\sigma_x^2}, \tag{3.46}$$

and

$$f(y) = \frac{1}{y\sqrt{2\pi\sigma}} \exp \frac{-(y - \theta_y)^2}{2\sigma_y^2} \tag{3.47}$$

where $-\infty < x, y, z = x - y, < \infty$.
Then

$$g(z) = \int_{-\infty}^{\infty} \frac{1}{2\pi\sigma_x\sigma_y} \exp \left[\frac{-(x-\theta_x)^2}{2\sigma_x^2} - \frac{(x-z-\theta_y)^2}{2\sigma_y^2} \right] dx. \tag{3.48}$$

By suitable algebraic steps, this can be changed to the form

$$g(z) = \int_{-\infty}^{\infty} \frac{1}{2\pi\sigma_x\sigma_y}$$

$$\exp \left\{ -\frac{\sigma_x + \sigma_y}{2\sigma_x^2\sigma_y^2} \left[x - \frac{\theta_x\sigma_y + (z+\theta_x)\sigma_x}{\sigma_x^2 + \sigma_y^2} \right]^2 - \frac{(z+\theta_y-\theta_x)^2}{2(\sigma_x^2+\sigma_y^2)} \right\} dx,$$

or

$$g(z) = \frac{1}{\sqrt{2\pi(\sigma_x^2 + \sigma_y^2)}} \exp\left\{-\frac{(z + \theta_y - \theta_x)^2}{2(\sigma_x^2 + \sigma_y^2)}\right\} I, \qquad (3.49)$$

where

$$I = \int_{-\infty}^{\infty} \frac{\sqrt{\sigma_x^2 + \sigma_y^2}}{\sigma_x \sigma_y \sqrt{2\pi}} \exp\left\{\frac{\sigma_x^2 + \sigma_y^2}{-2\sigma_x^2\sigma_y^2}\left[x - \frac{\theta_x\sigma_y + (z + \theta_x)\sigma_x}{\sigma_x^2 + \sigma_y^2}\right]^2\right\} dx = 1. \quad (3.50)$$

Hence

$$g(z) = \frac{1}{\sqrt{2\pi(\sigma_x^2 + \sigma_y^2)}} \exp\left\{-\frac{(z + \theta_y - \theta_x)^2}{2(\sigma_x^2 + \sigma_y^2)}\right\}, \qquad (3.51)$$

a normal function with

$$\text{mean} \quad \theta_z = \theta_x - \theta_y \qquad (3.52)$$

and

$$\text{standard deviation} \quad \sigma_z = \sqrt{\sigma_x^2 + \sigma_y^2}. \qquad (3.53)$$

Of course, the result in this particular case follows directly from the well-known theorem about the normality of the distribution of a linear function of normally distributed variables.

The probability that load exceeds strength is then

$$\int_0^{\infty} g(z)\,dz.$$

Let

$$t = \frac{z + \theta_y - \theta_x}{\sqrt{\sigma_x^2 + \sigma_y^2}}.$$

Then, the integral becomes

$$\int_{\frac{\theta_y - \theta_x}{\sqrt{\sigma_x^2 + \sigma_y^2}}}^{\infty} \frac{1}{\sqrt{2\pi}} e^{-t^2/2}\,dt,$$

the form in which most tables of the normal distribution are published. Such a table will be found in any standard text on statistics. A condensed version is presented as Table 3.5.

Consider the following numerical example. Let

$$\theta_x = 1, \quad n_x = n_y = 2, \quad \sigma_x = \sigma_y = 0.4, \quad m = 1.25.$$

Then the mean strength, θ_y, can be computed from the formula for m:

$$m = \frac{\theta_y - n_y\sigma_y}{\theta_x + n_x\sigma_x},$$

Table 3.5 CUMULATIVE NORMAL DISTRIBUTION

$$F(Y) = \int_Y^\infty \frac{1}{\sqrt{2\pi}}\, e^{-y^2/2}\, dy = P[y \geq Y]$$

y	F(Y)
0.0	0.5000
0.1	0.4602
0.2	0.4207
0.3	0.3821
0.4	0.3446
0.5	0.3085
0.6	0.2743
0.7	0.2420
0.8	0.2119
0.9	0.1841
1.0	0.1587
1.1	0.1357
1.2	0.1151
1.3	0.0968
1.4	0.0808
1.5	0.0668
1.6	0.0548
1.7	0.0446
1.8	0.0359
1.9	0.0287
2.0	0.0228
2.1	0.0179
2.2	0.0139
2.3	0.0107
2.4	0.0082
2.5	0.0062
2.6	0.0047
2.7	0.0035
2.8	0.0026
2.9	0.0019
3.0	0.00135
3.1	0.00097
3.2	0.00069
3.3	0.00048
3.4	0.00034
3.5	0.00023
3.6	0.00016
3.7	0.00011
3.8	0.00007
3.9	0.00005
4.0	0.00003

giving

$$1.25 = \frac{\theta_y - 2(0.4)}{1 + 2(0.4)}$$

or

$$\theta_y = (1.25)\,(1.8) + 0.8 = 3.05.$$

Then the lower limit of the integral is

$$\frac{\theta_y - \theta_x}{\sqrt{\sigma_x^2 + \sigma_y^2}} = \frac{3.05 - 1}{\sqrt{2(0.16)}} = 3.6.$$

The integral of the normal function from 3.6 to infinity is approximately 0.0001, indicating an extremely conservative design in which we might expect that load would exceed strength on the average of once every 10,000 times.

Moment Relationships for Specific Functions

It has been noted above that we can develop time-to-failure density functions from engineering statements about failure patterns. We also observed that the equations of these functions have certain constants which can be related to these engineering statements and also to the moments. Reference has been made to mean life and hazard rates in connection with previously discussed examples. We would now like to consider in some detail the relationship between moments and the constants in the density function for a few specific cases, and then follow this by an analysis of the moment values and associated properties of function graphs. This discussion should result in an understanding of the failure characteristics when we have available these numerical values, the density constants and moments. Specifically, this will yield an ability to compare different densities and reliabilities by comparing their moments, algebraic representations, and graphs.

We have observed that moment calculations can be done in many ways. It is likewise true that there are many ways in which moment formulas for specific functions can be developed. A number of these ways will be illustrated in the derivations to follow. The direct method is merely to use the moment definition, but this sometimes involves mathematical complexity. Hence it has been useful to develop other procedures, some of which are essentially trick procedures while others involve fundamental statistical techniques which are extremely important in a study of the theory of mathematical statistics. It should be noted that the mathematical material which follows is not essential to an understanding of the problems of reliability if one wishes to accept results without understanding the proofs. It is included for completeness and also to satisfy those readers who have an interest in the mathematical aspects of the subject.

Moments of the Exponential Function

We have made repeated references to the exponential time-to-failure density function, and it is natural to consider it first. It can be written in the form

$$f(t) = \frac{1}{\theta} e^{-t/\theta}, \quad 0 \leq t < \infty, \tag{3.54}$$

and the moments about the origin are

$$\nu_k = \int_0^\infty \frac{t^k}{\theta} e^{-t/\theta} \, dt, \tag{3.55}$$

It can easily be shown that

$$\nu_k = k\theta \nu_{k-1} = k!\theta^k, \tag{3.55a}$$

and we can prove that

$$\nu_1 = \theta, \quad \mu_2 = \theta^2, \quad \mu_3 = 2\theta^3, \quad \text{and} \quad \mu_4 = 9\theta^4. \tag{3.55b}$$

We can develop a recursion formula for the moments ν_k by integrating by parts.

Let

$$f = t^k \quad \text{and} \quad dv = \frac{1}{\theta} e^{-t/\theta} \, dt.$$

Then

$$df = kt^{k-1} \, dt \quad \text{and} \quad v = -e^{-t/\theta}$$

and

$$\nu_k = -t^k e^{-t/\theta} \Big|_0^\infty + k \int_0^\infty t^{k-1} e^{-t/\theta} \, dt,$$

or

$$\nu_k = k\theta \nu_{k-1}. \tag{3.56}$$

Now $\nu_0 = 1$ for any density. Hence, let $k = 1$, giving $\nu_1 = \theta$. By taking $k = 2, 3$, and 4, we obtain

$$\nu_2 = 2\theta^2, \nu_3 = 6\theta^3 \quad \text{and} \quad \nu_4 = 24\theta^4.$$

We have previously developed the following moment formulas:

$$\mu_2 = \nu_2 - \nu_1^2$$
$$\mu_3 = \nu_3 - 3\nu_2\nu_1 + 2\nu_1^3 \tag{3.57}$$
$$\mu_4 = \nu_4 - 4\nu_3\nu_1 + 6\nu_2\nu_1^2 - 3\nu_1^4, \quad \text{etc.}$$

Then for the exponential,

$$\sigma^2 = \mu_2 = 2\theta^2 - \theta^2 = \theta^2$$

$$\mu_3 = 6\theta^3 - 3(2\theta^2)\theta + 2\theta^3 = 2\theta^3 \tag{3.58}$$

$$\mu_4 = 24\theta^4 - 4(6\theta^3)\theta + 6(2\theta^2)\theta^2 - 3\theta^4 = 9\theta^4.$$

From the definition of the dimensionless moments, α_k, we obtain

$$\alpha_3 = \frac{2\theta^3}{\theta^3} = 2 \quad \text{and} \quad \alpha_4 = \frac{9\theta^4}{\theta^4} = 9. \tag{3.59}$$

Moments of the Normal Density Function

The moments of the normal density function are calculated from a recursion formula in a similar way. However, it is helpful to modify the function to give μ_k, moments about the mean, without bothering to write ν_k. This depends on the observation that the mean of a symmetrical density function is the center of symmetry. This can be expressed mathematically by use of Equation 3.16 and by taking moments about θ where

$$f(t) = \frac{1}{\sqrt{2\pi}\sigma} e^{-(t-\theta)^2/2\sigma^2} \tag{3.60}$$

and

$$\mu_k = \int_{-\infty}^{\infty} \frac{1}{\sqrt{2\pi}\sigma} (t - \theta)^k e^{-(t-\theta)^2/2\sigma^2} dt. \tag{3.61}$$

To prove that this is in fact μ_k, and a moment about the mean, it is only necessary to note the symmetry property. However, it is easier to see this by making the substitution $x = t - \theta$.
Then

$$f_k = \int_{-\infty}^{\infty} \frac{1}{\sqrt{2\pi}\sigma} x^k e^{-x^2/2\sigma^2} dx. \tag{3.62}$$

This is clearly the kth moment of a symmetrical density, since x occurs to an even power in the density function (excluding the x^k factor). It is also apparent by symmetry that all moments of odd order are zero, $\mu_1 = \mu_3 = \mu_5 = \ldots = 0$.

Again, we obtain the recursion formula by integrating by parts. Let

$$f = x^{k-1} \quad \text{and} \quad dv = \frac{1}{\sqrt{2\pi}\sigma} \times e^{-x^2/2\sigma^2} dx.$$

Then

$$df = (k - 1) x^{k-2} dx \quad \text{and} \quad v = \frac{-\sigma}{\sqrt{2\pi}} e^{x^2/2\sigma^2},$$

and

$$\mu^k = -x^{k-1}\frac{\sigma}{\sqrt{2\pi}}e^{-x^2/2^2}\Big]_{\infty}^{-\infty} + (k-1)\sigma^2 \int_{-\infty}^{\infty}\frac{1}{\sqrt{2\pi}\sigma}x^{k-2}e^{-x^2/2\sigma^2}\,dx \quad (3.63)$$

or
$$\mu_k = (k-1)\sigma^2\mu_{k-2}.$$

Let $k = 2$ and $k = 4$ to obtain $\mu_2 = \sigma^2$ and $\mu_4 = 3\sigma^4$. It is easy to check that $\alpha_3 = 0$, $\alpha_4 = 3$.

Moments of the Gamma and Weibull Density Functions

By similar methods, moments of the gamma and Weibull densities can be found to have the following forms (α_3 and α_4 for the Weibull density are too complex to be considered here):

	Gamma	*Weibull*
ν_1	$\beta(\alpha+1)$	$\alpha^{-1/\beta}\left(\frac{1}{\beta}\right)$
μ_2	$\beta^2(\alpha+1)$	$\alpha^{-2/\beta}\left[\left(\frac{2}{\beta}\right)! - \left\{\left(\frac{1}{\beta}\right)!\right\}^2\right]$
α_3	$2(\alpha+1)^{-1/2}$	—
α_4	$3\dfrac{(\alpha+3)}{\alpha+1}$	—

$$(3.64)$$

Moments of the Rectangular Density Function

The rectangular distribution in the continuous case is of interest since it illustrates a simple derivation of a formula for moments of all orders, in addition to being the equally likely density. Denote it by

$$f(x) = \frac{1}{2a}, \; -a \le x \le a. \quad (3.65)$$

In this form, it is obvious that the mean is zero because of symmetry. Hence the moments will be calculated immediately as moments about the mean, the μ_k moments.

$$\mu_k = \int_{-a}^{a}\frac{x^k}{2a}\,dx = \frac{x^{k+1}}{2a(k+1)}\Big]_{-a}^{a} = \frac{a^{k+1} - (-a)^{k+1}}{2a(k+1)}$$

$$= 0 \text{ if } k \text{ is odd}, \quad (3.66)$$

and
$$= \frac{a^k}{k+1} \text{ if } k \text{ is even}. \quad (3.67)$$

Thus

$$\mu_1 = \mu_3 = \ldots = 0,$$

$$\mu_2 = \frac{a^2}{3}, \; \mu_4 = \frac{a^4}{5}, \text{ etc.,} \tag{3.68}$$

$$\alpha_3 = 0 \quad \text{and} \quad \alpha_4 = 9/5.$$

Moments of Selected Discrete Density Functions

The binomial and Poisson density functions will be used to illustrate the computation of moments for discrete densities:

$$\text{Binomial:} \quad \mu(x) = \binom{n}{x} p^x q^{n-x}, \quad x = 0, 1, 2, \ldots, n \tag{3.69}$$

$$\text{Poisson:} \quad \mu(x) = \frac{e^{-m} m^x}{x!} \qquad x = 0, 1, 2, \ldots, \tag{3.70}$$

In the discrete case, x is used as the variable in place of t, which usually applies to a continuous time variable.

For the binomial density,

$$\nu_1 = \sum_{x=0}^{n} \binom{n}{x} p^x q^{n-x} x = \sum_{x=1}^{n} \frac{n! \, x}{x! \, (n-x)!} p^x q^{n-x}, \tag{3.71}$$

the index starting at $x = 1$, since the term for $x = 0$ vanishes. Then the formula can be written as

$$\nu_1 = np \sum_{x=1}^{n} \frac{(n-1)!}{(x-1)! \, (n-x)!} p^{x-1} q^{n-x}.$$

Let $y = x - 1$ and $m = n - 1$ in the summation. Then

$$\nu_1 = np \sum_{y=0}^{m} \frac{m!}{y! \, (m-y)!} p^y q^{m-y} = np(p+q)^m = np. \tag{3.72}$$

The higher-order moments can be derived by a slight modification of the method above. To find ν_2, the following procedure is used: First find $\nu_2 - \nu_1$, this being given by

$$\nu_2 - \nu_1 = \sum_{x=0}^{x=n} \binom{n}{x} p^x q^{n-x} (x^2 - x) = \sum_{x=0}^{x=n} \frac{n!}{x! \, (n-x)!} x(x-1) p^x q^{n-x}$$

$$= n(n-1)p^2 \sum_{x=2}^{x=n} \frac{(n-2)!}{(x-2)! \, (n-x)!} p^{x-2} q^{n-x}$$

$$= n(n-1)p^2 (p+q)^{n-2}$$

$$= n(n-1)p^2.$$

Then

$$v_2 = v_1 + (n^2 - n)p^2 = np + n^2p^2 - np^2, \tag{3.73}$$

and

$$\mu_2 = v_2 - v_1^2 = np + n^2p^2 - np^2 - n^2p^2$$
$$= np - np^2 = np(1 - p) = npq. \tag{3.74}$$

Higher-order moments can be computed in the same way. For the third moment the relationship desired is obtained by using

$$x(x - 1)(x - 2) = x^3 - 3x^2 + 2x,$$

meaning that we compute $v_3 - 3v_2 + 2v_1$. The answer will be $v_3 - 3v_2 + 2v_1 = n(n - 1)(n - 2)p^3$ and

$$v_3 = n(n - 1)(n - 2)p^3 + 3v_2 - 2v_1$$
$$= n(n - 1)(n - 2)p^3 + 3np^2 + 3n^2p^2 - 3np^2 - 2np$$
$$= n(n - 1)(n - 2)p^3 + 3np^2(n - 1) + np. \tag{3.75}$$

Then

$$\mu_3 = v_3 - 3v_2v_1 + 2v_1^3 = (n^3 - 2n^2 + 2n)p^3 + 3np^2(n - 1) + np$$
$$- 3n^2p^2 - 3n^3p^3 + 3n^2p^3 + 2n^3p^3 = np(2p^2 - 3p + 1)$$
$$= np(1 - p)(1 - 2p) = npq(q - p). \tag{3.76}$$

Then

$$\alpha_3 = \frac{\mu_3}{\mu_2^{3/2}} = \frac{npq(q - p)}{npq\sqrt{npq}} = \frac{q - p}{\sqrt{npq}}. \tag{3.77}$$

A similar system works for the computation of moments of the Poisson density:

$$v_1 = \sum_{x=0}^{\infty} \frac{xe^{-m}(m^x)}{x!} = m\sum_{x=1}^{\infty} \frac{e^{-m}(m^{x-1})}{(x - 1)!}.$$

Let $y = x - 1$. Then

$$v_1 = m\sum_{y=0}^{\infty} \frac{e^{-m}(m^y)}{y!} = m \tag{3.78}$$

$$v_2 - v_1 = \sum_{x=0}^{\infty} x(x - 1)\frac{e^{-m}(m^x)}{(x)!} = m^2\sum_{x=1}^{\infty} \frac{e^{-m}(m^{x-2})}{(x-2)!} = m^2.$$

Then

$$v_2 = m^2 + v_1 = m^2 + m, \tag{3.79}$$

and

$$\mu_2 = \nu_2 - \nu_1^2 = m. \tag{3.80}$$

$$\nu_3 - 3\nu_2 + 2\nu_1 = \sum_{x=0}^{\infty} x(x-1)(x-2)\frac{e^{-m}(m^x)}{x!} = m^3 \sum_{x=3}^{\infty} \frac{e^{-m}(m^{x-3})}{(x-3)!} = M^3,$$

$$\nu_3 = m^3 + 3m^2 + 3m - 2m = m^3 + 3m^2 + m, \tag{3.81}$$

$$\mu_3 = \nu_3 - 3\nu_2\nu_1 + 2\nu_1^3$$
$$= m^3 + 3m^2 + m - 3m^3 - 3m^2 + 2m^3 = m, \tag{3.82}$$

and

$$\alpha_3 = \frac{\mu_3}{\mu_2^{3/2}} = \frac{1}{\sqrt{m}}. \tag{3.83}$$

Moment Formulas Developed from Moment-Generating Functions

Because of the unusual importance of moment-generating functions in the mathematical development of theorems in sampling theory, we will illustrate their use in computing moments for certain density functions. (This section can be omitted without loss in continuity by readers not interested in the mathematical aspect of the subject.)

As the name implies, the *moment-generating function* is a function which can be used to generate moments of all orders for any specific density function. Many functions have this property. We shall consider only the most commonly used generator, defined by

$$g(x) = \int_a^b e^{tx} f(t)dt \tag{3.84}$$

where $g(x)$ is the generating function for density $f(t)$. It is understood that the limits of integration are the range of t used in defining the density.

Without bothering to state the precise mathematical assumptions involved, we can show that this is indeed a moment-generating function by the use of the series expansion of e^{tx} in the integral as follows:

$$g(x) = \int_a^b e^{tx} f(t)dt$$
$$= \int_a^b f(t)\sum_{i=0}^{i=\infty} \frac{(tx)^i}{i!} dt$$
$$= \sum_{i=0}^{i=\infty} \frac{x^i}{i!} \int_a^b t^i f(t) dt$$
$$= \sum_{i=0}^{i=\infty} \frac{x^i}{i!} \nu_i. \tag{3.85}$$

If ν_1 happens to be zero, the moments are μ_i.

In applying the moment-generating function to obtain moments, we would perform the integration first and then expand the integral as a series in powers of x. The coefficient of $x^i/i!$ is then the moment ν_i or μ_i if the first-order moment is zero. We shall apply the method to the exponential and the rectangular densities.

For the exponential density

$$g(x) = \int_0^\infty e^{tx} \frac{1}{\theta} e^{-t/\theta} \, dt \tag{3.86}$$

$$= \int_0^\infty \frac{1}{\theta} e^{(x-\frac{1}{\theta})t} \, dt$$

$$= \frac{1}{1 - \theta x} \quad \text{for } x < \frac{1}{\theta}. \tag{3.87}$$

Then

$$g(x) = 1 + \theta x + \theta^2 x^2 + \ldots + \theta^i x^i + \ldots$$

$$= 1 + \theta x + 2!\theta^2 \frac{(x^2)}{2!} + \ldots + i!\,\theta^i \frac{x^i}{i!} \ldots, \tag{3.88}$$

giving $\nu_k = k!\theta^k$ as before. $\tag{3.89}$
For the rectangular density

$$g(x) = \int_{-a}^a e^{tx} \frac{1}{2a} \, dt.$$

$$= \frac{1}{2ax} (e^{ax} - e^{-ax})$$

$$= 1 + \frac{a^2}{3}\left(\frac{x^2}{2!}\right) + \frac{a^4}{5}\left(\frac{x^4}{4!}\right) + \ldots, \tag{3.90}$$

giving $\nu_1 = 0$. Hence the function is a generator of μ moments and they are

$$\mu_k = 0 \quad \text{if } k \text{ is odd},$$

and

$$\mu_k = \frac{a^k}{k+1} \quad \text{if } k \text{ is even}, \tag{3.91}$$

in agreement with the answer obtained previously.

Some Important Probabilities Associated with the Normal and Exponential Functions

Because of the widespread use of the normal and exponential functions in reliability analysis, it is proper to examine some of the important probabilities associated with them. The normal function is of two-fold interest: it is not only encountered in describing failure patterns, but it is also fundamental in the more general field of sampling theory, including problems of estimation of population parameters. The latter point is discussed in this chapter and in subsequent chapters.

Figure 3.3 shows these two distributions in density and reliability forms. Since published tables of the normal and exponential functions are for distributions with unit variance, the functions are plotted in this form. This representation is not really a restriction since it merely reflects a choice of unit for plotting purposes. It is unusual to speak in terms of unit variance in the case of the exponential; the more common statement is *unit mean*. Actually the two statements are equivalent, since the variance is equal to the square of the mean for the exponential. It should also be observed that the usual representation of the normal function involves the assumption of a zero mean. This merely locates the vertical axis. Hence, for purposes of plotting, we can think of this axis as passing through the mean value, even though we may wish to consider a different axis position for other purposes.

The exponential has one unique property when the mean value is taken to be unity. The density function and the reliability function are identical in this and in no other case. Of course, this means that only one curve appears for the exponential in Figure 3.3.

Some of the most frequently encountered points are marked on the reliability curves, together with the numerical values of the reliability functions at these points. Consider first the normal function for which reliability values are shown for the mean and for 1, 2, and 3 units (standard deviations) above and below the mean. At the mean, the value of the reliability function is $R(0) = 0.5$. This says that half of the failures occur before the mean and half after the mean, a property shared by all other symmetrical functions.

The usual types of probability statements can be illustrated as follows: Consider the values of the reliability function at 1 unit below and 1 unit above the mean—that is, $R(-1)$ and $R(1)$, respectively. The value $R(-1) = 0.8413$ is the probability of survival until 1 unit of time (1 standard deviation) before the mean. Hence $1 - R(-1) = 0.1587$ is the probability of failure prior to this time. Similarly there is a probability of $R(1) = 0.1587$ of survival until 1 unit after the mean and a probability of $1 - R(1) = 0.8413$ of failure prior to this time. Hence, the probability of failure within the 1 unit of the mean is

$$R(-1) - R(1) = 0.8413 - 0.1587 = 0.6826,$$

which is roughly approximated as two-thirds. The probability of failure occurring outside this interval is

$$1 - 0.6826 = 0.3174,$$

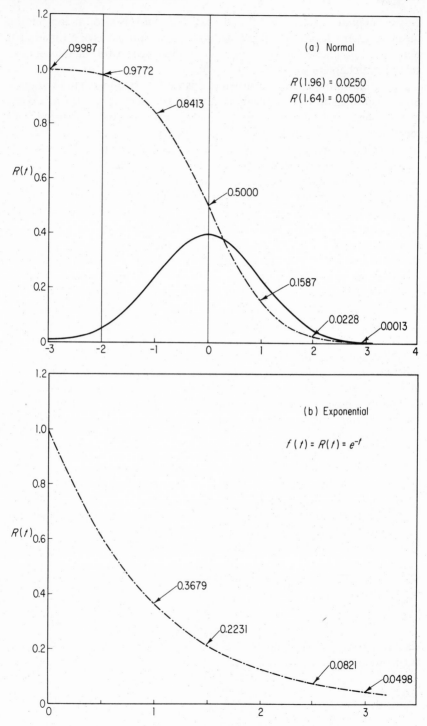

Fig. 3.3 Density and reliability functions for (a) normal and (b) exponential distributions.

or roughly one-third. Because of symmetry, the value 0.3174 could also have been computed as twice the reliability $R(1)$,

$$2R(1) = 2(0.1587) = 0.3174.$$

$R(1)$ is the area under the density curve to the right of the time $t = 1$, called the *tail* of the curve. We also speak of the tail on the left side of the curve, the area being $1 - R(-1)$ in the numerical example, which is also 0.1587 by symmetry. The probability of falling at least t units away from the mean is $2R(t)$, the area of two tails; while the probability of falling at least t units above the mean, $R(t)$, is the area of one tail. We are often interested in the following values of the reliability function in the normal case:

$$
\begin{array}{ll}
R(1) \quad = 0.1587 & 2R(1) \quad = 0.3174 \\
R(1.64) = 0.0505 & 2R(1.64) = 0.1010 \\
R(1.96) = 0.0250 & 2R(1.96) = 0.0500 \\
R(2) \quad = 0.0228 & 2R(2) \quad = 0.0456 \\
R(3) \quad = 0.0013 & 2R(3) \quad = 0.0026.
\end{array}
$$

The two-tailed area is about one-third for $t = 1$; 10 per cent for $t = 1.64$ or about $t =$ one and two-thirds; 5 per cent for $t = 1.96$ (often approximated by $t = 2$); and less than 3 in 1000 for $t = 3$.

The one-tailed area is 5 per cent at $t = 1.64$, or at about a t value of one and two-thirds; 2.5 per cent at $t = 1.96$, or at about $t = 2$; and 1 in 1000 at $t = 3$.

For most failure patterns which have the normal function as the time-to-failure density, the vertical axis passes through some point to the left of the mean. In the figure, this is shown by drawing the axis through the point two standard deviations to the left of the mean, making this point correspond to zero time. This implies negative time to failure for 2.28 per cent of the items, since the area of the tail is 0.0228 (obtained as one minus the value of the reliability function at that point). In practice we would actually observe these failures at installation, meaning at time zero, and they can therefore be viewed as defectives not caught by quality control. The probability of such an installation failure can be computed for a normal density if we know the coefficient of variation, which is the ratio of the standard deviation to the mean. A majority of the published tables of the normal function require us first to compute the reciprocal of the coefficient of variation. Thus, in the example, the coefficient of variation is 0.5, and the reciprocal, 2, is the value at which one enters the table to read the area of the tail, 0.0228.

In some respects, the interpretations are simpler for the exponential. We never speak of a two-tailed area, for obvious reasons. Furthermore, there are fewer critical probabilities to keep in mind. It is very important to realize that the value of the reliability at one mean life is 0.3679: the probability of lasting one mean life is only slightly larger than one-third for the exponential. It is difficult for some people to understand why this value is not 0.5 as in the case of the normal. The reason is that the mean is an *arithmetic mean* and not a median. In general, the two are different: the median will be equal to the arithmetic mean only if the density function is symmetrical. Nevertheless, it is striking to recognize that, for the exponential, one expects about two-thirds of the items to fail before one mean life.

3.4 The Sampling Problem

In the previous discussion, two fundamental purposes of statistical analysis have been mentioned. The first purpose is to describe a large mass of data by computing a small number of figures of merit which contain most of the important information included in the data. This deductive analysis is certainly *data reduction*. The second purpose is to determine from statistical analysis of the data some of the more important characteristics of the population of which it is but a sample. Hence, this purpose is often thought of as numerical inductive logic. It can be noted that deductive logical processes are involved when we assume that the population density is known and properties of samples are described in some sort of probability language. Thus, probability statements are involved in both deductive and inductive statistical reasoning. *Sampling theory* is that branch of mathematical statistics which deals with inductive logical processes associated with the description of a population from knowledge of the results of some experiment, usually in the form of test data.

The sampling problem can be considered in any of three distinct forms: point estimation, interval estimation, and hypothesis testing. We shall give short descriptions of each of these. At the outset, it is necessary to comment briefly on the terminology of sampling theory and thus to explain the framework within which this problem is being studied.

Following the usual practice, it is helpful to think in terms of a simple, well-understood characteristic of sample data and population densities—the arithmetic mean. Suppose we select a random sample of 100 parts from a production line, measure some characteristic of each of the parts, and compute the mean of the 100 measurements. Another sample of 100 parts would no doubt have a different mean. Indeed we can think of the distribution of all possible sample means which might arise from a continuation of this sampling procedure, that is, a density function describing the distribution of sample means. In turn, this density of sample means would have its own mean, variance, and all other properties associated with any density. Similarly, we could visualize a density of sample variances, standard deviations, medians, etc. The study of sampling theory requires the computation of the forms of these various densities from knowledge of, or assumptions about, the population. Of course, some properties of these densities are independent of the form of the density of the population from which samples are drawn, and these are referred to as either nonparametric or distribution-free. The following discussion illustrates both the derivation of a sampling distribution and a method of computation of moments in the solution of a specific problem.

An Example of the Distribution of the Mean

The concept of the density function of some sample characteristic can perhaps be clarified by developing the distributions of means of samples of various sizes for a simple case. To make the arithmetic manageable, consider a trivial urn example. Suppose that an urn contains three types of cards in equal numbers, identified by the numbers 1, 2, and 3 on the separate types. Cards are to be drawn, each being replaced before another is drawn. The parameter of concern is the arithmetic mean of the numbers on the cards which are drawn in this fashion.

Let x denote the variable of interest (the numbers on the cards) and let $f(x)$ denote the population density. Then the tabular representation of the population density is as follows:

x	$f(x)$
1	1/3
2	1/3
3	1/3

For the population to be sampled, the mean ($\bar{x}$) is given by

$$\bar{x} = \tfrac{1}{3}(1) + \tfrac{1}{3}(2) + \tfrac{1}{3}(3) = 2,$$

and the variance (σ^2) is

$$\sigma^2 = \tfrac{1}{3}(1-2)^2 + \tfrac{1}{3}(2-2)^2 + \tfrac{1}{3}(3-2)^2 = \tfrac{2}{3}.$$

Let $\bar{x}_i$ and σ_i^2 denote the mean and the variance, respectively, for distributions of means of samples of i drawn from this population with replacement as described. Of course, $\bar{x}_1 = \bar{x}$ and $\sigma_1^2 = \sigma^2$, since by definition the population density is exactly the same as the "sample of one" density. Consider samples of two. The sum of the numbers on the two cards can assume any of the values 2, 3, 4, 5, or 6 and the sample means can assume corresponding values, 1, 1.5, 2, 2.5, or 3. We can compute the probabilities of each of these values under the sampling procedure to find the density for means of samples of two, which is denoted by $f_2(\bar{x})$. This can be done by using an extension of the binomial distribution. In this case, we use a trinomial which gives the following simple probability generating function for the sample of two:

$$(\tfrac{1}{3}t + \tfrac{1}{3}t^2 + \tfrac{1}{3}t^3)^2 = \tfrac{1}{9}(t + t^2 + t^3)^2 = \tfrac{1}{9}(t^2 + 2t^3 + 3t^4 + 2t^5 + t^6).$$

The variable t in this expression is merely a *carrier* variable so arranged that the exponent indicates the sum of the numbers on the two cards. If $t^{1/2}$ had been used instead of t, the answer would have been

$$\tfrac{1}{9}(t + 2t^{1.5} + 3t^2 + 2t^{2.5} + t^3),$$

and the exponent on t would then have identified the mean of the sample instead of the sum of the numbers on the cards. The numerical coefficient is the probability associated with the sum in the first form and with the mean in the second form. This can be summarized in the following table:

Sum on Two Cards	$\bar{x}$	$f_2(\bar{x})$
2	1	1/9
3	1.5	2/9
4	2	3/9
5	2.5	2/9
6	3	1/9

The foregoing method for computing $f_2(\bar{x})$ is easy, but it is perhaps instructive to compute the probabilities directly without using the trinomial generating function. For this purpose, consider the calculation of probabilities of all possible outcomes of the two drawings. These outcomes are enumerated below.

First Card	Second Card	Sum	Mean
1	1	2	1
1	2	3	1.5
2	1	3	1.5
1	3	4	2
2	2	4	2
3	1	4	2
2	3	5	2.5
3	2	5	2.5
3	3	6	3

This tabulation shows that there are 9 possible outcomes of the trials: 1 trial yields a sum of 2 with a mean of 1; 2 trials yield a sum of 3 with a mean of 1.5; 3 trials yield a sum of 4 with a mean of 2; 2 trials yield a sum of 5 with a mean of 2.5; and 1 trial yields a sum of 6 with a mean of 3. Since the 9 cases are equally likely, the associated probabilities are (in order) $\frac{1}{9}$, $\frac{2}{9}$, $\frac{3}{9}$, $\frac{2}{9}$, and $\frac{1}{9}$, as before.

$$\bar{x}_2 = \tfrac{1}{9}[1 + 2(1.5) + 3(2) + 2(2.5) + 3] = \tfrac{1}{9}[1 + 3 + 6 + 5 + 3] = 2.$$

$$\sigma_2^2 = \tfrac{1}{9}[(1 - 2)^2 + 2(1.5 - 2)^2 + 3(2 - 2)^2 + 2(2.5 - 2)^2 + (3 - 2)^2]$$

$$= \tfrac{1}{9}[1 + 0.5 + 0 + 0.5 + 1] = \tfrac{1}{3}.$$

Note that the mean of $f_2(\bar{x}_2)$ is the same as the population mean. This is true for any size sample. However, the variance is not the same as the population variance; it is less, $\sigma_2^2 = \sigma^2/2$. In general, it will be seen that

$$\bar{x}_i = \bar{x}, \quad \text{and} \quad \sigma_i^2 = \frac{\sigma^2}{i}: \tag{3.92}$$

the mean of the samples is always distributed about the population mean, and the variance of the distribution of sample means is equal to the population variance divided by the sample size.

For samples of three cards, the distribution is developed in the same way. Using the trinomial method, the probabilities are generated by

$$\tfrac{1}{27}\,[t + t^2 + t^3]^3 = \tfrac{1}{27}\,[t^3 + 3t^4 + 6t^5 + 7t^6 + 6t^7 + 3t^8 + t^9],$$

where the exponent on t indicates the sum and the coefficient is the associated probability. The sample mean would be indicated if the exponent were divided by three. Thus, the density for means of samples of three is shown in the first two columns of the table below:

$\bar{x}$	$f_3(\bar{x})$	$f_3(\bar{x})$	$^2f_3(\bar{x})$
1	1/27	1/27	1/27
4/3	3/27	4/27	16/81
5/3	6/27	10/27	50/81
2	7/27	14/27	28/27
7/3	6/27	14/27	98/81
8/3	3/27	8/27	64/81
3	1/27	1/9	1/3
Sum		2	38/9

The last two columns show steps in the computation of the mean and variance of $f_3(\bar{x})$. They give

$$\bar{x}_3 = 2, \quad \sigma_3^2 = \frac{38}{9} - 2^2 = \frac{2}{9} = \frac{\sigma}{3}.$$

Using

$$\frac{1}{81} [t + t^2 + t^3]^4$$

it is easy to obtain $f_4(\bar{x})$ as

$\bar{x}$	$f_4(\bar{x})$
1	1/81
1.25	4/81
1.5	10/81
1.75	16/81
2	19/81
2.25	16/81
2.5	10/81
2.75	4/81
3	1/81

and to determine that

$$\bar{x}_4 = 2 \quad \text{and} \quad \sigma_4^2 = \frac{1}{6} = \frac{\sigma^2}{4}.$$

This process can be continued to obtain the distribution $f_i(\bar{x})$ for larger values of i—i.e., for larger samples—but naturally the arithmetic becomes quite burdensome.

It should be emphasized that the simplicity of this example should not mislead one into thinking that these properties of $f_i(\bar{x})$ are true only for such simple populations. The formulas (Equation 3.86)

$$\bar{x}_i = \bar{x} \quad \text{and} \quad \sigma_1^2 = \frac{\sigma^2}{i}$$

are valid for all sample sizes (i) and for both discrete and continuous populations. These characteristics are, therefore, distribution-free.

To show how $f_i(\bar{x})$ changes when i increases, the distributions are plotted in Figure 3.4 for $i = 1$, 2, 3, and 4. The larger the sample size, the more the means tend to concentrate around the true or population value (this is a reflection of the decreasing variance). To simplify the arithmetic, a symmetrical example was chosen. If the initial probabilities had not been symmetrical (that is, if the population distribution had been skewed), $f_i(\bar{x})$ would have become more symmetrical as the sample size increased. That is to say, the means of large samples tend to be symmetrically distributed even though the parent population is skewed. It can be shown that for any population, subject to very minor restrictions, the distribution of sample means approaches the normal or Gaussian distribution as sample size increases. It should be pointed out, however, that the sample means have a normal distribution for all sample sizes only if the parent population is normally distributed.

In the discrete case, to arrive at a form suitable for simple comparison of the distributions of sample means for various sample sizes, it is necessary to represent probabilities by areas rather than ordinates. This involves two related modifications in the computed probabilities. The first is to use touching columns to stand for probabilities which really apply only to the abscissa of the midpoint of the base of the column. The second is to adjust the height of the column to make the area equal the associated probability. To keep the figure readable, only the tops of the columns are shown.

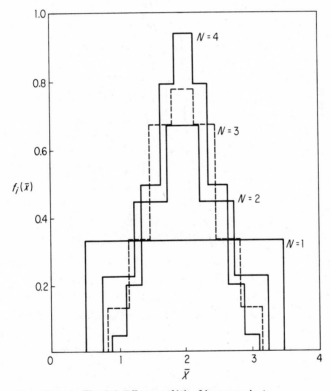

Fig. 3.4 Effect on $f_i(\bar{x})$ of increases in i.

To illustrate, the probability of a mean of 2.5 in a sample of 4 is 0.123. The width of the column is 0.25, since adjacent possible sample means differ by this amount. Hence, a column for this case extends from 2.375 to 2.625, or over a range of 0.25. The height of the column is 4(0.123) = 0.492, and the total area is equal to 0.123.

Suppose random samples of n are drawn from a normal parent population

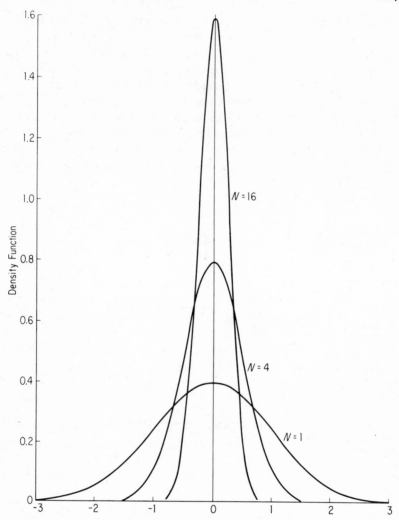

Fig. 3.5 Distributions of population ($n=1$) and means of samples ($n=4$, $n=16$).

with mean θ and variance σ^2. As stated above, it can be shown that sample means are distributed normally with mean σ and variance equal to σ^2/n. Figure 3.5 shows the population distribution, denoted by $n = 1$, and the distributions of sample means for sample sizes $n = 4$ and $n = 16$. It is readily observed that, as the sample size increases, the distribution of the mean becomes concentrated about the population mean. Thus, if we compute the mean of a sample of even moderate size, there is only a small chance that this mean will differ by any sizeable amount from that of the population. Thinking of the mean of a sample as a single item drawn from the distribution of the mean, we can compute the probability of a deviation from the true mean of at least a specified amount merely by computing the area of the tails of the curve of the distribution of the sample mean.

For example, suppose that a sample of 16 electron tubes is drawn at random and life tested. Suppose further that the population of tubes of this type has a mean life of 1200 hr and a standard deviation of 400 hr, and that the distribution of lives is normal. Then means of samples of 16 will be normally distributed about a mean of 1200 hr, with a standard deviation of $\dfrac{400}{\sqrt{16}} = 100$.

Therefore, about two-thirds of such samples will yield mean lives between 1100 and 1300 hr—i.e., within one standard deviation (of the distribution of means, 100) of the true mean. Stated in reverse, only one-third of such samples would have means outside this range. To get a 0.95 probability, we would take a range of 1.96 standard deviations, or 1004 to 1396 in this case, and then only 5 per cent of the samples would fall outside these limits.

Point Estimation

A constant computed from a set of sample measurements and used as an estimate of the value of a population characteristic is a point estimate. Thus, a *point estimate* is a single value computed from a sample and used as a "best" single guess as to the population value. The sample value used as the estimate is called a *statistic*, and the estimated population value is called a *parameter*. Note that any parameter can be estimated by many different such statistics. The parameter is unique but the estimating statistic is not. For example, the population mean can be estimated by computing a sample mean, median, geometric mean, the average of the lowest and highest sample values, etc., and we can find cases in which each of these sample statistics would be appropriate. We need to develop criteria for selecting a best statistic.

CRITERIA FOR A "BEST" ESTIMATE

The word "best" is usually defined in terms of many characteristics of the statistic used. It is natural to require it to have at least some of the following properties:

1. A best statistic is one which will, in the long run, tend to cluster around the true parameter value. It is *unbiased* if the mean of the statistic density function is the population parameter. The sample mean satisfies this requirement. However, the sample variance tends to be lower than the population variance, and therefore it is a biased statistic.

2. We would want a statistic to be stable under sampling, i.e., to have small sampling variation: The statistic density function should have a small variance. For example, the mean and median are two different

estimates of population mean. However, the median sampling density function has a variance larger than that of the mean. The mean is therefore a more *efficient* statistic than the median.

3. It is useful to have a *sufficient* statistic, which means that it uses all the information which can be furnished by the sample. Suppose we can estimate parameter θ by either of the statistics θ_1 and θ_2. If the estimate of θ based on both θ_1 and θ_2 is exactly the same as the estimate based on θ_1 alone, then θ_2 is really of no value, and θ_1 is said to be sufficient if this statement holds for all possible θ_2.

4. A fourth property, not identified in all cases with the best estimate but a frequent substitute for unbiased, is called consistent. An estimate is *consistent* if its bias approaches zero as the sample size approaches infinity. Sample variance is consistent even though it is biased.

5. Finally the term "best" is often defined as the statistic which is dictated by the method used to determine it. This may sound strange, but it is common. For example, the method of moments is one which uses unmodified sample moments as estimates of corresponding population moments. This is often used in curve-fitting problems, and the solution is accepted as best although it does not have all the properties listed above. A second common method is *maximum likelihood*, which determines the parameter estimate as that value (or values, if more than one are being estimated) which would make the observed sample more likely than any other parameter value (or values). This method also does not have all the properties listed above. For example, not all maximum-likelihood estimates are unbiased. Sometimes a modification of the method is used whereby the maximum-likelihood estimate is modified to make it unbiased.

In summary, a point estimate of a population parameter is a statistic computed from a sample—a set of data. It is rare that such an estimate is exactly equal to the parameter—i.e., that there is no error at all. If we think of the statistic as a single value—a sample of one—drawn from the density of the statistic (e.g., a mean drawn from the distribution of means of samples of a given size), it is possible to measure the likelihood of an error of some given magnitude. This is done by examining the variance of the density of the statistic, and using this variance to phrase some statement about the chance of an error of at least some selected magnitude. This procedure will be clarified by examples in later chapters.

A CLASS OF UNBIASED POINT ESTIMATES

(This section is of interest only to those who wish to study the mathematical aspects of the subject. It can be omitted without loss in continuity.)

Suppose $x_1, x_2, \ldots, x_n$ are a sample of n values drawn randomly from a population with density $f(x)$. Let

$$H(x_1, x_2, \ldots, x_n)$$

be any function defining a statistic which is considered as a point estimate of a population parameter. The estimate may or may not be biased depending on properties of both functions, $H(x_1, x_2, \ldots, x_n)$ and $f(x)$. However, if $H(x_1, x_2, \ldots, x_n)$ is of the form

$$H(x_1, x_2, \ldots, x_n) = \sum_{i=1}^{i=n} \frac{1}{n} g(x_i), \tag{3.93}$$

it can be shown that the statistic is an unbiased estimate regardless of the form of the population density, $f(x)$. This is particularly useful in view of the fact that sample moments about a constant are of this type. For example, sample moments about the origin are included by taking

$$g(x_i) = x_i^k. \tag{3.94}$$

The proof of this theorem follows.

The sample density in differential form is

$$f(x_1) \cdot f(x_2) \cdots f(x_n) \cdot dx_1 \cdot dx_2 \cdots dx_n.$$

Hence the expected value of the statistic is

$$E\left[\frac{1}{n}\sum_{i=1}^{n} g(x_i)\right] = \int\int \cdots \int \frac{1}{n}\sum_{i=1}^{n} g(x_i) \prod_{j=1}^{j=n} f(x_j) dx_j, \tag{3.95}$$

where the range of integration is the interval on x for which $f(x)$ is defined. Now this integral can be expressed as the sum of n integrals of the form

$$\frac{1}{n}\int\int \cdots \int g(x_i) \prod_{j=1}^{j=n} f(x_j) dx_i.$$

This can be expressed as the product of n integrals

$$\left[\frac{1}{n}\int g(x_i) f(x_i)\, dx_i\right]\left[\int f(x) dx\right]^{n-1} = \frac{1}{n} E\left[g(x_i)\right], \tag{3.96}$$

since $\int f(x)dx = 1$. Now the desired answer is the sum of n such terms, or

$$E\left[\frac{1}{n}\sum_{i=1}^{n} g(x_i)\right] = E\left[g(x)\right], \tag{3.97}$$

which was to be shown.

Although the proof above established that sample moments about zero are unbiased estimates of population moments about zero, it is not true that sample moments about the sample mean are unbiased estimates of population moments about the population mean. This can be proved by an example showing the biased character of sample variance. Let ν_1 and μ_2 be the population mean and variance, respectively. Now the sample variance is

$$\frac{1}{n}\sum_{i=1}^{i=n}(x_i - \bar{x})^2 \quad \text{where } \bar{x} = \frac{1}{n}\sum_{j=1}^{j=n} x_j.$$

Then the expected value of the sample variance is

$$
\begin{aligned}
E\left[\frac{1}{n}\sum_{i=1}^{i=n}(x_i - \bar{x})^2\right] &= E\left[\frac{1}{n}\sum_{i=1}^{i=n} x_i^2 - \left(\frac{\Sigma x j}{n}\right)^2\right] \\
&= E\left[\sum_{i=1}^{i=n}\left(\frac{1}{n} - \frac{1}{n^2}\right)x_i^2 - \frac{2}{n^2}\sum_{j=2}^{j=n}\sum_{k=1}^{k=j-1} x_j x_k\right] \\
&= \frac{n-1}{n^2}\sum_{i=1}^{i=n} E\left[x_i^2\right] - \frac{2}{n^2}\sum_{j=2}^{j=n}\sum_{k=1}^{k=j-1} E\left[x_j x_k\right] \\
&= n\,\frac{n-1}{n^2}(\mu_2 + \nu_1^2) - \frac{2}{n^2}\frac{(n-1)n}{2}\nu_1^2 \\
&= \frac{n-1}{n}\mu_2 \neq \mu_2.
\end{aligned}
\tag{3.98}
$$

The simplification from the summations to population moments needs explanation:

$$E[x_i^2] = [E\,x^2] = \mu_2 + \nu_1^2. \tag{3.99}$$

(The second moment about zero equals the sum of the variance and the square of the mean.)

$$E[x_j x_j] = E[x_j]\,E[x_k] = \{E[x]\}^2 = \nu_1^2. \tag{3.100}$$

Because of the independence, the expected value of the product is the product of the expected values. The change in constant coefficients is merely the counting of the number of equal terms involved.

An Example of a Maximum-Likelihood Estimate

Although the reader can disregard the proofs in this section, the results are of value in later chapters and they should be noted.

The binomial distribution provides a simple example of a maximum-likelihood estimate. In n repeated trials of an event in which there is a constant probability, p, of success in a single trial, the probability of c successes and n-x failures is

$$\frac{n!}{x!\,(n-x)!}\,p^x\,(1-p)^{n-x}.$$

If an experiment yields exactly x successes in n trials and we wish to estimate p by the method of maximum likelihood, it would be necessary to determine that value of p which would maximize the sample probability above. This says that we would have to think of x as being fixed (at the observed sample value) and view p as being the variable.

The problem is thus reduced to that of the simple *maximum-minimum* type of elementary calculus. We merely differentiate the function above with respect to p, equate the derivative to zero, and solve for p. The derivative is

$$\frac{n!}{x!\,(n-x)!}\,[xp^{x-1}(1-p)^{n-x} - (n-x)p^x(1-p)^{n-x-1}]$$

$$= \frac{n!}{x!\,(n-x)!}\,p^{x-1}(1-p)^{n-x-1}(x-np).$$

Setting this derivative equal to zero gives the solution

$$p = x/n.$$

Since the value of p thus obtained is only an estimate of the true or population value, it is customary to use some symbol other than p. A very common notation is

$$\hat{p} = x/n. \tag{3.101}$$

In practice it turns out that the mathematical steps are simplified by differentiating the logarithm of the function involved rather than the function itself. Of course, this will not change the solution, since the maximum of a function occurs for the same value of the variable as does the maximum of the logarithm of the function. In the example, the logarithm of the sample probability is

$$\log \frac{n!}{x!\,(n-x)!} + x \log p + (n-x) \log (1-p),$$

and the derivative of this with respect to p is

$$\frac{x}{p} - \frac{n-x}{1-p}.$$

Equating this to zero gives

$$\frac{x}{p} = \frac{n-x}{1-p} = 0$$

$$x - xp - np + xp = 0$$

$$x - p = 0$$

$$p = x/n \tag{3.102}$$

or

$$\hat{p} = x/n$$

as before.

Since the sample probability is called the likelihood function, we usually work with the logarithm of the likelihood function.

Interval Estimation

In the method known as *interval estimation*, we replace the single statistic which is used as a parameter estimate in point estimation by an interval within which we estimate that the parameter is located. For example, if a sample of electron tubes yielded 5 per cent defectives, then the point estimate might be 5 per cent, while an interval estimate might be less than 8 per cent or between 3 and 10 per cent. The exact determination of the interval depends on the assumptions we make—i.e., the kind of interval we wish to determine—and on a large amount of very complex mathematics.

These intervals are called *confidence intervals* because there is a serious problem in trying to use the word "probability" when speaking of the location of the parameter within or without the interval. There is a strong objection to speaking of "probability" unless we refer to a random event. In the present case, the parameter value is a fixed number, while the statistic is a random variable. Thus, for a given population density, we can logically speak of the probability that a statistic falls in a particular interval. On the other hand, it is illogical to speak of the probability that the parameter falls in a given interval. Indeed, in every case, it is true that the parameter either is or is not in the interval, and—by a stretch in usage—the probability might be termed one or zero for these two situations, respectively. We would be hard pressed to justify speaking in terms of any other value, say 0.9, as the probability that the parameter is in the interval.

To eliminate this semantic problem, alternative phraseology is used. We can speak of the degree of confidence that can be placed in the truth of the assertion that the parameter does in fact lie in the interval. Here confidence is measured by a positive proper fraction, although it is frequently translated into a percentage, and we then speak of a certain percentage confidence level. The meaning of such a statement is best explained by phrasing it in terms of the numbers involved.

Suppose a parameter ξ is estimated by a statistic $\hat{\xi}$ and the 90 per cent confidence interval is a to b. Then if we were to assert that ξ lies between a and b, $a \leq \hat{\xi} \leq b$, we would expect to be telling the truth 90 per cent of the time if we repeat the same procedure a large number of times. This, then, relates the 90 per cent to the confidence in the truth of the assertion, but not to the probability in the immediate case at hand.

In this case, a and b are called the 90 *per cent confidence limits*. Of course, they are computed from the sample data. In some instances they are selected in such a way that they are bisected by the point estimate; that is,

$$\hat{\xi} - a = b - \hat{\xi}.$$

This symmetrical arrangement is not necessary, of course. We could state a number of properties which a good confidence interval should have, but some might be contradictory to others. It is certainly true that, given a choice of two confidence intervals of the same level, we would be inclined to select the shorter of the two. The seriousness of making an error by over-estimating (or underestimating) the parameter value would also influence the type of interval selected.

3.5 Testing Hypotheses

In both point and interval estimation, the estimate of the population value follows data collection and is based on some computation using the sample data. By contrast, in hypothesis testing we usually state *before data collection* some hypothesis concerning parameters of one or more populations, then collect data, analyze the data, and arrive at a statement that the hypothesis is either accepted or rejected. Thus the conclusion of a hypothesis test is non-numerical, although the hypothesis itself may be a quantitative statement. This concept can be clarified if it is phrased in terms of an example.

In this example, it is desired to test whether or not a new method of sealing an electron tube will increase tube life. It is assumed that the mean life of tubes manufactured under the old process is known. The form of the density function is probably known or can be assumed. For tubes made by the new process, the form of the density function could be either assumed or estimated from sample data. Now, even though we would hope that the new process would improve mean life, the exact amount of improvement, if any, is unknown. We can formulate a hypothesis that the mean lives do indeed differ by some specific amount. If such a hypothesis is tested and rejected, we can say only that the size of the difference in mean lives was incorrectly estimated. A better conclusion can be reached by stating the hypothesis in a negative way. If the hypothesis states that there is no difference in mean lives (the so-called null hypothesis), then rejection results in a definite conclusion.

As in all types of estimation, there is a chance that the conclusion is incorrect. The hypothesis may be rejected as false when it is in fact true, or it may be accepted when it is actually false. These are termed Type I and Type II errors, respectively. Another kind of error is sometimes discussed. One might conclude from sample data that two production methods are different, but the whims of sampling might produce the result that the poorer method appears to be the better of the two. This Type III error is quite improbable and will not be considered at this time.

A complication which frequently enters into hypothesis testing results from consideration of more than one hypothesis at a time—the conclusion of the test being the acceptance of one hypothesis and rejection of the others.

Since it is possible to reject alternatives successively, we can restrict the discussion to choice between two hypotheses, say H_1 and H_2. Again there can be two types of errors: accepting H_1 when it should have been H_2, and the reverse.

Hypothesis testing is of extreme value in quality control and acceptance testing. These aspects will be covered in more detail in later chapters, and hence will not be discussed here.

3.6 Distributions Tabulated for Significance Testing

Among the many distributions which have been tabulated for use in all types of significance testing, we quickly think of the binomial, Poisson, normal, chi-square, t, and F distributions. Of course, there are others of great importance in point and interval estimation and in hypothesis testing. Complete discussion of these distributions and their uses is obviously beyond the scope of this chapter. It is possible here only to point out a few of the relationships between these distributions and to cite some of the historical events of interest in their development.

The binomial and normal distributions are basic in the theory of sampling. The binomial arose in connection with early gambling problems. The normal distribution, used in many applications in such fields as biometry and demography, then was applied as a continuous approximation of the binomial. It has been found that the normal distribution possesses two very important properties which enhance its use. First, on the basis of both theoretical considerations and observed data, it is believed that many population attributes probably follow the normal distribution. There are many illustrations of this in the life of electronic parts, product variation about bogie, and psychological variables such as intelligence quotients or grades in school. Second, it has been found that many other distributions approach the normal as a limiting case, thus making it useful as an approximation even when it is not the exact distribution.

The second point is of more interest in connection with sampling problems. The means of samples follow the normal distribution only if the parent population is normal. However, as the sample size increases, the distribution of sample means approaches normality for samples drawn from non-normal parent populations, subject only to very minor restrictions. Similar statements hold true for other parameters. Also, we can use the normal distribution for larger samples as an approximation for other distributions used in testing. For example, the chi-square and t distribution tests are replaceable by normal tests for large samples.

PROBLEMS

1. The 2 sets of data shown below were obtained with the use of tables of random variables. In each case the population mean was 100. (a) Make a frequency distribution for each set of data; (b) construct a histogram for each; (c) calculate the arithmetic mean of each; (d) compute the standard deviation of each; (e) calculate the arithmetic means of groups of 3 and 15, and construct a histogram of each group of means.

Set 1					Set 2				
87	179	78	60	117	176	43	70	229	471
133	134	73	126	52	74	184	106	36	84
52	115	60	118	140	26	83	89	18	281
118	106	69	87	46	38	483	10	11	5
85	96	96	122	21	27	274	244	2	37
53	101	111	69	102	9	69	147	102	13
48	108	92	26	107	27	201	12	171	44
123	149	71	51	81	17	15	58	48	112
89	93	119	91	110	19	154	128	182	41
138	132	75	48	160	167	143	10	9	58
58	111	144	105	82	72	2	51	35	5
96	96	125	90	105	40	37	141	86	2
63	98	84	41	99	310	20	92	255	127
130	117	163	73	102	3	121	56	32	190
121	87	160	71	79	42	4	107	11	32
46	112	93	129	77	51	113	127	182	100
70	64	71	136	111	132	140	129	74	27
62	62	120	120	134	181	19	193	46	412
70	134	102	144	119	215	8	64	25	38
109	42	120	85	142	124	60	92	99	95
74	130	150	86	106	51	32	41	76	19
145	117	99	15	71	28	141	14	10	106
94	76	66	85	112	330	332	62	106	25
105	70	102	161	64	83	16	6	262	9
145	99	90	102	107	128	41	391	128	64
64	118	139	100	128	22	45	106	230	78
97	96	116	115	104	11	49	248	17	106
71	95	139	158	117	45	17	159	59	10
66	124	107	84	106	43	158	85	101	91
87	100	65	111	109	38	89	72	114	37

2. Find the probability that load would exceed strength, assuming the load- and strength-density functions are normal and

$$\theta_y = 100$$
$$n_y = n_x = 2$$
$$\sigma_x = 10$$
$$\sigma_y = 15$$
$$m = 1.2.$$

3. Suppose the transconductance of a certain electronic tube is normally distributed with mean $\theta = 1600$ micromhos. The specification limits call for the product to lie between 1200 and 2000 micromhos with probability 0.80. (a) What is the maximum allowable standard deviation that the process can have and still maintain its quality? (b) If the process is at the value σ found in (a), what is the probability of sample average of 2 falling outside the specification limits? (c) Repeat (b) with sample average of 10.

4. Find the moment-generating function for the Poisson distribution (Equation 3.70). Calculate the mean and variance.

5. Suppose n tubes are put on life test. The test is terminated when all tubes have failed, the times of failures being $t_1, \ldots, t_n$. Assume that the time to failure has an exponential distribution with probability density function

$$f(t; \theta) = 1/\theta \, e^{-t/\theta}, \quad t > 0$$
$$= 0 \qquad\quad t \leq 0$$

Estimate θ by the method of maximum likelihood.

Data Collection Procedures

4.1 Introduction

The preceding chapters have dealt with concepts basic to the evaluation of system effectiveness. Factors affecting system effectiveness have been identified, defined, and charted to show their interrelationships, and mathematical methods have been introduced and justified on the basis of the necessity for precision in evaluating an individual system or comparing attributes of competing systems.

It should be recognized that there is a growing interest in all factors that comprise system effectiveness, both among military and commercial groups. Some of these factors are already specified in procurement documents; thus, both the manufacturer and user groups will share responsibility for product evaluation. In some cases impartial outside organizations may be brought in to conduct and monitor evaluation procedures. In any event, whether or not the concepts offered in this text undergo modification and regardless of who conducts evaluation studies, certain basic requirements must be met in order to conduct a meaningful evaluation. Thus, this chapter is concerned with problems pertinent to the organization and execution of measurement programs. For purposes of discussion, these problems may be classified in three categories, which are discussed in this order in the remainder of the chapter: (1) planning of the measurement program, (2) collection and reporting of measurement data, and (3) processing and handling of measurement data.

In the study of the material to follow, it should be remembered that basic measurement requirements and techniques are standard at any level of product division—for example, set, unit, assembly, subassembly, and part.

4.2 Planning a Measurement Program

Statement of Objectives

In planning a program to collect system measurement data, due consideration should be given to several factors that are important to the success of the study. First, as the purpose of the study is to collect data that will provide, upon analysis, an accurate evaluation of the study parameters, it is appropriate that a complete set of clearly stated technical objectives be established. These objectives determine the scope of the measurement program in terms of data to be collected and the best methods to be used for handling and processing them. Once the technical objectives of the study are determined, the planning and subsequent execution of the study must be directed specifically toward the realization of the objectives.

Personnel

Secondly, planning should consider the methods by which required data will be collected and processed. In general, one of two methods may be used, depending entirely upon the relative importance of accuracy versus cost. One method is to supply the operational, maintenance, or production personnel with data forms containing blanks for the desired information, and to ask that the forms be completed as directed. Forms containing the raw data are returned to a central collection point to be picked up periodically and processed. This method has the advantage of low cost, resulting from use of few personnel, but data so collected are invariably of questionable accuracy and completeness. Operational, maintenance, or production personnel, in general, tend to look upon data collection as just so much paper work, and in the press of more urgent responsibilities they tend to neglect it.

A second method—more costly but far more accurate—is to employ technical personnel who have the assigned responsibility for carrying out the measurement program. This method has many advantages. For example: (1) personnel concerned with the program can be given a thorough understanding of the objectives of the study; (2) a high interest in the study can be maintained at the source of the data; (3) as a result of their understanding of the study objectives and their close daily contacts with the system and with user or production personnel, the evaluation personnel can make decisions as necessary to keep the study on the right course; (4) data supplied under conditions of close monitoring and recheck require a negligible amount of rework and interpretation before final processing; (5) selective attention can be given to developing details or trends that are pertinent to the evaluation; and (6) inconsistencies and errors in the data can be detected through cursory checks and analyses, with corrections applied at or close to the time of occurrence.

The foregoing list of advantages is far from complete, and represents only the more important reasons for assigning the technical and administrative conduct of a measurement study to technical personnel, trained in measurement techniques, as their primary responsibility.

Development of a Study Plan

A third factor essential to the success of the measurement study is a detailed written document which is in effect a specification for the work to be done during the study. This study specification, if properly planned and written, is the primary medium for coordinating the plans of the project engineering group and the measurement, data-collection, and processing work of the measurement group. If the over-all evaluation program is of such magnitude that several people are assigned to the study, either in the same or in different locations—or if there is any possibility of personnel dislocation during the course of the study—a clearly written specification serves to preserve intact the original perspective and objectives of the study. The format of the specification should be determined by the nature and scope of the specific system evaluation to be conducted. However, in general, the information outlined below should be spelled out.

BACKGROUND

The specification should contain a brief, factual account of the development and objectives of the broad evaluation program of which the measurement study is a part. Although measurement personnel can make satisfactory observations and collect specified system data without knowing the background of the study, their engineering decisions and the quality of the data they collect will be improved if they understand the broad objectives of the program and know how their work affects the accomplishment of these objectives.

STUDY OBJECTIVES

The specification should clearly and concisely state the specific objectives to be realized during the course of the study. As previously stated, these objectives determine the scope of the work and provide the basic guidelines for decisions by measurement personnel during the course of the study.

DEFINITIONS OF TERMS

The specification should explicitly define the terms that are of interest to the study and that are used throughout the specification. Since the primary purpose of most studies is the measurement of some characteristic of a product, the characteristic must be defined in such a way that units of it can be measured. Many of the terms used in the evaluation of effectiveness factors are sufficiently well understood that confusion regarding their application now rarely occurs. The terms that are most likely to cause difficulty are those involving an opinion or a value judgment on the part of an observer. Terms such as *satisfactory performance* and *system failure* are of this nature. Unless they are carefully and explicitly described, there is danger that the data on a given characteristic will be confounded by the presence of data on other characteristics, and that the investigation will be delayed until the problem is corrected.

Probably the most important single item of data to be collected in any study is the time during which various events occur. In view of the wide variety of time measurements that are used, particular attention should be given to defining the time measurements required by the objectives of each study.

DATA REQUIREMENTS

The specification should indicate in detail each item of data to be collected and reported by the study group. The types of data required for final evaluation are determined by analysis of the study objectives, the product configuration, the usage environment, and the events of interest in the study. Each item of data should be defined in terms of type, criteria, and unit of measurement, as well as in terms of the time when the data must be reported.

DESCRIPTION OF THE PRODUCT AND SOURCES OF DATA

The specification should include a complete and detailed technical inventory of the product to be evaluated—with each subdivision of the product clearly identified by technical nomenclature, number, and function. It should also include a list of sources from which the items of information specified under Data Requirements will be extracted. The sources may include: production lines, operators, maintenance personnel at various echelons, supply depot records, and a wide variety of user organizational report forms, such as flight logs, maintenance records, UR reports, DD-787's, work orders, and weather reports.

DATA COLLECTION, PROCESSING, AND REPORTING METHODS

A detailed description of the methods to be used for the collection, processing, and reporting of data should also be included in the Data section of the specification. Data reporting forms, with detailed instructions for completing them, should be attached to the study specification as Appendices. Where raw data will be coded for integration into a mechanical data-processing system, the instructions must clearly explain the coding system.

REPORTS

The specification should include a requirement for the submission of periodic reports by the study group. The types of reports required and the times when they must be submitted should be specified.

AUTHORIZATION FOR THE STUDY

When a study is to be performed in cooperation with another group having a primary mission responsibility, the study group can operate most effectively if formal authorization and support of the study are obtained and distributed throughout all levels of authority concerned with the project. Authorization should be obtained in writing and should be referenced in the study specification.

MATERIAL AND FACILITIES

In order to carry out the study, the specification should list the material and facilities needed by the evaluation group, the sources from which material should be obtained, and the persons or groups responsible for procurement.

4.3 Collection and Reporting of System Data

So far the discussion has dealt largely with the planning of a measurement study. We have indicated why there must be personnel whose primary responsibility is the collection and reporting of data in a form most usable for analysis and why there must be a clearly written study specification.

From this point on, the chapter will describe several types of information of intrinsic value to a system study; some typical sources of the information; techniques for collecting and reporting the information; and certain typical problems which arise in the collection of data.

Data Forms

Figures 4.1, 4.2, 4.3, and 4.4 are representative data sheets used in studies of systems. Figure 4.1 represents a combined study of the reliability and maintainability of an airborne search radar system, with emphasis on the maintainability of the system. Figure 4.2 represents basically the same type of study on the AN/ARC-27 airborne communications system, but in addition it provides for reporting of special data relative to the measured performance parameters of the units being studied. Figures 4.3 and 4.4 also represent a combined reliability and maintainability study, this time on a complex airborne bombing/navigation system. The figures show that the study is an elaborate one, extending from the system level down through the subsystem, unit, and subunit levels to the individual part level.

The reporting forms are designed to facilitate machine processing of the information, which is transferred from the forms to 80-column punch cards. The amount and types of information reported can result in as many as six different types of data cards, as shown by Figures 4.3 and 4.4. The specific function of each card type is to expedite final analysis of data by permitting rapid and selective recovery of information pertaining to a specific system level or area of the study. To this end, personnel code all raw data in numerical groups for maximum usage of the 80-column card.

SYSTEM IDENTIFICATION AND CONTROL DATA

System identification and control data are those which will allow each defined event of the study, such as a malfunction, a repair action, a mission, a performance measurement, an installation, or a termination, to be related to a specific system and, if appropriate, to a specific subunit of that system. In Figure 4.1, column 80 indicates that card types 1 and 2 are used to record desired data at the system level. Columns 7 and 8 identify the specific system being studied. Column 6 identifies the group to which the system is assigned for purposes of statistical analysis. Note that columns 1 through 16 are the same for card types 1, 2, 3, and 4.

However, as the system of interest in Figure 4.1 is subject to interchange or substitution of components in the course of maintenance or repair, card type 3 incorporates information on the component type and serial number involved in the event of interest to the system. This information is carried over to card type 4, where part-symbol and part-type data associated with a specific event may be related to the component level as well as to the system level.

The same sort of analysis may be made of the form shown in Figure 4.2. The differences between the forms shown in Figures 4.1 and 4.2 are due to differences in test objectives. In the latter case, the system of interest is the aircraft communications system using the AN/ARC-27 receiver-transmitter, and the study objectives emphasize the reliability, maintainability, and performance of the RT unit. In view of these objectives, the form provides for specific identification of the aircraft number, the RT-unit group code, and the RT-unit number. However, assemblies within the RT unit are identified only by type.

The forms shown in Figures 4.3 and 4.4 were designed for reporting of data on a complex airborne weapons system being evaluated primarily from the points of view of reliability and maintainability. The same types of identification and control data are recorded on this form, but in considerably more detail because of the greater complexity of the system and of the study itself. For example, for each reported event, card type 4 requires notation of the unit type involved, the serial number of each unit installed or removed, and the coded location of this unit type in the system. Card type 6, at the part level, requires information on the symbol and serial number of the part, its position in the circuit, and its name and manufacturer.

EVENT IDENTIFICATION AND CONTROL DATA

Event identification and control data are those which uniquely relate items of system measurement data to a specific system event. Examination of the forms shown in Figures 4.1 and 4.2 reveals that each observed event is identified by a sequentially assigned event number and the date when the event occurred. In Figure 4.1, this information is given in columns 9 through 16, labeled MR# and Date. The abbreviation MR indicates that a Maintenance Report was the source of the information about the event. In Figures 4.3 and 4.4, event identification and control information is given in columns 7 through 15.

EVENT MEASUREMENT DATA

Event measurement data are defined in the study specification as data required in the quantification of attributes (such as reliability or maintainability) and in the determination of the interrelationships among attributes. As previously stated, the scope of such data is determined by the nature of the study objectives.

From Figures 4.1 through 4.4 (which illustrate most of the types of data collected and reported in a measurement study) it is apparent that many types of measurements are made. For example, performance and malfunction data are measured in terms of operator or maintenance judgments, symptoms, actions taken, parameter values, units or parts replaced, and times between various actions.

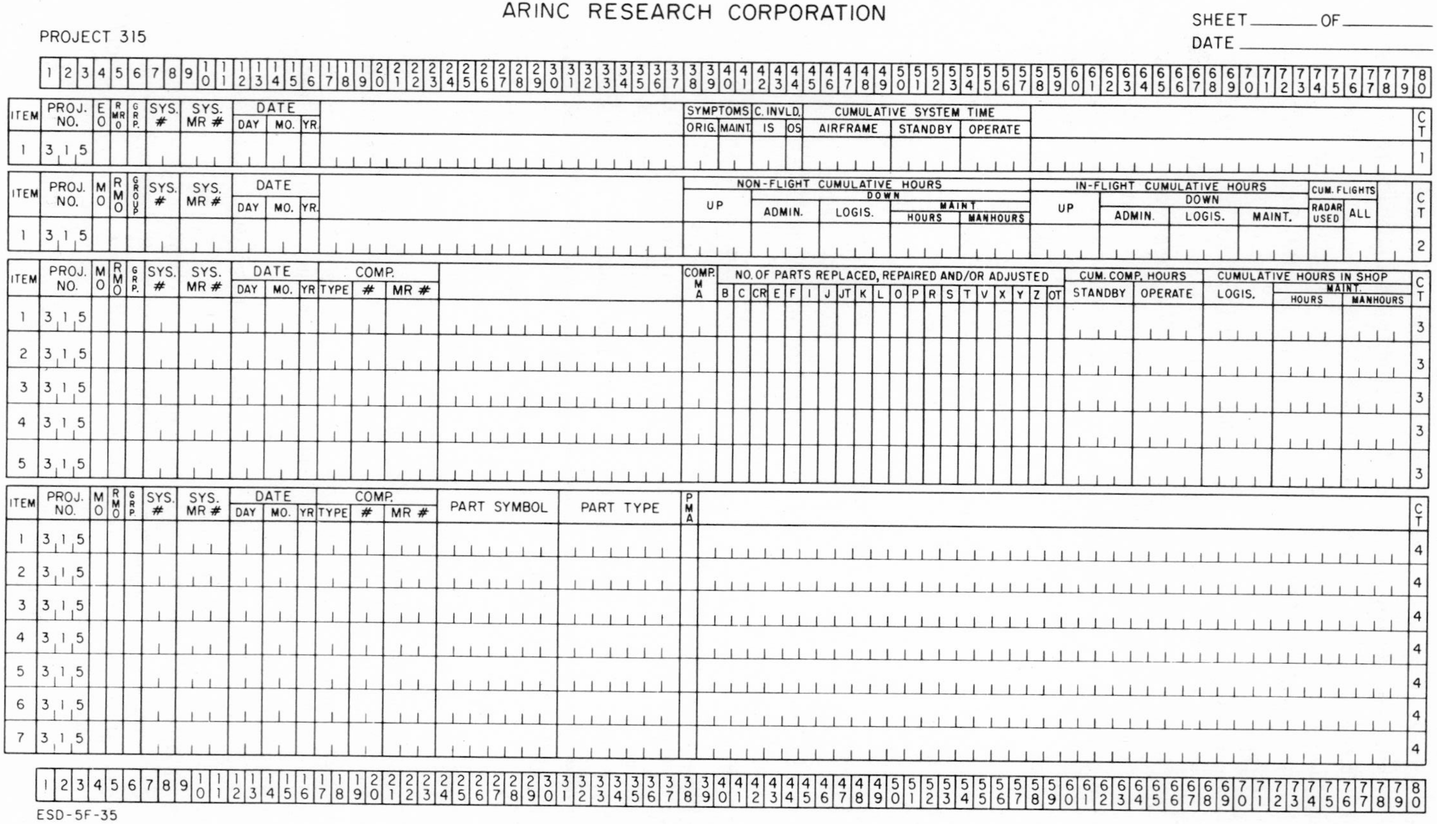

Fig. 4.1 Data reporting form, airborne search radar system.

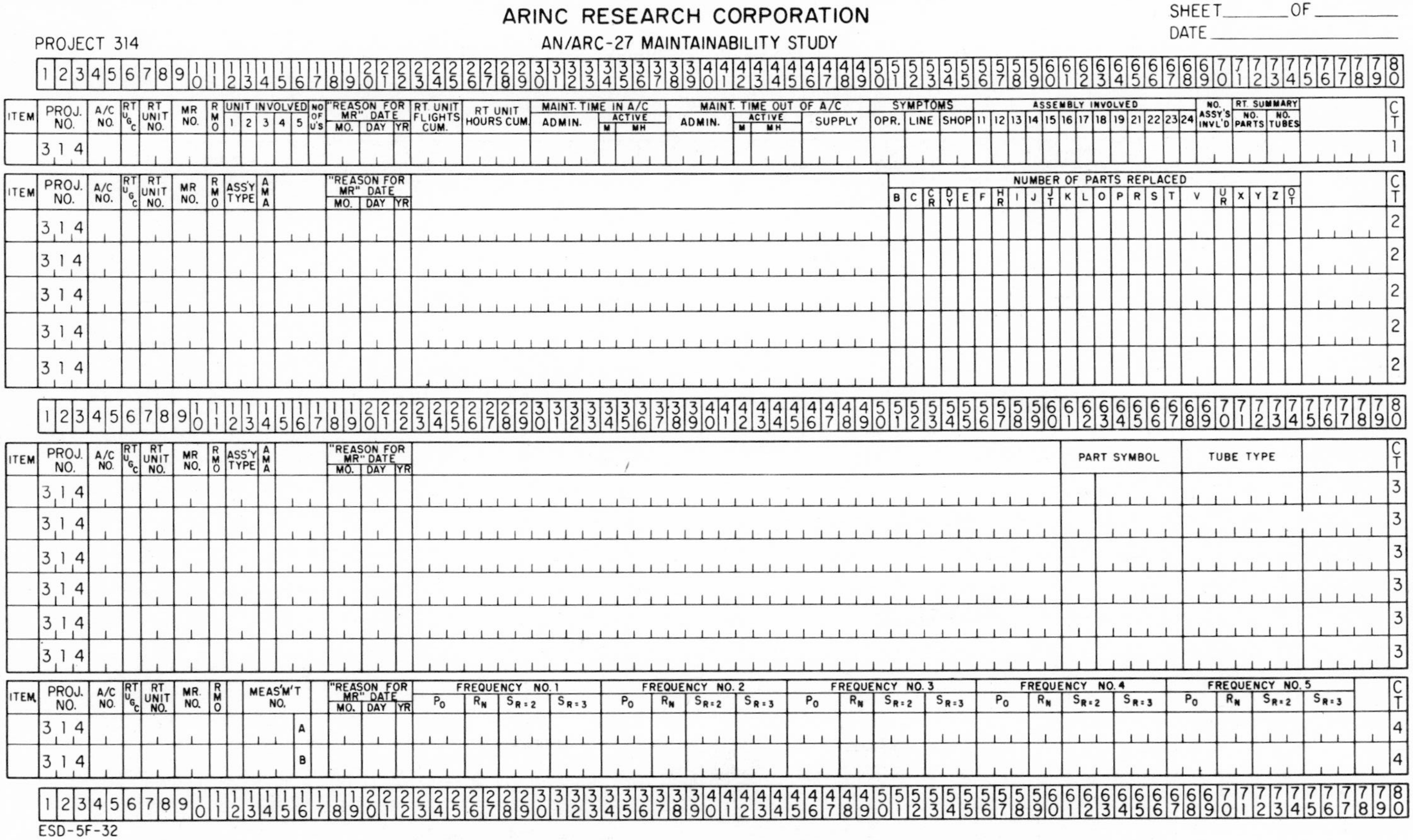

Fig. 4.2 Data reporting form, AN/ARC–27 communications system.

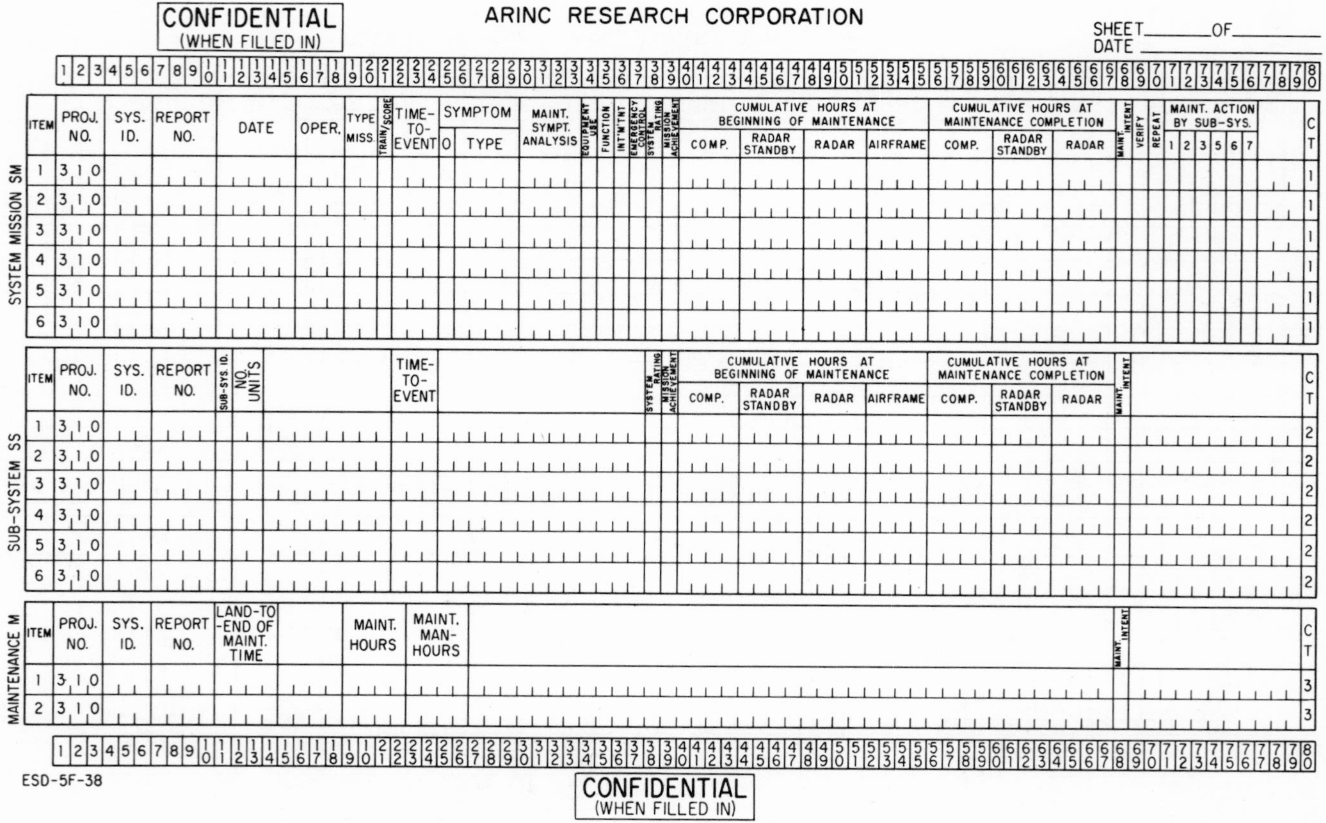

Fig. 4.3 Data reporting form, bombing/navigation system: System and subsystem levels.

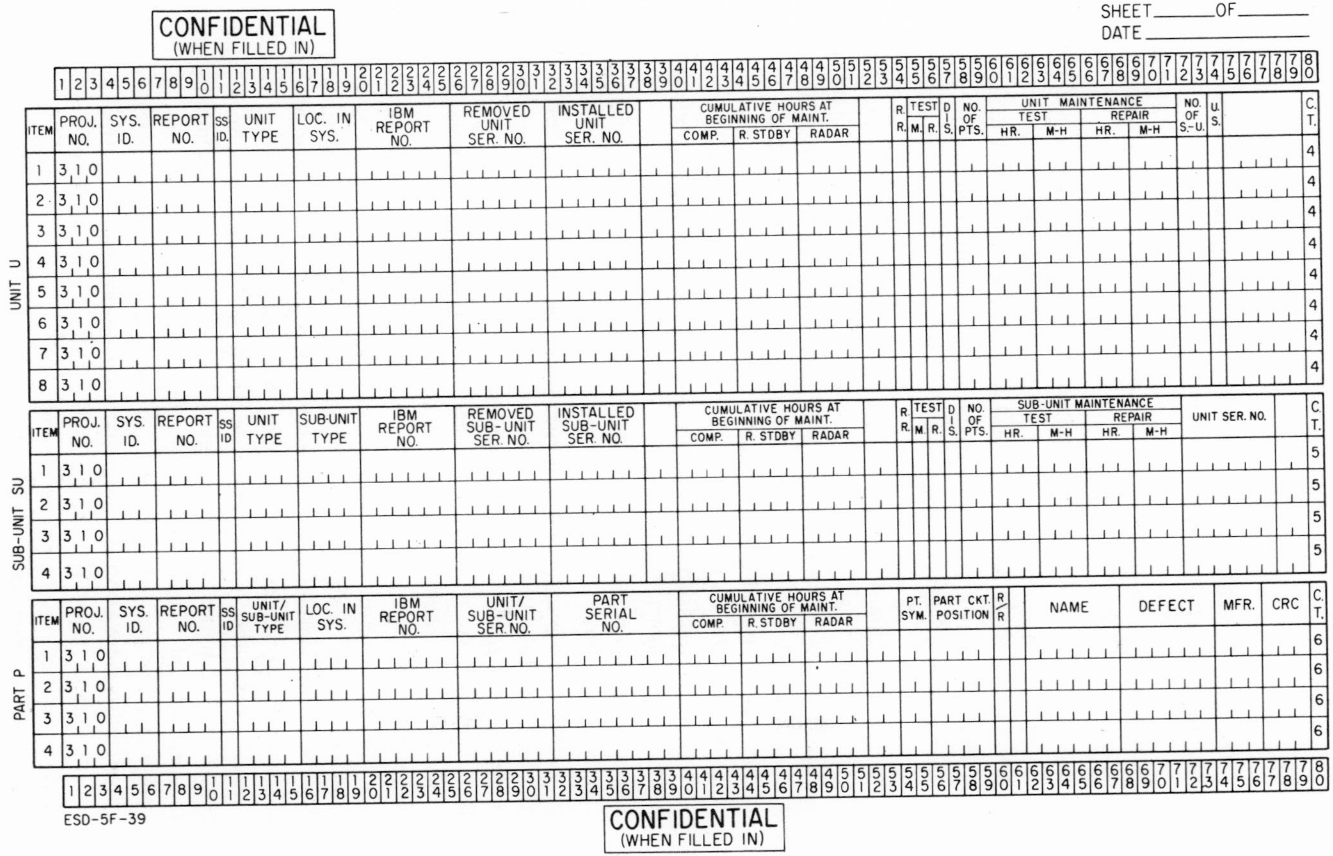

CONFIDENTIAL
(WHEN FILLED IN)

SHEET ______ OF ______
DATE ______

UNIT U

ITEM	PROJ. NO.	SYS. ID.	REPORT NO.	SS ID.	UNIT TYPE	LOC. IN SYS.	IBM REPORT NO.	REMOVED UNIT SER. NO.	INSTALLED UNIT SER. NO.		CUMULATIVE HOURS AT BEGINNING OF MAINT. COMP.	R. STDBY	RADAR		R. R.	TEST M.	TEST R.	D I S.	NO. OF PTS.	UNIT MAINTENANCE TEST HR.	TEST M-H	REPAIR HR.	REPAIR M-H	NO. OF S.-U.	U. S.		C. T.
1	3 1 0																										4
2	3 1 0																										4
3	3 1 0																										4
4	3 1 0																										4
5	3 1 0																										4
6	3 1 0																										4
7	3 1 0																										4
8	3 1 0																										4

SUB-UNIT SU

ITEM	PROJ. NO.	SYS. ID.	REPORT NO.	SS ID	UNIT TYPE	SUB-UNIT TYPE	IBM REPORT NO.	REMOVED SUB-UNIT SER. NO.	INSTALLED SUB-UNIT SER. NO.		CUMULATIVE HOURS AT BEGINNING OF MAINT. COMP.	R. STDBY	RADAR		R. R.	TEST M.	TEST R.	D I S.	NO. OF PTS.	SUB-UNIT MAINTENANCE TEST HR.	TEST M-H	REPAIR HR.	REPAIR M-H	UNIT SER. NO.		C. T.
1	3 1 0																									5
2	3 1 0																									5
3	3 1 0																									5
4	3 1 0																									5

PART P

ITEM	PROJ. NO.	SYS. ID.	REPORT NO.	SS ID	UNIT/ SUB-UNIT TYPE	LOC. IN SYS.	IBM REPORT NO.	UNIT/ SUB-UNIT SER. NO.	PART SERIAL NO.	CUMULATIVE HOURS AT BEGINNING OF MAINT. COMP.	R. STDBY	RADAR		PT. SYM.	PART CKT. POSITION	R/R		NAME	DEFECT	MFR.	CRC	C. T.
1	3 1 0																					6
2	3 1 0																					6
3	3 1 0																					6
4	3 1 0																					6

ESD-5F-39

CONFIDENTIAL
(WHEN FILLED IN)

Fig. 4.4 Data reporting form, bombing/navigation system: Unit, subunit, and part levels.

Sources of Raw Data

The raw data entered on the forms shown in the figures mentioned above come primarily from three sources:

1. Copies of standard operational, maintenance, or supply forms completed by users of the system
2. Special data forms originated by the study group to fill a specific data need or control function (these forms may be completed either by the users or by study group personnel in interviews with the users)
3. Direct instrumentation and engineering measurements performed on the system.

Examples of standard military forms are so numerous as to obviate detailed discussion. In the case of airborne systems, they may include pre-flight check lists filled out by line technical personnel, post-flight debriefing forms recording details of system performance and malfunctions, various types of flight records completed by pilot or crew members during operation or maintenance of the system, and work-order forms issued to maintenance personnel. Similar types of forms are used for fixed ground systems, mobile ground systems, and shipborne systems.

Special data forms are used in cases where the circumstances relating to a particular installation or end-use of the system require the development of special data-collection methods. Study group personnel must devise these methods on the basis of their understanding of the specification requirements and the special circumstances found in end-use.

Data derived from direct engineering instrumentation may include measurements of specified electrical or mechanical properties, special time measurements made by means of elapsed time indicators, measurements of system and subsystem usage modes as indicated by cycle counters, accelerometer recordings of shock and vibration, and temperature measurements.

The collection and use of the three types of raw data mentioned above are best illustrated by describing an actual field study conducted by ARINC Research Corporation. The study for which the form in Figure 4.1 was prepared was designed with the objective of evaluating the reliability and maintainability of the AN/APS-20E airborne search radar system. The system was to be evaluated in terms of the extent to which reliability and maintainability were affected by system modifications made to correct serious problems found in a previous study. Of the systems selected for investigation, half were modified and half were unmodified.

Ten reporting forms were used in the collection of raw field data. Four of these were standard military forms, and six were designed by the field study group to provide data not listed in the military forms and required for purposes of measuring the characteristics under study.

Part A of Figure 4.5 is an official U.S. Navy form used for reporting of events associated with a specific mission and aircraft. Part B, on maintenance, is used for reporting discrepancies observed on any part of the aircraft or its equipment during flight. Various time elements are also given. It is obvious, however, that even if the form is properly filled out, the amount of data provided is sketchy. Insofar as the study is concerned, the form serves only as the official record of the occurrence of an event of interest.

Figure 4.6 is a Radar Flight Report form designed by the study group to provide operational data on systems for which a pre-flight check was used as a maintenance device. Figure 4.7 presents a form which is identical to that given in Figure 4.6, except that the systems involved were subject to a post-flight check. The radar operator completed the appropriate form each time a flight using the system was made. The field man assigned to monitor each group of test aircraft collected the forms and reviewed them for accuracy of reporting. Any discrepancies found were checked during an interview with the operator, whose signature appeared on the form. The field man also checked the information on this form against that shown on the form in Figure 4.6.

Figure 4.8 is an official U.S. Navy Form originated by the maintenance organization to effect repair action in cases revealed by the discrepancies reported on Part B of OPNAV Form 3760-A (Figure 4.5). When filled out, the form is forwarded to the appropriate repair facility for action. Again, the information presented on the form is not adequate to describe the event of interest to the field study. The form was used primarily for control purposes in cross-checking the Radar Flight Reports (Figures 4.6 and 4.7) to ensure continuity in data collection.

Figure 4.9 (Maintenance Report JAX-315-2) is a data form designed by the field study group to provide a complete history of a system malfunction and its repair. Most of the final field data reported on the form shown in Figure 4.1 were coded and transcribed from this form. A study group member assigned to monitor a group of test aircraft was directly responsible for completion of this data form. He obtained the desired information either by direct observation, by personal interview of military personnel concerned with the event of interest, or by examination of the other forms discussed above.

Figure 4.10 is a special form for reporting of measurement data observed in a proposed engineering study of the magnetron used in the radar system. The requirement for these data was defined in the Study Specification. The data do not appear as system data in Figure 4.1, since they are treated separately as part of the special engineering study. Measurement was the direct responsibility of the field study group.

PART A – PREFLIGHT "K"
OPNAV FORM 3760-2 (Rev. 7-56)

NAVY AIRCRAFT
FLIGHT RECORD

(Retain at Place of take-off)

DATE

A/C MODEL	A/C SERIAL NO.	A/C REPORTING CUSTODIAN	KIND OF FLIGHT CODE

FUEL *(Gal./Lbs.)*	OIL *(Gal./Pts.)*	OXYGEN *(Lbs.)*	ORDNANCE *(Kind and quantity)*	DESTINATION

A/C LIMITATIONS

CERTIFICATION: I certify that this aircraft has been inspected this day in accordance with approved Preflight Instructions, is serviced as stated above, and is ready for flight.

ACCEPTANCE: I accept this aircraft for flight. I have examined the Last_____Discrepancy Reports on this aircraft. I certify that all requirements for weight and balance clearance, DD form 365F, on this aircraft have been fulfilled.

SIGNATURE *(Plane Captain)*	RATE	SIGNATURE *(Pilot)*	RANK

INSTRUCTIONS: List all personnel aboard on the REVERSE SIDE of PART A, if no other list is filed.
DETACH PART A when the rest of the form is taken into the A/C.
Always take PARTS B and C on NON-LOCAL FLIGHTS.

PART B – MAINTENANCE
OPNAV FORM 3760-2 (Rev. 7-56)

INSTRUCTIONS: Pilot will Promptly Complete and turn into A/C REPORTING CUSTODIAN of the A/C. Entries refer only to flights covered by this sheet. All time entries to be in HOURS and TENTHS.

AIRCRAFT DATA

DATE:

			LANDINGS		ENGINE TIME			LOCAL USE			LOGGED
			TYPE	NO.	ENG.	TOTAL TIME	COMBAT POWER AFTER BURNER TIME	1	2	3	*(Pilot leave blank)*
A/C MODEL		TOTAL A/C TIME	ARRESTED		1	.	.			A	A/C LOG
A/C SERIAL NO.			FCLP		2	.	.			B	ENG. LOG
A/C SIDE NO.		KIND OF FLIGHT CODE	OTHER FIELD OR SHIP		3	.	.			C	
			WATER		4	.	.			D	

AIRCRAFT DISCREPANCY REPORT

PILOT FILL OUT		MAINTENANCE FILL OUT	
▲ A/C CONDITION IS- *(Plainly Circle UP or DOWN)* ▼		CORRECTIVE ACTION TAKEN *(If the action corrects grounding discrepancies, circle Pilot's "X" which identifies it.)*	

ITEM NO.	X	DISCREPANCIES AND ITEMS TO BE CHECKED *(Place "X" opposite grounding discrepancies)*	ITEM NO.	WORK DONE	BY *(Initials)*	TIME AND DATE

Check here if continued on reverse side. □   Check here if continued on reverse side. □

SIGNATURE *(Pilot)*	RANK	UNIT	SIGNATURE *(Corrective action approved - approving authority.)*

PART C – OPERATIONS
OPNAV FORM 3760-2 (Rev. 7-56)

INSTRUCTIONS: All time entries to be in HOURS and TENTHS. Entries refer only to flights covered by this sheet. Use reverse for non-Pilot Personnel receiving flying time credit.

FLIGHT DATE

A/C MODEL	A/C SERIAL NO.	A/C REPORTING CUSTODIAN	KIND OF FLIGHT CODE	TOTAL A/C TIME

PILOT AND STUDENT PILOTS RECEIVING FLYING TIME CREDIT	UNIT 2.	KIND OF FLIGHT CODE 3.	PILOT TIME						INSTRUMENT TIME				LANDINGS							INSTRUMENT		CATAPULT "X"
			TOTAL	1ST PILOT	CO-PILOT	DUAL	SPE-CIAL CREW TIME	DAY	NIGHT	NIGHT VISUAL TIME	FIELD	CARRIER	OTHER-LAND OR SEA	ARRESTED	CARRIER ONLY TOUCH AND GO	HOLTER	NO.	TYPE				
NAME *(Pilot in Command)* 1.	1.			.	.	.	.	4.	4.		5.	5.		5.	5.	5.	NO.	TYPE 5.				
1.				.	.	.	.			.												
NAME 2.				.	.	.	.			.												
NAME 3.				.	.	.	.			.												
NAME 4.				.	.	.	.			.												
TOTAL COLUMNS IF MORE THAN ONE (1) PILOT			.																		

1. ONLY if A/C Commander time is earned, check the applicable box and time in this SPACE - - - - - - - - - - - - - - - - -
2. PILOT'S UNIT if different from that of A/C Reporting Custodian.
3. If different than kind of flight code entered in Parts A, B, and C for the A/C.
4. Enter "S" if Simulated.
5. Enter "N" if Night.

LOGGED *(Pilot leave blank)* 1.	FLIGHT 2.	PLACE 3.	TIME *(Zone)* 4.	
MASTER LOG	DEPARTURE			I certify that PARTS "B" and "C" are complete and correct.
AVIATORS LOG	STOPS ARR: / DEP:			
OPNAV C. Z	STOPS ARR: / DEP:			
	ARRIVAL			SIGNATURE *(Pilot)*

Fig. 4.5 Navy aircraft flight record (front of form).

PART A – BACK

INSTRUCTIONS: List all personnel aboard, if NO OTHER LIST HAS BEEN FILED.

NAME (Last - first - middle)	RANK OR GRADE "CIV" (if Civilian)	FILE OR SERVICE NUMBER	CREW POSITION "P" (if passenger)	HOME STATION

PART B – BACK   (CONTINUED)

ITEM NO.	X	DISCREPANCIES AND ITEMS TO BE CHECKED (Place "X" opposite grounding discrepancies)	ITEM NO.	WORK DONE	BY (Ini-tial)	TIME AND DATE

PART C – BACK

INSTRUCTIONS: List all non-pilot personnel receiving FLYING TIME CREDIT. (Exclude passengers)

NAME (Last - first - middle)	FILE OR SERVICE NUMBER	RANK OR RATE	REASON FOR *FLIGHT	TOTAL FLYING TIME	UNIT TO WHICH ATTACHED	REMARKS

* 1 CREW MEMBER

2 NON-CREW MEMBER - FORMAL SCHOOL TRNG.

3 NON-CREW MEMBER - CREW MEMBER TRNG.

4 NON-CREW MEMBER - AIRBORNE TECHNICAL SPECIALIST

Fig. 4.5 Navy aircraft flight record (reverse of form).

This report is required for EACH flight. When completed, turn in with the flight YELLOW SHEET

RADAR FLIGHT REPORT

PRE-FLIGHT CHECK

1. Phase I voltage _____ volts

2. Phase II   "   _____ volts

3. Phase III   "   _____ volts

4. Fixed 400   "   _____ volts

5. 28-V Aircraft
   Bus Voltage   _____ volts

   Steady? ☐ Erratic? ☐

6. Mag. Fil. V. _____ volts

7. Magnetron Curr._____ ma

8. Modulator Curr._____ ma

9. Range Marks normal?

   Yes ☐   No ☐

10. Target returns (check one)
    Good ☐ Fair ☐ Poor ☐
    None ☐

FROM OS-4 OSCILLOSCOPE

A. Mod Trigger normal? (Sharp, positive 40-volt pulse--minimum-- on 100 microsecond sweep is normal) Yes ☐ No ☐

B. Thyratron Trigger normal? (Sharp, positive 80-volt pulse on 100 microsecond sweep normal) Yes ☐ No ☐

C. Rad Video normal? (presence of grass is normal) Yes ☐ No ☐

D. Transmitter pulse normal? (Sharp 80-volt pulse on 100 microsecond sweep is normal) Yes ☐ No ☐

FILL IN FOR EVERY FLIGHT

Aircraft No. LF-____ Takeoff Date____ Time_____ (wrist watch)
                     Landing Date____ Time_____ (wrist watch)

Was Radar Used on Operate? Yes ☐ No ☐

If no trouble during flight, sign
Name & Rate_____

FILL IN ONLY IF MALFUNCTION OCCURS

FIRST symptoms of trouble_____

Time (wrist watch) of day when trouble detected:_____

Radar Timer readings when trouble detected: STANDBY_____OPERATE__

If Maintenance performed IN Flight--Time of day BEGUN____ ENDED____

Engine RPM at Time of Malfunction:____Bus Voltage_____volts
                                      Steady? ☐ Erratic? ☐

Any significant change in RPM just prior to malfunction? Yes ☐ No ☐

If yes, DURATION of High___ Low___ RPM

RPM changed from_____ RPM to_____ RPM

Reported by:

Note: An ARINC Research Corporation Representative will see you for additional information.

_____
Name/Rate

JAX 315-1

Fig. 4.6 Radar flight report, including pre-flight check.

This report is required for EACH flight. When completed, place in ARINC collection box in Electronics Shop.	RADAR FLIGHT REPORT

FILL IN FOR EVERY FLIGHT

Aircraft No. LG-____ Takeoff Date____ Time____ R (Roger time)

Was Radar used on OPERATE during this flight? Yes ☐ No ☐

Did any Radar trouble occur during OPERATE TIME? Yes ☐   No ☐

FILL IN ONLY IF MALFUNCTION OCCURS

FIRST symptoms of trouble:_____

Time of day (Roger time) when trouble detected:_____ R

Radar Timer readings when trouble detected: Standby___OPERATE___

If maintenance performed in flight, time of day Begun___Ended_____

Engine RPM at the moment malfunction occurred:_____RPM

Any significant change in RPM just prior to malfunction? Yes ☐ No ☐

    If yes, RPM changed from_____RPM to___RPM

                    Radar Operator_____
                                   Name/Rate

COMPLETE IN FLIGHT JUST BEFORE SECURING RADAR

1. Phase I Voltage_____volts		2. Phase II Voltage_____volts	
3. Phase III " _____volts		4. Fixed 400 " _____volts	
5. 28-V Bus " _____volts		6. Mag. Fil. " _____volts	
7. Mag. Current_____ma		8. Mod. Current_____ma	
9. Range marks normal Yes ☐ No ☐		10. Target returns (Check one) Good ☐ Fair ☐ Poor ☐ None ☐	

Note: Place all removed parts (tubes, resistors, capacitors, etc.) in ARINC collection box in Electronics Shop. ARINC envelopes may be used for these parts...for the purpose of identification. EFR's may be inserted in the envelope with the part. An ARINC Representative will see you for additional information concerning malfunction data.

JAX 315-1B (5-59)

Fig. 4.7 Radar flight report, including post-operative check.

WORK ORDER AND WORK ACCOMPLISHMENT RECORD
NAVAER-5230B

NOTE: If additional space is required, write "See attached" and use separate sheet.

1. AUTHORIZATION

REQUESTED BY	AIRCRAFT MODEL	WORK ORDER NO.	DATE
ACTIVITY	BUNO	WORK AUTHORIZED *(Signature)*	

2. WORK ORDER DATA

REFERENCE INSTRUCTION OR DIRECTIVE

WORK DESCRIPTION

ESTIMATED MATERIAL REQUIREMENTS

3. TECHNICAL DATA

PARTS REPLACED *(Stock Number and Name)*	SERIAL NUMBERS		REASON FOR REPLACEMENT	FUR/EFR REQ'D (✓)
	OLD PART	NEW PART		

OTHER COMMENTS AND RECOMMENDATIONS

DETAILED RESPONSIBILITIES

SHOP	WORK	REQUIRED COMPLETION DATE	ESTIMATED MAN-HOURS

SPECIAL INSTRUCTIONS

4. ACCOMPLISHMENT RECORD

WORK ACCOMPLISHMENT			INSPECTION	
DATE ACCOMPLISHED	ACTUAL MAN-HOURS	SUPERVISOR'S SIGNATURE	DATE	SIGNATURE OF AUTHORIZED INSPECTOR

5. MAINTENANCE OFFICE ACTION

LOG BOOK ENTRY REQUIRED *(Check appropriate box(es))*

☐ AIRCRAFT ☐ ENGINE ☐ ACCESSORIES & COMPONENTS ☐ WEIGHT & BALANCE ☐ OTHER *(Specify)*

LOG BOOK ENTRY (ENTRIES) MADE ________ *(Date)* ________ *(Initials)*

FOLLOW-UP DATA

NO FURTHER ACTION REQUIRED ____________________ *(Signature)*

(See instructions on reverse)

Fig. 4.8 Work order and work accomplishment record.

ARINC RESEARCH CORPORATION
Project 0315 -- Maintenance Report

System No._____

Group No._____

System MR No._____

Event Originator:

1. Operator.........................☐

2. Maintenance Man..............☐

3. ARC..............................☐

Reason for MR:

1. In-flight Malfunction.........☐

2. Swap.............................☐

3. Check............................☐

4. Delayed Repair................☐

5. Scheduled Maintenance......☐

6. Installation.....................☐

7. Termination....................☐

8. Modification...................☐

9. Other............................☐

Date and Time:

A/C Takeoff_____

Event Start_____

A/C Landing_____

Event End _____

Clock or Log Readings

Airframe_____

Standby_____

Operate_____

Symptoms:

Reported by operator_____
_____
_____
_____
_____

Name_____ Rate_____

Calendar Hour
_____
_____
_____
_____

Cum. System Time
_____
_____
_____

* *

MAINTENANCE ACTION

Symptoms:
Observed by maintenance man_____
_____
_____
_____

Name_____ Rate_____

	System Down Time	
Type of Time	In Flight	Non-Flight
Admin		
Logistic		
Maintenance		
Man Hours		

Item	Component Involved	Comp. Number	Comp MR#	Cum. Comp. Time		Time in Shop		
				Standby	Operate	Logistic	Maintenance Hours	Maintenance Man Hours
1								
2								
3								
4								

ACTION TAKEN: (Repórt adjustments, alignments, parts replaced, etc. and whether performed in A/C or shop)

Item_____ _____
_____

Item_____ _____
_____

Item_____ _____
_____

Item_____ _____
_____

JAX 315-2

Fig. 4.9 Maintenance report (front of form).

SYSTEM TIME

Time Type	Date and Hour Begins - Ends		Date and Hour Begins - Ends	
Admin.				
Logist.				
Maint.				
No. Men				

Admin.				
Logist.				
Maint.				
No. Men				

COMPONENT STATUS

Item	Removed from Aircraft Date - Hour		Returned to Aircraft Date - Hour	

COMPONENT TIME
(logistic--in shop only)

Item	Date and Hour Part Ordered - Received		Date and Hour Part Ordered - Received	

COMPONENT TIME
(maintenance--in shop only)

Item	Date and Hour Begins - Ends		Date and Hour Begins - Ends	

Fig. 4.9 Maintenance report (reverse of form).

ARINC Magnetron Filament Current Report

Syst. No.	Date and Hour When Checked	Mod Meter V When Checked	Mod Meter V At 85 Amp.	Clock Readings	
				St'by	Operate
001					
004					
029					
031					
032					
034					
037					
041					
042					
044					
046					
047					

Remarks: -

Note: An adjustment of the magnetron filament voltage is required if the Mod Meter voltage when checked is 2 or more scale divisions less or greater than the 85-ampere reference voltage.

Fig. 4.10 Magnetron filament current report.

The form shown in Figure 4.11 was developed by the field study group for reporting airframe time, usage of the system during flight, and operational status of the system during flight. The primary function of this form was to provide continuity of control in collection of data on system up time and down time during flight.

Figures 4.12 and 4.13 show the well-known FUR and DD-787 forms. The FUR (or Failure, Unsatisfactory, or Removal) report is submitted upon failure of a major nonexpendable item. The DD-787 form is used for reporting failures of expendable electronic parts. In the AN/APS-20E study, the function of these forms was to provide a cross-check on parts replaced during system maintenance events reported on the form shown in Figure 4.9.

It should not be inferred from the foregoing discussion that all the data-collection, control, and processing techniques referred to here would satisfactorily apply in all field studies. The procedures selected must be closely geared to the objectives and requirements set forth in the study specification, as well as to the conditions under which the system will be used in the field.

4.4 Storage and Recovery of Data

Data supplied by study group personnel are usually transcribed onto cards or tapes in order to facilitate storage, recovery, and analysis. The term *recovery* indicates the operations performed in retrieving data from coded form on storage devices and assembling it into listings, tabulations, computations, or other forms suitable for display and analysis.

The method of transcription is taken into consideration when the test is planned and is usually dictated by the amount of data anticipated. Machine methods have a tremendous advantage over manual methods if large quantities of data must be presented in several different ways, because—besides being faster—they are more repeatable and accurate than manual methods. The punch-card system is one well-established machine method used successfully in the studies cited in this discussion. Another large-volume process uses magnetic tape, as handled by typical commercial computers. The principal advantage of this latter method is that sorting, listing, and selective tabulation are readily and quickly handled.

From the point of view of flexibility, manual processing is more advantageous than machine processing, in that a program is less time-consuming and expensive to change once it is under way. Manual processing can be carried out by means of ledger books, file cards, key-sort cards, and other types of record forms, depending on the amount of data to be handled and the amount of time available. File cards and ledger books require that intermediate data forms be used in lieu of rapid-sort processes. The intermediate listings must be made manually and are time-consuming. Key-sort

A/C Buno_____   FLIGHT DATA   Month_____

Day	Flight Length	Takeoff Hour	Radar used--or Use attempted	Radar Down?

Note: Submit above data twice a month (1st and 15th).

Fig. 4.11 Flight data report.

PRESS HARD ☆ U. S. GOVERNMENT PRINTING OFFICE
1958 - 484137

YOU ARE MAKING OUT AN EIGHT PART FORM

PRESS HARD

1. REPORTING ACTIVITY	2. REPORT SERIAL	3. DATE OF TROUBLE	4. MAJOR COMMAND (Enter number in space at right)
			1 - LANT 3 - NABS 5 - FLAW 7 - NART 9 - MATS
			2 - PAC 4 - NATRA 6 - R & D 8 - BAR 0 - OTHER

5. ITEM IDENTIFICATION (Stock Number)	6. MFGR'S. CODE	7. ITEM PART NUMBER

8. ACCT. CODE	9. ITEM NOMENCLATURE	10. QUANTITY	11. OVERHAULED BY (Enter number in space at right)
			1 - ALAM 3 - CORP 5 - LAKE 7 - NORIS 9 - QUON.
			2 - CH. PT. 4 - JAX 6 - NORF 8 - PENS 0 - OTHER

12. AIRCRAFT/MISSILE/AG/CATAPULT	13. SYSTEM/ENGINE/ACCESSORY	14. AIRCRAFT/MISSILE/AG/CATAPULT	15. ENGINE/ACCESSORY
MODEL	MODEL	BUNO	SER. NO.

16. TIME (Hours)	17. OPERATING BASE	18. CONTRACT NUMBER	19. TROUBLE RESULTED IN (Check no more than two)
			0 AAR 11 FLIGA 2 FLAME-OUT 3 ENGINE FAILURE (Do not check if not removed)

20. HOW TROUBLE NOTICED
- 0 INOPERATIVE
- 1 INTERFER./BINDING
- 2 EXCESS. VIBRATION
- 3 UNSTABLE/SURGING
- 4 LEAKAGE
- 5 RPM OUT-OF-LIMITS
- 6 TEMP. OUT-OF-LMT
- 7 PRESS. OUT-OF-LMT
- 8 TROUBLE SHOOTING
- 9 PREVENTIVE MAINT.
- 11 OTHER (Amplify)
- 12 NOT REMOVED-UNSAT. (Amplify)

21. WHAT IS PART CONDITION
- 0 CHAFED
- 1 BROKEN
- 2 CRACKED
- 3 DISTORTED
- 4 SCORED
- 5 EXCESSIVE WEAR
- 6 DISCOLORED
- 7 OUT OF TOLERANCE
- 8 CORRODED
- 9 O.K.
- 11 CANNOT DETERMINE
- 12 OTHER (Amplify)

22. CAUSE OF TROUBLE
- 0 DESIGN DEFICIENCY
- 1 OP. TECH./ADJ.
- 2 NORMAL USE
- 3 FAULTY MFG /INSPEC.
- 4 DEFICIENT MAINT/O. H.
- 5 DAMAGED ON RECPT.
- 6 WEATHER CONDITION
- 7 FLUID CONTAMINATION
- 8 FOREIGN OBJ./COMBAT
- 9 OTHER PARTS
- 11 FAULTY PRESERV.
- 12 UNDETERMINED/OTHER (Amplify)

23. CIRCUMSTANCES
Special
- 0 FOLLOW-UP REPORT
- 11 HIGH TIME REMOVAL
- 12 MISSION ABORTED

Environment
- 1 SANDY/DUSTY
- 2 ARCTIC
- 3 TROPIC
- 4 ARID

Trouble Discovered During
- 5 FLIGHT OPS
- 6 GROUND OPS/TEST
- 7 MAINTENANCE
- 8 PRIOR PART INSTALL

24. DISPOSITION OF FAILED PART
- 0 RETURNED TO SUPPLY
- 1 REPAIRED/REINSTALLED
- 2 SURVEYED (Lost, Missing, or Destroyed)
- 3 HOLDING 30 DAYS (Show BASO or data returned to supply)
- 4 RELEASED FOR PRIORITY INVEST. _____ (Name of OO'R)
- PER: _____ (Ref. document specifying invest. OO'R)
- 5 TO CONTRACTOR: _____ (Name of contractor)
- VIA: _____ (Signature contractor's local rep.) (Date)
- FINAL DISPOSITION: _____ (Ref. doc. advising ultimate return to supply)

25. STATEMENT OF TROUBLE/CORRECTIVE ACTION (Check box only when publication as FUR Digest Phrase is desired)

26. AMPLIFYING REMARKS (Attach additional sheets, sketches, and photographs, as appropriate)

27. REPORT IS
- 0 FUR
- 1 AMPFUR
- 2 URGENT AMPFUR
- 3 FLIGHT SAFETY AMPFUR
- 4 PRIORITY DIR (OO'R use only)

FAILURE, UNSATISFACTORY OR REMOVAL REPORT	28. SIGNATURE	29. RANK/RATE	30. DATE
NAVAER - 3068 (REV. 8-58)			

FUR (Mail to FUR Center)

Fig. 4.12 Failure, unsatisfactory or removal (FUR) report.

REPORT THE FAILURE OF ONLY ONE PART OR TUBE ON THIS FORM

1. REPORT NO.	2. REPORTING ACTIVITY	3. REPAIRED OR REPORTED BY (NAME)	4. DATE OF FAILURE
5. EQUIPMENT INSTALLED IN (TYPE AND NO.)	6. TIME METER READING OR INSTALLATION LOG TIME	7. WAS MISSION ABORTED? ☐ YES ☐ NO	8. OPERATIONAL CONDITION

EQUIPMENT

9. MODEL DESIGNATION AND MOD. NO.	10. SERIAL NO.	11. CONTRACTOR	12. CONTRACT OR ORDER NO.

COMPONENT (MAJOR UNIT)

13. MODEL DESIGNATION AND MOD. NO.	14. SERIAL NO.	15. CONTRACTOR	16. CONTRACT OR ORDER NO.

ASSEMBLY OR SUBASSEMBLY

17. ASSEMBLY AND MOD. NO.	18. SERIAL NO.	19. MANUFACTURER	20. (LEAVE BLANK)

PART DATA

21. PART NAME OR TUBE TYPE	22. STOCK NO. (FAILED ITEM)	23. PART REF. DESIG. (V–101, R–101, ETC.)	24. REPAIR TIME (MAN-HOURS)
25. HOURS IN SERVICE	26. MANUFACTURER OF FAILED PART	27. SERIAL NO.	28. WAS REPLACEMENT PART AVAILABLE LOCALLY? ☐ YES ☐ NO

29. FIRST INDICATION OF TROUBLE

1	INOPERATIVE
2	INTERMITTENT
3	LOW PERFORMANCE
4	NOISY
5	OFF FREQUENCY
6	OUT OF ADJUSTMENT
7	OVERHEATING
8	UNSTABLE
9	OTHER

30. CHECK TYPE(S) OF TUBE OR PART FAILURE

Code		Code		Code	
007	ARCING	001	GASSY	790	OUT OF ADJUST.
710	BEARING FAILURE	300	GROUNDED	006	SHORTED
780	BENT	380	LEAKAGE	770	SLIP RING OR COMMUTATOR FAILURE
040	BINDING	730	LOOSE	018	TESTED OK. DID NOT WORK
070	BROKEN	004	LOW GM OR EMISSION		
720	BRUSH FAILURE	750	MISSING	020	WORN EXCESSIVELY
080	BURNED OUT	008	NOISY		SEE INSIDE FLAP FOR ADDITIONAL CODES
130	CHANGED VALUE	450	OPEN		
170	CORRODED	099	OTHER		

31. CAUSE OF FAILURE

2	FAULTY PACKAGING
5	MISHANDLING
6	INSPECTION OR TEST
1	NORMAL OPERATION
3	STORAGE
7	ASSOCIATED FAILURE—EXPLAIN
4	OTHER

32. WAS THE PART REPLACED DURING PREVENTIVE MAINTENANCE? ☐ YES ☐ NO

33. REMARKS (CONTINUE ON REVERSE SIDE IF NECESSARY)

DD (1 AUG 54) 787

U. S. GOVERNMENT PRINTING OFFICE 1957 O-F-431086

ELECTRONIC FAILURE REPORT

Fig. 4.13 Electronic failure report form DD 787.

cards are more readily handled, but they can accommodate a limited number of entries or separate pieces of information. In general, the key-sort feature of these cards is used for sorting only—not for coding or other purposes. The data are written on the cards, and listings are made manually after the cards are sorted.

Other methods of handling data can doubtless be made to work. However, in cases where the data comprise thousands of entries, it is recommended that machine processing be used.

4.5   Checking of Data for Evidence of Bias

An important function of the measurement analysis group, requiring constant and close liaison between this group and the measurement study group, is the continuous examination of final field data for cumulative errors in reporting and possible bias on the part of personnel observing the events. A considerable proportion of all system data, particularly data involving time measurements, is sequential or cumulative in nature, and any reporting errors which are allowed to go uncorrected may result in exasperating and costly delays during the analysis stage. It is well to devise a method whereby a running review of accumulating data can be made, and any errors detected can be quickly reported to measurement personnel for correction.

Study group personnel responsible for collection and monitoring of data at the field source should be alert for evidence of possible bias on the part of user personnel. Where possible, study personnel should separate final data into categories by collection period and location, and then run appropriate statistical tests for homogeneity. If some groups of data are significantly different from others and if the difference cannot be attributed to environmental factors, it may be that the problem is due to prejudice in reporting.

4.6   Summary

In this chapter on measurement problems, two major points have been emphasized. The first is that the success of a measurement study depends largely on good planning, including the careful formulation of objectives, the writing of a clear, comprehensive specification, and the establishment of collection and transcription methods which will provide greatest accuracy for least cost. The second major point is that measurement group personnel must exercise extreme care to see that the data are not biased as the result of prejudicial reporting. The methods used by ARINC Research Corporation have succeeded in a large number of field trials, but other methods can be used with equal success, depending on the situation. The principle to bear in mind is that the means must be suited to the purpose in view and must be rigidly adhered to throughout the course of the test.

PROBLEMS

1. What elements should be included in a field study specification?

2. What are possible sources of error in reported field data?

Data Analysis Procedures

5.1 Introduction

In Chapter 1 it was pointed out that there are several major attributes to be considered in an evaluation of the total value or worth of an electronic system —i.e., in an evaluation of system effectiveness. It was seen that effectiveness is influenced by factors in three broad areas: (1) reliability, or the time the system will operate failure-free; (2) maintainability, or the time and resources required to restore a system to satisfactory operation, once failure has occurred; and (3) design adequacy, or the ability of a system to accomplish its mission, given that it is operating within specifications.

In this chapter, some of the mathematical techniques required for the quantification of reliability and maintainability concepts are discussed in detail. Quantification and computation of system effectiveness are discussed in Chapter 11. It is natural to consider reliability first in this presentation, since this concept received concentrated study several years before similar attention was given to the subjects of maintainability and design adequacy. Theory and measurement techniques accordingly have reached a higher stage of development in the area of reliability. This does not imply that reliability is more important than the other attributes, but rather that it represents a convenient and logical starting point for the discussion of methods which are, in general, applicable to attributes other than reliability.

5.2 Survival Curves

Reliability was defined (in Chapter 1) as the probability that a system will perform satisfactorily for at least a given period of time when used under stated conditions. A reliability function was defined as this probability expressed as a function of the time period. A *reliability function* is, therefore, a mathematical formula relating the probability of satisfactory performance to time. The precise nature of this relationship is dependent on the distribution of times to failure for an item and theoretically could be of any form. However, it has been determined by experience that most failure patterns can be represented by a relatively small number of distribution types. The types most commonly encountered are (1) the normal or Gaussian, and (2) ⊼ the exponential, which is a special case of (3) the Weibull. These distributions are considered separately and in more detail later.

The concept of satisfactory performance is basic to the definition of reliability above. This fact implies that there are definite and accepted standards for determining when performance is satisfactory and when it is not. However, there is unfortunately too little general agreement concerning what constitutes satisfactory performance in an electronic system. The criteria in use range from the operator's judgment to the engineering-specification requirements to which the system was built. In evaluating reliability, all criteria of satisfactory performance should be considered, and the dominant criteria should be identified, so that proper emphasis can be placed on the various stresses causing deterioration in performance. The most comprehensive criteria of satisfactory performance are those used by personnel who operate the system in the field. Therefore, *unsatisfactory performance* will be defined as that system performance which some specified user regards as unsatisfactory. In the interval between two expressions of dissatisfaction on the part of the user, and after maintenance has been performed following the first complaint, the performance of the system is assumed to be satisfactory.

A *survival curve* is a graphic representation of the relationship between the probability of survival and time. Here probability of survival is synonymous with probability of nonfailure or probability of satisfactory performance. Three types of survival curves are of primary interest. The first is a discrete or point-type curve derived from observed data by nonparametric or distribution-free methods. The second type is a continuous curve based on an assumption as to the form of the distribution (Gaussian, exponential, etc.) and on values of the distribution parameters estimated from the observed data. The third type of curve is the true reliability function of the population from which the sample observations were drawn. This last function can only be estimated (i.e., not determined precisely), although the limits within which it will fall a given percentage of the time can be defined.

5.3   Mean Life

Mean life is the arithmetic average of the lifetimes of all items considered. A "lifetime" may consist of time between malfunctions, time between repairs, time to removal of tubes or other parts, or any other desired interval of observation.

Mean-life values have meaning only in relation to type of frequency distribution assumed by the data. For example, if a constant rate of malfunction is present in the system, the times between malfunctions will be exponentially distributed, and the mean life will occur at the point where there is a 36.8 per cent probability of survival. However, if the times between malfunctions are normally distributed, the rate of malfunction will increase with time and the mean life will occur at the point where there is a 50 per cent probability of survival.

The nature of the normal (Gaussian) distribution is such that the survival curve is not completely described unless the mean life and the standard deviation are specified. In the case of the exponential distribution, the survival curve is completely described when the mean life and the exponential distributional form are specified. This point is stressed because of a tendency on the part of some system investigators to give mean-life values without specifying the form of the distribution. While it is usual in system studies to assume that the data are exponentially distributed—in other words, that there is a constant rate of malfunction in the system—the distributional form must nevertheless be specified in order that the mean-life values can be properly interpreted.

A numerical example will emphasize the fact that mean life must be interpreted in relationship to the form of the distribution on which it is based. Assume that a system or subsystem, Equipment A, has a mean life of 100 hr and an exponential distribution of times to failure. Another system, Equipment B, exhibits a normal distribution of times to failure with a mean life of 100 hr and a standard deviation of 40 hr. Assume that one or the other of these systems is to be used for a mission of 10-hr duration and that the probability of nonfailure for at least 10 hr must be estimated for each system.

The reliability functions for the two systems may be written as follows:

$$R_A(t) = e^{-t/\theta} \tag{5.1}$$

and

$$R_B(t) = 1 - \left\{ \frac{1}{\sqrt{2\pi}\sigma} \int_{-\infty}^{t} e^{-(t-\xi)^2/2\sigma^2} dt \right\}. \tag{5.2}$$

From (5.1),

$$R_A(10) = e^{-10/100} = 0.905.$$

From (5.2),

$$R_B(10) = 1 - \left\{ \frac{1}{\sqrt{2\pi}\sigma} \int_{-\infty}^{t} e^{-(10-100)^2/2(40)^2} dt \right\}$$

$$= 1 - 0.0122 = 0.988.$$

Thus it is seen that the probability of nonfailure for 10 hr is significantly higher for Equipment B than it is for Equipment A, although the mean lives of the two equipments are equal.

When mean-life values, with no other information, are given as representative of equipment reliability, this is sometimes misinterpreted by the uninitiated to mean that the equipment will operate failure-free for a period of time equal to the mean life. The fallacy of this conclusion is evident from the example above, where 50 per cent of the B equipments and only 36.8 per cent of the A equipments could be expected to operate failure-free for 100 hr.

5.4 Use of Samples from a Population

It is essential to understand that the techniques of reliability measurement rest on statistical concepts. The reasons for this become apparent when it is understood that reliability measurements are made on samples representative of the total population of the item under consideration. The term *population* refers to all the items of a given type. For example, if the subject of investigation were the AN/ARC-27 communications equipment in Navy use, the population would be composed of all AN/ARC-27's used by the U.S. Navy. The term *sample* refers to a group of items taken from the population. In the case of the AN/ARC-27, the sample might consist of a group of equipments used by one squadron of naval aircraft at one base.

Since it is seldom feasible to make measurements on entire populations, the use of statistical techniques is necessary. Such techniques permit the extrapolation of results obtained from a sample to the population as a whole, and therefore to other similar populations.

The use of samples in the measurement of reliability requires that the final results be presented as an estimated value, with confidence limits to indicate the probable range within which the population mean will fall. Investigators usually select the 95 per cent confidence intervals—the range of values in which there is a 95 per cent probability that the population mean exists. The larger the size of the sample, the narrower the confidence interval becomes.

The normal and exponential survival curves are the basic types usually obtained in reliability studies. The normal curve presents the probability of survival of an item whose failure rate increases with time. The exponential curve presents the probability of survival of an item whose failure rate is constant with time.

5.5   Calculation of Survival Curves

Figure 5.1 presents a frequency distribution of failures in a fixed population of 90 items, over a 6-hr period. To obtain a survival curve from these data, the following simplified method is used:

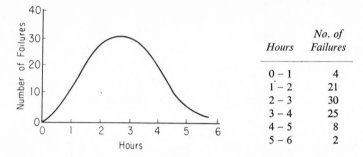

Hours	No. of Failures
0 – 1	4
1 – 2	21
2 – 3	30
3 – 4	25
4 – 5	8
5 – 6	2

Fig. 5.1 Normal distribution of failures in time.

During the first period of observation, from 0 to 1 hr, 4 of the original 90 items failed. The failure rate during this period was 4/90, or 0.0445, which is equivalent to a survival rate of 1 − 0.0445, or 0.9555. In the second period of observation, 21 of the 86 remaining items failed. The failure rate was 21/86, or 0.244, and the survival rate was 1 − 0.244, or 0.756. The tabulation above Figure 5.2 gives the failure rates and survival rates for the remaining periods of observation. It will be noted that the failure rate increases with time.

To obtain a survival curve, which is the cumulative probability of survival with time, the probability of survival in each time period is multiplied by the

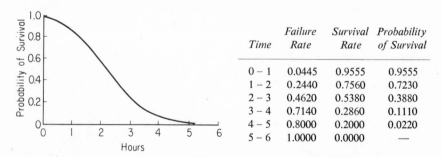

Time	Failure Rate	Survival Rate	Probability of Survival
0 – 1	0.0445	0.9555	0.9555
1 – 2	0.2440	0.7560	0.7230
2 – 3	0.4620	0.5380	0.3880
3 – 4	0.7140	0.2860	0.1110
4 – 5	0.8000	0.2000	0.0220
5 – 6	1.0000	0.0000	—

Fig. 5.2 Calculation and presentation of a normal survival curve.

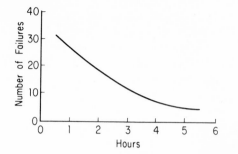

Hours	No. of Failures
0 – 1	30
1 – 2	20
2 – 3	13
3 – 4	9
4 – 5	6
5 – 6	4

Fig. 5.3 Exponential distribution of failures in time.

survival rate in the succeeding time period. Thus, $0.9555 \times 0.756 = 0.723$, $0.723 \times 0.538 = 0.388$, etc. The probability values are plotted versus the centers of the time periods as shown at the bottom of Figure 5.2.

Figure 5.3 presents a frequency distribution of failures for a population in which the removal rate is constant with time. The approach described in connection with the normal curve yields the tabulation and exponential survival curve shown in Figure 5.4.

Survival curves for most systems are of the exponential form. Survival curves for parts, on the other hand, are frequently of the normal form. As parts wear out, their failure rate increases and their probability of survival decreases. A large number of such parts, all having normal survival curves but each having a different mean life and variance, will produce a system malfunction rate which is essentially constant, since the mean lives of the parts will be randomly distributed.

To determine what type of population gives rise to a particular survival curve, the theoretical reliability function most closely resembling the curve is computed from sample parameters. The theoretical function is then matched to the observed curve by statistical techniques. If this procedure establishes that there is no significant difference between the observed and theoretical curves, the theoretical curve is usually employed for all additional calculations.

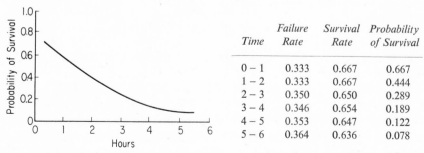

Time	Failure Rate	Survival Rate	Probability of Survival
0 – 1	0.333	0.667	0.667
1 – 2	0.333	0.667	0.444
2 – 3	0.350	0.650	0.289
3 – 4	0.346	0.654	0.189
4 – 5	0.353	0.647	0.122
5 – 6	0.364	0.636	0.078

Fig. 5.4 Calculation and presentation of an exponential survival curve.

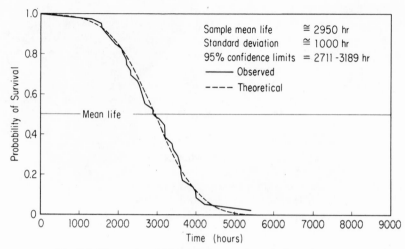

Fig. 5.5 Observed and theoretical normal survival curves.

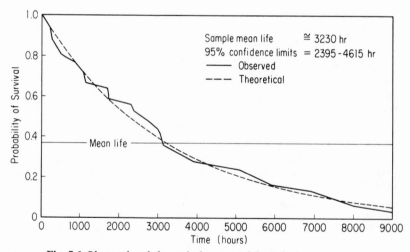

Fig. 5.6 Observed and theoretical exponential survival curves.

Figures 5.5 and 5.6 present survival curves estimated by nonparametric techniques, and corresponding theoretical reliability functions estimated from the parameters of the samples used. The data for both of the observed curves are empirical—i.e., they were obtained from tables of random numbers appropriately modified to provide the desired population. In each case, the population had a mean life of 3000 hr. Both samples were composed of 50 items, of which 35 were completed observations and 15 were censored observations.

The sample represented in Figure 5.5 had a mean life of approximately 2950 hr and a standard deviation of approximately 1000 hr. The curve shown in the figure is typical of reliability data associated with parts, such as electron tubes, which are operated under mild environmental conditions. The sample represented in Figure 5.6 had a mean life of approximately 3230 hr, with 95 per cent confidence limits of 2395 and 4615 hr. The results shown in this figure are typical of those obtained in system reliability studies.

5.6 Techniques Used in Computing Reliability Functions*

The Observed Data or the Sample

If a reliability study is concerned with the length of time a system (or any device) operates before it fails or requires repair, the observed data will consist of time intervals representing the lifetimes of the items being studied. For the sake of convenience, lifetime will be regarded as the time to failure or time between malfunctions of an item; however, lifetime can also be measured in terms of time between maintenance actions, number of times the product performs its function, or some other variable of interest.

In order to obtain a representative sample from a population of lifetimes, it is desirable that the item being observed be selected at random and that observations of time to failure be made under controlled or stable conditions. In most industrial or military equipment studies, the cost of obtaining a random sample of lifetimes is usually prohibitive because of the time involved in waiting for all the sampled items to fail. Furthermore, it may not be possible to control the conditions under which the items are observed, and the conditions often are not stable. Either or both of these considerations may cause the problem of estimating the reliability function to become rather complex.

If a sample contains both complete and incomplete lifetimes, the incomplete lifetimes are referred to as *censored* observations. These consist primarily of lifetimes which are too long to be observed completely (*terminated* observations) and lifetimes in which the item being observed is lost before completion of observation (*lost* observation). In the case of terminated observations, the length of observation time is controlled; in the case of lost observations, the length of observation time is not controlled. In either case, the investigator knows that the lifetime of the item exceeds the period of time during which the item was being observed. Terminated observations do not present a problem to the investigator other than to increase the complexity of his calculations, but lost observations may constitute a real problem because they may be associated with only a portion of the population.

*The material in this section is adapted from *Estimation of Reliability Functions*, by G. R. Herd, ARINC Research Monograph No. 3 (May 1, 1956).

Nonparametric Methods

Given a random sample of lifetimes, the reliability function may be estimated either (1) by noting the percentage or fraction of the sample which survives after a given time, T, or (2) by noting the time at which a given percentage or fraction of the sample still survives.

If the reliability function were presented graphically (see, for example, Figures 5.5 and 5.6), the first approach would be analogous to selecting a value on the abscissa (time) and determining the corresponding value on the ordinate (probability of survival). The second approach would be analogous to selecting a value on the ordinate and determining the corresponding value on the abscissa.

Although the two approaches appear to be identical, they are based upon different theoretical procedures. The first approach involves the estimation of the theoretical per cent or probability, $R(T)$. This procedure is equivalent to estimating the parameter of a binomial distribution, since if T is fixed, the number surviving T is a random variable. The appropriate estimate of $R(T)$ is given by

$$\hat{R}(T) = \frac{\text{number surviving time } T}{\text{initial sample size}} = \frac{N-k}{N}, \tag{5.3}$$

where k is the number which fail in time T.

The second approach involves a procedure in which the random variable is the time to failure or time between malfunctions. At the time of each failure or malfunction, the expected percentage surviving is known. This approach leads to an estimate of $\hat{R}(t)$ which is given by

$$\hat{R}(t_i) = \frac{N-i+1}{N+1}, \tag{5.4}$$

where i represents the ith ordered incident of failure, t_i represents the time to the ith failure, and r_i is the number of failures at time t_i. (The term *ordered* means that the times to failure are numbered in ascending order of magnitude; t_1 is the shortest lifetime and t_n is the longest lifetime.) The function $R(t_i)$ may be plotted by plotting the points

$$\left(\frac{N-i+1}{N+1}, t_i \right) \qquad i = 1, 2, \ldots, N.$$

Either of these procedures can be used to present the observed reliability function, since they are equivalent if used in connection with large samples. However, the second procedure is more advantageous than the first when the problem involves terminated observations.

If terminated or censored observations occur, and if the time of occurrence and the number of terminations are known, an estimate of $R(t)$ at the time of the kth failure is given by

$$\hat{R}(t_k) = \frac{N_1 - r_1 + 1}{N_1 + 1} \cdot \frac{N_2 - r_2 + 1}{N_2 + 1} \cdots \frac{N_i - r_i + 1}{N_i + 1} \cdots \frac{N_k - r_k + 1}{N_k + 1}$$

$$= \prod_{i=1}^{k} \left(\frac{N_i - r_i + 1}{N_i + 1} \right), \tag{5.5}$$

in which N_i is the number of survivors beginning the interval which precedes the ith failure and r_i is the number of failures occurring at the time of the ith failure. The reliability function is presented by plotting the expected fraction surviving as a function of the observed times.

When censorship occurs, the observed reliability function cannot be determined exactly. Instead, it is necessary to estimate this function by the method given in Equation 5.5. This is a nonparametric, or distribution-free, method of estimation. The variance of the estimate is

$$V[\hat{R}(t_k)] = \prod_{i=1}^{k} \left(\frac{N_i - r_i + 1}{N_i + 2} \right) - \prod_{i=1}^{k} \left(\frac{N_i - r_i + 1}{N_i + 1} \right)^2. \tag{5.6}$$

If the data-collection procedures are such that only the number of failures occurring within a time interval are reported—rather than the exact times at which failures occur—then the reliability function must be estimated for each interval rather than for each incident of failure. Equation 5.5 is still the basis for computing the observed reliability function, but r_i would be interpreted to mean the number of failures occurring in the interval. The computations will be illustrated in the following section.

Example of Nonparametric Computations

Table 5.1 presents the data obtained from a study of the reliability of vacuum tubes in a number of fixed ground-station communications receivers. The receivers operated continuously and were given an operational check once every 24 hr. When defective tubes were discovered during a check, these were reported as failures. In addition, some nonfailed tubes—i.e., tubes which had not caused an equipment malfunction—were sometimes replaced by the technician as a precautionary measure. All tubes, including replacements, carried a serial number decal so that operating time to failure or removal could be reported accurately to within 24 hr for each individual removal.

Although these data represent failure times for tubes, they might just as well have been times to system malfunction. In other words, the computational procedure is the same no matter whether the observations are times to failure or times between malfunctions.

Table 5.1 ORDERED TIME OBSERVATIONS FOR TUBES FROM A GROUP OF
FIXED GROUND-STATION COMMUNICATIONS RECEIVERS
(A) Times to Failure, (B) Censored Times

(A) *Time to Failure in Operating Hours*	(B) *Censored Operating Hours Following a Failure or a Replacement*
0	—
—	48,48
—	72,72,72
96	96,96
120	120
144,144	144,144
168,168	168,168,168
192	192
216	216
264,264	264
288	288,288,288
312,312	312,312,312
336	336
—	360,360
384,384,384	384
—	408
456	456,456
504	504,504
528	529,529,529
552,552	—
576	576
600,600	—
—	648
—	672,672,672,672,672
	672,672,672,672
696	—
720	—
744,744,744,744	744
768	768
792,792	792
816	—
—	840,840,840
—	864
888	—
—	912
—	936
960,960	960,960
—	1032
—	1152
1200,1200,1200,1200	1200
1248	1248
—	1320
—	1344,1344
—	1368
1416	1420
—	1512,1512
1584	—
1896	—
—	1968
2064	—
Total Time=31,920 hr Total No. of Observations = 48	Total Time=42,388 hr Total No. of Observations = 68

Table 5.2 COMPUTATION OF NONPARAMETRIC RELIABILITY FUNCTION

	Observed Data (from Table 5.1)			Computed Data $R(t_j) = \prod\limits_{i=1}^{k} \dfrac{N_i + 1 - r_i}{N_i + 1}$			
(1) t_i	(2) N_i	(3) r_i	(4) k_i	(5) $N_i + 1$	(6) $N_i + 1 - r_i$	(7) (6)/(5)	(8) $R(t_i)$
0	116	1		117	116	0.991	0.991
48			2				
72			3				
96	110	1	2	111	110	0.991	0.982
120	107	1	1	108	107	0.991	0.973
144	105	2	2	106	104	0.981	0.955
168	101	2	3	102	100	0.980	0.936
192	96	1	1	97	96	0.990	0.927
216	94	1	1	95	94	0.989	0.917
264	92	2	1	93	91	0.978	0.897
288	89	1	3	90	89	0.989	0.887
312	85	2	3	86	84	0.977	0.867
336	80	1	1	81	80	0.988	0.857
360			2				
384	76	3	1	77	74	0.961	0.824
408			1				
456	71	1	2	72	71	0.986	0.812
504	68	1	2	69	68	0.986	0.801
528	65	1	3	66	65	0.985	0.789
552	61	2		62	60	0.968	0.764
576	59	1	1	60	59	0.983	0.751
600	57	2		58	56	0.966	0.725
648			1				
672			9				
696	45	1		46	45	0.978	0.709
720	44	1		45	44	0.978	0.693
744	43	4	1	44	40	0.909	0.630
768	38	1	1	39	38	0.974	0.614
792	36	2	1	37	35	0.946	0.581
816	33	1		34	33	0.970	0.564
840			3				
864			1				
888	28	1		29	28	0.966	0.545
912			1				
936			1				
960	25	2	2	26	24	0.923	0.503
1032			1				
1152			1				
1200	19	4	1	20	16	0.800	0.402
1248	14	1	1	15	14	0.933	0.375
1320			1				
1344			2				
1368			1				
1416	8	1		9	8	0.889	0.333
1420			1				
1512			2				
1584	4	1		5	4	0.800	0.266
1896	3	1		4	3	0.750	0.200
1968			1				
2064	1	1		2	1	0.500	0.100
$\Sigma\, t_i r_i + \Sigma\, t_i k_i$ $= 74{,}208$		Total 48	Total 68				

Total Operating Time = 74,208 hr; Mean Time To Failure = θ = 74,208/48 = 1,546.0 hr (based on assumption of the exponential distribution).

Column A of Table 5.1 lists the times to failure for 48 individual tubes. Column B lists the times to removal for nonfailed tubes in the study, or what have been termed censored times. The data were observed over a period of approximately 6 months of continuous surveillance of the equipments. Table 5.1 presents the data in raw form, except that the times have been ordered for convenience.

Table 5.2 shows the form used for the computation. The first four columns were derived from the observed data in Table 5.1 by listing the number of failures and/or censored observations occurring at each reported time. Column 3 indicates how frequently the individual times to failure were observed. Column 4 indicates how frequently the individual times shown in Column 1 were censored rather than terminated by a failure. Column 2 gives the cumulative total, N_i, of actual and censored observations, beginning at the bottom of the table.

The nonparametric estimator for the reliability function is given by the equation at the top of columns 5 through 8. This is the same as Equation 5.5 in the previous section. In column 5 is entered $N_i + 1$, a value arrived at by adding 1 to the value given in column 2. However, an entry is made in column 5 only when there is an entry in column 3. To get the value given in column 6, r_i is subtracted from $N_i + 1$. The value in column 7 is obtained by dividing the value in column 6 by the value in column 5. Column 8, the nonparametric reliability function, is obtained by successive multiplication of the values given in column 7.

Computation of Theoretical Exponential Reliability Function

When the form of the distribution is sufficiently well defined, it is possible to estimate the reliability function in terms of the parameters of the distribution. This method has the advantage of permitting utilization of all the accumulated knowledge concerning the items in the population. In addition, the reliability function can be summarized by specifying the values of the parameters and can be compared with other reliability functions merely by comparing the values of the summarized data.

The exponential reliability function is a one-parameter distribution—i.e., it is completely described when the mean life, θ, is specified. The function is given by:

$$R(t) = e^{-t/\theta} \tag{5.7}$$

where θ is the mean life. The estimate of θ is given by

$$\theta = \frac{\sum t_i r_i + \sum t_i k_i}{r} \tag{5.8}$$

using the notation of Table 5.2.

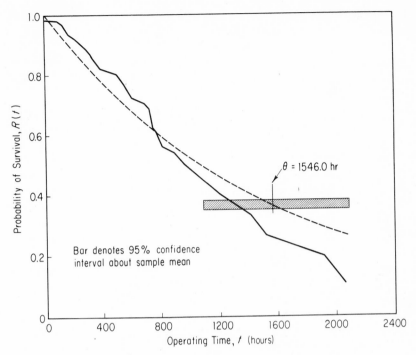

Fig. 5.7 Nonparametric reliability function and theoretical exponential reliability function.

As an example, the theoretical exponential reliability function corresponding to the data in Table 5.2 will be computed. The mean life, θ, computed at the bottom of Table 5.2, is 1546.0 hr.

Table 5.3 shows the computations for $R(t)$ for selected values of the t_i shown in Table 5.2. The values of e^{-x} can be obtained from tables of exponential functions or may be closely approximated by use of a log-log slide rule.*

Figure 5.7 presents a plot of the estimated nonparametric reliability function derived from column 8, Table 5.2, and a plot of the corresponding theoretical exponential function derived from column 3, Table 5.3. Determination of the confidence interval is discussed in Section 5.11.

Computation of Theoretical Normal Reliability Function

Censored observations cause more difficulty in estimating the parameters of a population whose underlying probability distribution is of the normal

---

*The nomograph on the inside back cover can also be used.

or Gaussian form,

$$u(t) = \frac{1}{\sigma\sqrt{2\pi}} e^{-(t-\mu)^2/2\sigma^2}$$ (5.9)

than they do in estimating the parameter of a population whose underlying probability distribution is of the simple exponential form.* A simple

Table 5.3 COMPUTATION OF THEORETICAL EXPONENTIAL RELIABILITY FUNCTION FOR $\theta = 1546$ HR

(1) t_i	(2) t_i/θ	(3) $e^{-t_i/\theta}$
0	0	1.000
96	0.0621	0.9398
216	0.1397	0.8696
312	0.2018	0.8173
456	0.2950	0.7445
552	0.3571	0.6997
696	0.4502	0.6375
792	0.5123	0.5991
888	0.5744	0.5630
960	0.6210	0.5374
1200	0.7762	0.4602
1416	0.9159	0.4002
1546	1.0000	0.3679
1896	1.2264	0.2933
2064	1.3351	0.2631

regression method for estimating the parameters μ (mean) and σ (standard deviation) of a normal distribution may be adequate in most practical situations. This method is described below.

Since $R(t_i)$ can be estimated, the normal deviate corresponding to the estimated probability $R(t_i)$—which will be designated as y_i—may be obtained from a table of normal probability areas. The observed time to failure, t_i, has the following linear relationship to y_i:

$$t_i = \mu + \sigma y_i.$$ (5.10)

The linear function of t may be determined by the method of least squares, in which the estimates of μ and σ are given as the solutions of the following equations:

$$\sum_i t_i = r\mu + \sigma \sum_i y_i$$ (5.11a)

*If there were no censored observations, the parameters of a normal population would be estimated by

$$\hat{\mu} = \frac{\sum t^i}{\sum r} \quad \text{and} \quad \hat{\sigma} = \sqrt{\frac{\sum r_i(\sum t_i^2) - (\sum t_i)^2}{\sum r_i(\sum r_i - 1)}}.$$

and

$$\sum_i t_i y_i = \mu \sum_i y_i + \sigma \sum_i y_i^2. \tag{5.11b}$$

Example of Computations for Normal Reliability Function

Table 5.4 presents some observed failure data for a sample of 50 vacuum tubes. At each instance of failure, three or four tubes were removed from the test. The operating environment was such that deterioration was believed to be the cause of failure, and a normal distribution was anticipated as the pattern for the distribution of times to failure. Table 5.4 lists the observed times when failures occurred and the number of censored observations at the time of each failure.

Table 5.4 OBSERVATIONS OF TIME TO FAILURE AND CENSORED TIME FOLLOWING A FAILURE IN A SAMPLE OF 50 TUBES

Hours Following a Failure, t_i	No. of Completed Observations, r_i	No. of Censored Observations, k_i
1,300	1	4
1,692	1	3
2,243	1	4
2,278	1	3
2,832	1	3
2,862	1	3
2,931	1	4
3,212	1	4
3,256	1	4
3,410	1	4
3,651	1	3
Total	$\Sigma r_i = 11$	$\Sigma k_i = 39$

Table 5.5 lists the hours to failure and the estimates of $R(t_i)$. The values of $R(t_i)$ are computed by the method illustrated in Table 5.2. The values of the normal deviate, y_i, were determined from a table of normal probability areas. The computations for the estimates of μ and σ, using Equations 5.11a and b, are shown at the bottom of Table 5.5.

These computations give as the estimates of μ and σ

$$\hat{\mu} = 3972 \text{ hr,}$$

and

$$\hat{\sigma} = 1208 \text{ hr.}$$

Table 5.5 Computational Procedure for Estimating the Mean and Standard Deviation of a Normal Distribution, Based on Ordered Observations Involving Censorship

Hours to Failure t_i	Reliability Function $R(t_i)$	Normal Deviate Corresponding to $R(t_i)$ y_i
1,300	0.980	-2.06
1,692	0.958	-1.73
2,243	0.935	-1.51
2,278	0.910	-1.34
2,832	0.883	-1.19
2,862	0.853	-1.05
2,931	0.819	-0.91
3,212	0.778	-0.77
3,256	0.726	-0.60
3,410	0.653	-0.39
3,651	0.522	-0.06
29,667		$\Sigma y_i = -11.61$

$$\Sigma t_i = 29,667 \qquad \Sigma y_i = -11.61 \qquad r = 11$$

$$\Sigma t_i y_i = -27,062.8 \qquad \Sigma y_i^2 = 15.77$$

$$29,667 = 11\mu - 11.61\sigma \qquad \hat{\sigma} = 1,208$$

$$-27,062.8 = -11.61\mu + 15.77\sigma \qquad \hat{\mu} = 3,972$$

The mean and standard deviation may be approximated by a graphical method which is often sufficiently accurate. The estimated values of $R(t_i)$ are plotted against t_i on probability paper. (The cumulative normal function plots as a straight line on this type of paper.) Then a line is "eye-fitted" to the points. The value of t at which the line crosses the 50 per cent probability point is the estimate of the mean, μ. The difference between this value of t and the value at which the line crosses the 84 per cent or the 16 per cent probability point provides an estimate of the standard deviation, σ.

Figure 5.8 is a plot of the values of $R(t_i)$ from Table 5.5 on probability paper. From the fitted straight line the estimates of μ and σ are

$$\hat{\mu} = 4000 \text{ hr},$$

and

$$\hat{\sigma} = 1260 \text{ hr}.$$

These values are in reasonably close agreement with the values obtained by an analytical solution of Equations 5.11a and b, and the process is much less time-consuming.

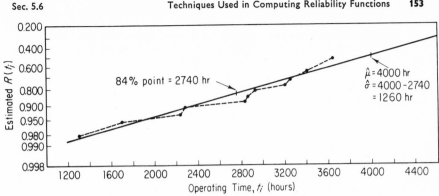

Fig. 5.8. Graphic estimation of the parameters of an assumed normal distribution.

When μ and σ have been determined, either from Equations 5.11a and b or graphically, 95 per cent confidence limits around the mean may be computed from the equation:

$$\text{confidence limits} = (\hat{\mu} \pm t\hat{\sigma}/\sqrt{n}), \qquad (5.12)$$

where t is obtained from a table of the upper percentage points of the t distribution (see any standard text on statistics) for $d = n - 1$ degrees of freedom. As n increases, t approaches the value 1.96, as shown below.

n	t
5	2.57
10	2.23
20	2.09
30	2.04
50	2.00
∞	1.96

For the estimates of μ and σ obtained in Table 5.5 the confidence limits are

$$(3972 \pm (2.20)\,(1208/\sqrt{11}) = 3972 \pm 801,$$

therefore

$$\text{upper confidence limit} = 4773 \text{ hr},$$

and

$$\text{lower confidence limit} = 3171 \text{ hr}.$$

Figure 5.9 presents a plot of the observed reliability function estimated by nonparametric methods, and the theoretical normal function based on the parameters μ and σ as estimated from Equations 5.11a and b. The 95 per cent confidence interval as calculated above is also shown on the figure.

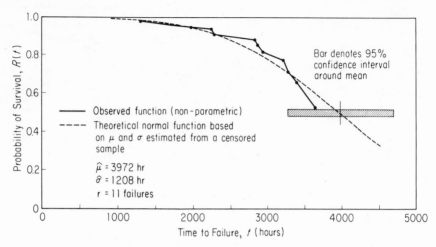

Fig. 5.9 Nonparametric and theoretical normal reliability functions.

5.7 Techniques Used in Computing Maintainability and Repairability Functions

The Observed Data

Maintainability was defined as the probability that, when maintenance action is initiated under stated conditions, a failed system will be restored to operable condition within a specified total down time. The *maintainability function* is this probability expressed as a function of the down time, t. The maintainability function may thus be written symbolically as

$$M(t) = f(t), \tag{5.13}$$

where $f(t)$ is the cumulative frequency distribution of the equipment down times. The observed data sample is then a series of observations of the length of time required to complete a maintenance action. The procedure to be described here is equally applicable to computations for repairability functions. The only difference is that the observed data consist of observations of active repair time rather than total maintenance time.

Since the maintainability function expresses the probability that a maintenance action is completed in time t or less, estimation of the function must take into account all the probabilities of completion in times less than t. The nonparametric estimator of $M(t)$ for time t_i is then

$$M(t_i) = \frac{\sum\limits_{j=1}^{i} n_j}{N + 1} \tag{5.14}$$

where n_j is the number of times the time t_j was observed and N is the total number of observations, or simply the sum of all the n_j. Equation 5.14 implies that the t_j have been arranged in order of ascending magnitude: i.e., t_1 is the smallest observed time and t_n is the largest.

Practical considerations rule out zero observations of t_i. In other words, a maintenance action always requires a minimum time for completion. It is possible to observe zero times for certain of the subdivisions of active repair time, such as preparation time and fault-location time, but it is assumed that a complete maintenance action can never occur in zero time. Furthermore, incomplete or censored observations do not present the problem in the computation of maintainability functions that they do in the computation of reliability functions. Although it is possible that if a maintenance study were terminated at a precise point in calendar time, some actions would be incomplete and therefore would constitute censored observations, this rarely occurs in practical field studies. The nature of maintenance is such that it is almost always possible to observe to completion the few actions that may be in progress when surveillance is terminated. This is in contrast to reliability surveillance, where it frequently is not possible to observe complete lifetimes for all items under surveillance.

The discussion above leads to the assumption that incomplete or censored observations do not occur in a sample of times required for maintenance. Departures from the validity of this assumption are of no real consequence in the estimation of maintainability and repairability functions.

Example of Computation of Repairability Functions

Columns 1 and 2 of Table 5.6 present a series of 46 observations of active repair time on an airborne communications transceiver. These times are shop repair times and do not involve any delay times, either logistic or administrative. In this particular sample, only one technician was involved in the repair so that the hours tabulated in Column 1 are also equivalent to the man-hours expended. Column 2, the n_i, shows how many times each t_i in Column 1 was observed. Since the data represent active repair times, the repairability function, as defined in Chapter 1, will be computed, but it should be noted that the computational procedure would be the same for a maintainability function.

Column 3 of Table 5.6 lists the estimated value of the repairability function, $M_R(t_i)$, as computed from Equation 5.14 for each value of t_i. It is seen that the computations for the nonparametric repairability function are somewhat simpler than those for the nonparametric reliability function.

Table 5.6 COMPUTATIONS FOR REPAIRABILITY FUNCTION
(A) NONPARAMETRIC, (B) THEORETICAL LOG-NORMAL

(A)			(B)			
(1)	(2)	(3)	(4)	(5)	(6)	(7)
Observed Data		Non-Parametric Function	$Log_{10}t_i$	$(Log_{10}t_i)^2$	$\dfrac{x_i - \hat{\xi}}{\hat{\sigma}}$	Theoretical Log-Normal
t_i	n_i	$\hat{M}_R(t_i)$	x_i	x_i^2		$M_R(t_i)$
0.2	1	0.021	−0.6990	0.4886	−2.04	0.021
0.3	1	0.043	−0.5229	0.2734	−1.67	0.048
0.5	4	0.128	−0.3010	0.0906	−1.21	0.113
0.6	2	0.170	−0.2218	0.0492	−1.05	0.147
0.7	3	0.234	−0.1549	0.0240	−0.91	0.181
0.8	2	0.277	−0.0969	0.0094	−0.79	0.215
1.0	4	0.362	0	0	−0.59	0.278
1.1	1	0.383	0.0414	0.0017	−0.51	0.305
1.3	1	0.404	0.1139	0.0130	−0.36	0.359
1.5	4	0.489	0.1761	0.0310	−0.23	0.409
2.0	2	0.532	0.3010	0.0906	0.03	0.512
2.2	1	0.553	0.3424	0.1172	0.12	0.548
2.5	1	0.574	0.3979	0.1583	0.23	0.591
2.7	1	0.596	0.4314	0.1861	0.30	0.618
3.0	2	0.638	0.4771	0.2276	0.39	0.652
3.3	2	0.681	0.5185	0.2688	0.48	0.684
4.0	2	0.723	0.6021	0.3625	0.65	0.742
4.5	1	0.745	0.6532	0.4267	0.76	0.776
4.7	1	0.766	0.6721	0.4517	0.80	0.788
5.0	1	0.787	0.6990	0.4886	0.85	0.802
5.4	1	0.808	0.7324	0.5364	0.92	0.821
5.5	1	0.830	0.7404	0.5482	0.94	0.826
7.0	1	0.851	0.8451	0.7142	1.15	0.875
7.5	1	0.872	0.8751	0.7658	1.22	0.889
8.8	1	0.894	0.9445	0.8921	1.36	0.913
9.0	1	0.915	0.9542	0.9105	1.38	0.916
10.3	1	0.936	1.0128	1.0258	1.50	0.933
22.0	1	0.957	1.3424	1.8020	2.18	0.985
24.5	1	0.979	1.3892	1.9299	2.28	0.989
$\Sigma n_i t_i = n = \Sigma n_i$ 166.0 hr $= 46$			$\Sigma n_i x_i =$ 13.1612	$\Sigma n_i x_i^2 =$ 14.3050		

$$\hat{\xi} = \frac{\Sigma n_i x_i}{\Sigma n_i} = \frac{13.1612}{46} = 0.2861$$

$$\hat{\sigma} = \sqrt{\frac{\Sigma n_i \, \Sigma n_i x_i^2 - (\Sigma n_i x_i)^2}{\Sigma n_i \, (\Sigma n_i - 1)}} = \sqrt{\frac{484.8128}{2070}} = \sqrt{0.234204} = 0.4840$$

$$\hat{\mu} = \text{antilog } \hat{\xi} = \text{antilog } (0.2861) = 1.932 \text{ hr}$$

$$\text{confidence limits} = \text{antilog } (\hat{\xi} \pm 2.00 \, \hat{\sigma}/\sqrt{n})$$

$$= \text{antilog } (0.2861 \pm 2.00 \times 0.4840/\sqrt{46}) = \text{antilog } (0.2861 \pm 0.1427)$$

Therefore, upper confidence limit = 2.684 hr, lower confidence limit = 1.391 hr

Estimation and Computation of Theoretical Repairability Functions

In order to determine a theoretical function that will describe the non-parametric function of Column 3, Table 5.6, it is necessary to make some decision regarding the form of the assumed theoretical function. Experimental data on maintenance or repair times obtained from a large number of diverse equipment types have indicated that the times follow the logarithmic-normal distribution. The *log-normal* distribution* is simply one in which the logarithms of the variable, rather then the variable itself, are normally distributed. In order to illustrate the reasoning which led to the use of the log-normal distribution, the underlying distribution, or density function, of the data in columns 1 and 2 of Table 5.6 will be examined.

Table 5.7 presents these data grouped in two different ways. In Part (A) of the table, the number of actions occurring in successive half-hour intervals are counted and tabulated. These grouped data are plotted as a histogram in Figure 5.10a. From this figure it is seen that a large number of observations cluster about a relatively low value of repair time, and there are relatively few observations of very long repair times. (It is important to note the scale break on the abscissa of Figure 5.10a.) In other words, the distribution is very much skewed to the left. Very frequently, a skewed distribution can be transformed to a normal distribution by a simple change of the random variable, substitution of log x for x. The nonexistence of zero repair times and the skewness of the distribution suggest the use of the log-normal distribution.

In Table 5.7(B) the observed data have been grouped in intervals whose width doubles with each successive interval. This type of interval might be called a *geometric interval* or a *logarithmic interval*, since this type of grouping is roughly equivalent to taking the logarithm to the base 2 of the times t_i. These grouped data are plotted as a histogram in Figure 5.10b. The change in the shape of the density function is immediately apparent from a comparison of Figures 5.10a and 5.10b. In the latter histogram most of the skewness has disappeared, and it appears that the distribution might be described by the normal function.

The foregoing discussion has been presented to lead the reader, by a somewhat intuitive approach, to the acceptance of the plausibility of the use of the log-normal distribution as a mathematical description of repair times. Various reasonable hypotheses regarding repair times will lead to the theoretical development of the log-normal distribution, but this development is beyond the scope of the present discussion.† It will be assumed here that the distribution of repair times is log-normal, and the remainder of this

---

*This is a standard abbreviation for logarithmic-normal.

†For further discussion, see *Yearly Review of Progress*, ARINC Research Corporation, Publication No. 101-28-166 (July 15, 1960).

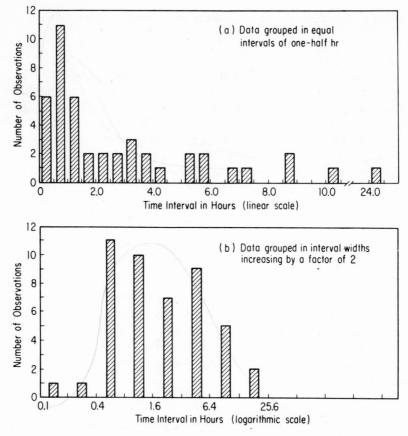

Fig. 5.10 Frequency distributions of active repair times (grouped data).

section will illustrate the procedures necessary to estimate the parameters of the log-normal function from the observed data.

Part (B) of Table 5.6 illustrates the procedure. Then $\hat{\xi}$ and $\hat{\sigma}$ are computed from the equations in the footnote on page 150, with t_i replaced by x_i and $\hat{\mu}$ replaced by $\hat{\xi}$. In this discussion it is necessary to distinguish between the mean of the distribution of the logarithms and the antilog of this value. (ξ is used to designate the former, and μ is used for the latter.) The reason for this is that μ turns out to be the median value of the distribution of the original variable, t, and it is desirable to express at least one of the parameters in terms of hours rather than logarithms of hours.

The solutions of Equation 5.11 for $\hat{\xi}$ and $\hat{\sigma}$ are given at the bottom of Table 5.6. The 95 per cent confidence limits about $\hat{\xi}$ are computed from Equation 5.12. The antilogs of these limits correspond to the confidence limits about the median value $\hat{\mu}$.

Table 5.7 FREQUENCY DISTRIBUTIONS OF ACTIVE REPAIR TIMES
(GROUPED DATA)

(A) EQUAL INTERVALS

Interval of t_i (hours)	n_i in Interval	Interval of t_i (hours)	n_i in Interval
0 – 0.5	6	—	—
0.51 – 1.0	11	6.6 – 7.0	1
1.1 – 1.5	6	7.1 – 7.5	1
1.6′ – 2.0	2	—	—
2.1 – 2.5	2	8.6 – 9.0	2
2.6 – 3.0	2	—	—
3.1 – 3.5	3	10.1 – 10.5	1
3.6 – 4.0	2	—	—
4.1 – 4.5	1	21.6 – 22.0	1
4.6 – 5.0	2	—	—
5.1 – 5.5	2	24.1 – 24.5	1

$$\Sigma n_i = 46$$

(B) LOGARITHMIC INTERVALS

Interval of t_i (hours)	n_i in Interval
0.1 – 0.2	1
0.21 – 0.4	1
0.41 – 0.8	11
0.81 – 1.6	10
1.61 – 3.2	7
3.21 – 6.4	9
6.41 – 12.8	5
12.9 – 25.6	2

$$\Sigma n_i = 46$$

Column 6 of Table 5.6 shows the computations of the values necessary to obtain column 7 from tables of the cumulative normal distribution. Column 7 then represents the theoretical log-normal repairability function corresponding to the nonparametric function of column 3. These two functions are plotted on semilog paper in Figure 5.11. It is evident from this figure that the fit of the theoretical to the observed curve is very good for this particular sample. A goodness-of-fit test shows no significant difference between the two. It is of interest to compare the arithmetic mean value of t with the median value obtained from the log-normal computations. The arithmetic mean is given by

$$\theta = \frac{\sum n_i t_i}{\sum n_i} = \frac{166.0}{46} = 3.61 \text{ hr},$$

as shown at the bottom of columns 1 and 2 of Table 5.6, whereas the median value is $\hat{\mu} = 1.93$ hr.

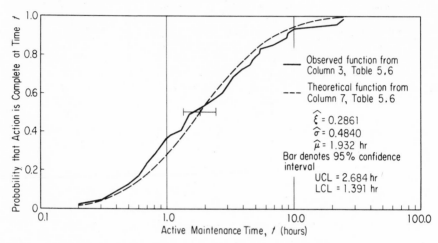

Fig. 5.11 Observed nonparametric and theoretical log-normal repairability functions.

It is evident that the arithmetic mean is unduly influenced by the large, infrequently occurring values of t_i, and that the median value is much more descriptive of the time required for a typical repair action.

The graphic method of estimating the parameters described in Section 5.6, page 152 may also be used for estimating $\hat{\xi}$ and $\hat{\sigma}$ for the log-normal distribution, and is adequate in many instances. For this method, the values of $\hat{M}_R(t_i)$ are plotted against t_i on log-probability paper, a straight line is drawn, and $\hat{\xi}$ and $\hat{\sigma}$ are estimated in the same manner as that used for the normal function.

5.8   Man-Hour Requirements

Man-Hour Ratio

For the maintainability and repairability functions discussed in the previous sections, the time base used was the total system down time or system active repair time. For simple systems, active repair time is frequently equivalent to active man-hours, since only one technician may be working.

For complex systems, this is usually not true, and it is useful to have some index which provides information regarding the average number of technicians required for maintenance. This index is termed the *Man-Hour Ratio* and is given by

$$R_{MH} = \frac{\text{total man-hours}}{\text{total active repair time}}.$$ (5.15)

The ratio can be used to convert the time scale of repairability functions to a man-hour scale.

Maintenance Support Index

Another index which is useful as a measure of maintenance man-power requirements is the so-called *Maintenance Support Index* (*M.S.I.*). This is defined as the number of active maintenance man-hours required per thousand system operating hours, as shown in Equation 5.16:

$$M.S.I. = \frac{(\text{total man-hours})\,(1000)}{\text{total operating hours}}.$$ (5.16)

This index reflects both the reliability and the repairability of a specific equipment and also provides an estimate of the manpower requirements.

5.9 Application of the Weibull Reliability Function

The *Weibull cumulative distribution function* (or *failure distribution*) is defined as

$$F(t) = 1 - e^{-(t-\gamma)^\beta/\alpha}, \quad \text{for} \quad \begin{aligned} t &\geq \gamma \\ a, \beta &> 0 \end{aligned}$$ (5.17)
$$= 0,$$

otherwise $F(t)$ is the probability that an item will fail at or before time t.

In Equation 5.17, the three unknown parameters are

$$\alpha = \text{scale parameter}$$
$$\beta = \text{shape parameter}$$
$$\gamma = \text{location parameter}$$

Upon setting $\beta = 1$, Equation 5.17 becomes the exponential distribution with delay (i.e., γ can be thought of as a guarantee period within which no failures can occur, or as a minimum life).

However, it is generally reasonable to assume that an item is subject to failure from the instant it is put into use. In this event, $\gamma = 0$ and Equation

5.17 reduces to the more common form,

$$F(t) = 1 - e^{-t^{\beta}/\alpha}, \quad \text{for} \quad t \geq 0 \tag{5.18}$$
$$\alpha, \beta \geq 0.$$

Equation 5.18, when $\beta = 1$, thus becomes the one-parameter exponential with $\alpha = $ mean life. It should be stressed that $\alpha = $ mean life only when $\beta = 1$.

The reliability function, $R(t) = 1 - F(t)$, for the two-parameter Weibull distribution is

$$R(t) = e^{-t^{\beta}/\alpha} \tag{5.19}$$

where $R(t)$ is the probability that no item will fail at or before time t.

The Weibull probability density function is obtained by taking the first derivative of Equation 5.18, the Weibull failure distribution. Thus

$$f(t) = \frac{\beta t^{\beta-1}}{\alpha} \cdot e^{-t^{\beta}/\alpha} \tag{5.20}$$

which, for $\beta = 1$, is identical to the exponential density function with $\alpha = \theta$.

When dealing with failure rates, the Weibull shape parameter β is of especial importance because it describes the mode of failure. Thus, for $\beta = 1$, as implied above, the failure rate is constant over time as it is for the exponential distribution. A value for $\beta < 1$ indicates that the failure rate is a decreasing function of time, while a $\beta > 1$ means that the failure rate is increasing with time.

The hazard rate function, $Z(t)$, also known as the instantaneous failure rate at time t, is defined for the Weibull distribution as

$$Z(t) = \frac{\beta t^{\beta-1}}{\alpha}. \tag{5.21}$$

Figure 3.1a (Chapter 3), shows for the Weibull distribution the probability density function, $f(t)$, the reliability function, $R(t)$, and the hazard rate function, $Z(t)$, for values of $\beta = 1, 2, 3,$ and 4 and fixed $\alpha = 1$.

When α and β are known, the mean life, μ, can be obtained from the equation

$$\mu = \alpha^{1/\beta} \Gamma(1/\beta+1), \tag{5.22}$$

where $\Gamma 1/\beta + 1$ is interpreted as the gamma function of $1/\beta + 1$. From Equation 5.22, it is easily verified that mean life is equal to α only when $\beta = 1$. For values of $\beta < 1$, $\alpha < t$. Conversely, when $\beta > 1$, $\alpha > t$.

Thus, if one assumes the Weibull distribution as a functional form of the failure distribution, the problem of estimating the parameters α and β must be solved. Two methods for solving this problem—the maximum-likelihood and minimum-chi-square methods—are quite involved and for a high degree of accuracy require a highspeed electronic computer.

Fortunately, through the use of Weibull probability graph paper, a simple method is available for obtaining estimates of the parameters β and α of the Weibull distribution. (This graph paper, developed by Professor J. H. K. Kao of Cornell University, can be obtained from Cornell University for a small fee.)

To understand how these estimates can be determined on this type of graph, note that in Figure 5.12 the Weibull probability paper that is depicted has four scales, one at each of the four sides of the graph. The upper and right scales are referred to as the principal scales and are linear; the lower and left scales are auxiliary and nonlinear. During the process of estimation, the raw failure data are plotted with reference to the auxiliary scales. Then estimates of β and α are obtained with reference to the principal scales. The principal scales are related to the principal ordinate and the principal abscissa, which are represented on the graph in Figure 5.12 by the two dark lines that intersect at point 0, 0.

To obtain estimates of β and α on the Weibull probability graph, the accumulated percentage of failure is plotted versus failure age. The per cent failure is plotted with reference to the left scale (auxiliary ordinate), and failure age with reference to the bottom scale (auxiliary abscissa). Then a straight line is fitted to the plotted points (usually by eye). At this stage of the process, the estimates of β and $\ln \alpha$ (natural logarithm of α) can be read directly from the right scale (labeled m in Figure 5.12, the scale for the principal ordinate).

In certain instances, it is necessary to obtain the estimate of $\ln \alpha$ from the upper m scale; this occurs, for example, when the straight line fitted to the plotted data (known as the Weibull plot) intersects the principal ordinate *outside* the graph.

Note in Figure 5.12 that the estimate of β (slope of the Weibull plot) is obtained by drawing a line (parallel to the Weibull plot) through point 1,0, located on the principal axis, so that the line intersects the principal ordinate. The point of intersection of the Weibull plot with the principal ordinate is the estimate of β, and it is read from the right scale, m.

To obtain the estimate of $\ln \alpha$, read from the right m scale the value that corresponds to the intersection of the Weibull plot with the principal ordinate. Find the antilogarithm of $\ln \alpha$, using as an aid a table of natural logarithms. When the Weibull plot intersects the principal ordinate outside the graph, to obtain an alternate estimate of $\ln \alpha$ read from the upper m scale that value which corresponds to the intersection of the Weibull plot with the principal abscissa. This value is the estimate of $\beta^{-1} \ln \alpha$; and from it—knowing the estimated value of β—we can readily obtain $\ln \alpha$ and, subsequently, α.

This technique, as illustrated in Figure 5.12, employs data presented in Table 5.8, which lists accumulated percentage of failures versus selected time

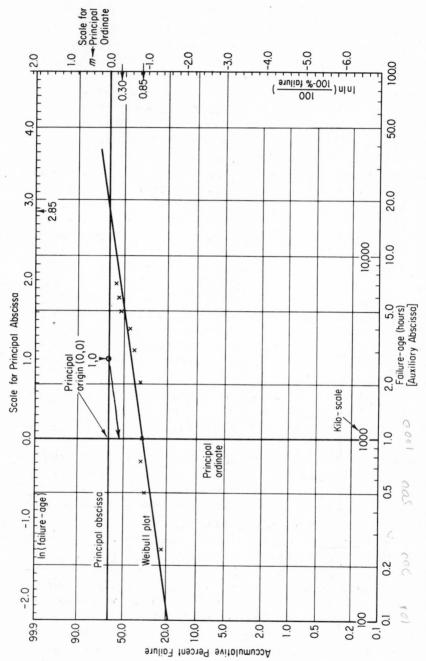

Fig. 5.12 A graphic procedure for obtaining estimates of parameters associated with the Weibull distribution.

intervals of a germanium power transistor. Note that column 2 lists the number of new failures observed during the hours (given in column 1) of a 7000-hr storage life-test run. Column 3 shows the accumulative number of failures per original sample size. Column 4 gives the accumulative percentage of failures of the transistor. From column 4 the accumulative per cent-failure values were plotted versus the failure-age values in column 1.

Table 5.8 GERMANIUM POWER TRANSISTOR: ACCUMULATIVE
PER CENT FAILURE VS. SELECTED TIME INTERVALS

(1)	(2)	(3)	(4)
Failure-Age (hours)	Failures	Accumulative Failures/Sample Size of 75	Accumulative Per Cent Failure
250	17	17/75	22.7
500	8	25/75	33.3
750	1	26/75	34.7
1000	1	27/75	36.0
2000	0	27/75	36.0
3000	5	32/75	42.7
4000	3	35/75	46.7
5000	4	39/75	52.0
6000	3	42/75	56.0
7000	2	44/75	58.7

The next step is to modify the auxiliary abscissa scale by multiplying it by 1000 (kilo-scale) to plot the point and to fit a straight line to the points. In this case, a relatively good fit results. The following step is to obtain an estimate of β. This is done by drawing a line—parallel to the fitted line (Weibull plot)—through the point $(1,0)$ on the principal abscissa to the point of intersection with the principal ordinate. If we read the value determined at the point of intersection from the right, or m scale, we obtain $\beta = 0.30$.

The estimated value of $\ln \alpha$ is obtained by reading from the same m scale the value that corresponds to the intersection of the Weibull plot with the principal ordinate. The value so obtained is $\ln \alpha = 0.85$. Using this value of $\ln \alpha$, we find—with the aid of a table of natural logarithms—that the anti-logarithm is $\alpha = 2.34$ (in the kilo-scale). However, if the Weibull plot does not intersect the principal ordinate on the graph, the estimate of $\beta^{-1} \ln \alpha$ can be obtained by reading this estimate from the upper m scale that corresponds to the intersection of the Weibull plot with the principal abscissa. The latter procedure, in this example, gives a value of $\beta^{-1} \ln \alpha = 2.85$. Since β was estimated above as being equal to 0.30, the result is

$$\ln \alpha = 2.85(0.30) = 0.855,$$

which compares quite favorably with the previous estimate of $\ln \alpha = 0.85$.

To determine the unscaled value of α, we multiply the value of α above (obtained in the kilo-scale) by $10^{3\beta}$. Thus

$$\alpha = 2.34(10^{3\beta}) = 2.34(10^{0.9}) = 2.34(7.943) = 18.6.$$

However, it is usually more convenient to retain a certain chosen time unit, say kilo-hour, throughout the analysis. It is essential to note that the estimate of α is in the kilo-scale, or whatever scale is being used.

After we determine estimates for α and β, the reliability function, $R(t)$, is easy to obtain. By the use of Equation 5.19, we determine the reliability function that corresponds to the failure data listed in Table 5.6 and shown in Figure 5.13. The computations which lead to this function appear in Table 5.9.

Similarly, the hazard rate function, $Z(t)$, is obtained through use of Equation 5.21 and is shown in Figure 5.14. The underlying calculations are listed in Table 5.10. It is interesting to note, with reference to Figure 5.14, that the resulting hazard rate curve is a decreasing function of time over the 7000-hr time period during which the life test took place. This, of course, is as expected since the estimated value of the shape parameter β is less than 1.0.

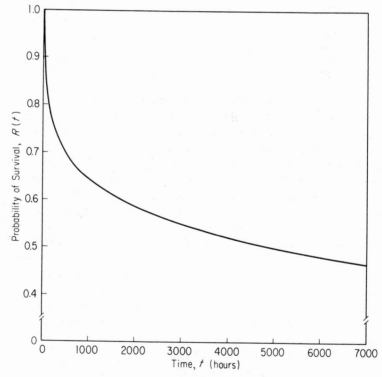

Fig. 5.13 Theoretical Weibull reliability function, $R(t)$, for $\beta = 0.30$ and $\alpha = 2.34$ (kilo-scale).

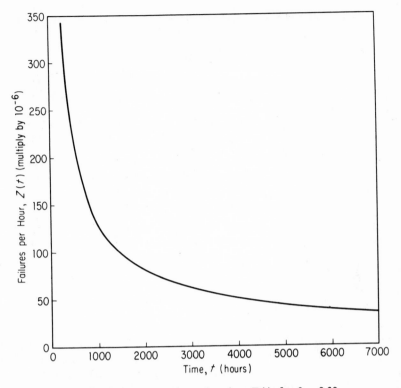

Fig. 5.14 Weibull hazard-rate function, $Z(t)$, for $\beta = 0.30$ and $\alpha = 2.34$ (kilo-scale).

Table 5.9 COMPUTATION OF THEORETICAL WEIBULL RELIABILITY FUNCTION FOR $\beta = 0.30$ AND $\alpha = 2.34$ (KILO-SCALE)

t_i	t_i (Kilo-Scale)	t_i^β	$\dfrac{t_i^\beta}{\alpha}$	$R(t_i) = e^{-t_i^\beta/\alpha}$
0	0	0	0	1.00
250	0.25	0.66	0.28	0.76
500	0.5	0.81	0.35	0.70
750	0.75	0.92	0.39	0.68
1000	1.0	1.00	0.43	0.65
2000	2.0	1.23	0.53	0.59
3000	3.0	1.39	0.59	0.55
4000	4.0	1.52	0.65	0.52
5000	5.0	1.62	0.69	0.50
6000	6.0	1.71	0.73	0.48
7000	7.0	1.79	0.76	0.47

Table 5.10 COMPUTATION OF HAZARD-RATE FUNCTION FOR
$\beta = 0.30$ AND $\alpha = 2.34$ (KILO-SCALE)

(1)	(2)	(3)	(4)		
t_i	t_i (Kilo-Scale)	$t_i^{\beta-1}$	$\beta t_i^{\beta-1}$	$Z(t_i) = \dfrac{\beta t_i^{\beta-1}}{\alpha}$	Failures per Hour $Z(t_i)$ (Multiply by 10^{-6})
250	0.25	2.64	0.79	0.34	340
500	0.50	1.62	0.49	0.21	210
750	0.75	1.22	0.37	0.16	160
1000	1.0	1.00	0.30	0.13	130
2000	2.0	0.62	0.19	0.08	80
3000	3.0	0.46	0.14	0.06	60
4000	4.0	0.38	0.11	0.05	50
5000	5.0	0.32	0.10	0.043	43
6000	6.0	0.29	0.09	0.038	38
7000	7.0	0.26	0.08	0.034	34

5.10 Tests for Validity of the Assumption of a Theoretical Reliability Function

The validity of many statistical techniques used in the calculation, analysis, or prediction of reliability depends on the distribution of the failure times. Many techniques are based on specific assumptions about the probability distribution and are often sensitive to departures from the assumed distributions.* That is, if the actual distribution differs from that assumed, these methods sometimes yield seriously wrong results. Therefore, in order to determine whether or not certain techniques are applicable to a particular situation, some judgment must be made as to the underlying probability distribution of the failure times.

Graphical Techniques

Some theoretical reliability functions, such as those based on the exponential, normal (Gaussian), log-normal, and Weibull distributions, will plot as straight lines on special types of graph paper. For example, Figure 5.15 shows a theoretical exponential reliability function, and Figure 5.8 show as theoretical normal reliability function as straight lines on semilog and on normal probability paper, respectively.

If it cannot be determined visually that the reliability function follows a straight line when plotted on some special graph paper, a graphical test for goodness of fit can be a valuable aid in making this judgment. The Kolmogorov-Smirnov or "d-test" for goodness of fit is one of many tests designed

*That is, not robust.

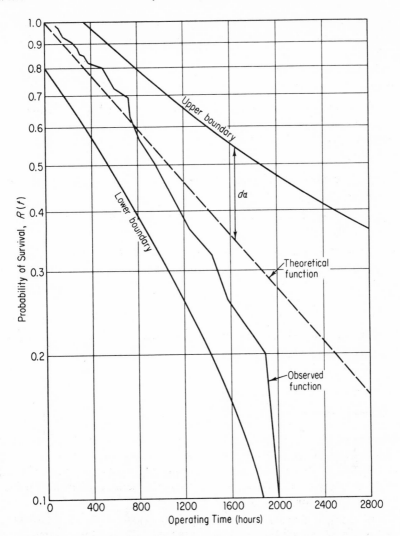

Fig. 5.15 Example of the application of the *d*-test.

for the purpose of testing whether or not the assumptions made about distributions of failure times are reasonable.

This test is based upon the fact that the observed cumulative distribution of a sample is expected to be fairly close to the true distribution. The goodness of fit is measured by finding the point at which the sample and the population are farthest apart and comparing this distance with the entry in a table of critical values, Table 5.11, which will then indicate whether such a large distance is likely to occur. If the distance is too large, the chance that the observations actually come from a population with the specified distribution is very small. This is evidence that the specified distribution is not the correct one.

Using the data in Table 5.2, the graphical procedure for applying the test would be as follows:

1. Draw the curve for the theoretical distribution which is assumed to be an exponential with an $MTBF = 1546.0$ hr.

2. Find the value, d, $(1.36/\sqrt{48} = 0.196)$ from Table 5.11 which corresponds to sample size, $n = 48$, and level of significance,† $\alpha = 0.05$.

3. Draw curves at a distance $d = 0.196$ above and below the theoretical curve drawn in step 1.

4. On the same graph draw the curve corresponding to the observed function.

5. If the observed function were to pass outside the band bounded by the two curves above and below the theoretical curve, there would be about a 5 per cent chance that the sample came from an exponential population with a mean life of 1546 hr.

6. If the observed function remains inside the band, as it does in the example, this *does not prove* that the assumed distribution is exactly right, but only that it might be correct and it is not unreasonable to assume that it is correct.

Chi-Square Goodness-of-Fit Test

The standard chi-square test considers a random sample of size n grouped into class intervals, and the observed frequencies for each class interval, with the restriction that the total expected frequencies equal the observed sample size. If m parameters are estimated from the sample, then m additional restrictions are imposed on the chi-square values.

When censorship occurs, the observed frequencies in each class interval are more highly correlated than they are in the case of a random sample because the sample size varies for each interval. However, for any interval, it is possible to consider only the binomial (failure or nonfailure); therefore,

$$(\chi_i)^2 = \frac{(r_i - n_i p_i)^2}{n_i p_i (1 - p_i)}, \tag{5.23}$$

where p_i is the conditional probability of failure in the ith interval, given nonfailure in the previous intervals $(i-1, i-2, \ldots, 1)$.

---

†The level of significance is approximate because of the estimating of parameters from the sample and the use of censored values.

Table 5.11 CRITICAL VALUES $d\alpha(n)$ OF THE MAXIMUM ABSOLUTE
DIFFERENCE BETWEEN SAMPLE AND POPULATION
RELIABILITY FUNCTIONS

Sample Size (N)	*Level of Significance (α)*				
	0.20	0.15	0.10	0.05	0.01
3	0.565	0.597	0.642	0.708	0.828
4	0.494	0.525	0.564	0.624	0.733
5	0.446	0.474	0.474	0.565	0.669
10	0.322	0.342	0.368	0.410	0.490
15	0.266	0.283	0.304	0.338	0.404
20	0.231	0.246	0.264	0.294	0.356
25	0.21	0.22	0.24	0.27	0.32
30	0.19	0.20	0.22	0.24	0.29
35	0.18	0.19	0.21	0.23	0.27
40	0.17	0.18	0.19	0.21	0.25
45	0.16	0.17	0.18	0.20	0.24
50	0.15	0.16	0.17	0.19	0.23
over 50 }	$\dfrac{1.07}{\sqrt{n}}$	$\dfrac{1.14}{\sqrt{n}}$	$\dfrac{1.22}{\sqrt{n}}$	$\dfrac{1.36}{\sqrt{n}}$	$\dfrac{1.63}{\sqrt{n}}$

If there are k class intervals excluding the last one (the interval between the last failure and infinity, which contains only censored observations), then

$$(\chi_{k-m})^2 = \sum_{i=1}^{k} \frac{(r_i - n_i p_i)^2}{n_i p_i (1 - p_i)}, \tag{5.24}$$

where r_i is the number of observed frequencies in the ith interval, and

$$p_i = 1 - \frac{R(t_i)}{R(t_{i-1})}, \tag{5.25}$$

where t_i is the right-hand end point of the ith interval.

Since the p_i's are conditional, each term represented by Equation 5.23—if the p's are known—is a chi-square variate with one degree of freedom; therefore, the sum represented by Equation 5.24 has k degrees of freedom. However, if the p's must be estimated through the parameters of $R(t)$, then the k degrees of freedom are reduced by the number, m, of parameters estimated.

Comparison of χ^2 (chi-square) and d-Tests

The d-test is superior to χ^2 in the following ways:

1. The d-test requires only the assumption of a continuous distribution, while the chi-square test requires the assumption that observed frequencies are normally distributed about their expected frequencies.

2. The exact distribution of d is known and tabled for small sample sizes, while the exact distribution of chi-square is known and tabled only for infinite sized samples.

3. The d-test can be used to test for deviations in a given direction, while chi-square can be used only for a two-sided test.

4. The d-test uses ungrouped data so that every observation represents a point of comparison, while the chi-square test requires the data to be grouped into cells with arbitrary choice of interval, size, and selection of starting point.

5. The d-test can be used in a sequential test where data become available from smallest to largest, computations being continued only up to the point at which rejection occurs.

The chi-square test is superior to the d-test in the following ways:

1. Chi-square does not require that the hypothesized population parameters be completely known in advance.

2. Chi-square can be partitioned and added.

3. Chi-square can be applied to discrete populations.

For example, using the data in Tables 5.2 and 5.3 and applying Equation 5.24, the value of 9.94 is obtained in the following manner:

Hours following a malfunction	r_i	n_i	$R(t_i)$	P_i	$n_i P_i$	$n_i P_i (1 - P_i)$	$(r_i - n_{Pi})^2$	χ^2
0–215	8	116	0.87	0.13	15.08	13.1196	50.1264	3.82
216–455	10	94	0.74	0.15	14.10	11.9850	16.8100	1.40
456–695	8	71	0.64	0.14	9.94	8.5484	3.7636	0.44
696–959	11	45	0.54	0.16	7.20	6.0480	14.4400	2.39
960–1199	6	25	0.46	0.15	3.75	3.1875	5.6625	1.59
1200–2064	5	14	0.26	0.43	6.02	3.4314	1.0404	0.30

$$\Sigma \chi^2 \quad = 9.94$$

This value can then be compared to the distribution of χ^2 for five degrees of freedom (since to compute $R(t_i)$ the estimated value of θ was used). The tabular value at the 5 per cent significance level is 11.07 (from Table 5.12). Therefore, the null hypothesis that the observed failure density can be represented by a theoretical exponential distribution has not been rejected.

5.11 Confidence-Interval Estimates for Parameters of Exponential Distribution

Two situations have to be considered for estimating confidence intervals: one in which the test is run until a preassigned number of failures ($r*$) occurs, and one in which the test is stopped after a preassigned number of test hours ($t*$) is accumulated. The formula for the confidence interval employs the χ^2 (chi-square) distribution. A short table of χ^2 values is given as Table 5.12. The general notation used is

$$\chi^2(p, d),$$

where p and d are two constants used to choose the correct value from the table.

The quantity p is a function of the confidence coefficient; d, known as the degrees of freedom, is a function of the number of failures. Equations 5.26 and 5.27 are for one-sided or two-sided $100(1-\alpha)$ per cent confidence intervals. For nonreplacement tests with a fixed truncation time, the limits are only approximate.

EQUATIONS FOR CONFIDENCE LIMITS ON MEAN LIFE

Type of Confidence Limits	Fixed Number of Failures, $r*$	Fixed Truncation† Time, $t*$	
One-Sided (Lower Limit)	$\left(\dfrac{2T}{\chi^2\alpha,\,2r},\infty\right)$	$\left(\dfrac{2T}{\chi^2(\alpha,\,2r+2)},\infty\right)$	(5.26)
Two-Sided (Upper and Lower Limits)	$\left(\dfrac{2T}{\chi^2\left(\frac{\alpha}{2},2r\right)},\dfrac{2T}{\chi^2\left(1-\frac{\alpha}{2},2r\right)}\right)$	$\left(\dfrac{2T}{\chi^2\left(\frac{\alpha}{2},2r+2\right)},\dfrac{2T}{\chi^2\left(1-\frac{\alpha}{2},2r\right)}\right)$	(5.27)

†For non-replacement tests, only one-sided intervals are possible for $r = 0$. Use $2n$ degrees of freedom for the lower limit if $r = n$.

The terms used are identified as follows:

n = number of items placed on test at time $t = 0$,

$t*$ = time at which the life test is terminated,

θ = mean life,

$\chi^2_{\frac{\alpha}{2}}(2r+2)$ = for example, the $\alpha/2$ percentage point of the chi-square distribution for $(2r+2)$ degrees of freedom,

r = number of failures accumulated to time $t*$,

$r*$ = preassigned number of failures,

α = acceptable risk of error,

$1-\alpha$ = confidence level.

Note that T is computed as follows, depending on the type of test procedure:

Probability

DF	0.99	0.975	0.95	0.90	0.80	0.75	0.50	0.25	0.20	0.10	0.05	0.025	0.01	0.001
1	0.000157	0.000983	0.00393	0.0158	0.0642	0.10153	0.455	1.323	1.642	2.706	3.841	5.024	6.635	10.827
2	0.0201	0.0506	0.103	0.211	0.446	0.5753	1.386	2.772	3.219	4.605	5.991	7.377	9.210	13.815
3	0.115	0.216	0.352	0.584	1.005	1.2125	2.366	4.108	4.642	6.251	7.815	9.348	11.341	16.268
4	0.297	0.484	0.711	1.064	1.649	1.9225	3.357	5.385	5.989	7.779	9.448	11.143	13.277	18.465
5	0.554	0.831	1.145	1.610	2.343	2.674	4.351	6.625	7.289	9.236	11.070	12.832	15.086	20.517
6	0.872	1.237	1.635	2.204	3.070	3.454	5.348	7.840	8.558	10.645	12.592	14.449	16.812	22.457
7	1.239	1.689	2.167	2.833	3.822	4.254	6.346	9.037	9.803	12.017	14.067	16.013	18.475	24.322
8	1.646	2.179	2.733	3.490	4.594	5.070	7.344	10.218	11.030	13.362	15.507	17.534	20.090	26.125
9	2.088	2.700	3.325	4.168	5.380	5.898	8.343	11.388	12.242	14.684	16.919	19.023	21.666	27.877
10	2.558	3.247	3.940	4.865	6.179	6.737	9.342	12.548	13.442	15.987	18.307	20.483	23.209	29.588
11	3.053	3.816	4.575	5.578	6.989	7.584	10.341	13.701	14.631	17.275	19.675	21.920	24.725	31.264
12	3.571	4.404	5.226	6.304	7.807	8.438	11.340	14.845	15.812	18.549	21.026	23.336	26.217	32.909
13	4.107	5.008	5.892	7.042	8.634	9.299	12.340	15.984	16.985	19.812	22.362	24.735	27.688	34.528
14	4.660	5.628	6.571	7.790	9.467	10.165	13.339	17.117	18.151	21.064	23.685	26.119	29.141	36.123
15	5.229	6.262	7.261	8.547	10.307	11.036	14.339	18.245	19.311	22.307	24.996	27.488	30.578	37.697
16	5.812	6.907	7.962	9.312	11.152	11.912	15.338	19.368	20.465	23.542	26.296	28.845	32.000	39.252
17	6.408	7.564	8.672	10.085	12.002	12.791	16.338	20.488	21.615	24.769	27.587	30.191	33.409	40.790
18	7.015	8.231	9.390	10.865	12.857	13.675	17.338	21.605	22.760	25.989	28.869	31.526	34.805	42.312
19	7.633	8.906	10.117	11.651	13.716	14.562	18.338	22.717	23.900	27.204	30.144	32.852	36.191	43.820
20	8.260	9.591	10.851	12.443	14.578	15.452	19.337	23.827	25.038	28.412	31.410	34.169	37.566	45.315
21	8.897	10.283	11.591	13.240	15.445	16.344	20.337	24.935	26.171	29.615	32.671	35.479	38.932	46.797
22	9.542	10.982	12.338	14.041	16.314	17.239	21.337	26.039	27.301	30.813	33.924	36.780	40.289	48.268
23	10.196	11.688	13.091	14.848	17.187	18.137	22.337	27.141	28.429	32.007	35.172	38.075	41.638	49.728
24	10.856	12.400	13.848	15.659	18.062	19.037	23.337	28.241	29.553	33.196	36.415	39.364	42.980	51.179
25	11.524	13.119	14.611	16.473	18.940	19.939	24.337	29.339	30.675	34.382	37.652	40.646	44.314	52.620
26	12.198	13.844	15.379	17.292	19.820	20.843	25.336	30.434	31.795	35.563	38.885	41.923	45.642	54.052
27	12.879	14.573	16.151	18.114	20.703	21.749	26.336	31.528	32.912	36.741	40.113	43.194	46.963	55.476
28	13.565	15.308	16.928	18.933	21.588	22.657	27.336	32.620	34.027	37.916	41.337	44.460	48.278	56.893
29	14.256	16.047	17.708	19.768	22.475	23.566	28.336	33.711	35.139	39.087	42.557	45.722	49.588	58.302
30	14.953	16.791	18.493	20.599	23.364	24.476	29.336	34.799	36.250	40.256	43.773	46.98	50.892	59.703

Table 5.12 DISTRIBUTION OF χ^2

Replacement Tests (failure replaced or repaired) $T = nt^*$ $\qquad$ (5.28)

Nonreplacement Tests $= \sum\limits_{i=1}^{r} t_i + (n-r)t^*$ $\qquad$ (5.29)

where t_i = time of the ith failure.

Censored Items (withdrawal or loss of items which have not failed)

$\quad$ (a) If failures are replaced and censored items are not replaced

$$T = \sum_{j=1}^{r} t_j + (n-c)\, t^* \qquad (5.30)$$

$\qquad$ where t_j = time of censorship
$\qquad\qquad\quad c$ = number of censored items.

$\quad$ (b) If failures are not replaced

$$T = \sum_{i=1}^{r} t_i + \sum_{j=1}^{c} t_j + (n-r-c)\, t^*. \qquad (5.31)$$

Example 1. Twenty items undergo a replacement test. Testing continues until ten failures are observed. The tenth failure occurs at 80 hr. Determine (1) the mean life of the items, and (2) the one-sided and two-sided 95 per cent confidence intervals.

(1) From Equation 5.28 and

$$\hat{\theta} = \frac{nt^*}{r} = \frac{(20)\,(80)}{10} = 160 \text{ hr.}$$

(2) From Equation 5.26, where

$\alpha = 1 - \text{confidence level} - 1 - 0.95 = 0.05$
$2r = 2\,(\text{number of failures}) = 2(10) = 20$

$$\frac{2T}{\chi^2(\alpha, 2r)}, \infty = \frac{2(1600)}{\chi^2(0.05,20)}, \infty = \frac{3200}{31.41}, \infty$$

$\qquad\qquad\qquad$ = 101.88 hr for the lower (one-sided) 95 per cent confidence level.

(3) From Equation 5.27,

$$\left(\frac{2T}{\chi^2\left(\frac{\alpha}{2}, 2r\right)}, \frac{2T}{\chi^2\left(1-\frac{\alpha}{2}, 2r\right)} \right) = \left(\frac{3200}{34.17}, \frac{3200}{9.591} \right)$$

$\qquad\qquad$ = 93.65 hr for the lower (two-sided) 95 per cent confidence interval

and $\qquad$ = 333.65 hr for the upper (two-sided) 95 per cent confidence interval.

(Assumption of Exponential Distribution)

Degrees of Freedom	Lower Limit						Upper Limit					
	99% Two-Sided / 99-1/2% One-Sided	98% Two-Sided / 99% One-Sided	95% Two-Sided / 97-1/2% One-Sided	90% Two-Sided / 95% One-Sided	80% Two-Sided / 90% One-Sided	60% Two-Sided / 80% One-Sided	60% Two-Sided / 80% One-Sided	80% Two-Sided / 90% One-Sided	90% Two-Sided / 95% One-Sided	95% Two-Sided / 97-1/2% One-Sided	98% Two-Sided / 99% One-Sided	99% Two-Sided / 99-1/2% One-Sided
2	0.185	0.217	0.272	0.333	0.433	0.619	4.47	9.462	19.388	39.58	100.0	200.0
4	0.135	0.151	0.180	0.210	0.257	0.334	1.21	1.882	2.826	4.102	6.667	10.00
6	0.108	0.119	0.139	0.159	0.188	0.234	0.652	0.909	1.221	1.613	2.3077	3.007
8	0.0909	0.100	0.114	0.129	0.150	0.181	0.437	0.573	0.733	0.921	1.212	1.481
10	0.0800	0.0857	0.0976	0.109	0.125	0.149	0.324	0.411	0.508	0.600	0.789	0.909
12	0.0702	0.0759	0.0856	0.0952	0.107	0.126	0.256	0.317	0.383	0.454	0.555	0.645
14	0.0635	0.0690	0.0765	0.0843	0.0948	0.109	0.211	0.257	0.305	0.355	0.431	0.500
16	0.0588	0.0625	0.0693	0.0760	0.0848	0.0976	0.179	0.215	0.251	0.290	0.345	0.385
18	0.0536	0.0571	0.0633	0.0693	0.0769	0.0878	0.156	0.184	0.213	0.243	0.286	0.322
20	0.0500	0.0531	0.0585	0.0635	0.0703	0.0799	0.137	0.158	0.184	0.208	0.242	0.270
22	0.0465	0.0495	0.0543	0.0589	0.0648	0.0732	0.123	0.142	0.162	0.182	0.208	0.232
24	0.0439	0.0463	0.0507	0.0548	0.0601	0.0676	0.111	0.128	0.144	0.161	0.185	0.200
26	0.0417	0.0438	0.0476	0.0513	0.0561	0.0629	0.101	0.116	0.130	0.144	0.164	0.178
28	0.0392	0.0413	0.0449	0.0483	0.0527	0.0588	0.0927	0.106	0.118	0.131	0.147	0.161
30	0.0373	0.0393	0.0425	0.0456	0.0496	0.0551	0.0856	0.0971	0.108	0.119	0.133	0.145
32	0.0355	0.0374	0.0404	0.0433	0.0469	0.0519	0.0795	0.0899	0.0997	0.109	0.122	0.131
32	0.0339	0.0357	0.0385	0.0411	0.0445	0.0491	0.0742	0.0834	0.0925	0.101	0.113	0.122
36	0.0325	0.0342	0.0367	0.0392	0.0423	0.0466	0.0696	0.0781	0.0899	0.0939	0.104	0.111
38	0.0311	0.0327	0.0351	0.0375	0.0404	0.0443	0.0656	0.0732	0.0804	0.0874	0.0971	0.103
40	0.0299	0.0314	0.0337	0.0359	0.0386	0.0423	0.0619	0.0689	0.0756	0.0820	0.0901	0.0968

To Use: Multiply value shown by total test hours to get upper and lower confidence limits in hours.

Note: $d = 2r$, except for the lower limit on tests truncated at a fixed time and where $r < n$. In such cases, use $d = 2(r+1)$.

Table 5.13 FACTORS FOR CALCULATION OF MEAN LIFE
CONFIDENCE INTERVALS FROM TEST DATA [FACTORS $= 2/\chi^2(p,d)$]

Example 2. Twenty items undergo a nonreplacement test, which is terminated at 100 hr. Failure times observed were 10, 16, 17, 25, 31, 46, and 65 hr. Calculate (1) the one-sided approximate 90 per cent confidence interval ($\alpha = 0.10$), and (2) the two-sided approximate 90 per cent confidence limits.

(1) From Equation 5.26,

$$\left(\frac{2T}{\chi^2(\alpha,\, 2r+2)},\, \infty \right) = \left(\frac{2\left[\sum_{i=1}^{7} t_i + (20-7)\,(100) \right]}{\chi^2(0.10,\, 16)},\, \infty \right)$$

$$= \left(\frac{30.20}{23.54},\, \infty \right)$$

$$= 128.28 \text{ hr for the lower single-sided 90} \\ \text{per cent confidence interval.}$$

(2) From Equation 5.27,

$$\left(\frac{2T}{\chi^2\left(\frac{\alpha}{2},\, 2r \pm 2\right)},\, \frac{2T}{\chi^2\left(1-\frac{\alpha}{2},\, 2r\right)} \right) = \left(\frac{3020}{26.30},\, \frac{3020}{6.57} \right)$$

$$= 114.83 \text{ hr for the lower (two-sided) 90 per cent confidence interval}$$

and 459.67 hr for the upper (two-sided) 90 per cent confidence interval.

Table 5.13, extracted from the RADC Reliability Notebook, presents the factor $2/\chi^2(p,d)$ for one-sided and two-sided confidence limits, at six confidence levels of each. Multiplying the appropriate factor by the observed total life T gives a confidence limit on $\hat{\theta}$. Figure 5.16 presents a graphical technique for determining upper and lower confidence limits for tests truncated at a fixed time, when the number of failures is known.

The probability of (or proportion of items) surviving t hours is found by

$$\hat{R}(t) = e^{-t/\hat{\theta}}. \tag{5.32}$$

The confidence interval on $R(t)$ is

$$(e^{-t/\hat{\theta}}L < R(t) < e^{-t/\hat{\theta}}U), \tag{5.33}$$

where

$\hat{\theta}_L$ and $\hat{\theta}_U$ are the lower and upper confidence limits on $\hat{\theta}$.

Example 3. Based on the data of Example 1, (1) what is the probability of an item's surviving 100 hr? (2) what are the two-sided 95 per cent confidence limits on this probability?

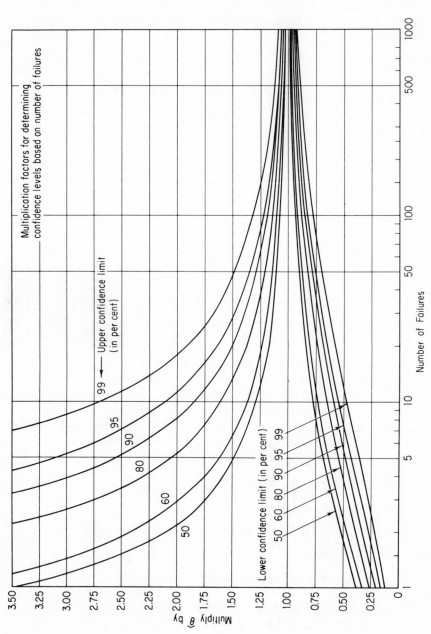

Fig. 5.16 Multiplication ratios for determining upper and lower confidence limits vs. number of failures for tests truncated at a fixed time.

(1) From Equation 5.32,

$$\hat{R}(100) = e^{-100/\theta} = e^{-100/160} = 0.535.$$

(2) The two-sided 95 per cent confidence limits $= (e^{-100/93.65}, e^{-100/333.65})$
$$= (0.344, 0.741).$$

To estimate the time period, $\hat{T}_R$, for which there is a reliability of R, the estimate $\hat{T}_R$, is

$$\hat{T}_R = \theta \log \frac{1}{R}. \tag{5.34}$$

The confidence limits on T_R define a tolerance interval, since these limits permit the statement that there is 100 (1-α) per cent confidence that R per cent or more of the items in the population will survive T_R or more time units. The 100 (1-α) per cent confidence limits on T_R or more time units are given below.

EQUATIONS FOR CONFIDENCE LIMITS ON RELIABILITY

Type of Confidence Limits	Fixed Number of Failures, r^*	Fixed Truncation Time, t^*	
One-Sided (Lower Limit)	$\left(\dfrac{2T \log 1/R}{\chi^2(\alpha,2r)}, \infty\right)$	$\left(\dfrac{2T \log 1/R}{\chi^2(\alpha,2r+2)}, \infty\right)$	(5.35)
Two-Sided (Upper and Lower Limits)	$\left(\dfrac{2T \log 1/R}{\chi^2\left(\frac{\alpha}{2},2r\right)}, \dfrac{2T \log 1/R}{\chi^2\left(1-\frac{\alpha}{2},2r\right)}\right)$	$\left(\dfrac{2T \log 1/R}{\chi^2\left(\frac{\alpha}{2},2r+2\right)}, \dfrac{2T \log 1/R}{\chi^2\left(1-\frac{\alpha}{2},2r\right)}\right)$	(5.36)

Example 4. For Example 1, what is the estimated time period for which the reliability is 0.80? What are the 95 per cent one- and two-sided confidence limits on T_R?

Since $\theta = 160$ hr (from Example 1), Equation 5.34 yields $T_{0.80} = \theta \log \dfrac{1}{0.8} = (160)(0.22314) = 35.70$ hr.

The 95 per cent one-sided and two-sided confidence limits on T_R, from Equations 5.35 and 5.36, are

$$\left(\frac{2(1600)(0.22314)}{31.41}, \infty\right) = \begin{array}{l} 22.73 \text{ hr for the lower} \\ \text{(one-sided) 95 per cent confidence limit} \end{array}$$

and

$$\left(\frac{2(1600)(0.22314)}{34.17}, \frac{2(1600)(0.22314)}{9.591}\right)$$

$= 20.90$ hr for the lower (two-sided) 95 per cent confidence limit

and 74.45 hr for the upper (two-sided) 95 per cent confidence limit.

5.12   Confidence-Interval Estimates for the Binomial Distribution

For situations where reliability is measured as a ratio of the number of successes to the total number of trials, the confidence interval is determined by consideration of the binomial distribution. Table XI of Hald's *Statistical Tables and Formulas* (John Wiley & Sons, Inc., New York, 1952) gives 95 per cent and 99 per cent confidence limits for a wide range of values. Figure 5.17 allows a rough estimate to be made when the number of successes (S) and the number of trials (N) are known.

Example 5. $S = 8$; $N = 10$. (1) What is the reliability estimate? (2) What are the two-sided upper and lower 95 per cent confidence limits? Answers: (1) 0.80; (2) 0.98 and 0.43.

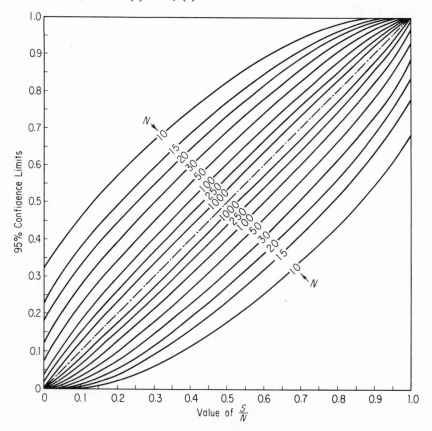

Fig. 5.17 Chart for 95 per cent confidence limits on the probability $\frac{S}{N}$. [From C. J. Clopper and E. S. Pearson, "The use of confidence or fiducial limits illustrated in the case of the binomial," *Biometrika*, Vol. 26 (1934), p. 410. Reprinted with permission.]

PROBLEMS

1. The following data are field observations of the removal times of a specific tube type.

Time (in hours)	Number of Removals	Number Censored
0	1	
120	1	
280		2
300		2
310		2
330	1	
340	1	
360	2	
470	2	
490	2	
550		2
560	2	
570	1	
580	2	
610	2	
650	1	
690	1	
780	1	
800	2	
860	1	
880	1	
1000	1	
1020	1	
1040	1	
1050	1	
1200	2	
1260	1	
1310	1	
1320	1	
1650	1	
1670	1	
1800		1
	35	9

(a) Using the data above, construct an observed reliability function and a theoretical normal reliability function. Plot both functions. (b) Determine the two-sided 95 per cent confidence interval for the mean.

2. The following data are field observations of the repair times of a specific equipment type.

Time (in hours)	Number of Observations
0.1	2
0.2	3
0.3	3
0.4	4
0.5	9
0.6	3
0.7	2
0.8	2
1.0	7
1.1	1
1.2	1
1.5	4
2.0	3
2.1	1
2.2	2
2.5	2
3.0	1
3.4	1
3.8	1
4.0	1
4.4	1
4.7	1
5.0	1
5.2	1
7.0	2
7.5	1
11.6	1
17.0	1
45.7	1
	63

(a) Using the data above, construct an observed repairability function and a theoretical log-normal repairability function. (b) Plot both functions. (c) Determine the two-sided 95 per cent confidence interval for the median.

3. The following are field observations of the failure times of a radar system.

Time (in hours)	Number of Failures
3.0	1
11.8	1
15.0	1
21.2	1
24.0	1
29.0	1
50.0	1
54.0	1
70.0	1
84.5	2
91.0	1
104.0	1
122.0	1
153.0	1
166.0	2
202.0	1
255.0	1
280.0	1
345.1	1
405.0	1
425.0	1
493.0	1
565.0	1
650.0	1

(a) Using the data above, construct an observed reliability function and a theoretical exponential reliability function. Plot both functions. Determine the two-sided 95 per cent confidence interval for the mean. (b) Using the data above, construct and plot a theoretical Weibull reliability function.

Reliability Allocation

6.1 Introduction

System design engineers must translate over-all system characteristics, including reliability, into detailed specifications for the numerous units that make up the system.† The process of assigning reliability requirements to individual units to attain the desired system reliability is known as *reliability allocation*. More is involved, however, than a simple mathematical equality. The reliability of an individual unit varies with the type of function to be performed, the complexity of the unit, and the method of accomplishing the function, to name a few of the more important factors. The role a unit plays in a particular system also enters into consideration.

The various factors influencing unit reliability must be considered if the most economical and realistic requirements are to be specified. Today, the designer and his reliability specialists recognize these influences but generally have no quantitative method of relating the various factors. The problem of realistic allocation is further complicated by the lack of detailed information on many of these factors early in the system design and analysis phases.‡

---

†The term *unit* as used herein represents that level of the system at which the system requirement is to be allocated.

‡When detailed design information is available, the methods described in Chapter 9 are used.

Consequently, reliability requirements for units, equipments, and subsystems are often assigned on the basis of past performance data, and are subjectively adjusted to make the system reliability equal to the combined reliabilities of the system sublevels.

Mathematical Basis of Reliability Allocation

The allocation of system reliability involves solving the basic inequality

$$f(\hat{R}_1 \cdot \hat{R}_2 \cdots \hat{R}_n) \geq R^* \tag{6.1}$$

where

$\hat{R}_i =$ the allocated reliability parameter for the ith unit,

$R^* =$ the system reliability requirement parameter, and

$f =$ the functional relationship between unit and system reliability.

For a simple series system in which the R's represent probability of survival for t hours, Equation 6.1 becomes

$$\hat{R}_1(t) \cdot \hat{R}_2(t) \cdots \hat{R}_n(t) \geq R^*(t). \tag{6.2}$$

Theoretically, Equation 6.2 has an infinite number of solutions, assuming no restrictions on the allocation. The problem is to establish a procedure that yields a unique or limited number of solutions, by which consistent and reasonable reliabilities may be allocated. For example, the allocated reliability for a simple unit of demonstrated high reliability should be greater than for a complex unit whose observed reliability has always been low.

The Value of an Allocation Program

Although several methods for attacking the problem have appeared in the literature (some of which are reviewed in Section 6.2), a standardized program for reliability allocation is still needed. Such a program would offer a number of benefits:

1. The well-meaning but ineffectual philosophy often applied to reliability —"we will do the best we can"—would be replaced by a contractual obligation in the form of quantitative reliability requirements that force contractors to consider reliability equally with other system parameters, such as performance, weight, and cost.

2. Since an allocation forces contractors to plan on meeting specified reliability goals, then improved design, procurement, manufacturing, and testing procedures would result. This situation would not only ensure a reliable system, but should improve the state of the art of all facets of the associated program.

3. Reliability allocation focuses attention on the relationships between component, equipment, subsystem, and system reliability, leading to a more complete understanding of the basic reliability problems inherent in the design.

4. Requirements determined through an allocation procedure would be more realistic, consistent, and economically attained than those obtained through subjective or haphazard methods, or those resulting from crash programs initiated after unfortunate field experiences.

5. A reliability allocation program could lead to optimum system reliability since the program would provide for handling such factors as essentiality, cost, maintenance, weight, and space.

6. The over-all cost of such a program would be more than counterbalanced by a saving in the amount of time and money usually expended in meeting specified reliability goals; in addition, substantial reductions of operational, maintenance, and management costs would be realized.

Implementation of an allocation program requires a quantitative contractual reliability requirement at the system level. Mutual acceptance by customers and their contractors is also required for (1) definitions of success and failure, and (2) the criticality of various failure modes, along with procedures and criteria for establishing methods for demonstrating conformance to the allocated unit requirements.

Moreover, the program and methods presented in this chapter can apply to the suballocation of reliability within the various units. The allocation program is necessarily one of continual refinement. Original requirements determined at the design stage should be critically examined and revised as more experience, knowledge, and test data become available during the advance of the system life cycle through the design, development, and production phases.

6.2 Feasibility Prediction

A *feasibility prediction* is an initial estimate of the feasibility of building a system having a certain operational reliability. This prediction is usually based on design concepts and estimated complexity. Complexity is perhaps the one, single factor most responsible for poor reliability, especially for systems having no provision for immediate adjustments or corrective maintenance, such as missiles and satellites. Early in the system-planning phase, a keen awareness of the relationship between complexity and reliability will serve to keep a degree of realism in preliminary estimates of "assurable" reliability for a given design configuration.

A review of the field reliability achievements of other systems can be of benefit in evaluating the magnitude of the reliability problem which confronts the proposed new system. To facilitate such a review, Figure 6.1 gives a log-plot of system mean time between failures versus nonredundant system complexity. Complexity is measured in terms of the number of active element groups (*AEG*'s) making up the system. An *AEG* is defined as an electron tube or transistor, and its associated circuitry.† The average or

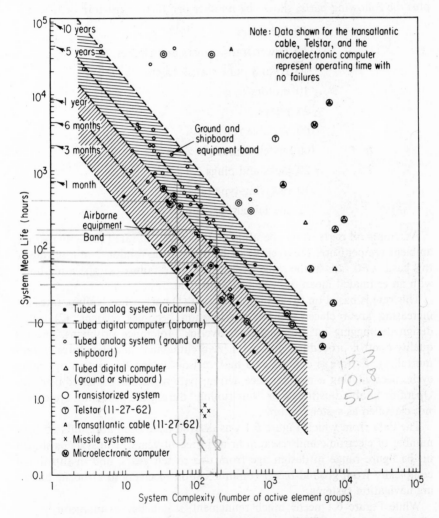

Fig. 6.1 Effects of system complexity on reliability.

†Mechanical parts such as motors, gyros, relays, etc., are included in the associated circuitry.

typical *AEG* for analog transistorized systems consists of:

> 1 silicon transistor
> 1 silicon diode†
> 3 composition resistors
> 2 paper capacitors

plus the following parts, shown by number per 1000 transistors or tubes:

> 50 inductors
> 10 power transformers and chokes
> 10 IF and pulse transformers
> 10 motors
> 25 relays
> 100 variable resistors
> 100 switch contacts
> 20 jacks and plugs
> 10 receiving-type tubes
> 2 special-purpose tubes

Assuming all parts to be operating at 25 per cent of rated capacity in 25°C ambient temperature, the average inherent or "true random" failure rate of this basic *AEG* falls in the range of 1-to-3 $\times$ 10^{-6} failure per operating hour, with an estimated mean of 1.2×10^{-6} per hour.

This rate is based on the assumptions that (1) parts whose failure rates are increasing are replaced before deterioration becomes evident, (2) initial design de-bugging has eliminated "maverick" tolerance problems, and (3) quality-control procedures for parts have eliminated initial failures (infant mortality). Failure is defined as a malfunction during the system operating cycle, necessitating a maintenance action, part replacement, or adjustment. Operator knob-adjustments or "fine tuning" during the operating cycle were not classified as system failures.

The data from which Figure 6.1 was derived have been accumulated by a number of electronic and research firms in recent years. The systems shown in the figure range in design age from two to ten years, and in functional complexity from ground-based communication receivers to airborne bombing/navigation systems.

While Figure 6.1 needs much refinement, it can be advantageously employed for evaluation of design feasibility and for "ballpark" predictions of system reliability. For example, on the basis of the data plotted in the figure,

---

†For digital systems, a tentative allocation is 10 diodes per *AEG*.

we question the feasibility of an airborne equipment with 1000 *AEG*'s having a mean life of 100 hr; at least it is likely that a unique approach or state-of-the-art breakthrough would be needed to meet such a requirement. The figure indicates a mean life of 1–10 hr would be more appropriate for this equipment.

For gross estimation purposes, the upper boundary of the ground equipment based on Figure 6.1 can be used as a tentative midline for satellites. This estimate, which is based on a limited amount of data,† is equivalent to using a factor of 0.5 to describe the satellite-to-ground failure rate relationship.

Basic Allocation Methods‡

Basic reliability allocation methods are discussed in this section. For a straightforward presentation of fundamentals, the discussion excludes several related but complex factors: multimodal design, feasibility of the over-all requirement, and unit redundancy.

The procedures are based on part failure (hazard) rates, which are assumed to be constant. The other necessary assumptions are that unit failures are independent, and that failure of any unit will result in system failure; i.e., the system is composed of units in series.

These assumptions lead to the following equations:

Let

$$R_j(t) = \text{reliability of the } j\text{th unit over } t \text{ operating hours,}$$

and

$$R(t) = \text{reliability of the system over } t \text{ operating hours.}$$

Then

$$R(t) = R_1(t) \, R_2(t) \dots R_n(t). \tag{6.3}$$

If

$$\lambda_j = \text{failure rate of the } j\text{th unit,}$$

and

$$\lambda_s = \text{failure rate of the system,}$$

†*Satellite Reliability Spectrum*, ARINC Research Corporation, Publication No. 173-5-280 (January, 1962).

‡More detailed descriptions of the methods appear in the following references:
Electronic Industries Association (formerly RETMA), "Determination of Permissible Component Part Failure Rates," *Electronic Applications Review*, Vol. 4, No. 1 (September, 1956), 10-12.
H. E. Frederick, "A Reliability Allocation Technique," *Proceedings of the Fourth National Symposium on Reliability and Quality Control* (January, 1958), pp. 314-317.
H. S. Balaban, H. R. Jeffers, and D. O. Baechler, *The Allocation of System Reliability*, ARINC Research Corporation, Publication No. 152-2-274 (November 30, 1961).

Equation 6.3 becomes

$$e^{-\lambda_s t} = e^{-\lambda_1 t} e^{-\lambda_2 t} \ldots e^{-\lambda_n t}. \tag{6.4}$$

A reliability allocation for an over-all requirement expressed in terms of $R(t)$ or λ can be performed by an essentially identical approach. Allocations based on a system failure rate requirement will be discussed first. Since a series system with constant failure rates is assumed, the method is identical for requirements stated in terms of mean life or MTBF, which is the reciprocal of failure rate.

The allocation method comprises the following steps:

1. Given a series system with n units, the system failure rate is equal to the sum of the unit failure rates. If λ^* is the system failure rate requirement, allocated unit failure rates $\hat{\lambda}_j$ must be chosen so that

$$\hat{\lambda}_1 + \hat{\lambda}_2 + \ldots + \hat{\lambda}_n \leq \lambda^*.$$

2. Observed or estimated unit failure rates are obtained:

$$\lambda_1, \lambda_2, \ldots, \lambda_n.$$

3. Relative unit weights are computed from the equation

$$w_j = \frac{\lambda_j}{\sum\limits_{k=1}^{n} \lambda_k}.$$

4. Since w_j represents the relative failure vulnerability of the jth unit and $\sum w_j = 1.0$,

the system failure rate requirement λ^* can be apportioned over the units by the formula

$$\hat{\lambda} \leq w_j \lambda^*. \tag{6.5}$$

If the equality sign holds, maximum allocated failure rates result which satisfy the requirement

$$\sum_{j=1}^{n} \hat{\lambda}_j = \lambda^*.$$

For a system requirement expressed in terms of probability of survival over t hours, i.e., $R^*(t)$, the same formulas can be used, since it is required that

$$R^*(t) = e^{-\lambda^* t} \leq e^{-\hat{\lambda}_1 t} e^{-\hat{\lambda}_2 t} \ldots e^{-\hat{\lambda}_n t}$$

Substituting Equation 6.5,

$$R^*(t) \leq e^{-w_1 \lambda^* t} e^{-w_2 \lambda^* t} \ldots e^{-w_n \lambda^* t}$$

leading to the basic allocation formula

$$\hat{R}_j(t) > [R^*(t)]^{w_j}. \tag{6.6}$$

Example 1. Assume, for a particular system, a maximum failure rate requirement of $\lambda^* = 0.010$ (equivalent to a specified mean life of 100 hr). This failure rate is to be allocated to the four serial units comprising the system.

(1) Estimated unit failure rates are

$$\lambda_1 = 0.002, \ \lambda_2 = 0.003, \ \lambda_3 = 0.004, \ \lambda_4 = 0.007$$

$$\sum_{k=1}^{n} \lambda_k = 0.002 + 0.003 + 0.004 + 0.007 = 0.016.$$

(2) Relative unit weights

$$w_1 = \frac{0.002}{0.016} = 0.1250, \ w_2 = \frac{0.003}{0.016} = 0.1875,$$

$$w_3 = \frac{0.004}{0.016} = 0.2500, \ w_4 = \frac{0.007}{0.016} = 0.4375.$$

(3) Allocated unit failure rates are then

$$\hat{\lambda}_1 = (0.1250)\,(0.010) = 0.001250;$$
$$\hat{\lambda}_2 = (0.1875)\,(0.010) = 0.001875;$$
$$\hat{\lambda}_3 = (0.2500)\,(0.010) = 0.002500;$$
$$\hat{\lambda}_4 = (0.4375)\,(0.010) = 0.004375.$$

Example 2. Assume for the same system a reliability requirement of 0.90 for 10 hr of operation. Unit reliabilities for 10 hr are to be determined which are consistent with this requirement.

(1) Unit failure rates are

$$\lambda_1 = 0.002, \ \lambda_2 = 0.003, \ \lambda_3 = 0.004 = \lambda_4 = 0.007.$$

(2) Relative unit weights are

$$w_1 = 0.1250, \ w_2 = 0.1875, \ w_3 = 0.2500, \ w_4 = 0.4375.$$

(3) Allocated 10-hr reliabilities (Equation 6.6) are

$$R_1(10) = [R^*(10)]^{w_1} = (0.90)^{0.1250}$$

$$= \text{antilog}\ [(0.1250)\ \log\ (0.90)]$$

$$= 0.987.$$

$$R_2(10) = (0.90)^{0.1875} = 0.980$$

$$R_3(10) = (0.90)^{0.2500} = 0.974$$

$$R_4(10) = (0.90)^{0.4375} = 0.955$$

Check $(0.987)\,(0.980)\,(0.974)\,(0.955) = 0.8997.$

The system reliability requirement $R^*(T)$ can be converted to a failure rate requirement by the formula

$$\lambda^* = \frac{-\log R^*(T)}{T}.$$ (6.7)

Allocated unit failure rates can then be used to obtain the allocated unit reliabilities by the formula

$$\hat{R}_j(T) = e^{-\hat{\lambda}_j T}.$$ (6.8)

6.3   The AGREE Allocation Method

The reliability allocation method described in the AGREE Report† is somewhat more sophisticated than the methods discussed in the previous sections. The principal difference is that the AGREE method is based on unit complexity rather than unit failure rates. Also, unit importance or essentiality—the relationship between unit and system failure—is considered explicitly in the AGREE allocation formula.

The allocation formula is used to determine a minimum acceptable mean life for each unit (AGREE uses the term "equipment") to satisfy a minimum acceptable system reliability. The assumption is made that units within the system have independent failure rates and operate in series with respect to their effect on mission success.

Unit complexity is defined in terms of modules and their associated circuitry, where a module is an electron tube, a transistor, or a magnetic amplifier. Diodes represent a half-module. AGREE states that for digital computers, where the module count is high, the module count should be reduced to allow for the fact that failure rates for digital parts are generally far lower than for radio-radar types.

The importance factor for a unit is defined in terms of the probability of system failure if that particular unit fails. If the importance factor of a unit equals one, the unit must operate satisfactorily for the system to operate satisfactorily; if the factor equals zero, then failure of the unit has no effect on system operation.

The specific basis of the allocation is to require that each module make an equal contribution to system success. An equivalent requirement is that each module have the same mean life or failure rate. With the approximation that $e^{-x} \approx 1 - x$, where x is small, the allocated failure rate of the jth unit is shown in the AGREE report to be

$$\hat{\lambda}_j = \frac{n_j[-\log R^*(T)]}{E_j t_j N}$$ (6.9)

†*Reliability of Military Electronic Equipment*, Advisory Group on Reliability of Electronic Equipment, Office of the Assistant Secretary of Defense (June 4, 1957), 52–57.

where

n_j = the number of modules in the jth unit,

E_j = the importance factor of the jth unit,

t_j = the number of hours the jth unit will be required to operate in T system hours $(0 < t_j \leq T)$, and

N = total number of modules in the system.

The allocated reliability for the jth unit for t_j unit operating hours, $\hat{R}(t_j)$, is

$$\hat{R}(t_j) = 1 - \frac{1 - [R^*(T)]^{n_j/N}}{E_j}. \qquad (6.10)$$

The AGREE report cautions against use of the allocation formula for units of very low importance, which, if included, will distort the allocation. The report also briefly discusses allocation when redundancy exists, but the formulas given are very poor approximations, at best.

Example 3. Assume a reliability requirement of 0.90 for 12 hr of continuous operation for an early-warning radar set. The units of the set are

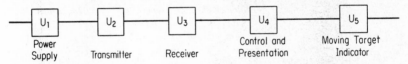

U_1	U_2	U_3	U_4	U_5
Power Supply	Transmitter	Receiver	Control and Presentation	Moving Target Indicator

Allocate the minimum acceptable unit failure rates and unit reliabilities. The step-by-step solution follows:

(1) Determine unit complexity.

Unit	Number of Modules $-n_j$
1	35
2	91
3	88
4	231
5	88

$$N = \sum n_j = 533$$

(2) Determine unit operating hours, t_j.

Each unit except the moving-target indicator (U_5) has a 100 per cent duty cycle. This indicator will be required for approximately 6 of the 12 hr of operation. Hence $t_1 = t_2 = t_3 = t_4 = 12$; $t_5 = 6$.

(3) Determine unit importance, E_j.

From the operating requirements on the set, it is determined that all units except the moving-target indicator are required. If the moving-target indicator fails, an estimated 25 per cent of the targets may be lost in ground clutter. Since unit importance is defined as the probability that the system will fail to accomplish its mission if the unit fails while all others are satisfactory, then

$$E_1 = E_2 = E_3 = E_4 = 1.00$$

$$E_5 = 0.25.$$

(4) Determine the unit allocations.

Table 6.1 summarizes the necessary inputs to the formulas for allocated minimum acceptable failure rate or for allocated reliability.

Table 6.1 ALLOCATION INPUTS FOR EXAMPLE 3

System: AEW Radar Set Primary Mission: Detection of Aircraft		Reliability Requirement: $R^*(12 \text{ hr}) = 0.90$	
Unit	Number of Modules n_j	Required Operating Time (hr), t_j	Importance E_j
1—Power supply	35	12	1.00
2—Transmitter	91	12	1.00
3—Receiver	88	12	1.00
4—Display and Control	231	12	1.00
5—Moving-Target Indicator	88	6	0.25
	$N = 533$		

(5) The allocated minimum acceptable unit failure rates, from Equation 6.9, are

$$\hat{\lambda}_1 = \frac{35 \, [-\ln 0.90]}{(533) \, (1.00) \, (12)} = 575 \times 10^{-6} \qquad 576.6$$

$$\hat{\lambda}_2 = \frac{91 \, [-\ln 0.90]}{(533) \, (1.00) \, (12)} = 1495 \times 10^{-6} \qquad 1499$$

$$\hat{\lambda}_3 = \frac{88 \, [-\ln 0.90]}{(533) \, (1.00) \, (12)} = 1445 \times 10^{-6}$$

$$\hat{\lambda}_4 = \frac{231 - \ln 0.90}{(533)(1.00)(12)} = 3790 \times 10^{-6}$$

$$\hat{\lambda}_5 = \frac{88 - \ln 0.90}{(533)(0.25)(6)} = 11{,}560 \times 10^{-6}.$$

(6) Allocated unit reliabilities, from Equation 6.10:

$$\text{Unit 1: } R_1(12) = \frac{1 - [0.90]^{35/533}}{1.0}$$

$$= (0.90)^{35/533} = 0.993$$

Unit 2: $\hat{R}_2(12) = (0.90)^{91/533} = 0.982$

Unit 3: $\hat{R}_3(12) = (0.90)^{88/533} = 0.983$

Unit 4: $\hat{R}_4(12) = (0.90)^{231/533} = 0.955$

$$\text{Unit 5: } \hat{R}_5(6) = 1 - \frac{1 - [0.90]^{88/533}}{0.25} = 0.932.$$

$$\hat{R}_5(12) = \hat{R}_5(6) + [1 - \hat{R}_5(6)][1 - E_5] = 0.983.$$

As a check:

$$R_{\text{set}} \ (12 \text{ hr}) = \hat{R}_1 \hat{R}_2 \hat{R}_3 \hat{R}_4 [\hat{R}_5 \ (12)]$$

$$= 0.8998.$$

6.4 Other Considerations

The allocation procedures described herein are limited in certain respects, such as the influence of maintenance, costs, weight, space, redundancy, and development time. Although several procedures are available for handling these factors,† they are too detailed for inclusion in this text.

PROBLEMS

1. A proposed airborne analog system is to consist of three serial units. The estimated *AEG* allocation for each unit is: A = 100; B = 150; C = 50.
    (a) Determine the median, upper, and lower *MTBF* estimates for this system.
    (b) Determine the equivalent probability of survival [$R(t)$] for each *MTBF* estimate if the mission time is 8 hr.

†The following reference considers many of the factors above and provides much of the required input data: *The Allocation of System Reliability*, ARINC Research Corporation, Publication No. 152-2-274 (November, 1961).

2. Assume in Problem 1 that $R(t)$ must equal 0.9 for the 8-hr mission. (a) What is the equivalent $MTBF$? (b) Assume that a detailed reliability prediction indicates the following $MTBF$ values for each unit: A = 200 hr; B = 80 hr; C = 300 hr. What is the estimated $MTBF$ of the system? (c) Determine the required reliability allocation for each unit if the system $MTBF$ found in (a) is to be achieved.

3. In order to improve the reliability of the system outlined in Problem 1, it is decided to add another unit similar to B to be used as a standby to be switched into service in the event that B fails. Assuming 100 per cent reliability of the switching mechanism, determine the new system $MTBF$, using the data given in 2(b).

4. Assume that the essentiality of Unit A of the system outlined in Problem 1 is 0.5 and that it would normally be used for approximately 2 of the 8 hr of operation. Using the AGREE procedure, determine the reliability levels that should be allocated to each unit. Assume that the system reliability requirement is 0.9 for an 8-hr mission.

Redundancy:

Situations Not Dependent On Time

7.1 Introduction

In reliability engineering, *redundancy* can be defined as the existence of more than one means for accomplishing a given task. In general, all means must fail before there is a system failure. This chapter presents methods for predicting and evaluating some of the effects of redundancy on system reliability. Three basic steps are involved: (1) determination of the paths for successful system operation through reliability block diagrams; (2) determination of the mathematical model for individual path reliability and overall system reliability through the use of probability theory and part-failure information; (3) extension of these results to time-dependent situations.

This chapter deals with the first two of the steps above. Chapter 8 will extend the discussion on redundancy to include situations where time-to-failure densities have to be considered.

Terminology

To keep the discussion general, the following terminology is used:

Path: a physical means for accomplishing a given task. A redundant path has at least one functional duplicate; therefore, failure of a redundant path will not necessarily result in system failure.

Element: the basic system level under consideration. An element may be a component, a circuit, an equipment, or even a system. Each redundant element has at least one functional duplicate.

Unit: the next higher level of system assembly. A unit may consist of redundant elements, nonredundant elements, or both. If the elements are components, the unit comprising these elements is a circuit; if the elements are circuits, the units are equipments or subsystems, etc.

Block: a group of units that perform an integrated function. A block may be an equipment, a subsystem, or a system, depending on the basic level of redundancy.

Notation

In connection with probability statements about redundant configurations, the following notation will generally be used:

R = probability of success, or reliability, of a system or block;

$\bar{R}$ = probability of failure, or unreliability, of a system or block $(R + \bar{R} = 1.0)$;

r = probability of success, or reliability, of a unit or path;

$\bar{r}$ = probability of failure, or unreliability, of a unit or path $(r + \bar{r} = 1.0)$;

p_i = probability of success, or reliability, of element i;

q_i = probability of failure, or unreliability, of element i $(p_i + q_i = 1.0)$.

Elements of a system will usually be denoted by A, B, C, etc. For probability statements concerning element success or failure, the following notation will be used:

A = the event: success of element A;

$\bar{A}$ = the event: failure of element A;

$P(A)$ = probability that event A occurs (A operates) = p_a;

$P(\bar{A})$ = probability that event $\bar{A}$ occurs (A fails) = q_a.

As an aid in understanding this material, it is suggested that the reader review the discussion of basic probability theory given in Chapter 2.

7.2 Types of Redundancy

Redundancy can be categorized according to three classification criteria: the system level at which it is applied; the state of the redundant elements while the system is in operation (i.e., duplicate elements may both be active or one may be on a standby basis); and the existence or nonexistence of decision and switching (DS) devices.

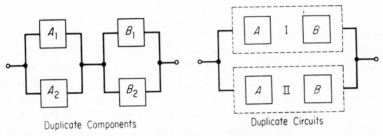

Duplicate Components   Duplicate Circuits

Fig. 7.1 Basic types of redundancy.

The basic types of redundancy may be illustrated by means of a simple example. Consider a two-component circuit within a system. The designer has the choice of duplicating components or duplicating circuits. These alternative levels are shown in Figure 7.1. If the redundant elements (say, components A_2 and B_2, or circuit II) are continuously energized and are employed in performing the circuit function, there is *active* redundancy. If the duplicate elements do not perform any function unless the primary element fails, there is *standby* redundancy. A standby element may be fully or partially energized, or completely inactive. If it is assumed that circuit II is on a standby basis, then it is necessary to have some device that will detect a failure in primary circuit I and switch the load to circuit II, as shown in Figure 7.2.

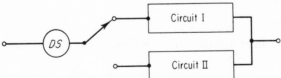

Fig. 7.2 Redundant circuits with switching device.

The four engines of a bomber are an example of active redundancy, in that failure of one or more engines will not necessarily result in a plane crash. A spare tire in the trunk of a car is a standby redundant element. If it is inflated, it can be considered to be partially energized. The decision and switching device in this system is, of course, the driver.

The great majority of redundant configurations are combinations or modifications of the basic types illustrated; i.e., most redundancies can be diagrammed as some combination of series and parallel elements, with any necessary switching devices included.*

7.3   Evaluation of Simple Redundant Circuits

Basic Assumptions

The units under consideration are those whose elements are assumed to be independent and whose operation can be described in discrete terms as either a "success" or a "failure" over a specified time interval. For the present, it will be assumed that all elements are continuously energized and that switching devices are unnecessary or failure-free; it will also be assumed that path failure of any type has no adverse effects on the operation of the surviving parallel paths—e.g., precautions such as use of a fuse have been taken to avoid unit failure through shorting if a parallel path shorts.

Evaluating the reliability of redundant configurations under the assumptions above requires no more than basic probability theory. The first step is to draw a *reliability block diagram*, showing in sequence those elements or units which must function for successful operation. From the reliability block diagram and the definition of block or system success, the various paths for successful operation can be determined. When any element ceases to function properly, the path containing that element is no longer usable. Since redundancy implies more than one path for system success, reliability is mathematically defined as the probability that at least one path exists between the block input and the block output.

Basic Configurations

Series System

A series system of n elements can be represented by the following diagram:

The reliability of this "n-element system," under the basic assumptions of independent element failures and the necessity of successful operation of all

---

*In another type of configuration, known as *voting redundancy*, three or more elements operate in conjunction with a switch which selects the elements with agreeing outputs if they constitute a majority. This type of redundancy is often used in computer applications.

elements for system success, is

$$R = P(A_1), \ldots, P(A_2), \ldots, P(A_n)$$

$$= p_1, p_2, \ldots, p_n. \tag{7.1}$$

If all elements are identical with reliability p,

$$R = p^n. \tag{7.2}$$

If all elements have a high reliability (say $p_i > 0.95$), a useful approximate formula for R is

$$R = 1 - \sum_{i=1}^{n} q_i \tag{7.3}$$

or, for identical elements each with unreliability q,

$$R = 1 - nq. \tag{7.4}$$

For example, 5 series elements each with a reliability of 0.98 give an exact system reliability of $(0.98)^5 = 0.9039$. From Equation 7.4,

$$R = 1 - 5\,(0.02) = 0.90.$$

In general, for all the following formulas which contain terms of the form x^k, and

$$0 < x < 1.0, k = 2, 3, 4, \ldots$$

the approximate value of this term for x close to 1.0 is

$$1 - k\,(1 - x). \tag{7.5}$$

Similarly, terms which have the form

$$x_1, x_2, x_3, \ldots, x_k$$

can be approximated by

$$1 - [(1-x_1) + (1-x_2) + \ldots + (1-x_k)] \tag{7.6}$$

if all or most of the x_i are close to 1.0.

Parallel System

Figure 7.3 represents the basic redundant unit of two elements in parallel. There are two possible paths for successful signal flow—$I - A_1 - O$ or $I - A_2 - O$. If the probability of A_1 operating successfully over the specified time interval is p_1, and if p_2 is the corresponding probability for element A_2, the probability of success for the unit can be found by either the additive or the multiplicative rule of probability.

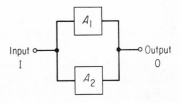

Fig. 7.3 Duplicate parallel elements.

Additive Rule: the unit is successful if at least A_1 or A_2 is operable. Since both elements are energized, the events A_1 and A_2 (the symbol A_1 now representing the event—success of element A_1) are not mutually exclusive—i.e., both A_1 and A_2 can occur. Therefore, the probability of success using the additive rule for nonmutually exclusive events is

$$R = p_1 + p_2 - p_1 p_2. \tag{7.7}$$

Multiplicative Rule: the only way the unit can fail is through failure of both elements. Since A_1 and A_2 are assumed to be independent, the probability that both elements fail is the product of their unreliabilities. One minus this probability is therefore the reliability or

$$\bar{R} = q_1 q_2$$
$$R = 1 - \bar{R} = 1 - q_1 q_2,$$

where

$$q_1 = 1 - p_1$$
$$q_2 = 1 - p_2.$$

Example. If $p_1 = p_2 = 0.90$,

by the additive rule, $R = 0.90 + 0.90 - 0.81 = 0.99$.

By the multiplicative rule,

$$\bar{R} = (0.10)(0.10) = 0.01$$
$$R = 1 - 0.01 = 0.99.$$

By simple extension of these results, the reliability of a unit with m parallel elements is

$$R = 1 - q_1 q_2 \ldots q_m; \tag{7.8}$$

or, if all elements are identical with unreliability q,

$$R = 1 - q^m. \tag{7.9}$$

Series-Parallel Configurations

If we have a series of n basic parallel units, commonly called a *series-parallel configuration*, the reliability block diagram will have the form shown in Figure 7.4.

This is the series-parallel redundant counterpart of an n-element non-redundant system. For each unit there are two possible paths for unit success; hence, there are a total of 2^n possible paths for block or system success. For unit j to be successful, either A_{1j} or A_{2j} must be operating. The probability of

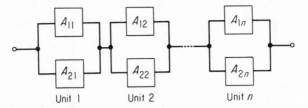

Fig. 7.4 Series-parallel configuration.

success for the unit is therefore

$$r_j = P(A_{1j} \text{ or } A_{2j})$$

$$= p_{1j} + p_{2j} - p_{1j}p_{2j} \qquad \text{(additive rule)}$$

$$= 1 - q_{1j}q_{2j}. \qquad \text{(multiplicative rule)}$$

Assuming that unit failure probabilities are independent, the system reliability is the product of unit reliabilities, or

$$R = (1-q_{11}q_{21})(1-q_{12}q_{22})\ldots(1-q_{1n}q_{2n})$$

$$= \prod_{j=1}^{n} (1-q_{1j}q_{2j}), \qquad (7.10)$$

where the symbol $\prod$ represents the product. If all elements are identical, with a reliability of p and an unreliability of $q = 1 - p$, then

$$R = (1-q^2)^n. \qquad (7.11)$$

If q is small, then

$$R \approx 1 - nq^2. \qquad (7.12)$$

To illustrate the possible reliability advantages of redundancy for this simple model, assume a basic circuit of three identical elements, each with a reliability of 0.8 over the time period of interest. Without redundancy,

$$R = (0.8)^3 = 0.512.$$

When each element is duplicated by a series-parallel arrangement,

$$R = (1-q^2)^3 = (1-0.2^2)^3 = 0.885.$$

A series of multiple parallel elements rather than two parallel elements is shown in Figure 7.5.

By extending the results given in the previous section,

$$R = \prod_{j=1}^{n} (1-q_{1j} \cdot q_{2j} \cdots q_{mj}), \qquad (7.13)$$

where q_{ij} is the failure probability of the ith element in the jth unit. If the elements in a unit are identical (e.g., $A_{11} = A_{21} = \ldots = A_{m1}$), then

$$R = \prod_{j=1}^{n} (1-q_j^m), \qquad (7.14)$$

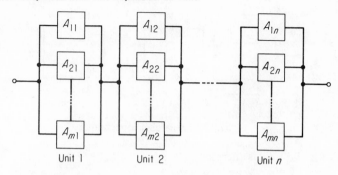

Fig. 7.5 General representation of a series of multiple parallel elements.

where q_j is the unreliability of an element in the jth unit. If all elements are identical with reliability p,

$$R = [1-(1-p)^m]^n = (1-q^m)^n, \tag{7.15}$$

and for small q,

$$R \approx 1 - nq^m. \tag{7.16}$$

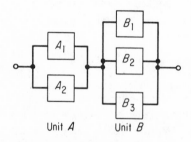

Fig. 7.6 Nonsymmetric series-parallel configuration.

The examples above represent symmetric configurations, in that each unit has the same number of elements. An example of a nonsymmetric configuration is given in Figure 7.6.

For Unit A to be successful, at least A_1 or A_2 must be operating satisfactorily. Similarly, any B-element is necessary for success of Unit B. The unit reliabilities are therefore

$$r_A = 1 - q_{a_1} q_{a_2}$$

$$r_B = 1 - q_{b_1} q_{b_2} q_{b_3}.$$

Since both units are necessary for successful operation,

$$R = r_A r_B.$$

In general, for series-parallel configurations where the elements are in parallel and the units are in series, the first step is to evaluate the reliability of each unit. The product of the unit reliabilities is equal to the system reliability.

Parallel-Series Configurations

A parallel-series configuration for an n-element basic system is shown in Figure 7.7.

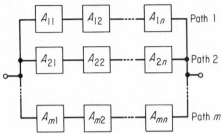

Fig. 7.7 Parallel-series configuration.

The reliability of each path is found by multiplying the reliabilities of its elements; i.e.,

$$r_i = P(A_{i1}) \, P(A_{i2}) \ldots P(A_{in}). \tag{7.17}$$

Performing this operation for each of the m paths will reduce the Figure 7.7 diagram to the one shown below:

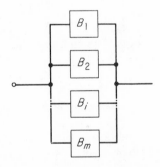

B_i now represents the series of elements $A_{i1}, A_{i2}, \ldots, A_{in}$, and the probability that B_i is successful is r_i. The system will fail only if all B's fail, or

$$R = 1 - (1-r_1)(1-r_2) \ldots (1-r_m)$$

$$= 1 - \prod_{i=1}^{m} (1-r_i). \tag{7.18}$$

If all paths are identical (i.e., $r_1 = r_2 = r_m = r$), then

$$R = 1 - (1-r)^m. \tag{7.19}$$

In terms of element reliabilities where p_{ij} is the reliability of the jth series element in the ith path,

$$R = 1 - (1-p_{11}\,p_{12} \cdots p_{1n})\,(1-p_{21}\,p_{22} \cdots p_{2n}) \cdots$$

$$(1 - p_{m1}\,p_{m2} \cdots p_{mn})$$

$$= 1 - \prod_{i=1}^{m} (1 - p_{i1}\,p_{i2} \cdots p_{in}). \tag{7.20}$$

If all paths are identical (i.e., $p_{11}=p_{21}= \ldots =p_{mi}=p_1, p_{12}=p_{22}= \ldots =p_{m2} =p_2$, etc.),

$$R = 1 - (1-p_1\,p_2 \ldots p_n)^m. \tag{7.21}$$

If all elements are identical with reliability p,

$$R = 1 - (1-p^n)^m. \tag{7.22}$$

To illustrate the application of these formulas and the reliability advantages of this type of redundancy under the basic assumptions, assume a basic three-element system with reliabilities of p_1, p_2, and p_3, as shown in the following diagram:

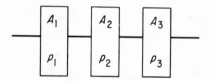

The system reliability is p_1, p_2, p_3. When an identical parallel path is added, the system reliability becomes

$$1 - (1-p_1\,p_2\,p_3)^2.$$

If $p_1 = p_2 = p_3 = p$, the ratio of reliabilities of the two competing systems is

$$\frac{R_{\text{redundant}}}{R_{\text{nonredundant}}} = 2 - p^3.$$

If $p = 0.80$, the redundant system is approximately 1.5 times more reliable than the nonredundant system; for $p = 0.90$, it is approximately 1.3 times more reliable; and for $p = 0.95$, it is approximately 1.15 times more reliable.

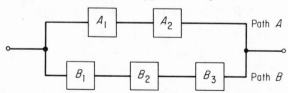

Fig. 7.8 Nonsymmetric parallel-series configuration.

For nonsymmetric configurations, the procedure is identical. This type of parallel-series redundancy is shown in Figure 7.8.

The reliability of each path is simply the product of the reliabilities of the elements in the path. Hence,

$$r_A = p_{a1} p_{a2}$$

$$r_B = p_{b1} p_{b2} p_{b3}.$$

Since the system is successful if either path is successful, the addition rule of probability is used, giving

$$R = r_A + r_B - r_A r_B$$

or

$$R = 1 - (1-r_A)(1-r_B)$$

$$= 1 - \bar{r}_A \bar{r}_B.$$

In general, for parallel-series configurations where the elements are in series and the paths are in parallel, the first step is to obtain the path reliabilities by calculating the product of the element reliabilities. One minus the product of the path unreliabilities is equal to the system reliability.

Partial Redundancy

In the previous examples, the block or system was successful if at least one of several possible paths was successful. There may be configurations where at least k out of m $(k<m)$ possible paths must be successful. These configurations exhibit partial redundancy. (If $k = m$, the system is a nonredundant or series one, since all elements are required for successful operation.) An example of partial redundancy is a four-engine plane in which at least two engines are required for mission success. Another example is the Saturn system, in which clustered rocket boosters are used.

Consider the simple case of three parallel elements (paths) shown in the diagram below:

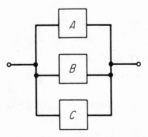

Assume that the system is successful if at least two of the three elements are successful $(m=3, k=2)$. If the three elements are different and have reliabilities of p_a, p_b, and p_c, then by straightforward application of basic probability theory involving combination of events,

$$R = p_a p_b + p_a p_c + p_b p_c - 2p_a p_b p_c.$$

(Derivation: The following events lead to system success: ABC, $\bar{A}BC$, $A\bar{B}C$, $AB\bar{C}$. Since these four events are mutually exclusive,

$$R = P(\bar{A}BC) + P(A\bar{B}C) + P(AB\bar{C})$$
$$= p_a p_b p_c + (1-p_a)\, p_b p_c + p_a(1-p_b)\, p_c + p_a p_b(1-p_c). \quad (7.23)$$

After simple algebra, this equation is equivalent to that given above).

A general method for determining the reliability function of systems with partial redundancy is now described. Let A_i represent success of the ith element and $\bar{A}_i$ represent failure of the ith element. Then $(A_i + \bar{A}_i)$ represents all possibilities for the element—namely, success or failure. For two elements, A_1 and A_2, we can generate all possible system states by forming the product

$$(A_1 + \bar{A}_1)(A_2 + \bar{A}_2) = A_1 A_2 + A_1 \bar{A}_2 + \bar{A}_1 A_2 + \bar{A}_1 \bar{A}_2. \quad (7.24)$$

Each term represents a particular combination of element states. For example, $A_1 A_2$ represents success of both A_1 and A_2; $\bar{A}_1 A_2$ represents failure of A_1 and success of A_2. The states that lead to system success can be determined from the block diagram. Thus, for a two-element series system, only state $A_1 A_2$ will lead to success; for a two-element parallel system, the first three states represent a successful mission.

System reliability is simply the sum of the probabilities of those states that lead to system success. Since independent element failures have been assumed, it is possible to make use of the fact that

$$P(A_i A_j) = P(A_i)\, P(A_j) = p_{ai}\, p_{aj}$$
$$P(\bar{A}_i A_j) = P(\bar{A}_i)\, P(A_j) = q_{ai}\, p_{aj}.$$

For n total system elements, the results above are easily extended by finding the product

$$(A_1 + \bar{A}_1)(A_2 + \bar{A}_2) \dots (A_n + \bar{A}_n)$$

and obtaining the probabilities of the states representing system success. For complex configurations, it is sometimes easier to define the terms which represent system failure and subtract the probabilities of these states from one.

The following illustrates the method used when two elements out of three are required:

$$(A + \bar{A})(B + \bar{B})(C + \bar{C}) = ABC + AB\bar{C} + A\bar{B}C + A\bar{B}\bar{C}$$
$$+ \bar{A}BC + \bar{A}B\bar{C} + \bar{A}\bar{B}C + \bar{A}\bar{B}\bar{C}. \quad (7.25)$$

All triplets which have two or more letters with no bars above them represent system outcomes with at least two successes. Hence,

$$R = P[ABC] + P[AB\bar{C}] + P[A\bar{B}C + P[\bar{A}BC]$$

$$= P(A)P(B)P(C) + P(A)P(B) - P(A)P(B)P(C) + P(A)P(C)$$

$$\qquad - P(A)P(B)P(C) + P(B)P(C) - P(C)P(B)P(C)$$

$$= P(A)P(B) + P(A)P(C) + P(B)P(C) - 2P(A)P(B)P(C)$$

$$= p_a p_b + p_a p_c + p_b p_c - 2p_a p_b p_c.$$

If $p_a = p_b = p_c$, then

$$R = 3p^2 - 2p^3.$$

This result for identical elements could have been obtained directly from the binomial distribution. If p is the constant probability of a success, the probability of at least k successes out of m trials or possibilities is

$$P(x \geq k; m) = \sum_{x=k}^{m} \frac{m!}{x!(m-x)!} p^x (1-p)^{m-x}, \qquad (7.26)$$

where $x! = x(x-1) \ldots 3 \cdot 2 \cdot 1$ and $0! = 1$.

For three parallel elements where at least two are required for system success,

$$R = P(x \geq 2; n=3) = \sum_{x=2}^{3} \frac{3!}{x!(3-x)!} p^x (1-p)^{3-x}$$

$$= \frac{3!}{2!1!} p^2(1-p) + \frac{3!}{3!0!} p^3$$

$$= 3p^2 - 2p^3.$$

If each element in the configuration above has a reliability of 0.90 for a specified time interval, then for situations in which

(a) all three elements are required,

$$R = p^3 = 0.729;$$

(b) at least two out of three elements are required,

$$R = 3p^2 - 2p^3 = 0.972;$$

(c) at least one element is required,

$$R = 1 - (1-p)^3 = 0.999.$$

7.4 Redundancy Effectiveness for Simple Models

Limitations of the Models and Measures of Effectiveness

The tables and graphs presented in this section show the improvements that can be obtained by introducing redundant elements into a system. Only the simple redundant models discussed in Section 7.3 are considered. The reader must bear in mind that these models are based on assumptions that are not always valid in actual system design. The need for switching devices, the increase in complexity and weight, the dependence between the failure probabilities of redundant paths, the possibility of shorting of parallel paths, and the like may reduce the effectiveness of the redundancy application. On the other hand, if maintenance is performed at periodic intervals and failed redundant elements are replaced, then the redundant application is more beneficial than indicated.

To simplify the presentation it will be assumed that all elements are identical. Since the purpose of this section is to show the relative efficacy of various configurations rather than the actual numerical value of reliability, this assumption is not too serious. It is reasonable to conjecture that the redundant elements are not more reliable than the basic elements. The only plausible situation in which this would not be true would be one in which the redundant elements were of such a nature that system effectiveness was reduced; for example, a slide rule might be a redundant element to a bombing/navigation computer: the slide rule is more reliable but not as effective or as accurate as the computer. Therefore, it is safe to state that, by assuming identical duplicate elements, the resulting numerical value of system reliability is a maximum.

One further point to consider is the measure used to evaluate reliability improvement for fixed time intervals. If R_1 is the reliability of System 1(S_1), and R_2 is the reliability of System 2(S_2), three apparent measures, assuming S_2 is the more reliable, are

$$\text{(a)} \quad R_2/R_1, \quad \text{(b)} \quad \bar{R}_1/\bar{R}_2, \quad \text{(c)} \quad R_2 - R_1.$$

To be more specific, assume that $R_1 = 0.50$ and $R_2 = 0.90$. Then from (a) we can say that S_2 is 1.8 times (or 80 per cent) more reliable than S_1. From (b) we can say that S_1 is 5 times or 500 per cent more unreliable than S_2. From (c) we can say that the reliability of S_2 is 0.40 greater than that of S_1.

Each of these measures has some usefulness. However, an 80 per cent increase in reliability means little if the value is still far below the requirement. For example, the value for measure (a) will be identical to that given above if $R_1 = 0.05$ and $R_2 = 0.09$. If the system were required to have a reliability of, say, 0.85 or greater, we would not be satisfied with the improvement, even though reliability would be increased by 80 per cent.

It is apparent that no one measure can be effectively used to evaluate reliability improvement. It is felt that the best approach for measuring the effectiveness of redundancy is to evaluate the dollar value of improved reliability against the cost of achieving the improvement. This approach will lead to optimum methods for improving reliability, as it will inevitably take account of such factors as redundancy, design simplification, improved components, and better quality control.

Methods for incorporating the considerations above into meaningful measures are beyond the scope of this chapter. Therefore, reliability comparisons will be made either by representing the difference $(R_2 - R_1)$ graphically or by giving the ratio of reliabilities R_2/R_1. If any two of five reliability values—R_1, R_2, R_2/R_1, $\bar{R}_1/\bar{R}_2$, $R_2 - R_1$—are known, then the other three values are easily found through simple algebra.

Basic Parallel Configurations

The reliability of a block with m parallel elements is given by the equation

$$R = 1 - (1-p)^m. \tag{7.27}$$

In Figure 7.9 are plotted the solutions to this equation for values of m ranging

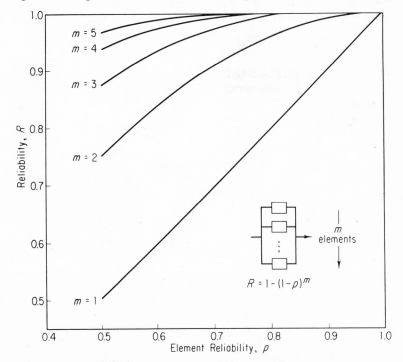

Fig. 7.9 Reliability of parallel configurations.

from 1 through 5, and values of p ranging from 0.5 to 1.0. The curve for $m = 1$ represents the nonredundant configuration. The significant point illustrated by the figure is that for any value of p, the greatest increase in reliability occurs when m is increased from 1 to 2. For $p > 0.90$, there is usually little practical advantage to having more than two parallel elements. Similarly, if $0.8 < p < 0.9$, it is questionable if more than three redundant elements are of practical utility. Again, it is cautioned that these conclusions are based only on reliabilities; such factors as criticality, cost, weight, and maintenance are not considered.

Series-Parallel Configurations (see Figure 7.5)

The reliability of a block of n series units with m parallel elements in each unit is

$$R = [1 - (1-p)^m]^n. \tag{7.28}$$

In Figure 7.10 are plotted the solutions to this equation for all combinations of: $n = 1, 2, 3, 4, 5$; $m = 1, 2, 3$; and $p = 0.5$ and 0.9. When $n = m = 1$, the configuration is nonredundant. Note that when n and p are fixed, R increases as m increases; and when m and p are fixed, R decreases as n increases. Although it is not very apparent from the figure, the rate of reliability improvement increases rather slowly for $p > 0.80$ as m increases beyond 3. Again, for most practical situations, values of m greater than 3 are usually unwarranted.

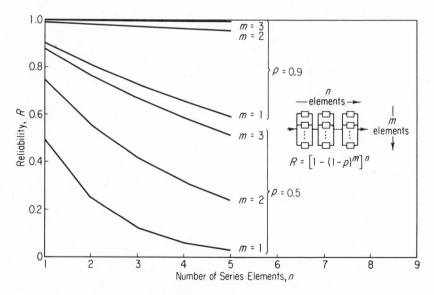

Fig. 7.10 Reliability of series-parallel configurations.

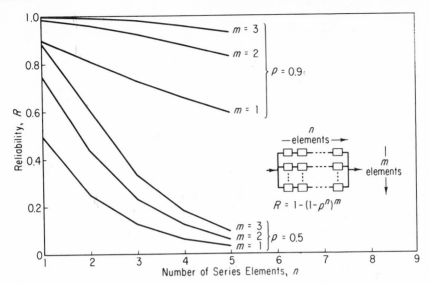

Fig. 7.11 Reliability of parallel-series configurations.

Parallel-Series Configurations (see Figure 7.6)

The reliability of a block with m parallel paths of n elements each is

$$R = 1 - (1-p^n)^m. \tag{7.29}$$

Figure 7.11 gives the solutions to this equation for all combinations of $n = 1$, 2, 3, 4, 5; $m = 1, 2, 3$; and $p = 0.5$ and 0.9. When $m = 1$, the configuration is nonredundant. When n and p are fixed, R increases as m increases; and when m and p are fixed, R decreases as n increases. Note that when $n \geq 5$ and $p \leq 0.5$, an increase in m results in a negligible increase in R.

Partial Redundancy

The reliability of a partially redundant unit of m parallel paths where at least k ($k<m$) paths must be satisfactory for system success is

$$R = \sum_{x=k}^{m} \frac{m!}{x!\,(m-x)!} p^x (1-p)^{m-x}. \tag{7.30}$$

In Figure 7.12, the solutions to this equation are plotted for all combinations of $m = 5$; $p = 0.5, 0.7, 0.9$; and $k = 1, 2, 3, 4, 5$. Obviously, for a fixed p, R decreases as k increases. In fact, R approaches the series reliability as k approaches m (this is shown by the sharp decline in R as k increases).

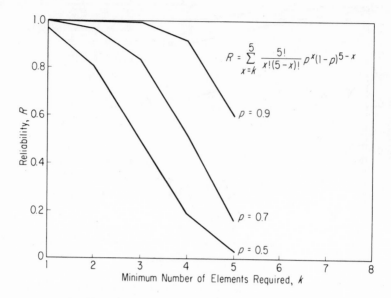

Fig. 7.12 Reliability of partially redundant configurations (five elements in parallel).

Series-Parallel Versus Parallel-Series Configurations

Comparison of Figures 7.10 and 7.11 indicates that series-parallel configurations result in higher reliability than equivalent parallel-series configurations. This is to be expected, since a basic n-element system replicated m times has m^n possible paths for system success for the series-parallel configuration; in the parallel-series configuration there are only m possible paths.

To illustrate the comparison, consider a three-element system in which each element has a reliability of 0.9. If the system is nonredundant, reliability is $0.9^3 = 0.729$. If the system is duplicated by adding one parallel path, there are three possible redundant configurations involving a total of six elements:

A. *Parallel-Series*

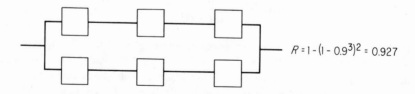

$$R = 1 - (1 - 0.9^3)^2 = 0.927$$

B. *Mixed-Parallel*

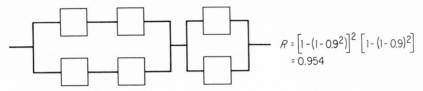

$$R = \left[1-(1-0.9^2)\right]^2 \left[1-(1-0.9)^2\right]$$
$$= 0.954$$

C. *Series-Parallel*

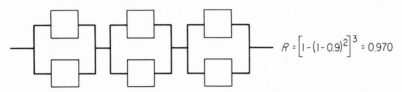

$$R = \left[1-(1-0.9)^2\right]^3 = 0.970$$

Case A can be considered to be system duplication; Case B represents unit duplication; and Case C represents element duplication. With the assumptions stated previously, it is easily shown that the lower the system level at which redundancy is introduced, the more effective it is—i.e., series-parallel redundancy is more effective than parallel-series redundancy.

Since it is usually more difficult to duplicate elements than to duplicate units or systems, care should be taken in determining the best level of redundancy. In terms of reliability alone, element duplication may be unwarranted in that it yields only a slightly higher reliability than, say, duplication of pairs of elements. Table 7.1 gives the value of the ratio R_{sp}/R_{ps} for all combinations of $n = 1, 2, 3, 4, 5$; $m = 1, 2, 3, 4, 5$; and $p = 0.5$ and 0.9. The special case of $n = m$ is plotted in Figure 7.13. Note that for $p = 0.9$, the ratio is very close to one for all values of $n = m$, while for low element reliability ($p=0.5$) the ratio increases very rapidly as $n = m$ increases. This indicates that, as p increases, the difference between series-parallel and parallel-series reliability decreases.

Table 7.1 RATIO OF SERIES-PARALLEL RELIABILITY TO PARALLEL-SERIES RELIABILITY FOR VARIOUS COMBINATIONS OF m, n, AND p

Number of Parallel Elements (m)	Number of Series Elements (n)							
	2		3		4		5	
	$p = 0.5$	$p = 0.9$	$p = 0.5$	$p = 0.9$	$p = 0.5$	$p = 0.9$	$p = 0.5$	$p = 0.9$
1	1.00	1.00	1.00	1.00	1.00	1.00	1.00	1.00
2	1.29	1.02	1.80	1.05	2.61	1.09	3.85	1.14
3	1.32	1.00	2.03	1.02	3.33	1.04	5.65	1.07
4	1.29	1.00	1.99	1.01	3.40	1.01	6.07	1.03
5	1.23	1.00	1.87	1.00	3.19	1.00	5.81	1.01

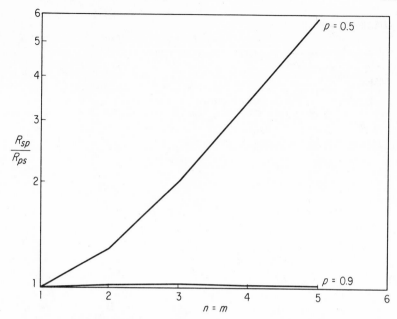

Fig. 7.13 Ratio of series-parallel reliability to parallel-series reliability when $n=m$.

7.5   Open- and Short-Circuit Failures

The previous redundant models were based on the assumption that individual element or path failure has no effect on the operation of the surviving paths. Consider a simple parallel unit composed of two elements, A and B, each of which can fail in either of two ways—open failure or short-circuit failure.* Since a short in either of the two elements will result in unit failure, the assumption that individual path failure does not result in unit failure is not always true.

Failure Probability Model

The probabilities that are necessary to describe element failure can best be illustrated by a simple example. Assume that 100 randomly selected items are tested for a prescribed time to determine the failure probabilities. The results are as follows:

100 Items Tested
   80 items experienced no failure;
   15 items experienced an open failure;
   5 items experienced a short-circuit failure.

*Diodes are a good example of elements which can fail in either mode.

Therefore, the estimated probability of success is $80/100 = 0.80$. The estimated probability of an open failure is 0.15, and the estimated probability of a short-circuit failure is 0.05. The sum of these latter two probabilities (opens and shorts are mutually exclusive events) is the probability of element failure $= 0.15 + 0.05 = 0.20$, which, of course, could have been obtained by subtracting the probability of element success from 1.

The following notation will be used to represent these probabilities:

$p =$ probability of success (0.80);

$q =$ probability of failure either by opening or by shorting (0.20);

$P(O) = q_o =$ probability of an open failure (0.05);

$P(S) = q_s =$ probability of a short-circuit failure (0.05).

Then,

$$q_o + q_s = q = 1 - p. \tag{7.31}$$

The conditional probabilities of short and open failures are sometimes used to represent element failure probabilities. The data indicate that 15 of the 20 failures that occurred were due to opens. Therefore, the conditional probability of an open failure is 15/20 or 0.75. Similarly, the conditional probability of a short-circuit failure is 5/20 or 0.25. If

$q_o' =$ conditional probability of an open $= P(O|F) = q_o/q$,

$q_s' =$ conditional probability of a short $= P(S|F) = q_s/q$,

then the following relationships hold true:

$$q_o' + q_s' = 1.0$$
$$q_o'q = q_o$$
$$q_s'q = q_s.$$

Reliability of Basic Parallel Configurations

DEFINITION OF FAILURE

For two elements in the active-parallel redundant configuration below,

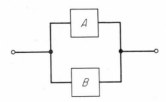

the unit will fail if

(1) either A or B shorts, or
(2) both A and B open.

The respective probabilities of these two events are

(1)
$$P_{a \text{ or } b}(S) = P_a(S) + P_b(S) - P_a(S)\,P_b(S)$$
$$= 1 - [1 - P_a(S)]\,[1 - P_b(S)]$$
$$= 1 - (1-q_{sa})\,(1-q_{sb}).$$

(2)
$$P_{ab}(O) = P_a(O)\,P_b(O)$$
$$= q_{oa}\,q_{ob},$$

where $P_i(O)$ is the probability that element i opens and $P_i(S)$ is the probability that element i shorts. Since events (1) and (2) are mutually exclusive, the probability of unit failure is the sum of the two event probabilities, or,

$$P(F) = \bar{R} = P_{a \text{ or } b}(S) + P_{ab}(O)$$
$$= 1 - (1-q_{sa})\,(1-q_{sb}) + q_{oa}\,q_{ob}. \tag{7.32}$$

In general, if there are m parallel elements,

$$\bar{R} = 1 - \prod_{i=1}^{m}(1-q_{si}) + \prod_{i=1}^{m} q_{oi}; \tag{7.33}$$

and the reliability is, of course, equal to $1 - \bar{R}$, or

$$R = \prod_{i=1}^{m}(1-q_{si}) - \prod_{i=1}^{m} q_{oi}. \tag{7.34}$$

If all elements are identical, the reliability of the unit is

$$R = (1-q_s)^m - q_o^m. \tag{7.35}$$

Note that if short failures are impossible ($q_s=0$), the reliability equations are the same as those given in the discussion of "Parallel System", page 201.

OPTIMUM REDUNDANCY LEVEL*

It is apparent that by introducing the possibility of short-circuit failures, unit reliability may be decreased by adding parallel elements. For example, if

---

*The following two papers give more detailed treatment of this subject:
H. W. Price, "Reliability of Parallel Electronic Components," *Symposium on Military Electronics Reliability and Maintainability*, Rome Air Development Center, Vol. 3 (November, 1958), 69–76.

R. E. Barlow and L. C. Hunter, *Mathematical Models for System Reliability*, Sylvania Electronic Defense Laboratory Report No. EOL-E35 (ASTIA No. 228–131) (August, 1959), 65–69.

(a) $q_s = 0, q_o = 0.10$ (c) $q_s = 0.10, q_o = 0.10$

  $m = 1: R = 0.90$ $m = 1: R = 0.80$

  $m = 2: R = 0.99$ $m = 2: R = 0.80$

  $m = 3: R = 0.999$ $m = 3: R = 0.73$

(b) $q_s = 0.05, q_o = 0.10$ (d) $q_s = 0.20, q_o = 0.10$

  $m = 1: R = 0.85$ $m = 1: R = 0.70$

  $m = 2: R = 0.89$ $m = 2: R = 0.63$

  $m = 3: R = 0.86$ $m = 3: R = 0.51.$

For cases (a) and (b), the addition of one parallel element ($m=2$) increases unit reliability. For (a), the reliability increases as m increases and, in fact, approaches 1 as m approaches infinity. However, for (b), increasing m from 2 to 3 results in decreased reliability and, in fact, the reliability continues to decrease as m gets larger. Therefore, for case (b), the optimum number of parallel elements for maximum reliability is two. For case (c), R is the same for $m = 1$ and 2 but decreases for $m > 3$. For case (d), the maximum reliability occurs for $m = 1$, the nonredundant configuration.

For any range of q_o and q_s, *the optimum number of parallel elements is one if $q_s > q_o$.* For most practical values of q_o and q_s, the optimum number turns out to be two. In general, for a given q_s and q_o, the reliability as a function of m would have the form shown below.

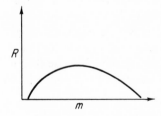

Therefore, by taking the derivative of R with respect to m, we can find the optimum number of parallel elements for maximizing reliability. If m' is equal to the optimum number, the solution of the equation $\partial R/\partial m = 0$ for m will yield a unique m', which is derived in the following manner:

$$\frac{\partial R}{\partial m} = \frac{\partial[(1-q_s)^m - q_o^m]}{\partial m}.$$

As the derivative of a^x with respect to x is equal to $a^x \log_e a$,

$$\frac{\partial R}{\partial m} = (1-q_s)^m \log_e (1-q_s) - q_o^m \log_e q_o.$$

Equating to zero and solving for m gives

$$m' = \frac{\log_e \left[\dfrac{\log_e q_o}{\log_e (1-q_s)} \right]}{\log_e \left[\dfrac{1-q_s}{q_o} \right]}. \tag{7.36}$$

Figure 7.14, which is based on this equation, gives the optimum number of parallel elements for values of q_o ranging from 0.001 to 0.5 and for the ratio q_s/q_o ranging from 0.001 to 1.0 (use left-hand and bottom axes). A designer who has knowledge of the general range of element failure probabilities, and possibly of the ratio of short to open probabilities can use this chart to determine the optimum number of parallel elements. For example, if it is believed that over-all element reliability is somewhere between 0.8 and 0.9 and that opens are likely to occur about twice as often as shorts, then

$$0.1 \le q \le 0.2, \quad \text{and} \quad q_s/q_o \approx 0.5.$$

Since $q_o + q_s = q$,

$$0.1 \le \tfrac{3}{2} q_o \le 0.2,$$

or

$$0.07 \le q_o \le 0.13.$$

For each of the values of q_o between 0.07 and 0.13, the optimum number is determined at $q_s/q_o = 0.5$. If this number is the same for all or nearly all possible values of q_o, then the optimum design is fairly well established. In this case, it is seen that two is the optimum number of parallel elements. If an optimum number boundary line is crossed somewhere in the interval of possible values of q_o, then it will be necessary to narrow the length of this interval, either through reappraisal of existing failure data or through tests specifically designed to yield relatively precise information.

If optimum values are desired for $q_o \le 0.001$, the following method can be used:

$$R_{(m)} = (1-q_s)^m - q_o^m.$$

Since $q_o < 0.001$, then for $m > 1, q_s < q_o < 0.001$; and therefore a very good approximate formula is

$$R_{(m)} = 1 - mq_s - q_o^m.$$

A boundary line between two optimum numbers, m' and $m' + 1$, for given values of q_o and q_s can be determined from the solution of the equation

$$R_{(m')} = R_{(m'+1)},$$

or

$$1 - m' q_s - q_o^{m'} = 1 - (m'+1) q_s - q_o^{(m'+1)}.$$

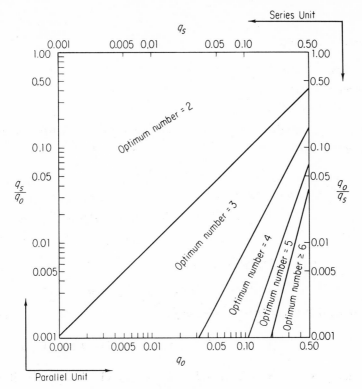

Fig. 7.14 Optimum number of elements for parallel or series units whose elements can short and open.

After some algebra,

$$q_s/q_o = q_o^{m'-1} - q_o^{m'}.$$

The value q_s/q_o is the vertical axis of Figure 7.14. Therefore, for a given value of q_s and m', the corresponding q_s/q_o can be found from the equation above. Solution of the equation for various values of m and q_o will generate enough points to extend the figure.

Reliability of Basic Series Configurations

The results given in the previous section indicate that if $q_s > q_o$, the optimum number of parallel paths is one. However, addition of an element in *series* will result in an increase in reliability if q_s is much greater than q_o. The reliability of a series system in which both short-circuit and open failures are possible is estimated below, with a two-element series unit used for illustration:

DEFINITION OF FAILURE

The unit will fail if

(1) both A and B short, or
(2) either A or B opens.

Note that the definition of failure for a two-element parallel unit is exactly the opposite. The respective probabilities of these two events are

(1) $$P_{ab}(S) = P_a(S) P_b(S) = q_{sa} q_{sb}.$$

(2) $$P_a(O) \text{ or } P_b(O) = P_a(O) + P_b(O) - P_a(O) P_b(O)$$
$$= 1 - [1 - P_a(O)] [1 - P_b(O)]$$
$$= 1 - (1 - q_{oa}) (1 - q_{ob}).$$

Since events (1) and (2) are mutually exclusive,

$$P(F) = \bar{R} = 1 - (1 - q_{oa}) (1 - q_{ob}) + q_{sa} q_{sb}.$$

In general, if there are n series elements,

$$\bar{R} = 1 - \prod_{i=1}^{n} (1 - q_{oi}) + \prod_{i=1}^{n} q_{si} \text{ and } R = \prod_{i=1}^{n} (1 - q_{oi}) - \prod_{i=1}^{n} q_{si}.$$

If all elements are identical, the reliability of an n-element series unit is

$$R = (1 - q_o)^n - q_s^n.$$

Note that n replaces m in the equation for a parallel unit and the positions of q_o and q_s are reversed. If $q_s = 0$, the equations for R reduce to those given in Section 7.3, pages 200-202.

OPTIMUM REDUNDANCY LEVEL

Assume that $q_s = 0.20$ and $q_o = 0.10$. Section 7.5 on page 218 indicates that the optimum number of parallel elements is one, yielding a reliability of $1 - 0.20 - 0.10 = 0.70$. If another identical series element is added, then

$$R = (1 - 0.10)^2 - 0.20^2 = 0.77.$$

If a third series element is added,

$$R = (1 - 0.10)^3 - 0.20^3 = 0.73.$$

Since the plot of reliability against values of n for a fixed q_o and q_s is concave downward with a single maximum, it is seen that $n = 2$ is the optimum choice of this set of q_o and q_s.

Using the same approach as that for the parallel configuration case, it is easily shown that the optimum number of series elements for a given q_o and q_s is

$$n' = \frac{\log_e\left[\frac{\log_e q_s}{\log_e(1-q_o)}\right]}{\log_e\left[\frac{1-q_o}{q_s}\right]}. \tag{7.37}$$

Figure 7.14 can also be used to determine the optimum number of series elements by using the upper and right-hand axes. As in the parallel system, if $q_o > q_s$, the optimum number of series elements is one.

The method for obtaining the boundary line for $[n', (n'+1)]$ for values of $q_s < 0.001$ is analogous to the method described for parallel configurations for $q_o < 0.001$. The equation

$$\frac{q_o}{q_s} = q_s^{n'-1} - q_s^{n'}$$

is used.

Series-Parallel and Parallel-Series Configurations

SERIES-PARALLEL

Consider four elements arranged as follows:

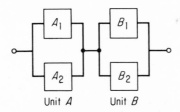

Unit A Unit B

Each element performs the same function. Success is defined as an output from at least one element. Therefore, the block is successful if

(1) both units have less than two opens, and
(2) at least one unit has no shorts.

Let

$\alpha =$ probability that at least one unit has two opens

$=$ one minus the probability that both units have at least one "no open"

$= 1 - [1 - P_{a_1}(O)P_{a_2}(O)][1 - P_{b_1}(O)P_{b_2}(O)].$

β = probability that at least one element in each unit shorts

$$= \{1 - [1 - P_{a_1}(S)(1 - P_{a_2}(S)]\}\{1 - [1 - P_{b_1}(S)][1 - P_{b_2}(S)]\}.$$

Then,

probability of block failure $= \alpha + \beta$;

reliability of block $= 1 - \alpha - \beta$.

Since

$$P_i(O) = q_{oi} \quad \text{and} \quad P_i(S) = q_{si},$$

$$R_{sp} = [1 - q_{oa_1} q_{oa_2}][1 - q_{ob_1} q_{ob_2}] - [1 - (1 - q_{sa_1})(1 - q_{sa_2})]$$
$$\times [1 - (1 - q_{sb_1})(1 - q_{sb_2})].$$

When $A_1 = B_1$ and $A_2 = B_2$ (identical units) and all components perform the same function,

$$R_{sp} = [1 - q_{oa} q_{ob}]^2 - [1 - (1 - q_{sa})(1 - q_{sb})]^2.$$

For n identical units each containing m elements,

$$R_{sp} = [1 - \prod_{i=1}^{m} q_{oi}]^n - [1 - \prod_{i=1}^{m}(1 - q_{si})]^n; \tag{7.38}$$

and if all elements are identical,

$$R_{sp} = [1 - q_o^m]^n - [1 - (1 - q_s)^m]^n. \tag{7.39}$$

If q_s and q_o are small,

$$R_{sp} \approx 1 - n q_o^m - (m q_s)^n. \tag{7.40}$$

PARALLEL-SERIES

Consider four elements arranged as follows:

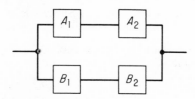

Each element performs the same function. Success is defined as an output from at least one element. Therefore, the block is successful if,

(1) at least one path has no opens, and

(2) both paths have less than two shorts.

Let

γ = probability that at least one element in each path has opened

$$= \{1-[1-P_{a_1}(O)]\,[1-P_{a_2}(O)]\}\{1-[1-P_{b_1}(O)]\,[1-P_{b_2}(O)]\}$$

δ = probability that at least one path has two shorts

= one minus the probability that both paths have at least one "no short"

$$= 1 - [1-P_{a_1}(S)\,P_{a_2}(S)]\,[1-P_{b_1}(S)\,P_{b_2}(S)].$$

Then,

probability of block failure $= \gamma + \delta$.
reliability of block $= 1 - \gamma - \delta$.

Since

$$P_i(O) = q_{oi} \quad \text{and} \quad P_i(S) = q_{si},$$

$$R_{ps} = [1-q_{sa_1}\,q_{sa_2}]\,[1-q_{sb_1}\,q_{sb_2}] - [1-(1-q_{oa_1})\,(1-q_{oa_2})]$$
$$\times\,[1-(1-q_{ob_1})\,(1-q_{ob_2})].$$

If the paths are identical ($A_1=B_1$, $A_2=B_2$) and all components perform the same function,

$$R_{ps} = [1-q_{sa}\,q_{sb}]^2 - [1-(1-q_{oa})\,(1-q_{ob})]^2. \tag{7.41}$$

For m identical paths each containing n elements,

$$R_{ps} = [1- \prod_{i=1}^{n} q_{si}]^m - [1- \prod_{i=1}^{m} (1-q_{oi})]^m. \tag{7.42}$$

If all elements are identical, the reliability of an $n \cdot m$ parallel-series unit is

$$R_{ps} = [1-q_s^n]^m - [1-(1-q_o)^n]^m. \tag{7.43}$$

The approximate formula for small q_o and q_s is

$$R_{ps} = 1 - mq_s^n - (nq_o)^m. \tag{7.44}$$

OPTIMUM REDUNDANCY CONFIGURATION

The rewriting of equations for series-parallel and parallel-series configurations of an $n \cdot m$ unit with identical elements gives

$$R_{sp}(n \cdot m) = [1-q_o^m]^n - [1-(1-q_s)^m]^n$$

$$R_{ps}(n \cdot m) = [1-q_s^n]^m - [1-(1-q_o)^n]^m.$$

To find the optimum configuration for either type of configuration, set the partials of R with respect to n and m equal to zero, and then solve simultaneously for n' and m'. This can be done for both the series-parallel and the parallel-series configurations. Call the optimum combinations $(n'_{sp} \cdot m'_{sp})$ and $(n'_{ps} \cdot m'_{ps})$, respectively. Then, the optimum configuration is the one which represents the maximum of

$$[R_{sp}(n'_{sp} \cdot m'_{sp}), \; R_{ps}(n'_{ps} \cdot m'_{ps})].$$

Note that if $q_o = q_s$, $R_{sp}(m \cdot n) = R_{ps}(n \cdot m)$; therefore,

(a) $\qquad\qquad R_{sp}(n'_{sp} \cdot m'_{sp}) \geq R_{sp}(m'_{ps} \cdot n'_{ps}) = R_{ps}(n'_{ps} \cdot m'_{ps})$

(b) $\qquad\qquad R_{ps}(n'_{ps} \cdot m'_{ps}) \geq R_{ps}(m'_{ps} \cdot n'_{ps}) = R_{sp}(n'_{sp} \cdot m'_{sp}).$

The only way both of the equations above can hold is for

$$R_{sp}(n'_{sp} \cdot m'_{sp}) \text{ to equal } R_{ps}(n'_{ps} \cdot m'_{ps}).$$

Hence, if $q_o = q_s$, maximum reliability can be obtained with either a series-parallel or a parallel-series configuration.

There are, of course, numerous other types of redundancy networks (e.g., bridge networks) which are essentially combinations of series-parallel and parallel-series networks. These types of redundant configurations are not discussed because of space limitations. Two good references on the reliability evaluation and optimum design of such redundancies are

Samuel E. Estes, *Methods of Determining Effects of Component Redundancy on Reliability*, Report No. 7849-R-3, Massachusetts Institute of Technology (ASTIA No. 205965) (August, 1958).

C. R. Gates, *The Reliability of Redundant Systems*, Memo 20-76, Jet Propulsion Laboratories, California Institute of Technology (August, 1952).

7.6 Redundancy Involving Switching

Up to this point, it has been assumed that devices for detecting element failure and switching-in redundant elements are either unnecessary or failure-free. Time-dependent situations must be discussed before switching reliability can be treated fully; however, a simple example will prove fruitful in illustrating the effects of switching failure on redundancy applications. The following types of switching failure will be considered:

(a) Dynamic failure—failure to switch when required

(b) Static failure—inadvertent or premature switching

(c) Contact failure—inability of the switch to maintain a good connection.

Dynamic switching failure always causes system failure. Static failure causes system failure (a) if the duplicate element has failed (assuming switching is only in one direction); or (b) assuming that a system failure indicator is present, if this indicator is energized by the switch although the duplicate element is operational. Contact failure is always assumed to cause system failure.

It is important to note that in this model, a "switch" includes both the decision *and* the switching device—e.g., a failure to switch may result from the inability of the decision device to detect a failed element or from the inability of the switching device to obey a correct decision and switch to the duplicate element.

It is assumed that compensating errors cannot occur—e.g., if the decision device does not detect a failure, the switching device does not operate inadvertently in a fashion resulting in correct operation. The probabilities associated with each of the three types of switching failure are conditional; since a dynamic failure can occur only if the energized element fails, a static failure can occur only if the energized element is satisfactory, and a good connection has to exist before a contact failure can occur.

Consider a two-path parallel system which requires a decision and switching device and which remains latched to B once B is energized by the switch.

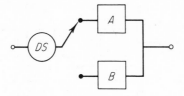

Three possible states that may lead to system success are

State 1: A and B are successful (AB).

State 2: A succeeds, B fails ($A\bar{B}$).

State 3: A fails, B succeeds ($\bar{A}B$).

State 1 requires no contact failure (dynamic failure can occur only if A fails and a static failure in this case does not result in system failure).

State 2 requires no contact failure and no static failure.

State 3 requires no contact failure and no dynamic failure (static failure cannot occur if A fails).

Let

p_i = element reliability ($i = a, b; q_i = 1 - p_i$);

p_d = conditional dynamic reliability (switching when required);

p_t = conditional static reliability (no switching when not required);

p_c = contact reliability.

Then the probability of success is

$$R = \underset{\text{State 1}}{p_a\,p_b\,p_c} + \underset{\text{State 2}}{p_a\,q_b\,p_c\,p_t} + \underset{\text{State 3}}{q_a\,p_b\,p_c\,p_d}$$

$$= p_c[p_a\,p_b + p_a\,q_b\,p_t + q_a\,p_b\,p_d].$$

For simplicity, assume that $p_a = p_b = p; p_d = p_t = p'$; then,

$$R = p_c[p^2 + 2pqp'].$$

For failure-free switching $(p_c = p' = 1.0)$, $R = p^2 + 2pq$. This is greater than nonredundant reliability, p, and the redundancy application therefore increases reliability. However, if p_c and p' are less than 1.0, the relationships between $p, p_c,$ and p' become important for determining if redundancy is beneficial and at what level it should be introduced.

As an example, assume a four-element system in which each element at time T has a reliability of 0.8 $(p=0.8)$; therefore, the system reliability, calculated by using the product rule, is $(0.8^4) = 0.41$. Also assume that $p_c = 0.9$ and $p_d = p_t = p' = 0.95$. If the system is duplicated as shown below,

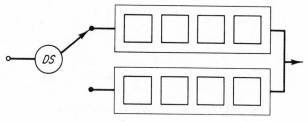

then

$$R = 0.9[(0.41)^2 + 2(0.41)\,(0.59)\,(0.95)] = 0.56.$$

If the system is split into halves and each half is duplicated,

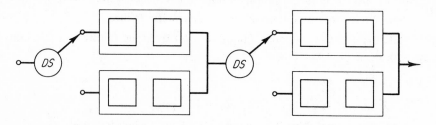

then

$$R = 0.9^2\,[(0.64)^2 + 2(0.64)\,(0.36)\,(0.95)]^2 = 0.58.$$

Duplication of each element

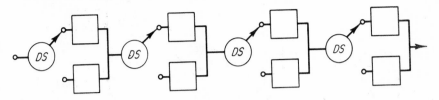

yields
$$R = 0.9^4 [(0.8)^2 + 2(0.8)(0.2)(0.95)]^4 = 0.52.$$

For these fictitious probabilities, the lowest level of redundancy is not the best approach either in terms of reliability or in terms of complexity. It has been shown that the optimum level of redundancy is a function of the non-redundant system reliability and the switching reliabilities.* Replication of more elements than this optimum number will not result in maximum reliability improvement.

Series-Parallel Arrangements

Figure 7.15 gives reliability block diagrams of a system with a series-parallel arrangement of units with decision and switching devices. The first step in evaluating the reliability of this system is to predict the probability of successful outcome for each redundant path $(A, B, \ldots, G)$ within each unit. Assuming that all elements are independent, this is accomplished by using the multiplicative rule; for example, the reliability of path A in unit 1 is $r_A = P(A_1) P(A_2) P(A_3)$. If this is done for all paths, the system block diagram reduces to the form shown in the second part of Figure 7.15. By the method described in Section 7.6, the reliability of each of the three units may be evaluated. This will simplify the diagram to the form shown in the bottom part of the figure.

The combination of DS_2 and units 2 and 3 can again be evaluated by the method described in Section 7.6. As unit 1 is in series with this combination, the system reliability is the product of the reliability of unit 1 and that of the combination of DS_2, unit 2, and unit 3.

Further Consideration of Switching Reliability

The switching model used in the previous section is a simplified representation of an actual system configuration. A more sophisticated configuration is shown in Figure 7.16.

*For example, see a paper "Reliability Improvement Through Redundancy at Various Levels", by B. J. Flehinger, *IRE National Convention Record*, Part 6 (1958).

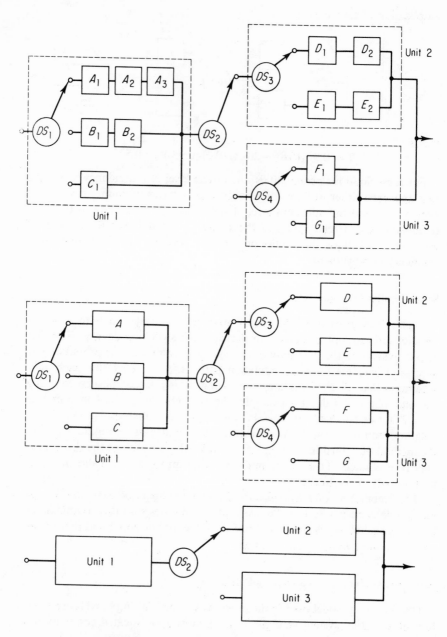

Fig. 7.15 Reliability block diagram for three-unit series-parallel system with *DS* devices.

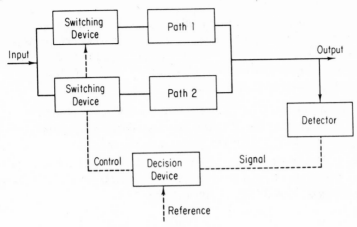

Fig. 7.16 More complex system with *DS* devices.

In this diagram, the blocks labelled "switching device" refer to groups of switching mechanisms—specifically, to all such mechanisms as are necessary to perform switching from path 1 to path 2. Therefore, if the switches are actually relays, there are multiple contacts involved, since the input, power, and output signals have to be transferred. The detector measures the output of the energized paths; the decision device, using some reference, determines if a failure has occurred within the energized path and, if so, energizes the switching devices. It is assumed that if any of the contacts of an energized path experience a contact failure, correct operation requires a detection of this event and appropriate response by the decision device.

For reliability purposes, the decision and detection devices can be lumped into one unit provided that the probability of compensating errors is assumed to be negligible (e.g., the detector fails to detect an error but the decision device operates inadvertently).

More detailed discussion of this model is given in Chapter 8. However, it is worthwhile to point out here that the switching devices used for redundant switching can, themselves, be made redundant. It is shown in Chapter 8 that if the failure probabilities of the path, the switching device, and the detector/decision device are equal, then the reliability of a system with two parallel paths is less than that of a nonredundant system which does not require switching. In fact, more than three parallel paths are required to achieve higher reliability than that of the nonredundant system. Rather than to add a fourth or fifth parallel path, it may be more economical in terms of space, weight, and cost to increase the reliability of the switching function by adding redundant switches.

Space limitations prohibit discussion of methods for devising optimum redundant configurations and predicting their reliability. In general, the methods are very similar to those used for elements which can open or short—an open corresponding to a failure of the contact to switch paths (dynamic failure), and a short corresponding to the failure of the contact to remain latched to a satisfactory path (static failure). The following two references give excellent treatments of this subject:

E. F. Moore and C. E. Shannon, "Reliable Circuits Using Less Reliable Relays," *Journal of the Franklin Institute*, Part I (September, 1956), pp. 191-208, and Part II (October, 1956), pp. 281-297.

W. E. Dickinson and R. M. Walker, "Reliability Improvement by the Use of Multiple-Element Switching Circuits," *IBM Journal of Research and Development*, Vol. 2, No. 2 (April, 1958).

PROBLEMS

1. The reliability block diagram of a two-element system with the associated element reliabilities is shown below.

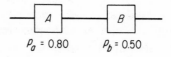

$$p_a = 0.80 \qquad p_b = 0.50$$

Assume independent element operation and equivalent cost, weight, and space factors for the two elements. In order to increase system reliability, two additional elements can be added to form a redundant design. Describe the reliability characteristics of various arrangements under the assumption of failure-free switching. Which arrangement is most desirable? If switching is not failure-free, what further considerations are necessary?

2. Derive the reliability equation in terms of element reliabilities, p_a, p_b, etc., for the following redundant design. (Assume independent element operation with no short possibilities.)

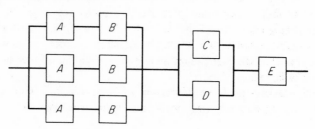

3. Which of the following three competing systems would you select if reliability were the prime consideration?

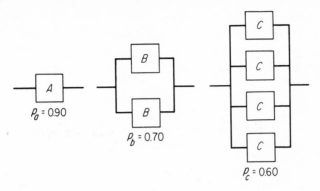

4. If the three systems in Problem 3 are equally effective, which offers the highest reliability per dollar if the costs of the elements are

$$C_a = \$1000, \; C_b = \$500, \; C_c = \$200?$$

Redundancy In Time-Dependent Situations

8.1 Introduction

The previous chapter discussed system reliability at a given point in time, and it was assumed that the reliability of each element was known for the specific time interval under consideration. This chapter considers the *functional relationship* between element reliability and time. Therefore, the evaluation of system reliability will generate a system *reliability function* that is time-dependent.

If the functional relationship between reliability and time is known, inclusion of the time factor does not introduce changes in the basic principles used to evaluate system reliability. As an example, if there is a parallel arrangement of two elements, A and B, with respective probabilities, p_a and p_b, for a specific time interval, then

$$R = p_a + p_b - p_a p_b.$$

In determining the system reliability function for any time period, the element reliabilities are expressed as functions of time; hence,

$$R(t) = p_a(t) + p_b(t) - p_a(t)p_b(t) = 1 - q_a(t)q_b(t),$$

where $R(t)$ is system reliability at time t, and $p_i(t) = 1 - q_i(t)$ is the reliability of the ith element at time t.

The following notation will generally be used in the ensuing discussion:

$f(t)$ = the time-to-failure density function.

$F(t)$ = the system or block cumulative time-to-failure density function, which is the unreliability at time t. That is,

$$F(t) = \int_0^t f(t)dt.$$

$R(t)$ = the system or block reliability function or the reliability at time t:

$$R(t) = 1 - F(t) = \int_t^\infty f(t)dt.$$

$\lambda(t)$ = the instantaneous failure rate (hazard rate) at time t.

θ = mean life, or the average time to failure for nonrepairable items, or the average time between failures for repairable items.

The notation for element, path, and unit reliabilities is the same as that for the non-time-dependent situations, except that the time dependency is indicated —e.g., r becomes $r(t)$.

8.2 Estimation of Mean Life

The mean life represents the average failure time. If $f_s(t)$ represents the time-to-failure density of a system (with or without redundant elements), then the system mean life can be calculated from

$$\theta_s = \int_0^\infty t f_s(t)dt.$$

Often it is easier to obtain the system reliability function, $R_s(t)$, directly from the element reliability functions rather than to obtain $f_s(t)$ from the element time-to-failure densities. If this is so, then the equation below is the easiest to use for estimating θ_s:

$$\theta_s = \int_0^\infty R_s(t)dt.$$

This equation holds for the usual condition in which the hazard rate, $\lambda(t)$, does not approach zero as t gets infinitely large. The derivation will be shown, since it illustrates some of the relationships between $R(t)$, $f(t)$, and $\lambda(t)$.

$$\lambda(t) = \lim_{h \to 0} \frac{R(t) - R(t+h)}{hR(t)}.$$

By l'Hospital's rule, differentiate both numerator and denominator with respect to h:

$$\lambda(t) = \lim_{h \to 0} \frac{f(t+h)}{R(t)}$$

$$= \frac{f(t)}{R(t)} \quad \text{since} \quad \frac{dR(x)}{dx} = -f(x)$$

Hence,

$$f(t) = \lambda(t)\, R(t) = -\frac{dR(t)}{dt}$$

$$\theta = \int_0^\infty tf(t)dt = \int_0^\infty t\left(-\frac{dR(t)}{dt}\right)dt.$$

Let

$$u = t \quad dv = -\frac{dR(t)}{dt}\, dt \quad du = dt \quad v = -R(t).$$

Then, integration by parts gives

$$\theta = -tR(t)\Big|_0^\infty + \int_0^\infty R(t)dt.$$

For the first term on the right, the integral is obviously equal to 0 at $t = 0$, since $0 \leq R(t) \leq 1$ for all t. For the upper limit,

$$\lim_{t \to \infty}\left[-\frac{t}{R(t)^{-1}}\right] = \lim_{t \to \infty}\left[-\frac{1}{R(t)^{-2}\lambda(t)R(t)}\right]$$

$$= \lim_{t \to \infty}\left[-\frac{R(t)}{\lambda(t)}\right] = 0,$$

provided that $\lambda(t)$ does not approach 0 as t approaches infinity. Hence, the first term on the right is equal to zero and, therefore,

$$\theta = \int_0^\infty R(t)dt.$$

8.3 Active-Parallel Redundancy

Basic Configuration

From m active-parallel elements in which failure probabilities are independent and failure of one parallel path has no effect on the operation of surviving paths, reliability—in situations that are not time-dependent—is

$$R = 1 - q_1 q_2 \ldots q_m;$$

or

$$R = 1 - q^m$$

if all elements are identical. In the time-dependent case, where

$$q_i(t) = 1 - p_i(t) = 1 - \int_t^\infty f(t)dt = \int_0^\infty f(t)dt,$$

reliability is

$$R(t) = 1 - q_1(t)\,q_2(t)\ldots q_m(t)$$

or

$$R(t) = 1 - [q(t)]^m$$

if $[q_1(t) = q_2(t) = \ldots = q_m(t) = q(t)]$. The mean life of the system θ_s is

$$\theta_s = \int_0^\infty R(t)dt.$$

THE EXPONENTIAL DISTRIBUTION

The time-to-failure density of the exponential distribution is

$$f(t) = \lambda e^{-\lambda t},$$

where λ is the constant failure (hazard) rate, or

$$f(t) = 1/\theta \; e^{-t/\theta},$$

where θ is the mean life and is equal to $1/\lambda$. The reliability equation for an element with an exponential density function is

$$p(t) = \int_t^\infty \lambda e^{-\lambda t} \, dt = e^{-\lambda t}.$$

If two elements in parallel have constant failure rates, λ_a and λ_b, then

$$R(t) = 1 - q_a(t)\,q_b(t) = 1 - (1 - e^{-\lambda_a t})\,(1 - e^{-\lambda_b t})$$
$$= e^{-\lambda_a t} + e^{-\lambda_b t} - e^{-(\lambda_a + \lambda_b)t}.$$

If $\lambda_a = \lambda_b = \lambda$ (identical elements),

$$R(t) = 1 - [q(t)]^2 = 1 - [1 - e^{-\lambda t}]^2 = 2e^{-\lambda t} - e^{-2\lambda t}.$$

The constant failure rates of the elements in the redundant configuration above cannot be combined in the usual manner (addition) to obtain the system failure rate, since the system failure rate is not constant but increases with time. This is true since the number of paths for successful operation decreases because of individual path failure. However, system mean life can be found by integrating the reliability function over the range of t from 0 to ∞. Therefore,

$$\theta_s = \int_0^\infty R(t)dt$$

$$\theta_s = \begin{cases} \dfrac{1}{\lambda_a} + \dfrac{1}{\lambda_b} - \dfrac{1}{\lambda_a + \lambda_b} & \lambda_a \neq \lambda_b \\[3mm] \dfrac{3}{2\lambda} & \lambda_a = \lambda_b = \lambda. \end{cases}$$

In general, for m active-parallel elements,

$$R(t) = 1 - \prod_{i=1}^{m} [1 - e^{-\lambda_i t}]$$

$$\theta_s = \sum_{i=1}^{m} \frac{1}{\lambda_i} - \sum_{\substack{i,j=1 \\ i<j}}^{m} \frac{1}{\lambda_i + \lambda_j} + \sum_{\substack{i,j,k=1 \\ i<j<k}}^{m} \frac{1}{\lambda_i + \lambda_j + \lambda_k} - \text{etc.}$$

If all elements are identical with constant failure rate λ,

$$R(t) = 1 - [1 - e^{-\lambda t}]^m$$

$$\theta_s = \frac{1}{\lambda} \sum_{i=1}^{m} \frac{1}{i}.$$

Example. An element has a constant failure rate of 100 failures per million hr (100×10^{-6}) or a mean life of 10,000 hr. If an identical element is to be added in an active-parallel configuration, obtain (a) the reliability function, (b) the reliability at $t = 1,000$ hr, and (c) the mean life.

(a) *Reliability Function*

$$R(t) = 2e^{-\lambda t} - e^{-2\lambda t} = 2e^{-0.0001t} - e^{-0.0002t}.$$

(b) *Reliability at 1000 hr*

$$R(1000) = 2e^{-(0.0001)\ (1000)} - e^{-(0.0002)\ (1000)}$$

$$= 2e^{-0.1} - e^{-0.2} = 0.991.$$

(c) *Mean Life*

$$\theta_s = \int_0^\infty (2e^{-0.0001t} - e^{-0.0002t})\, dt$$

$$= \frac{3}{2(0.0001)} = 15,000 \text{ hr.}$$

RELIABILITY IMPROVEMENT

In Figure 8.1 are plotted the reliability functions for a system with m parallel elements ($m = 1, 2, 3, 4, 5$) where the time axis is measured in terms of (λt) units. For any given time interval, the numerical difference in reliability between two configurations can easily be estimated from the graph. It is seen that this difference is small for times close to 0 and for long time intervals, since the reliability functions start at 1.0 and approach 0 asymptotically.

Another method of measuring reliability improvement is to calculate the ratios (or differences) in mean lives of two configurations. Table 8.1 gives the ratios of mean lives for (a) $\dfrac{\theta(m)}{\theta(1)}$, and (b) $\dfrac{\theta(m)}{\theta(m-1)}$, for $m = 1, 2, 3, 4, 5$ (based on the formula

$$\theta(m) = \theta \sum_{i=1}^{m} \frac{1}{i}$$

where θ is the mean life of one element and is equal to $1/\lambda$).

From this table it is seen that the maximum improvement in mean life occurs when $m = 2$. The reason for this is that the addition of one parallel element to a group of $(k - 1)$ parallel elements ($k > 1$) increases the mean life by $1/\lambda k = \theta/k$, which is greatest when $k = 2$.

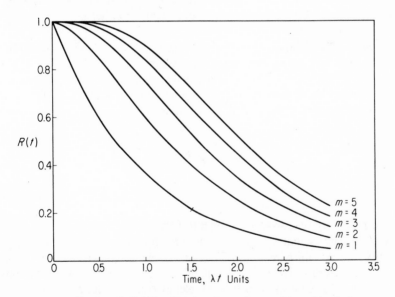

Fig. 8.1 Reliability function for systems with m identical parallel elements, each with failure rate λ.

Table 8.1 RATIOS OF MEAN LIVES
FOR m ACTIVE-PARALLEL ELEMENTS

m	$\dfrac{\theta(m)}{\theta(1)}$	$\dfrac{\theta(m)}{\theta(m-1)}$
1	1.00	—
2	1.50	1.50
3	1.83	1.22
4	2.08	1.14
5	2.28	1.10

The indicated improvements are, in most cases, the maximum that can be achieved. For example, if the parallel elements can experience short and open failures and/or if switching is necessary, the effectiveness of the redundancy application is reduced. These and other factors will be considered in later sections.

REDUNDANCY VERSUS IMPROVED ELEMENTS

A system designer may have the option of adding redundant elements or using improved elements in a nonredundant configuration to increase reliability. If more reliable elements can be developed, the design chosen will depend to a large extent on effectiveness, cost, weight, maintenance, and other related considerations. For the purpose of this discussion, it will be assumed that these limiting factors are equivalent for either of the two types of design.

If there are two standard elements in active-parallel, each with a mean life of θ, it is seen that, under the exponential assumption, the mean life of the system (θ_s) is equal to $(3/2)\theta$. To achieve the same mean life with a single improved element, an element mean life of $(3/2)\theta$ is required. In terms of standard element mean life units, $y = t/\theta$, this gives:

Redundant standard elements

$$R(y) = 2e^{-y} - e^{-2y}, \quad \theta_s = (3/2)$$

Improved element

$$R(y) = e^{-2y/3}, \quad \theta_s = (3/2).$$

The two reliability functions are shown in Figure 8.2.

The figure shows that in early life the redundant system has the greater reliability; after approximately 1.75 θ units, the improved single-element system is the more reliable. The point of intersection of the two functions will of course change if more redundant elements are added, if the degree of element improvement varies, if standby redundancy is used, etc.

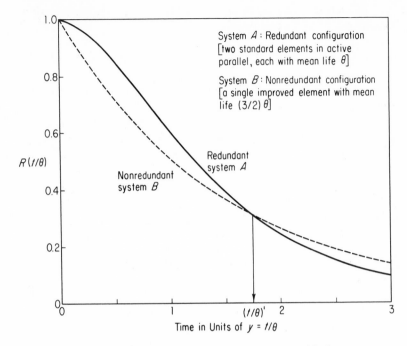

Fig. 8.2 Reliability functions for two systems with the same mean life.

In redundancy applications, there is usually one time, say t', when the reliability of a nonredundant system with improved elements is equal to the reliability of a redundant system with less reliable elements. When $t < t'$, the redundant system has the greater reliability. When $t > t'$, the improved-element system is superior. The choice of the system configuration depends on a very important parameter—the required time of operation (RTO) of the system. The RTO refers to the average operating time that will be accumulated before scheduled maintenance, complete system overhaul, or system scrapping. For nonrepairable systems such as missiles or satellites, the RTO is defined by the mission requirements. If the RTO is less than t', a system employing redundancy is more reliable than one employing improved elements. If continued satisfactory operation is required for a period greater than t', then improved elements lead to greater reliability.

THE NORMAL DISTRIBUTION

When deterioration is the dominant cause of failure and the ratio of the standard deviation to the mean life (coefficient of variation) is small, the normal distribution may appropriately describe the time-to-failure distribution of an element. The normal density function is

$$f(t) = \frac{1}{\sigma\sqrt{2\pi}} \exp\left[-\frac{(t-\mu)^2}{2\sigma^2}\right] \qquad \text{for } t, \mu, \text{ and } \sigma > 0,$$

when t = length of life, μ = mean life, and σ = standard deviation.

It will be assumed that μ and σ are known. Methods for estimating these parameters are presented in the references below.* Since the normal density is not directly integrable, reliability equations have to be shown in integral form. A normal time-to-failure density with mean μ and standard deviation σ will be denoted by

$$n(t; \mu, \sigma).$$

The reliability equation for two elements in parallel is

$$R(t) = 1 - q_a(t)\, q_b(t),$$

where $q_i(t)$ is the unreliability at time t of the ith element. Substitution of the notation for normal densities gives

$$R(t) = 1 - \int_0^t n(t; \mu_a, \sigma_a)dt \int_0^t n(t; \mu_b, \sigma_b)dt.$$

If the following notation is used,

$$z_a = \frac{t - \mu_a}{\sigma_a}, \qquad z_b = \frac{t - \mu_b}{\sigma_b},$$

then both $n(t; \mu_a, \sigma_a)$ and $n(t; \mu_b, \sigma_b)$ become normal densities with a mean of 0 and a standard deviation of 1, or

$$n(t; \mu_a, \sigma_a) \rightarrow n(z_a; 0, 1)$$

$$n(t; \mu_b, \sigma_b) \rightarrow n(z_b; 0, 1).$$

Also let

$$F(Z_a) = \int_0^{Z_a} n(z_a; 0, 1)dz_a = \int_0^{Z_a} \frac{1}{\sqrt{2\pi}} \exp\left[-\frac{(z_a)^2}{2}\right] dz_a$$

and

$$F(Z_b) = \int_0^{Z_b} n(z_b; 0, 1)dz_b = \int_0^{Z_b} \frac{1}{\sqrt{2\pi}} \exp\left[-\frac{(z_b)^2}{2}\right] dz_b$$

where Z_a and Z_b refer to values of z_a and z_b for a specific t. Then

$$R(t) = 1 - F(Z_a)\, F(Z_b).$$

*ARINC Research Corporation, *General Report No. 2* (July, 1957), 247–250.

G. R. Herd, *Estimation of Reliability Functions*, ARINC Research Monograph No. 3 (May, 1956), 13–14, 23–25.

Numerical values for

$$F = \int_0^Z \frac{1}{\sqrt{2\pi}} \exp\left[-\frac{z^2}{2}\right] dz$$

are found in a table of the cumulative normal distribution, such as Table 3.5 (p. 80).

To illustrate the procedure, it is assumed that two parallel components, A and B, have failure times that are approximately normally distributed, and that the distribution has the following parameters:

$$\mu_a = 300 \qquad \mu_b = 400$$
$$\sigma_a = 40 \qquad \sigma_b = 60.$$

In order to evaluate the reliability of this redundant unit at, say, 350 hr, the following computation is performed:

$$Z_a = \frac{350-300}{40} = 1.25, \qquad Z_b = \frac{350-400}{60} = -0.833$$

and

$$R(350) = 1 - F(1.25)\ F(-0.833) = 1 - (0.894)\ (0.202) = 0.819.$$

If the elements are identical, with mean life μ and standard deviation σ,

$$R(t) = 1 - [F(X)]^2,$$

where

$$X = \frac{t - \mu}{\sigma}.$$

Other Configurations

Reliability functions for series-parallel, parallel-series, and partially redundant configurations are easily found by extending the results given in the previous chapter to time-dependent situations. Table 8.2 (pp. 245-7) gives the reliability functions for various configurations in terms of element reliabilities, $p(t)$, and unreliabilities, $q(t)$, where

$$p(t) = \int_t^\infty f(t)dt$$

$$q(t) = 1 - p(t) = \int_0^t f(t)dt.$$

When $f_{ij}(t) = 1/\lambda_{ij}\, e^{-\lambda_{ij}t}$ (i.e., the element in the ith row and jth column of the reliability block diagram has an exponential density with constant failure rate λ_{ij}), then

$$p_{ij}(t) = e^{-\lambda_{ij}t}$$
$$q_{ij}(t) = 1 - e^{-\lambda_{ij}t}.$$

Table 8.2 also gives the equations for system mean life when all elements are assumed to have identical exponential densities with constant failure rate λ. For complex redundant configurations where elements have constant but different failure rates, evaluation of the mean life equation, $\theta_s = \int_0^\infty R(t)dt$, can be quite involved. A reasonable approximation for θ_s can often be obtained by plotting the reliability function and estimating that the mean life is between the time periods t_1 and t_2, where $R(t_1) = 0.45$ and $R(t_2) = 0.40$. As an example, assume the following redundant system:

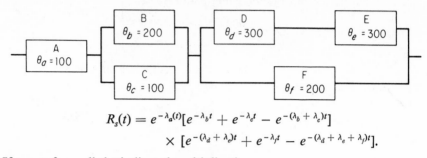

$$R_s(t) = e^{-\lambda_a(t)}[e^{-\lambda_b t} + e^{-\lambda_c t} - e^{-(\lambda_b + \lambda_c)t}]$$
$$\times [e^{-(\lambda_d + \lambda_e)t} + e^{-\lambda_f t} - e^{-(\lambda_d + \lambda_e + \lambda_f)t}].$$

If we perform all the indicated multiplications,

$$R_s(t) = \sum_{i=1}^{5} e^{-\lambda_i t} - \sum_{i=6}^{9} e^{-\lambda_i t},$$

where

$$\lambda_1 = \lambda_a + \lambda_b + \lambda_f = 0.020$$
$$\lambda_2 = \lambda_a + \lambda_b + \lambda_d + \lambda_e = 0.022$$
$$\lambda_3 = \lambda_a + \lambda_c + \lambda_d + \lambda_e = 0.027$$
$$\lambda_4 = \lambda_a + \lambda_c + \lambda_f = 0.025$$
$$\lambda_5 = \lambda_a + \lambda_b + \lambda_c + \lambda_d + \lambda_e + \lambda_f = 0.037$$
$$\lambda_6 = \lambda_a + \lambda_b + \lambda_d + \lambda_e + \lambda_f = 0.027$$
$$\lambda_7 = \lambda_a + \lambda_c + \lambda_d + \lambda_e + \lambda_f = 0.032$$
$$\lambda_8 = \lambda_a + \lambda_b + \lambda_c + \lambda_d + \lambda_e = 0.032$$
$$\lambda_9 = \lambda_a + \lambda_b + \lambda_c + \lambda_f = 0.030.$$

By integrating the reliability function, we find that

$$\theta_s = \sum_{i=1}^{5} \frac{1}{\lambda_i} - \sum_{i=6}^{9} \frac{1}{\lambda_i} = 199.5 - 132.9 = 66.6.$$

If the reliability function is plotted, we find that at $t = 58.5$, $R(t) = 0.45$, and at $t = 67$, $R(t) = 0.40$. Therefore, it would be estimated that

$$58.5 < \theta < 67,$$

which agrees with the exact value given above.

Table 8.2 Reliability Functions for Various Active-Parallel Configurations

Reliability Block Diagram	Configuration	Reliability Function, $R(t)$	Exponential Mean Life, θ
	1. Basic Series (a) General case	$p_{11}(t)\,p_{12}(t)\cdots p_{1n}(t)$	
	(b) Identical elements	$p(t)^n$	$\dfrac{1}{n\lambda}$
	2. Basic Parallel (a) General case	$1 - q_{11}(t)\,q_{21}(t)\cdots q_{m1}(t)$	
	(b) Identical elements	$1 - q(t)^m$	$\dfrac{1}{\lambda}\sum_{i=1}^{m}\dfrac{1}{i}$
	3. Series-Parallel 2×2 (a) General case	$[1 - q_{11}(t)q_{21}(t)]\,[1 - q_{12}(t)q_{22}(t)]$	
	(b) Identical elements in units	$[1 - q_1(t)^2]\,[1 - q_2(t)^2]$	
	(c) Identical elements	$[1 - q(t)^2]^2$	$\dfrac{11}{12\lambda}$

(continued)

Table 8.2 (continued) RELIABILITY FUNCTIONS FOR VARIOUS ACTIVE-PARALLEL CONFIGURATIONS

4. Series-Parallel $m \times n$

(a) General case

$$\prod_{j=1}^{n}[1 - q_{1j}(t)q_{2j}(t)\cdots q_{mj}(t)]$$

(b) Identical elements in units

$$\prod_{j=1}^{n}[1 - q_j(t)^m]$$

(c) Identical elements

$$[1 - q(t)^m]^n$$

$$\frac{1}{\lambda}\sum_{j=1}^{n}\left[(-1)^{j+1}\binom{n}{j}\sum_{i=1}^{jm}\frac{1}{i}\right]$$

$$\frac{3}{4\lambda}$$

5. Parallel-Series 2×2

(a) General case

$$1 - [1-p_{11}(t)p_{12}(t)][1-p_{21}(t)p_{22}(t)]$$

(b) Identical elements in paths

$$1 - [1-p_1(t)^2][1-p_2(t)^2]$$

(c) Identical elements

$$1 - [1-p(t)^2]^2$$

Table 8.2 (continued) RELIABILITY FUNCTIONS FOR VARIOUS ACTIVE-PARALLEL CONFIGURATIONS

6. Parallel-Series $m \times n$

Path 1

Path 2

Path m

At least k elements required

(a) General case

$$1 - \prod_{i=1}^{m} [1 - p_{i1}(t) p_{i2}(t) \cdots p_{in}(t)]$$

(b) Identical elements in paths

$$1 - [1 - p_1(t) p_2(t) \cdots p_n(t)]^m$$

(c) Identical elements

$$1 - [1 - p(t)^n]^m \qquad \frac{1}{n\lambda} \sum_{i=1}^{m} \frac{1}{i}$$

7. Partial Redundancy (require at least k satisfactory elements)

(a) Identical elements

$$\sum_{i=k}^{m} \binom{m}{i} p(t)^i [1 - p(t)]^{m-i} \qquad \frac{1}{\lambda} \sum_{i=k}^{m} \frac{1}{i}$$

Notation

$p(t) =$ element reliability function
$\quad = \int_t^\infty f(t)\,dt.$

$q(t) = 1 - p(t)$
$\quad =$ element unreliability function.

Element ij refers to the element in the ith row and jth column:

$i = 1, 2, \ldots, m; \quad j = 1, 2, \ldots, n.$

When elements have exponential failure density with failure rate λ,

$$p(t) = e^{-\lambda t}, \quad q(t) = 1 - e^{-\lambda t}.$$

Open- and Short-Circuit Failures

The following notation will be used in the ensuing discussion of basic active-parallel and series configurations containing elements which can fail by either open-circuit or short-circuit failures.*

$f_i(t)$ = time-to-failure density of the ith element

q'_{oi} = conditional probability of an open failure given that a failure has occurred†

$q'_{si} = 1 - q'_{oi}$ = conditional probability of a short-circuit failure given that a failure has occurred.

Therefore, the *probability of an open before time t is* $q'_{oi} q_i (t)$ and the *probability of a short before time t is* $q'_{si} q_i(t)$, where $q_i(t)$ is equal to the probability of failure before time t, or

$$q_i(t) = \int_0^t f_i(t)dt.$$

Following the same approach used for the non-time-dependent configurations, the reliability function for m active-parallel elements is

$$R(t) = \prod_{i=1}^m [1 - q'_{si}q_i(t)] - \prod_{i=1}^m q'_{oi}q_i(t)$$

or, if all elements are identical,

$$R(t) = [1 - q'_s q(t)]^m - [q'_o q(t)]^m.$$

For the range of t where $q'_s q(t)$ is small,

$$R(t) \approx 1 - mq'_s q(t) - [q'_o q(t)]^m.$$

A system of n *series* elements each of which can either open or short has as its reliability function

$$R(t) = \prod_{j=1}^n [1 - q'_{oj}q_j(t)] - \prod_{j=1}^n q'_{sj}q_j(t).$$

For identical elements,

$$R(t) = [1 - q'_o q(t)]^n - [q'_s q(t)]^n,$$

---

*The assumptions have been made that for a two-element parallel unit ⌐□□⌐; a short of either element will lead to failure, while a single open will not; in a two-element series system –□–□–, an open in either element will lead to system failure, while a single short will not.

†This notation implies that the conditional probabilities of open- and short-circuit failures are not time-dependent. If this is invalid, the q'_{oi} should be written as $q'_{oi}(t)$; and q'_{si} as $q'_{si}(t)$.

which is approximately equal to

$$1 - n\, q_o' q(t) - [q_s' q(t)]^n$$

for small $q_o' q(t)$.

The mean lives of the parallel or series configurations can be found by integrating the reliability function over the range from 0 to ∞. It has been shown that this operation will lead to the following equations for identical elements with constant failure rates:*

Parallel Configuration:

$$\theta_p = \frac{1}{\lambda} \sum_{i=1}^{m} \frac{q_o'^{\,m-i}}{i}$$

Series Configuration:

$$\theta_s = \frac{1}{\lambda} \sum_{j=1}^{n} \frac{q_s'^{\,n-j}}{j}.$$

When $\theta_p(m)$ is equated to $\theta_p(m-1)$, the solution for various values of q_o' will yield the critical interval that is optimal for the value of m used. Because of the symmetry between parallel and series circuits, the optimal n's are easily found by substituting the complements (q_s') for the critical intervals of q_o'. This calculation has been performed for values of m_{opt} and n_{opt} from 1 through 10. The results are shown in Table 8.3, which appeared originally in the referenced report by Barlow and Hunter.

Dependent Failure Probabilities

Up to this point, it has been assumed that failure of an active redundant element has no effect on the other active elements. This might occur, for example, in a two-parallel-element system where both elements carry the full load, although only one of the two elements is actually performing the circuit function at any given time. An example of conditional or dependent events is illustrated by the following block diagram:

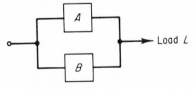

A and B are both fully energized, and normally share or carry half the load—$L/2$. If either A or B fails, the survivor must then carry the full load.

*R. E. Barlow and L. C. Hunter, *Mathematical Models for System Reliability*, Sylvania Electronic Defense Laboratory Report Number EDL-E35 (ASTIA No. 228-131) (August 11, 1959), 65–69.

Table 8.3 PARALLEL AND SERIES CIRCUIT PARAMETERS
WHICH MAXIMIZE MEAN LIFE

Optimal m for a Parallel Circuit	Optimal n for a Series Circuit	Critical Intervals for q_o'
1	more than 10	$q_o' < 0.040$
1	10	$0.040 \le q_o' < 0.046$
1	9	$0.046 \le q_o' < 0.054$
1	8	$0.054 \le q_o' < 0.063$
1	7	$0.063 \le q_o' < 0.077$
1	6	$0.077 \le q_o' < 0.096$
1	5	$0.096 \le q_o' < 0.125$
1	4	$0.125 \le q_o' < 0.175$
1	3	$0.175 \le q_o' < 0.272$
1	2	$0.272 \le q_o' < 0.500$
1	1	$q_o' = 0.5$
2	1	$0.500 < q_o' \le 0.728$
3	1	$0.728 < q_o' \le 0.825$
4	1	$0.825 < q_o' \le 0.875$
5	1	$0.875 < q_o' \le 0.904$
6	1	$0.904 < q_o' \le 0.923$
7	1	$0.923 < q_o' \le 0.937$
8	1	$0.937 < q_o' \le 0.946$
9	1	$0.946 < q_o' \le 0.954$
10	1	$0.954 < q_o' \le 0.960$
more than 10	1	$q_o' > 0.960$

Hence, the probability that one (say B) fails is dependent on the state of the other, if failure probability is related to load or stress. A simple example would be a two-engine plane which, if one engine fails, can still keep flying. However, the surviving engine now has to carry the full load and thus will have a higher probability of failing.

For this relatively simple example, a direct derivation of the reliability function can be performed by considering all possible ways for the system to be successful, as given in the following diagram.

Success = Conditions (1), (2), or (3)

(The bar above a letter—e.g., $\bar{A}$—represents failure of that element. The prime—e.g., A'—represents operation of that element under full load; absence of a prime represents operation under half load.)

To show the derivation, let

$f(t)$ represent the failure-time density under partial or half load (i.e., when both elements are operating);

$g(t)$ represent the element failure-time density under full load (i.e., when one element has failed); and

$t < t$ represent some fixed point in time.

For identical elements

$$f_a(t) = f_b(t) = f(t)$$

$$g_a(t) = g_b(t) = g(t).$$

The system is operating satisfactorily at time t if either A or B or both are operating successfully. Under the assumption that the elements are independent if both are operating, the probability that both will operate until time t is

$$\left[\int_t^\infty f(t)dt \right]^2 = [R_f(t)]^2.$$

The density for one element failing at time t_1 and the other surviving to t_1 under $L/2$ and from t_1 to t under L is

$$f(t_1) \int_{t_1}^\infty f(t)dt \int_{t-t_1}^\infty g(t)dt = f(t_1)\, R_f(t_1)\, R_g(t-t_1).$$

Since t_1 can range from 0 to t, this density is integrated over that range, and the resulting probability is doubled because the event can occur in either of two ways. Hence,

$$R(t) = [R_f(t)]^2 + 2 \int_0^t f(t_1)\, R_f(t_1)\, R_g(t-t_1)dt_1.$$

For the case of the two different elements with half-load densities, $f_a(t)$ and $f_b(t)$, and full-load densities, $g_a(t)$ and $g_b(t)$,

$$R(t) = R_{f_a}(t)R_{f_b}(t) + \int_0^t f_a(t_1)R_{f_b}(t_1)R_{g_b}(t-t_1)dt_1$$

$$+ \int_0^t f_b(t_1)R_{f_a}(t_1)R_{g_a}(t-t_1)dt_1.$$

If the elements' failure times are exponentially distributed and each has a mean life of θ under load $L/2$, and $\theta' = \theta/k$ under load $L(k > 0)$, solution of the equation above gives

$$R(t) = \frac{2\theta'}{2\theta' - \theta} e^{-t/\theta'} - \frac{\theta}{2\theta' - \theta} e^{-2t/\theta}$$

when $k \neq 2$. When $k = 2$,

$$R(t) = \left(\frac{2t + \theta}{\theta}\right) e^{-2t/\theta}.$$

The system mean life can be shown to be equal to

$$\theta_s = \frac{\theta}{k} + \frac{\theta}{2}.$$

Note that when $k = 1$, the system is one in which load-sharing is not present or increased load does not affect the element failure probability. This is exactly the assumption made for previous discussions of active-parallel redundancy, and therefore θ_s for $k = 1$ is equal to $(3/2)\theta = 3/2\lambda$. If there were only one element, it would be operating under full load; therefore the system mean life would be $\theta_s = \theta/k$.

Hence, addition of a load-sharing element increases the system mean life by $\theta/2$. This gain in mean life is equivalent to that gained when the elements are independent, but the over-all system reliability is usually less because θ' is usually less than θ ($k > 1$), and therefore

$$\theta' + \theta/2 < \theta + \theta/2 = 3/2\,\theta.$$

The use of a single improved element when this dependent model is assumed is of interest. To illustrate the effects of using improved single elements or redundant standard elements, the following configurations will be considered:

A: Single standard element; $\theta = 50$

B: Single improved element; $\theta = 100$

C: Dependent model, standard elements; θ (half load) $= 100$, θ' (full load) $= 50$.

The mean lives and reliability functions of these three configurations are

A: $\theta_A = 50$, $R_A(t) = e^{-t/50}$

B: $\theta_B = 100$, $R_B(t) = e^{-t/100}$

C: $\theta_C = 100$, $R_C(t) = e^{-t/50}(1 + t/50)$.

The reliability functions are shown in Figure 8.3. Note that although systems B and C have the same mean life, the redundant system has greater reliability in early life. After approximately 125 hr, the improved single-element system is superior. If such factors as effectiveness, cost, weight, and complexity are approximately equivalent for systems B and C, the choice would relate to the Required Time of Operation (RTO) concept discussed on page 241.

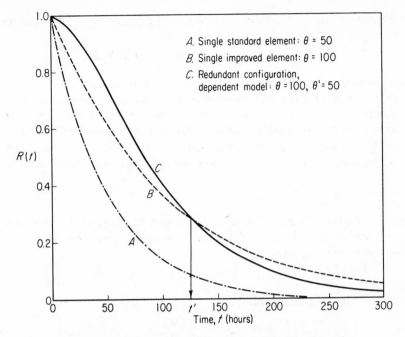

Fig. 8.3 Reliability functions for redundant configuration (dependent model) and nonredundant configurations.

8.4 Standby Redundancy

In a system with redundant elements that are completely on standby, no time is accumulated on standby elements until the primary element fails.* For simplicity, it will first be assumed that standby elements cannot fail if de-energized, and that the necessary switching devices are failure-free. If the system contains two elements, A and B, the reliability function can be found directly as follows: The system will be successful at time t if either of the following two conditions holds (letting A be the primary element):

(1) A succeeds up to time t, or

(2) A fails at time $t_1 < t$ and B operates from t_1 to t.

The following diagram represents these two conditions:

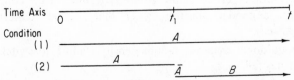

*If standby elements are partially or wholly energized, then one has a dependent model similar to that discussed in Section 8.3, pages 250-252.

Translation of the two conditions above to time-dependent probabilities gives

$$R(t) = \int_t^\infty f_a(t)dt + \int_0^t \Big[f_a(t_1) \int_{t-t_1}^\infty f_b(t)dt \Big] dt_1,$$

where $f(t)$ is the time-to-failure density function of an element.

The first term of this equation represents the probability that element A will succeed until time t. The second term, excluding the outside integral, is the density function of A failing exactly at t_1 and B succeeding for the remaining $(t-t_1)$ hr. Since t_1 can range from 0 to t, t_1 is integrated over that range.

For the exponential case where the element failure rates are λ_a and λ_b,

$$R(t) = \frac{\lambda_b}{\lambda_b - \lambda_a} e^{-\lambda_a t} - \frac{\lambda_a}{\lambda_b - \lambda_a} e^{-\lambda_b t},$$

which is a form of the mixed exponential. It can be shown that it does not matter whether the more reliable element is used as the primary or the standby element. If $\lambda_a = \lambda_b = \lambda$,

$$R(t) = e^{-\lambda t} (1 + \lambda t).$$

For n elements of equal reliability,

$$R(t) = e^{-\lambda t} \sum_{r=0}^{n-1} \frac{(\lambda t)^r}{r!}.$$

The system mean life evaluated by integrating the reliability function is

$$M = \sum_{i=1}^{n} \theta_i \qquad (\theta_1 \neq \theta_2 \neq \ldots \neq \theta_n)$$

or

$$M = n\theta \qquad (\theta_1 = \theta_2 = \ldots = \theta_n).$$

Switching Failures

The following notation will be used with reference to a two-element standby redundant unit requiring a decision-and-switching device that switches in one direction only:

$f_a(t) =$ failure density of element A,

$f_b(t) =$ failure density of element B,

$f_b'(t) =$ failure density of element B when on standby,

$f_x(t) =$ conditional contact failure density (failure of the contact to maintain a good connection, given that a good connection initially existed),

$f_y(t) =$ conditional dynamic failure density (failure to switch, given that A has failed),

$f_z(t) =$ conditional static failure density (switching when not required).

Note that $f_x(t)$, $f_y(t)$, and $f_z(t)$ refer to decision-and-switching device failures which, in some cases, may not be time-dependent. If such failures are not time-dependent, the appropriate failure density is replaced by a constant probability of failure.

When all ways that the system can be successful are considered,

$$R(t) = R_x(t) \left\{ R_a(t) R_z(t) + \int_{t_1=0}^{t} [R_a(t_1)f_z(t_1)R_z(t-t_1)R'_b(t_1)R_b(t-t_1)]dt_1 \right.$$

$$\left. + \int_{t_2=0}^{t} [f_a(t_2)R_z(t_2)R_y(t_2)R'_b(t_2)R_z(t-t_2)R_b(t-t_2)]dt_2 \right\}.$$

$R_x(t)$ is the probability that a good connection exists at least to time t. The first term inside the brackets represents the probability that A operates to t without any premature switching. The second term represents the probability that a static failure occurs at time $t_1 < t$ but B operates to t. The last term represents the probability that A fails at time $t_2 < t$ and the decision-and-switching device switches to B (no dynamic failure), which operates to t.

This equation represents a general case in that the following possibilities are included:

1. A and B can be different elements.
2. A static failure can occur if B is energized, resulting in no output or a false indication of system failure. If a static failure cannot occur when B is energized, then $R_z(t-t_1) = R_z(t-t_2) = 1.0$.
3. B can fail while on standby, and its failure density may be different from that when B is energized. If B is a "cold" rather than a "warm" or "hot" reserve, $f_b(t)$ may equal zero and therefore $R'_b(t) = 1.0$.

If we assume identical elements with constant failure rate, λ, and if $\lambda = 0$ when an element is not energized, then the equation above becomes

$$R(t) = e^{-(\lambda + \lambda_z + \lambda_x)t} \left[1 + \lambda_z t + \frac{\lambda}{\lambda_y}(1 - e^{-\lambda_y t}) \right].$$

If we assume in addition that once B is energized no further switching can occur, the equation becomes

$$R(t) = e^{-(\lambda + \lambda_x)t} \left\{ 1 + \frac{\lambda}{\lambda_z + \lambda_y}[1 - e^{-(\lambda_z + \lambda_y)t}] \right\}.$$

If we further assume that $\lambda_x = \lambda_y = \lambda_z = 0$, then, since

$$\lim_{\alpha \to 0} \frac{\lambda}{\alpha}(1 - e^{-\alpha t}) = \lambda t, \qquad (\alpha = \lambda_z + \lambda_y)$$

we have

$$R(t) = e^{-\lambda t}(1 + \lambda t), \text{ which was found on page 254.}$$

Optimum Design: General Model

If there are n redundant paths for successful operation [$(n-1)$ standby paths], and if each path requires a switching device, the reliability block diagram of a general model can be shown as follows:

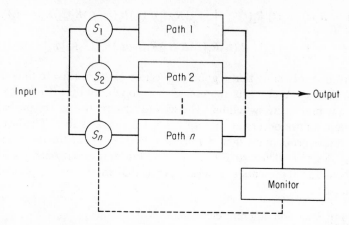

In this model, the monitor represents the failure-detection and switching-control functions. These two functions can be considered as one for reliability purposes, if it is assumed that the probability of compensating errors is negligible.

RELIABILITY ASPECTS OF THE SYSTEM

The following assumptions concerning the reliability aspects of the system are made:

1. Switching is in one direction only.
2. Standby or reserve paths cannot fail if not energized.
3. Switching devices should respond only when directed to switch by the monitor; false switching operation (static failure) is detected by the monitor as a path failure, and switching is initiated.
4. Switching devices do not fail if not energized.
5. Monitor failure includes both dynamic and static failures.

If the following constant failure rates are assumed,

λ = total failure rate of the series elements in a path,

λ_s = failure rate of the switching device (includes contact failure),

λ_m = failure rate of the monitor,

then for n total paths,

$$R_n(t) = e^{-(\lambda + \lambda_s + \lambda_m)t} \sum_{i=0}^{n-1} \frac{[(\lambda+\lambda_s)t]^i}{i!}.$$

To illustrate the reliability aspects of this model, it will be assumed that the system specification requires a high reliability for a mission of T hr. A nonredundant system would therefore have a reliability of

$$R_1(T) = e^{-\lambda T},$$

since no switching is required. The redundant system would have a reliability of

$$R_n(T) = e^{-(\lambda + \lambda_s + \lambda_m)T} \sum_{i=0}^{n-1} \frac{[(\lambda+\lambda_s)T]^i}{i!}.$$

$R_n(T)$ can be rewritten in terms of failure-rate ratios by using the fact that

$$\frac{|\log_e R_1(T)|}{\lambda} = \frac{|\log_e e^{-\lambda T}|}{\lambda} = T$$

and therefore

$$R_n(T) = \exp\left[\frac{-(\lambda+\lambda_s+\lambda_m)|\log_e R_1(T)|}{\lambda}\right] \sum_{i=0}^{n-1} \frac{\left[(\lambda+\lambda_s)\dfrac{|\log_e R_1(T)|}{\lambda}\right]^i}{i!}.$$

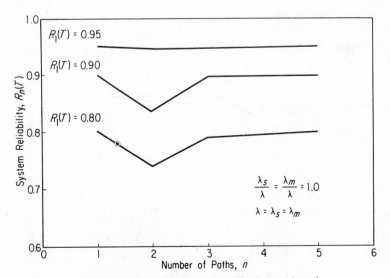

Fig. 8.4 Mission reliability for n redundant paths, when $\lambda_s/\lambda = \lambda_m/\lambda = 1.0$

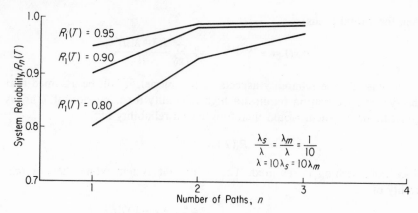

Fig. 8.5 Mission reliability for n redundant paths, when $\lambda_s/\lambda = \lambda_m/\lambda = 1/10$.

By letting $K = |\log_e R_1(T)|$, a constant, we have

$$R_n(T) = \exp\left[-(1+\lambda_s/\lambda + \lambda_m/\lambda)\,K\right]\sum_{i=0}^{n-1}\frac{[(1+\lambda_s/\lambda\,K]^i}{i!}.$$

It is easily shown that the maximum reliability at time T that can be achieved, as n approaches infinity, is $e^{-\lambda_m T}$. *Therefore, if $\lambda_m \geq \lambda$, the optimum design is that containing one element which does not require switching.*

Since the equation above is a function of the failure-rate ratios λ_s/λ and λ_m/λ, and some constant, $K = |\log_e R_1(T)|$, the mission reliability of the redundant system can be plotted as a function of the total number of redundant paths. In the figures that follow, the reliability for $n = 1$ is that of a nonredundant system $[R_1(T)]$ where no switching or monitoring is necessary. In each figure λ has been established as a reference, and λ_s and λ_m are compared to λ.

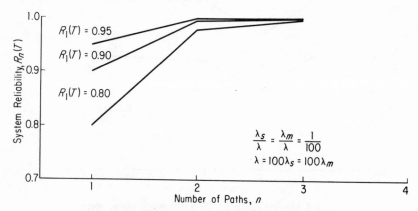

Fig. 8.6 Mission reliability for n redundant paths, when $\lambda_s/\lambda = \lambda_m/\lambda = 1/100$.

In Figures 8.4, 8.5, and 8.6, λ_s is assumed to be equal to λ_m. These three graphs show how system reliability is influenced by the number of paths if the switching device and the monitor have failure rates that are 1, 1/10, and 1/100 as great as the path failure rate.

Figure 8.7 gives the reliability of the redundant system as a function of the number of paths for various ratios of λ_m/λ when $R\,(T) = 0.80$. For illustrative purposes, λ_s/λ has been arbitrarily selected to be 1/1000 in computing the results shown in the figure. Figure 8.8 is similar except that λ_m/λ is assumed to be 1/1000 and λ_s/λ is allowed to vary.

GENERAL CONCLUSIONS

The following general conclusions can be drawn from the material developed and displayed above:

1. As the number of redundant paths increases, the mission reliability approaches the reliability of the monitor.

2. When the failure rates of the path, the switching devices, and the monitor are equal, standby redundancy with two paths results in a mission reliability considerably less than that of a single nonredundant path (Figure 8.4).

3. For systems where the switching-device and monitor failure rates are less than the path failure rate, the greatest increase in reliability occurs when one redundant path is added to a single path (Figures 8.5 and 8.6).

4. For a given path and switching-device failure rate, reliability improvement increases rapidly as the monitor failure rate decreases and the number of redundant paths increases (Figure 8.7). The same is true if the monitor failure rate is held constant and the switching-device failure rate decreases (Figure 8.8).

5. Significant improvement in mission reliability through redundancy results from the utilization of switching devices and monitors that are much more reliable than the path being switched (Figures 8.7 and 8.8).

8.5 Active-Parallel Versus Standby Redundancy

For the basic models discussed previously (independent elements, failure-free switching, and perfect reliability of de-energized elements), the reliability equations indicate that standby redundancy is preferable to active

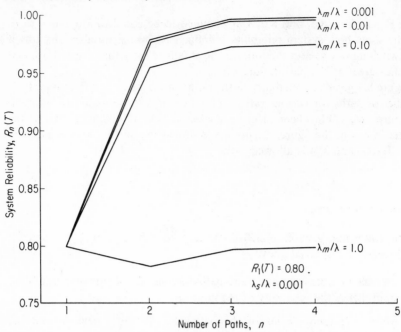

Fig. 8.7 Mission reliability for n redundant paths, when
$R_1(T) = 0.80$ and $\lambda_s/\lambda = 0.001$.

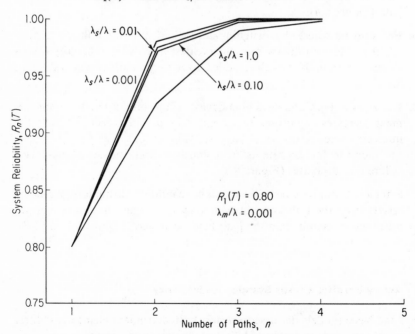

Fig. 8.8 Mission reliability for n redundant paths, when
$R_1(T) = 0.80$ and $\lambda_s/\lambda = 0.001, 0.01, 0.10,$ and 1.0.

redundancy. For example, two elements, each with an exponential mean life of θ, give a system mean life of $(3/2)\theta$ when there is active redundancy and 2θ when there is standby redundancy.

However, elements are not always independent; switching is rarely failure-free; and certain parts and components can fail without being energized—e.g., due to effects of radiation. Therefore, if there are two elements with exponential failure densities, it is most unlikely that the standby system will exhibit a mean life that is $\theta/2$ hr greater than that of an equivalent active-parallel system.

The following considerations are important in comparing the reliability aspects of the two systems:

1. For active-parallel systems, elements which can short as well as open will result in system failure; an element short in a standby system will not usually cause a system failure. If both elements are functioning, there is also the possibility of detrimental interaction between the elements, of which the short-circuit situation is a special case.

2. If load-sharing exists, the active-parallel elements have less stress imposed upon them for that period of time when more than one is operating. In standby systems, the operating element carries the full load. It can be shown that for a two-element redundant system where θ is the mean life under half load and $\theta' = \theta/k$ is the mean life under full load, if switching is failure-free, then for

$$k < 2, \text{ standby is preferable}$$

$$k = 2, \text{ there is no preference}$$

$$k > 2, \text{ active-parallel is preferable.}$$

3. Switching is always required in standby systems and is often very unreliable. Switching may also be required in active-parallel systems (e.g., when both elements are fully energized but only one is actually performing its function). However, active-parallel switching is usually less complex and therefore more reliable.

4. A system may be more effective in one of the two modes. For example, a generator is most efficient at rated load. If two generators are put in active-parallel redundancy, they have to share the load and therefore are not operating at maximum efficiency. If one is placed on standby, then the operating generator is under rated load. Note that in this example the active-parallel configuration results in load-sharing; hence, the *reliability of the system* may be greater than that of the standby configuration, but the *effectiveness of the system* may be less.

8.6 Maintenance Considerations

Periodic Maintenance*

In the previous sections, the reliability improvements achieved through redundancy were based on the assumption of unattended system operation. If periodic maintenance is performed, however, these improvements can be much greater. This section gives general reliability and mean-life equations for situations in which the following maintenance procedure is used:

Periodic maintenance is performed every T hr, starting at time 0. Every element is checked, and any one which has failed is replaced by a new and statistically identical component. If the exponential failure law is assumed for all components, the system is restored to new condition after each maintenance action, since no deterioration has taken place.

To derive the reliability function, a time period of t hr can be written as

$$t = jT + \tau \qquad j = 0, 1, 2, \ldots; \quad 0 \leq \tau < T.$$

Let $R_T(t)$ denote the reliability function of a redundant system in which maintenance is performed every T hr. For a time period such that $j = 1$ and $\tau = 0$,

$$R_T(t = T) = R(T).$$

If $j = 2$ and $\tau = 0$, the system has to operate the first T hr without failure of any redundant configuration. After replacement of all failed elements, another T hr of failure-free system operation is required; hence

$$R_T(t=2T) = [R(T)]^2.$$

If $0 < \tau < T$, then an additional τ hr of failure-free system operation are required, and

$$R_T(t=2T+\tau) = [R(T)]^2 \, R(\tau).$$

In general,

$$R_T(t=jT+\tau) = [R(T)]^j \, R(\tau) \qquad j = 0, 1, 2, \ldots; \quad 0 \leq \tau < T.$$

The mean life of a redundant system in which periodic maintenance is performed is

$$\theta_s = \int_0^\infty R_T(t)dt.$$

*See an article "Analysis of Reliability Improvement Through Redundancy," by Donald E. Rosenheim, *Proceedings of the New York University Conference on Reliability Theory* (June, 1958).

The integral over the range $0 < t < \infty$ can be expressed as the sum of integrals over intervals of T, or

$$\theta_s = \sum_{j=0}^{\infty} \int_{jT}^{(j+1)T} R_T(t)dt.$$

If $t = jT + \tau$, then $dt = d\tau$, and the limits of the integral become 0 to T. Hence,

$$\theta_s = \sum_{j=0}^{\infty} \int_0^T R_T(t)d\tau = \sum_{j=0}^{\infty} \int_0^T [R(T)]^j R(\tau)d\tau.$$

Since $1/1 - x = \sum_{j=0}^{\infty} x^j$, replacement of x with $R(T)$ gives

$$\theta_s = \frac{\int_0^T R(\tau)d\tau}{1 - R(T)}.$$

Example. To an element with an exponential mean life of 100 hr is added an identical element in an active-parallel configuration. Compare the reliability functions and mean lives for situations in which $T = 150$, 100, 50, and 10 hr, and for situations in which there is no maintenance ($T = \infty$).

(A) *Reliability Functions*

No maintenance: $(T = \infty)$

$$R(t) = 2e^{-t/100} - e^{-t/50}$$

Maintenance: $(T = jT + \tau; 0 \leq \tau < T)$

$T = 150$: $R_T(t) = [2e^{-1.5} - e^{-3}]^j [2e^{-\tau/100} - e^{-\tau/50}]$

$T = 100$: $R_T(t) = [2e^{-1} - e^{-2}]^j [2e^{-\tau/100} - e^{-\tau/50}]$

$T = 50$: $R_T(t) = [2e^{-0.5} - e^{-1}]^j [2e^{-\tau/100} - e^{-\tau/50}]$

$T = 10$: $R_T(t) = [2e^{-0.1} - e^{-0.2}]^j [2e^{-\tau/100} - e^{-\tau/50}]$.

The functions are plotted in Figure 8.9. Note that from 0 to 10 hr, all five functions are identical since $j = 0$ over this period for each of the four maintained systems.

(B) *Mean Life*

No maintenance:

$$\theta_s = \tfrac{3}{2}\theta = \tfrac{3}{2}(100) = 150$$

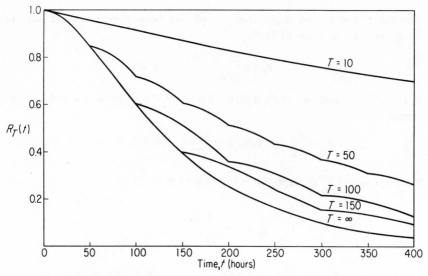

Fig. 8.9 Reliability functions for active-parallel configuration on which maintenance is performed every T hours.

Maintenance:

$$\theta_s = \frac{\int_0^T R(\tau)d\tau}{1 - R(T)} = \frac{\int_0^T [2e^{-\tau/100} - e^{-\tau/50}]d\tau}{1 - 2e^{T/100} + e^{-T/50}}$$

$$= \frac{150 + 50e^{-T/50} - 200e^{-T/100}}{1 - 2e^{-T/100} + e^{-T/50}}$$

$T = \infty : \quad \theta_s = 150$

$T = 150: \quad \theta_s = 179$

$T = 100: \quad \theta_s = 208$

$T = 50: \quad \theta_s = 304$

$T = 10: \quad \theta_s = 1097.$

Note the increases in mean life (and reliability) that can be achieved by a preventive maintenance policy. The common statement that maintenance is wasteful when all elements have constant failure rates is certainly not true if redundancy exists.

Maintenance Immediately Following Failure*

Reliability functions for some two-unit redundant designs, for which repair of a failed unit is possible, have been developed by Epstein and Hosford and are presented below.

*B. Epstein and J. Hosford, "Reliability of Some Two-Unit Redundant Systems," *Proceedings of the Sixth National Symposium on Reliability and Quality Control* (January, 1960), 469–488.

It is assumed that, at $t = 0$, all elements are in an operable condition. Repair starts immediately upon failure of a unit or element. The rate of repair (analogous to the rate of failure) is assumed to be a constant, independent of time.

Design A: Two units in active redundancy. The constant failure rate of each unit is λ and the constant repair rate is μ. For the reliability function,

$$R(t) = \frac{s_1 e^{s_2 t} - s_2 e^{s_1 t}}{s_1 - s_2}$$

where

$$s_1 = - \tfrac{1}{2}\left[3(\lambda+\mu) - \sqrt{\lambda^2 + 6\lambda\mu + \mu^2}\right]$$

$$s_2 = - \tfrac{1}{2}\left[3(\lambda+\mu) + \sqrt{\lambda^2 + 6\lambda\mu + \mu^2}\right]$$

$$\text{Mean life} = \frac{3\lambda + \mu}{2\lambda^2}.$$

Design B: Two units in standby redundancy. Constant unit failure rate $= \lambda$; constant unit repair rate $= \mu$.

$$R(t) = \frac{s_1' e^{s_2' t} - s_2' e^{s_1' t}}{s_1' - s_2'}$$

where

$$s_1' = - \tfrac{1}{2}\left(2\lambda+\mu-\sqrt{\mu^2+4\lambda\mu}\right)$$

$$s_2' = - \tfrac{1}{2}\left(2\lambda+\mu+\sqrt{\mu^2+4\lambda\mu}\right)$$

$$\text{Mean life} = \frac{2\lambda + \mu}{\lambda^2}.$$

Design C: Two units in a standby redundant design. It takes exactly τ time units to repair a failed unit.

$$R(t) = \sum_{i=0}^{[t/\tau]+1} \frac{e^{-\lambda t}(\lambda t)^i}{i!} \, [1-(i-1)\tau/t]^i$$

where $[t/\tau]$ is the greatest integer $\leq t/\tau$.

Figure 8.10 is a plot of the examples given by Epstein and Hosford for the following designs and parameters:

Type A: $\lambda = 0.1/\text{hr}$, $\mu = 2/\text{hr}$

Type B: $\lambda = 0.1/\text{hr}$, $\mu = 2/\text{hr}$

Type C: $\lambda = 0.1/\text{hr}$, $\tau = 0.5$ hr.

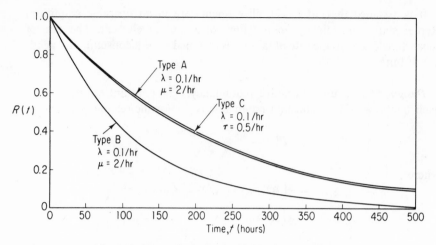

Fig. 8.10 Comparison of reliability functions for three main-tenance situations.

8.7 Summary

Chapters 7 and 8 have presented methods for evaluating the reliability of systems with various types of redundancy. This evaluation is performed by determining the paths for successful system operation through reliability block diagrams. Probability theory is then utilized to determine the system reliability or the system reliability function for time-dependent operation.

An important area that has not been considered is the economics of redundancy. All redundancy applications are governed by such limiting factors as feasibility, cost, weight, and complexity. These factors can be grouped under one heading—reliability per dollar. Redundancy will yield the maximum reliability per dollar if we take the trouble to understand basic redundancy concepts, to perform detailed analyses of the system and its requirements, and to give careful consideration to any factors that will limit the effectiveness of redundancy.

PROBLEMS

1. The estimated failure probability for an element that can short or open is 0.15. The ratio of short to open failure probabilities is known to be approximately 0.05. What is the optimum number of parallel elements to use? The optimum number of series elements? If shorts were more likely to occur, say, $q_o/q_s = 0.50$, what would be the optimum design?

2. What is the optimum design for conditions given in Problem 1 if the mean life is to be maximized? (Use Table 8.3.)

3. A designer has the choice of the following two designs:

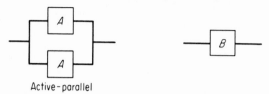

Active-parallel

Elements A and B have exponential failure densities with mean lives of 10 hr and 15 hr, respectively. Which of the two designs should be chosen if the mission time was (a) 5 hr, (b) 17.5 hr, (c) 30 hr?

Fundamentals of Reliability Prediction

NINE

During the past several years reliability prediction techniques have been developed to a fine art in the field of electronics and to a lesser extent for propulsion systems. The methods used in evaluating reliability in mechanical systems, however, need further study. Control of failure frequency for any type of system can be improved with more complete knowledge of failure modes and failure mechanisms. The purpose of this chapter is to review some of the fundamental bases and underlying philosophy of prediction as applied to electronics and propulsion systems.

The methods and data presented in subsequent sections of this chapter are as detailed and explicit as possible. It is recognized, however, that there is always some degree of subjectivity in reliability prediction. Consequently, it is not expected that accurate predictions can be made without the application of good judgment—prediction is still both an art and a science.

9.1 Purposes of Reliability Predictions

The real value of any numeric expression lies not in the number itself, but in the information it conveys and the use made of that information. Reliability predictions do not, in themselves, contribute significantly to the reliability of a system. Rather, they constitute criteria for selecting courses of action that affect reliability. Persons making reliability predictions should be aware of the role of predictions in improving and assuring reliability. Such awareness will promote the full exploitation of the potential value of the predictions. Accordingly, Section 9.1 discusses the primary purposes of reliability predictions.

Feasibility Evaluation

Reliability requirements cannot always be met simply by designing circuits that will perform the necessary functions. When an initial design concept has been proposed, it is important to determine how consistent it is with the desired reliability—before time and funds are spent in detailed design and development. The primary means for establishing reliability feasibility at this stage is the comparison of predesign predictions with requirements. Consideration of the direction, magnitude, and causes of discrepancies between predictions and requirements plays a major role in determining proper courses of action.

If a reliability requirement greatly exceeds the predicted value, it is logical to decide that the design concept is not feasible. A new design concept has to be developed (possibly entailing an entirely different functional approach, inclusion of redundancy, etc.); the requirements have to be changed; or perhaps program concepts have to be completely revised.

On the other hand, if the predicted reliability is well above the requirement, considerable confidence in the adequacy of the design concept is gained. This confidence might justify modification of the scope or depth of the planned reliability program and permit reduction of over-all program costs.

Where there is no great difference between predicted and required reliabilities, it is logical to proceed with design and development in accordance with the initial concept. In this situation, considerable effort should be devoted to reliability testing, design reviews, and all of the other elements of a comprehensive reliability program to assure that the requirement will, in fact, be met.

Comparison of Alternate Configurations

Comparison of alternate configurations or design concepts is closely associated with, or is even an integral part of, feasibility evaluation. But evaluation of alternate configurations goes beyond feasibility evaluation, in that several configurations might be capable of achieving the required reliability. Where this is so, reliability predictions, along with estimates of cost, weight, performance, etc., serve as a basis for selecting the optimum configuration.

Comparison of alternate configurations should be considered a process apart from feasibility evaluation in still another sense: The choice among different designs or configurations is a decision process that should extend, in time, beyond feasibility studies *per se*. A continuous program of evaluation and comparison of configurations often discloses areas wherein relatively minor changes would have a significant effect on reliability. The numerical estimates of reliability are indispensable sources of information to be considered, along with other pertinent factors, in deciding whether to make the indicated changes.

Usually it is not practical to make extensive configuration changes in the midst of a development program. However, if it becomes apparent that a reliability requirement cannot be met with a particular configuration, while it could be readily achieved with an alternate configuration, extensive redesign might be the only practical solution.

Allocation of Requirements

Reliability requirements must be established for all items of equipment within a system. These requirements are determined in consideration of the reliability required of the system, the relative complexity of the equipments, and, sometimes, the relative importance of the functions performed by the equipments. The process whereby the requirements are established is known as *reliability allocation*.

Identification of Reliability Problems

Predictions contribute significantly to a system's reliability through their ability to identify sources and causes of unreliability. The process of compiling parts lists, performing stress analyses, and constructing reliability block diagrams frequently discloses misapplication and overstressing of parts, less-than-optimum selection of parts, and incompatibility among input- and output-characteristic tolerances. In this sense, predictions should be considered a part of the over-all design-review function and their importance as such should not be underestimated.

In addition to discovering specific design errors, reliability predictions permit the identification, and facilitate the solution, of problems that are broader in scope and more general in nature. This is accomplished through tabulation and grouping of predicted reliabilities, or unreliabilities, for specific part types, part classes, and equipment types, and for operation in various modes or during different phases of a mission. Knowledge of the relative contribution of various items of equipment and modes of operation to the system's unreliability constitutes a sound basis for determining the need for, and the expected benefits of, part improvement programs, circuit

and equipment redesign, inclusion of redundancy, reallocation of requirements, and other courses of action.

Maintenance and Logistics Planning

If a system contains redundant elements, its reliability can be increased by
replacing failed redundant elements before the redundant group fails.
Periodic replacement (prior to failure) of items whose hazard rates increase
with time also improves reliability. For a system with no redundancy and no
items that have increasing hazard rates, the capability of repair has no direct
effect on reliability. Even for such systems, however, the ability to repair
upon failure may have a great effect on the probability of mission success.

Decisions as to what items should be made repairable during a mission
must be based on the expected frequency or probabilities of failure. Predictions are also used for making initial estimates of requirements for test
equipments and other physical facilities, maintenance personnel, spare parts
and equipments, optimum placement of test points, and accessibility for
maintenance.

Determining Data Deficiencies

The process of making reliability predictions focuses attention on items for
which adequate data are not available. Stress analysis requires knowledge of
part ratings, which occasionally may not be included in a specification.
Frequently, also, it will be found that the necessary failure information is not
available, particularly for new parts and parts peculiar to a specific application. Discovering these data deficiencies in the process of predicting reliability allows early corrective action. Often the corrective action will consist
of revising a specification or selecting a different part. In other cases it may
be necessary to initiate a data-collection program or to perform special tests.

Development of Trade-Offs with Other System-Worth Parameters

Reliability, of course, is only one of the factors which determine the over-
all worth of a system. The design of a system with the highest possible
reliability would be expected to differ from that of a system with the least
weight, the highest performance capability, the lowest cost, the highest
maintainability, or the shortest development time. It is, therefore, necessary
to make trade-offs among these attributes to develop a system which has the
maximum over-all worth. These trade-offs require quantitative estimates of
reliability.

Measuring Progress

Finally, predictions are the means whereby progress toward achieving the reliability goal is measured. This is an obvious purpose of predictions and, of course, it is associated with most of the purposes already discussed. Measurement of progress may be considered a purpose apart from the others, however, in the sense that it occupies a different position in the logical process of decision making.

If comparison of predictions with previous predictions and requirements indicates that a program is progressing satisfactorily, it may be decided that activities should continue as planned. If, on the other hand, predictions indicate that sufficient progress has not been made, the predictions can be further employed, as described in the preceding sections, to determine what action should be taken. In either event (progress is satisfactory or unsatisfactory), the measure of progress afforded by the predictions does serve a unique purpose.

Another purpose of reliability prediction closely related to measuring progress is *demonstration* of compliance with the requirement. In programs where few systems are produced and time is short, predictions are often accepted as evidence of compliance in lieu of demonstration tests.

9.2   Predictions at Various Stages of Development

Predesign Predictions

Predictions used in feasibility studies,* evaluation of alternate configurations, and allocation of reliability obviously must be made prior to, or during, the feasibility-study and early system-engineering states. These initial predictions will necessarily be based on little or no detailed design information. Predictions made at this very early stage of development are referred to in this manual as *predesign predictions.*

Because of the limited information on which they must be based, predesign predictions are not expected to be as precise as later predictions. They cannot contribute appreciably to identifying specific reliability problems and pointing out data deficiencies, or to certain other objectives of reliability predictions. However, predesign predictions, by influencing decisions on design concepts and the scope of the reliability program, can have a substantial impact on system reliability and system development.

*See Chapter 6.

Interim Predictions

To facilitate timely decisions on design details and other elements of a program and thus to enhance the ultimate reliability of a system, predesign predictions should be updated periodically. As the design progresses, successive predictions are based on accumulating knowledge of the parts complement, the application stresses, the manner in which circuit functions are accomplished, and the environmental conditions to which the parts will be subjected. Thus, consecutive predictions can be made with increasing precision. Predictions made after the predesign prediction, but prior to design completion, are designated as *interim predictions.*

Final Design Predictions

When an equipment's design is completed, a *final design prediction* is made. Because it reflects complete design information, the final design prediction is the best reliability estimate possible before actual reliability measurement through operational testing.

It should not be inferred from the term *final design prediction* that no more than one such prediction will be required for an equipment. If the design is changed, as designs frequently are, the final design prediction should be revised accordingly. (When used in this sense the word "final" means "complete" or "detailed.")

Other " Predictions "

A laboratory test of an equipment may provide an exact measure of the reliability level during that specific test. However, this measure itself has little value compared with its implications about the reliability of the same or similar equipments in future tests or end-use. Similarly, reliability measurement of a group of operational systems is primarily an assessment of the expected reliability of the group, or of an entire system population, in future operation.

The drawing of inferences from measurements, as described above, can be interpreted quite correctly as a predictive process. In fact, it can be argued with considerable justification that almost any method of assessing reliability, at any stage of development, is predictive in nature.

9.3 Approach to Prediction

Relationship between Parts and Systems

An *electronic system* is a collection of parts physically and electrically joined together in such a manner that, collectively, they perform a desired function or functions. If a system is capable of satisfactorily performing* its functions at some point in time, it will continue to have that capability until a significant change occurs in the operating characteristics of some part, or a group of parts. If part failure is said to occur when the characteristics of a part, or group of parts, have changed to the point where they exceed the limits† within which the system functions are satisfactorily performed,† it is apparent that the system will fail whenever a part fails and, conversely, whenever the system fails, one or more parts must have failed.

The relationship between part and system failure having been established, it is evident that the reliability of a system is determined by the number of parts it comprises and by the reliabilities of these individual parts. One of the basic problems in predicting the reliability of a system, then, is determining the expected reliabilities of the individual parts—as they are applied in the system.

Within the definition of part failure stated previously, the reliability of a part is determined by three factors: (1) the characteristics of the part at the beginning of the operating period of interest, (2) the characteristic limits which constitute failure, and (3) the magnitude of the changes occurring in characteristics during the period of operation. If these three factors could be determined for every part in a system prior to each period of operation, reliability could be predicted precisely. In fact, the ability to predict individual failures, as implied here, suggests that all failures could be prevented, with a resulting system reliability of 1.0.

Unfortunately, it is not possible to determine these factors for individual parts. To do so would require complete knowledge of the physics and chemistry of all failure modes and, in addition, would involve a monumental task of analysis and computation. Therefore, the usual approach to predicting reliability, and the approach used herein, is necessarily simpler and less rigorous than that suggested by the preceding discussion.

---

*For purposes of this discussion, *satisfactory performance* of the system is considered to be operation within specified functional characteristic limits. Also, satisfactory performance is considered synonymous with success or nonfailure, and unsatisfactory performance constitutes failure.

†Part failure is discussed further, and this definition is modified, in the section on "Clarification of Part-System Failure Relationships", page 278.

Types of Failures

To examine the manner in which part failures occur, failures are divided here into two categories. In the first category are parts whose measurable functional characteristics change rather gradually to the point where failure occurs. The second category comprises parts whose functional characteristics change abruptly and drastically. An example of the first type is an electron tube whose transconductance has diminished to the point of failure from a build-up of interface resistance. The second type of failure is exemplified by tubes which have become inoperative because their heaters have opened.

These two categories of failure are often called *tolerance failures* and *catastrophic failures*, respectively. This is, perhaps, an unfortunate choice of terminology since it often leads to incorrect inferences about the effects of the two types of failure on circuit or system performance. It should be understood that, within the meaning of the terms as used here, catastrophic part failures need not have catastrophic effects on the circuit or system performance, and, conversely, the fact that certain failures are referred to as tolerance failures does not preclude their having catastrophic effects on the system's performance.

Another popular misconception concerning tolerance and catastrophic failures is that the former type occurs only after parts have accumulated a great deal of operating time, while the latter type are typically "early life" failures. Although tolerance failures are likely to predominate after parts have operated a long time, it should be recognized that failures of either type can, and do, occur at any time. The electron tube with an open heater may very well have arrived at that state as a result of evaporation and crystallization of heater wire materials over a long period of time. Shorted capacitors, another classic example of catastrophic failures, may have become shorted because of an equally lengthy process of physical and chemical reaction to thermal, electrical, and mechanical stresses. On the other hand, tolerance failures may occur in practically new parts because their initial characteristics are marginal or the circuits in which they are employed will tolerate very little degradation of characteristics.

Although tolerance and catastrophic failures may be similar in their effects on system performance and in the time-dependence of their underlying causes, or mechanisms of failure, there is a significant difference between them: The frequency of occurrence of catastrophic failures tends to be independent of initial part characteristics and circuit tolerances, while tolerance failures are very much dependent on these factors. This difference between catastrophic and tolerance failures implies that the approach to predicting the two types of failures and the problems involved might be different.

Basis for Prediction of Catastrophic Failures

Prediction of catastrophic failures is a straightforward process that remains basically the same regardless of the data or detailed procedures used. The fundamental premise of catastrophic-failure prediction is that like parts have approximately the same reliability in one system as in any other system, if they are subjected to the same stresses. This premise permits the application of reliability data obtained from prior operation of parts to predict their reliabilities in new systems.

There is no available reliability data from application of parts under all possible combinations of stresses that might prevail in a system. However, large quantities of data have been obtained from the operation of parts under widely varied stress conditions. These data, in combination with failure mechanism analyses where necessary, have been used to establish relationships between reliability and several of the pertinent stresses. The relationships thus established are usually depicted as trade-off curves that allow, through interpolation and extrapolation, prediction of reliability under specific stress conditions for which observed data may not be available in significant quantities.*

Most data sources present reliability vs. stress trade-offs for temperature, power dissipation, or voltages, depending on the part type. Although their individual effects have not been established, part data taken from system operation also reflect the combined effect of other stresses—shock, vibration, pressure, humidity, and cycling—which are present in the system. These stresses, as opposed to the internally generated stresses for which trade-offs have been established, are determined, in large part, by the external environment to which a system is subjected. Collectively, they are often referred to as *use conditions*.

There is only a limited amount of information—from both theory and test results—to explain the fundamental relationships between external stresses, or use conditions, and failures. The information available consists primarily of predicted and observed reliabilities of systems whose use conditions are different. If system use conditions are categorized according to whether they represent digital or analog circuitry in ground, shipborne, airborne, missile, or spacecraft applications, the data indicate that an approximately constant ratio may exist between predictions from a single source of data and observed reliabilities of systems within a use-condition category. These observed ratios are used to adjust data for prediction of systems whose use conditions differ from those reflected in the data.

*See Section 9.6 for a discussion of data sources.

Prediction of Tolerance Failures

Tolerance failures are a function of essentially the same stresses that produce catastrophic failures. Therefore, much of the foregoing discussion of catastrophic failures is applicable to prediction of tolerance failures. However, tolerance failures are also directly affected by initial functional characteristics of parts and circuit tolerances.

Sophisticated techniques, such as Monte Carlo and regression analysis, have been used occasionally to predict tolerance failures. These techniques, although different in many respects, have an important similarity; i.e., they consider the functional relationships between individual circuit performance and part characteristics, and they apply expected distributions of part characteristics at particular points in time to estimate reliability. These techniques have been used with some success on relatively simple systems. However, they have not been developed to the point where it is feasible to use them on a wide scale in complex systems.

There are two alternatives to the use of detailed analytical techniques for predicting tolerance failures. The first is to ignore them and assume they are a negligible portion of total failures. Proponents of this approach point out that tolerance failures are individually predictable and that, in a well-designed system, they can be eliminated by preventive maintenance. This implies that new systems will be so designed that the pertinent characteristics of individual parts can be readily measured in the system and that levels of characteristics which constitute failure can be easily established.

Data indicate that tolerance failures have contributed a sizeable portion of past system failures. Further, there is no indication from available data that these failures occur less frequently in new systems than in older ones. In addition, preventive maintenance is impossible in many applications. Therefore, it is imperative that predictions include tolerance failures.

The other approach is to assume that in new systems tolerance failures will represent the same proportion of total failures as they did in previous systems. This is tantamount to assuming that, on the average, circuit tolerances and distributions of part characteristics will be the same as they have been in the past.

Many data sources include tolerance failures.* In such instances the data are applied in predictions without adjustment for tolerance failures. Where the prediction data are limited to catastrophic failures, they must be adjusted. Since a constant ratio of tolerance to catastrophic failures is assumed, the adjustment may be combined with the adjustment employed to correct for use conditions.

*See Section 9.6 for a discussion of data sources.

Clarification of Part-System Failure Relationships

In Section 9.3 on page 274 a simple relationship was established between part failure and system failure by use of a definition for part failure. From a theoretical standpoint, this definition is adequate. For practical reasons, however, the relationship between part failure and system failure must be further clarified and the definition of part failure must be modified.

A part used in a redundant element of a system cannot cause the system to fail unless the other redundant element fails. Therefore, according to the definition of part failure given in Section 9.3, page 274 the reliability of a part in a redundant element would be a complex function of the performance capability of all the parts in the parallel path. This concept of part failure is incompatible with the data used in prediction.

Part reliability data are collected and computed in such a manner that *part failure* can reasonably be assumed to cause system failure only when the part is used in a nonredundant configuration. Therefore, for parts in a redundant element, *failure* is redefined as having occurred when the characteristics of a part, or group of parts, exceed the limits within which the system's functions would be satisfactorily performed if the parts were not in a redundant path.

Given this modified definition of part failure, and reliability data derived in accordance therewith, the reliability of a system which contains redundant elements is not simply the product of the reliabilities of its parts. Therefore, a somewhat more complex formula relating system reliability to part reliabilities must be developed in each case for predicting the reliability of a system which includes redundancy. Development of the reliability formula is discussed further and illustrated in the following section.

Further clarification of the relationship between part failure and system failure is needed inasmuch as, even in series configurations, several parts often cause a system failure. Conversely, some system failures are repaired by adjustments without part replacement. Each of the parts that has to be replaced to correct a failure might be considered a failure itself according to almost any definition of part failure. However, if part failure data were simply computed on the basis of the number of part failures observed, their use in prediction would yield biased estimates of system reliability. Therefore, most sources present data that have been adjusted to reflect the contribution of each part type to system failure, rather than the total number involved in system failure. Thus the data provide an unbiased estimate of system reliability. But they may yield a biased estimate of the expected distribution of part failures.

Derivation of System Reliability Formula

The starting point in the derivation of the reliability formula is a functional diagram of the system showing the flow of inputs and outputs as required for system operation. This diagram, which is also called the *reliability block diagram*, may look entirely different from the engineering drawing that shows fabrication relationships (although perhaps not geometry). The purpose of the reliability diagram is to show how critical the system elements are in system operation—to show the operational modes and redundancy (when it exists) of the elements.

Of course, the system operation may vary during the mission (or specified operating time), and this variation must be considered in determining the mission failure pattern. For this purpose, the mission or operating time can be divided into time intervals during which system operation is dependent on a single functional diagram; that is, black box requirements do not change during any one interval. Then, the entire mission failure pattern can be synthesized by combining the patterns for all intervals.

Example 1. For an interval in which black box operational requirements do not change, consider that system operation is dependent on a relatively simple reliability block diagram involving six functional units, as shown in Figure 9.1a, with system operational output coming from Unit 6.

The probability of system success, p_s, can be computed if the probability of success for each unit, $p_i (i = 1, 2, \ldots, 6)$, is known. System operation does not require all units to be operating properly; the following combinations of units are acceptable: 1–4–6, 2–4–6, and 3–5–6. Then, assuming independence of unit failures, usual probability theory gives

$$p_s = [(p_1 + p_2 - p_1 p_2)p_4 (1 - p_3 p_5) + p_3 p_5]p_6$$
$$= [p_1 p_4 + p_2 p_4 + p_3 p_5 + p_1 p_2 p_3 p_4 p_5$$
$$- p_1 p_2 p_4 - p_1 p_3 p_4 p_5 - p_2 p_3 p_4 p_5]p_6.$$

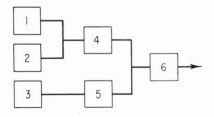 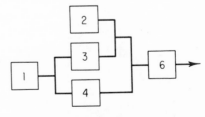

Fig. 9.1a Reliability block diagram for six-unit system.

Fig. 9.1b Reliability block diagram for a complex six-unit system.

Example 2. Figure 9.lb gives an example in which units are allowed to interact in more ways than in Example 1. The derivation of the failure probability is more difficult for this situation, but the underlying method is the same. Several approaches may be used, two of which are illustrated here.

A rather simple approach is the technique of "counting cases," that is, considering paths that can be followed to obtain system operation. The sequences of units that can produce system operation are 2–5–6, 1–3–5–6, and 1–4–6. The probability of one or more of these paths existing without failure must be computed. To do this, let the probabilities of successful operation of the units be denoted by p_i; the probabilities of failure by $q_i (i=1, 2, \ldots, 6)$; and the probability of successful system operation by P. Then, count all possible unit success-failure combinations (cases) that lead to system operation and identify their associated probabilities.

1. First, consider the case in which Unit 5 fails and path 1–4–6 must therefore be followed for system operation. The probability of system success is $q_5 p_1 p_4 p_6$.

2. In all other cases, Unit 5 must operate. The functional diagram is then effectively that shown in Figure 9.1c. (Unit 5 can be omitted because its outcome is set with probability p_5.) If Units 2 and 6 operate properly, system operation is assured. For this case, the probability of system success is $p_2 p_5 p_6$.

3. If Unit 2 fails in any case having Unit 5 operating, the functional diagram is further simplified as shown in Figure 9.1d. The probability of system success is $q_2 p_5 p_1 p_6 (p_3 + p_4 - p_3 p_4)$.

The probability of system success, considering all ways in which the units can interact, is the sum of the three results given above:

$$P = p_1 p_4 q_5 p_6 + p_2 p_5 p_6 + p_1 q_2 p_5 p_6 (p_3 + p_4 - p_3 p_4).$$

Substituting $(1 - p_i)$ for q_i gives

$$P = p_6[p_1 p_4 + p_2 p_5 + p_1 p_3 p_5 - p_1 p_3 p_4 p_5 - p_1 p_2 p_3 p_5 - p_1 p_2 p_4 p_5 + p_1 p_2 p_3 p_4 p_5]$$
$$= p_6[p_1 p_4 + p_2 p_5 + p_1 p_3 p_5 + p_1 p_2 p_3 p_4 p_5 - p_1 p_2 p_3 p_5 - p_1 p_2 p_4 p_5 - p_1 p_3 p_4 p_5].$$

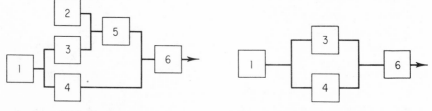

Fig. 9.1c Simplification of the reliability block diagram of figure 9.1b.

Fig. 9.1d Second simplification of the reliability block diagram of figure 9.1b.

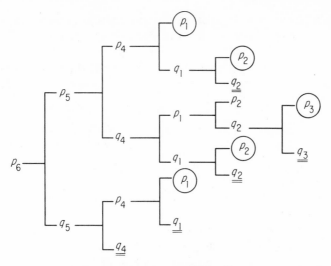

Fig. 9.2 Output success/failure diagram.

The probability of successful operation for the system shown in Figure 9.1a can also be determined by drawing a diagram that identifies unit outputs by the associated success (p_i) and failure (q_i) probabilities. The diagram is constructed by starting with Unit 6, which gives the system output, and selecting orders of units which lead to system success or failure. Each branch of the diagram is terminated when system success or failure is determined; failure is denoted by the double line under the terminal probability, and success is indicated by a circle about the terminal probability. (See Figure 9.2.)

The probability of system success, considering all ways in which the units can interact, is obtained by adding probabilities for all branches of the diagram leading to system success. Thus,

$$P = p_6[p_1p_4p_5 + q_1p_2p_4p_5 + p_1p_2q_4p_5 + p_1q_2p_3q_4p_5 + q_1p_2q_4p_5 + p_1p_4q_5].$$

Substituting $(1-p_i)$ for q_i gives the same formula obtained by the counting cases approach:

$$P = p_6[p_1p_4 + p_2p_5 + p_1p_3p_5 + p_1p_2p_3p_4p_5 - p_1p_2p_3p_5 - p_1p_2p_4p_5 - p_1p_3p_4p_5].$$

More complicated types of redundancy can be handled in similar ways. For example, system success might depend on three out of four paths (or units) operating properly. Sometimes designs are set up to determine operation according to a majority rule of a set of units. Since modifications in theory are not involved, these more complicated reliability diagrams are not considered here.

In the preceding discussions, it was implied that success probability was constant for each unit and hence for the system. This is true for a specified mission of a particular time length. However, the derivation of a formula for the probability of system success is not based on this assumption, and all steps in the derivation still hold if the probability is time-dependent. Time dependence can be indicated by substituting $R_i(t)$ for p_i in the success probability formula for any particular case. For example, this substitution in the formula derived in Example 2 gives the following system reliability function:

$$R(t) = R_6(t)\,[R_1(t)R_4(t) + R_2(t)R_5(t) + R_1(t)R_3(t)R_5(t)$$
$$+ R_1(t)R_2(t)R_3(t)R_4(t)R_5(t) - R_1(t)R_2(t)R_3(t)R_5(t)$$
$$- R_1(t)R_2(t)R_4(t)R_5(t) - R_1(t)R_3(t)R_4(t)R_5(t)].$$

The derivation further holds whether the reliability function is exponential, Gaussian, gamma, or any other form. The only critical assumption involved is the independence of the failures of the units that make up the system under consideration.

The same method illustrated here for arriving at a reliability formula for system units can also be applied at lower levels of the system. Each unit reliability function can be expressed in terms of subunits, or even parts. Hence, the system reliability function itself can be expressed in terms of part reliability functions. The system level for which the reliability function is derived depends on the specific data available for study. Computation work loads are minimized when failure patterns for larger subsystems are available, and this consideration influences the choice of the reliability test program.

Hazard Rates and Failure Rates

Chapters 2 and 3 discussed many of the probability and statistical concepts pertinent to reliability prediction. That discussion will not be repeated here. However, the discussion of failure rates and hazard rates has implications particularly pertinent to prediction of electronic equipments and the part reliability data used for such predictions. These concepts are reviewed and further discussed below.

Failure rate was defined as the rate of failure expected in the time interval t to $t + h$, expressed on a per-equipment-per-unit-time basis, considering only the proportion of the equipments operable at time t, the start of the interval. The failure rate formula is

$$f(t, h) = \frac{R(t) - R(t+h)}{hR(t)} \tag{9.1}$$

where

$R =$ the probability of survival from time 0 to the time specified in parentheses,

$t =$ the beginning of the time period of interest, and

$h =$ the length of the time interval.

The *hazard rate function*, $Z(t)$, is defined as the limit of the failure rate as the interval length, h, approaches zero. It is given by the equation

$$Z(t) = \frac{u(t)}{R(t)} \qquad (9.2)$$

where

$u(t) =$ the negative of the rate of change of the reliability function with time, or the density function.

Thus, the hazard rate is the rate of change of unreliability with time, adjusted for the proportion surviving up to that time. Hazard rate, then, is a *conditional instantaneous rate*. In this same sense, failure rate is a *conditional average rate over a finite period of time*.

The hazard rate function has an interesting property with regard to the relationship between system or equipment reliability and the reliabilities of parts, where the parts are nonredundant and fail independently. That relationship is

$$Z_s(t) = Z_1(t) + Z_2(t) + \ldots + Z_n(t) \qquad (9.3)$$

where

$Z_s(t) =$ the system hazard rate at time t, and

$Z_i(t)$, $i = 1, 2, \ldots, n$ are the hazard rates of the n parts at time t.

This additive property of hazard rates is independent of the functional form of the reliability functions. However, when the part reliability functions are exponential, hazard rates are especially convenient expressions of reliability data because hazard rates are then constant and, thus,

$$R_s(t) = e^{-\sum\limits_{i=i}^{n} \lambda_i t_i} \qquad (9.4)$$

where

$\lambda_i =$ the constant hazard rate of the ith part, and

$t_i =$ the operating time of the ith part in the time interval from 0 to t of system operation.

Therefore, it has become almost universal practice to compile part reliability data in the form of hazard rates when the failure times are exponentially distributed.

It should be noted that hazard rates are equivalent to failure rates only when the interval over which failure rate is computed approaches zero. In spite of this restriction, the term "failure rate" is widely used to describe measures of reliability which are really hazard rates.

Actual failure rates are used infrequently in reliability engineering. There-fore, misuse of the term "failure rates" to describe data that are, in fact, hazard rates usually is not of serious consequence. The term "failure rate" will be considered synonymous with "hazard rate" and the two terms will be used interchangeably in the remainder of this chapter, unless otherwise noted.

Implications of Constant Failure Rate Assumption

In a nonredundant situation it is a very simple task to compute system reliability from part reliability data when part failure rates are constant with time. If part failure rates are not constant, the task is only slightly more difficult and more time-consuming, provided that the part reliability functions are known. The prediction methods described in Section 9.4 are not dependent upon the part failure rates being constant.

However, it has been found that the failure rates of many electronic parts are reasonably constant. For many other part types, even though it is known that their failure rates are a function of time, it has been found that they change very little over periods that are long in comparison with most mission lengths of interest. Where either of these situations exists, very little error is caused by use of constant failure rate approximations.

Even where part failure rates change greatly with time, the use of an equivalent average failure rate will result in very little error in prediction of a system's reliability for a specific time, if that time coincides with the time over which the failure rates are averaged. However, if these time periods do not coincide, very large errors may result.

Part reliability data from most sources are presented as constant failure rates. For a few part types, some source documents give a time limit beyond which it should not be assumed that failure rates are constant. Many of these sources assume that the failure rates given are reasonable approximations of the rates expected over operating periods of any length, provided that normal preventive maintenance is performed.

9.4   Final Design Prediction Procedures

The procedures applicable to final design predictions consist of the 11 basic steps listed below. These steps are thereafter discussed under separate headings.

Step-By-Step Procedures

1. Define the System.
2. Define Failure.
3. Define Operating and Maintenance Conditions.
4. Construct Reliability Block Diagram(s).
5. Develop Reliability Formula.
6. Compile Parts Lists.

7. Perform Stress Analyses.
8. Assign Failure Rates or Probabilities of Survival.
9. Combine Part Failure Rates (or Probabilities of Survival).
10. Modify Preliminary Block Failure Rates or Reliabilities.
11. Compute System Reliability.

Define the System

The initial step in a reliability prediction is to define the system. The term "system" is used here to denote the particular collection of items to which a prediction pertains.

If a prediction is to be made for an entire subsystem, that subdivision constitutes a system. Likewise, if a prediction of the probability of success for a single function is required, the collection of parts which perform that function is referred to as a system. It follows, of course, that if an item of equipment performs several functions, for which separate predictions are required, it comprises several systems.

The task of defining the system, then, consists of explicitly describing the functions and physical boundaries of the items that constitute the system. Particular attention must be given to interfaces among systems so that all pertinent items will be considered in a prediction and there will be no unwanted duplication of coverage in predictions for adjacent systems.

Define Failure

Unless otherwise specified, failure is the occurrence of any condition which renders the system incapable of operating within its specified performance parameter limits. In this situation, the task of defining failure consists of listing or referencing the appropriate limits. Whenever this concept of failure is not being used, the modified definition of failure must be explicitly given in accordance with the terms of the appropriate exhibit. For example, the data presented in Table 9.4 are based on operator complaints.

Define Operating and Maintenance Conditions

Operating conditions include the system's operational profile and the environmental conditions prevailing during the various periods of operation. The operational profile is defined in terms of the elapsed mission times or mission phases at which the system is turned on, and the duration of operation during each phase. The sequence of functions necessary for success and the duty cycles of items within the system are also elements of the operational profile that must be defined.

During each period of operation the pertinent environmental conditions must be established by test, reference, or assumption. Definition of environmental conditions should encompass all the factors which might affect reliability, whether or not their effects can be quantitatively assessed. Complete definition of the environmental conditions establishes the conditions under which the system is designed to operate and, thus, the conditions to which the prediction pertains.

The maintenance conditions expected to affect reliability must be established. Pertinent items include: replacement schedules for parts with known or estimated limited lives; other preventive maintenance schedules; identification of items which may be replaced or repaired during a mission; requirement for special equipments or facilities. Also, the reliabilities of some items that include redundancy may be a direct function of repair or replacement times and monitoring schedules. These items should be identified and, depending on the terms of pertinent contract exhibits, maintainability predictions may be required as inputs to the reliability prediction.

Construct Reliability Block Diagram

A reliability block diagram may be considered a logic chart which, by means of the arrangement of blocks and lines, depicts the effect of failure of items of equipment on the system's functional capability. Items whose failure causes system failure are shown in series with other items. Items whose failure causes system failure only when some other item has also failed are drawn in parallel with the other items.

Reliability block diagrams often look very much like functional block diagrams. In fact, they are developed through analysis of the functional relationships among items of equipment as shown by functional block diagrams and circuit schematics. However, the purpose of a functional block diagram is quite different from that of a reliability block diagram, and the two usually are not interchangeable. Functional block diagrams show functional sequences and signal paths, and items which are wired in parallel are drawn in parallel. While it may be convenient to arrange blocks on a reliability diagram in the same order in which their functions are performed, it is not necessary that they be so arranged. The reliability block diagram does not necessarily show signal paths, and items wired in parallel are not shown in parallel if failure of one causes system failure.

One of the first tasks in constructing a reliability block diagram is to determine the complexity levels of equipment items which are to be shown as separate blocks. For a complex system it is often convenient to use several block diagrams. The first would usually be a simple diagram showing the first-order subdivision breakdown of the system. Separate block diagrams are then constructed for each of the first-order subdivisions. This process of diagramming goes on until individual blocks represent complexity of such an order that their reliabilities, or failure rates, can be readily estimated from part level data.

Although the level of complexity to be depicted by individual blocks cannot be explicitly defined without considerable knowledge of each specific system, there are some generally applicable limitations and guides which should be observed:

1. Individual blocks should not include items from more than one "black box."

2. Items (other than at the part level) developed or produced by different design groups or other organizations should not be included within a single block.

3. One-shot devices, and other items whose reliabilities are not time-dependent, and items that do not have constant failure rates should be shown as separate blocks.

4. Blocks should not include redundant groups of items.

5. Blocks should include only items that have the same duty cycle.

It is usually convenient and advantageous to arrange blocks on a reliability diagram in the same sequence as that in which their functions are performed. Blocks representing redundant items and alternate modes of operation should be shown in parallel. Other blocks are placed in series.

On a two-dimensional diagram it is frequently not possible to convey all of the pertinent information merely by the arrangement of blocks and interconnecting lines. Therefore, appropriate notation should be included on the diagram or in accompanying verbal descriptions. The notation should describe types of redundancy, where it is not obvious from the diagram. Where failure of a redundant element degrades performance or places additional stresses on the items in alternate paths, it should be so noted. Also, operating times, or cycles, of the individual blocks should be noted, if different from system operating time. In addition, items which may be repaired or replaced during a mission should be identified, and monitoring intervals for those items should be stated.

If system operation varies during a mission (or other specified operating time), this variation must be considered in determining the system failure pattern. For this purpose, mission time should be divided into intervals during which the system configuration is constant, and separate diagrams, or sets of diagrams, should be developed for each interval.

Example. An example of the reliability block diagramming process is given in Figures 9.3, 9.4, and 9.5. Figure 9.3 is a functional block

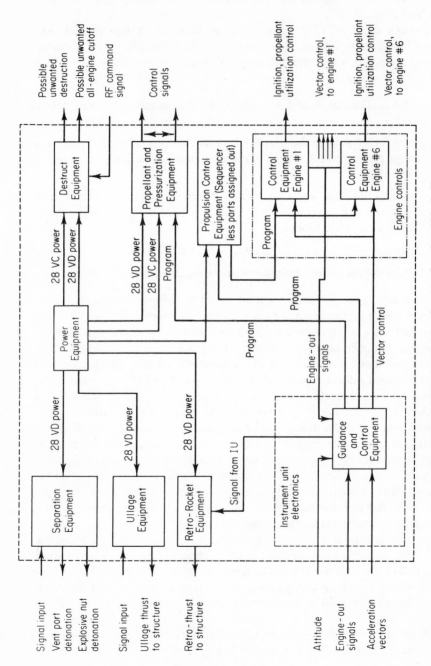

Fig. 9.3 A missile electronics subsystem functional block diagram.

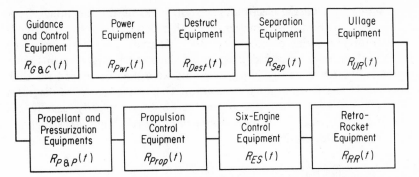

Fig. 9.4 A missile electronics subsystem — simplified reliability block diagram.

$$R_{\text{electronics}}(t) = R_{G\&C}(t) \times R_{Pwr}(t) \times R_{Dest}(t)$$
$$\times R_{Sep}(t) \times R_{UR}(t) \times R_{P\&P}(t)$$
$$\times R_{Prop}(t) \times R_{ES}(t) \times R_{RR}(t)$$

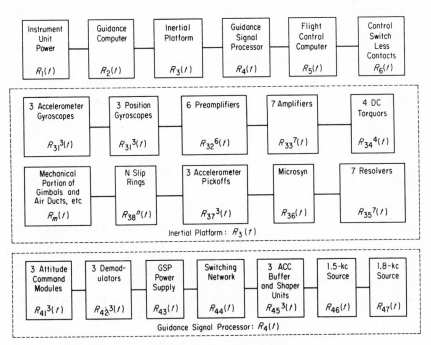

Fig. 9.5 A missile guidance and control equipment—simplified reliability block diagram.

$$R_{G\&C}(t) = R_1(t)\ R_2(t)\ R_3(t)\ R_4(t)\ R_5(t)\ R_6(t)$$
$$= R_1(t)\ R_2(t)\ R_{31}^6(t)\ R_{32}^6(t)\ R_{33}^7(t)\ R_{34}^4(t)\ R_{35}^7(t)\ R_{36}(t)$$
$$\times R_{37}^3(t)\ R_{38}^n(t)\ R_m(t)\ R_{41}^3(t)\ R_{42}^3(t)\ R_{43}(t)\ R_{44}(t)$$
$$\times R_{45}^3(t)\ R_{46}(t)\ R_{47}(t)\ R_5(t)\ R_6(t)$$

diagram of a missile electronics subsystem. From the functional block diagram, analysis of schematics, and other descriptions of the system's operation, a series of reliability block diagrams were developed. The first of these, a simplified diagram of the system, is shown in Figure 9.4. Detailed reliability block diagrams were then developed for each of the blocks in the simplified diagram. One of these detailed diagrams was selected as illustrative of the range of complexity usually encountered. Figure 9.5 is the simple form of a reliability block diagram where all items are in series.

Develop Reliability Formula

The reliability formula expresses mathematically the relationship of system reliability to the reliabilities of the equipment items depicted as blocks on the reliability diagram. In the simple cases where all blocks are in series, the formula may be merely

$$R_s = R_1 R_2 \ldots R_n, \tag{9.5}$$

where

$$R_s = \text{system reliability, and}$$

$$R_1, R_2, \ldots, R_n = \text{reliabilities of blocks 1, 2, ..., } n, \text{ respectively.}$$

Where there are no items whose hazard rates change with time, one-shot devices, etc., the formula may be even simpler; i.e.,

$$R_s(t) = \exp\left[-\sum_{i=1}^{n} \lambda_i t_i \right] \tag{9.6}$$

where

$\lambda_i =$ the failure rate of the ith block, and

$t_i =$ the operating time of the ith block in system operating time t.

If a system includes redundant items, items whose failure rates change with time, one-shot devices, provision for replacement of failed redundant elements, etc., the reliability formula becomes more complex. Development of reliability formulas was discussed in some detail in Section 9.3, and several illustrations are given in the example on page 287. Therefore, the subject will not be treated further here.

Compile Parts Lists

For each block on the reliability block diagram, individual parts should be listed in some convenient order. If the block contains several subdivisions, such as assemblies or subassemblies, it is usually advantageous to list parts by circuit symbol within part type within block subdivision. Only parts that can cause the block to fail are considered in the prediction. However, all parts within a block should usually be listed, with appropriate notation to identify those whose failure will not cause the block to fail. If, for some parts, only particular failure modes will affect the block, those parts and the pertinent failure modes should also be identified.

In addition to identifying the parts within blocks, parts lists serve as basic work sheets for determining stresses and part failure rates or reliability estimates. Therefore, parts lists should include part descriptions, pertinent ratings, and space for entering operating voltages, currents, power dissipation, stress indices, and failure rates or probabilities of survival.

Perform Stress Analyses

The first step in a stress analysis is to determine from design analysis, or measurements where possible, pertinent operating voltages, currents, power dissipation, etc., for each part. Stress indices are then calculated through comparison of operating conditions with rated values. If operating conditions change during a mission, several stress indices may have to be calculated for a part.

The pertinent voltages, currents, and other characteristics used in calculating stress indices, and the stress indices themselves, should be recorded on the parts lists.

Assign Part Failure Rates or Probabilities of Survival

The next step in predicting reliability is to assign failure rates, or other measures of reliability, to the individual parts. For most part types the measure of reliability will be obtained from a dependable data source and, thus, they will usually be in the form of constant failure rates. The stress indices determined in the preceding step, ambient temperatures, and other pertinent information are used to obtain these failure rates in the manner described in Section 9.7. If the stress indices or ambient temperatures vary during a mission, a separate failure rate for each mission phase should be assigned to each affected part.

Data that apply to parts whose failure rates change with time, to one-shot devices, or to other parts whose probabilities of survival are not time-dependent, are usually expressed in the form of probabilities and should be recorded as such on the parts lists. If the probabilities of survival are time-dependent, appropriate values should be recorded for each of the time periods of interest.

Combine Part Failure Rates or Probabilities of Survival to Obtain Preliminary Block Failure Rates or Reliabilities

For blocks which contain only series parts with constant failure rates, the part failure rates are added together to obtain the preliminary block failure rate. If a block includes redundant groups of parts, or parts which do not have constant failure rates, the part failure rates or probabilities are substituted in the block reliability formula to obtain the preliminary block reliability.

If part failure rates or probabilities of survival vary during a mission, they should be appropriately combined to obtain a preliminary block failure rate for each mission phase.

Modify Preliminary Block Failure Rates or Reliabilities

Block failure rates or probabilities of survival are next modified to account for tolerance failures and use conditions. As an example, typical adjustment factors to be used with MIL-HDBK-217 data* are listed in Table 9.1. All of

Table 9.1 ADJUSTMENT FACTORS FOR ELECTRONIC
SYSTEMS DATA

	Adjustment Factor	
Use Conditions	*Digital Circuitry*	*Analog Circuitry*
Ground and Shipboard Equipments	0.5	1.0
Airborne Equipments	5.0	7.5
Spacecraft and Booster Equipments Operation During Boost and Retro Firing	45.0	150.0
In-Orbit or Coast Operation	0.5	1.0
Operation on Lunar Surface	0.5	1.0

the factors except those for lunar surface operation were obtained by comparing observed reliabilities with predictions based on MIL-HDBK-217 data.† Insofar as the observed data are indicative, the values selected are

*See "Correction Factors for Stress", page 318, for discussion of adjustment factors for data from sources other than MIL-HDBK-217.

†Some of the factors listed (particularly those for in-orbit operation) were obtained from very limited data. It is interesting to note, however, that these adjustment factors are very similar to those which have been independently generated by several organizations for use with MIL-HDBK-217 data. As more data become available, it is expected that these factors will be revised as necessary.

proper adjustments for prediction of reliabilities that can reasonably be expected within the "design state of the art" and with the use of parts comparable to those presently available.

Of course, there are no data upon which to base adjustments for lunar operation. In the absence of better information, it was assumed that equipments would be operated in a controlled environment, or so designed that they would withstand stresses peculiar to the lunar environment without degradation of reliability below that experienced in ground equipment. Therefore, the adjustment factors for lunar operation are the same as those for ground and shipboard equipments.

The final block failure rate for each mission phase is the product of the appropriate factor from Table 9.1 and the preliminary block failure rate for that phase. Block reliabilities are adjusted by

$$R_f = R_p^k \qquad (9.7)$$

where

R_f = the final block reliability,

R_p = preliminary block reliability, and

k = the appropriate adjustment factor from Table 9.1.

It is common to compute the value of R_p^k by natural logarithms, that is, by using tables of the exponential function.

Compute System Reliability

System reliability is computed by entering the block reliabilities and failure rates in the system reliability formula and solving for the time periods or mission phases of interest. In addition to computing reliability at certain specified time periods, or for various mission phases, it is necessary to compute the system's reliability at a sufficient number of points in time to enable reliability functions to be plotted for each of the applicable operating conditions. Also, the reliability estimates for the various mission phases should be combined to show system reliability for the entire mission.

If the mean time between failures ($MTBF$) is a parameter of interest, it may, of course, be calculated by evaluating the integral of the reliability function from 0 to ∞. If the system reliability function is very complex, it may be a tedious task to obtain the integral. In this case, reasonable estimates of $MTBF$ may be made by use of a planimeter or by adding blocks to estimate the area under the reliability function curve.

9.5 Predesign and Interim Predictions

The procedures for making predesign and interim predictions are basically the same as those for final design predictions, as discussed in Section 9.4. The differences among the predictions lie in the degree of precision with which the 11 basic steps may be implemented.*

For predictions made at any stage of development, each of the steps should be carried out to the maximum extent. The system, failure, and operating and maintenance conditions should be defined as explicitly as possible. Reliability block diagrams are constructed to the lowest identifiable function, and appropriate system reliability formulas are established.

Precise parts lists, of course, cannot be compiled prior to design of an equipment. It is necessary, however, to make an estimate of the parts complements of the various system subdivisions (blocks on the reliability diagram). The parts complement estimates should include the number of parts by type. If this is not possible, predictions are based on the number of active elements, i.e., tubes or transistors, and an assumed distribution of other parts by type per active element.

Stress analyses obviously cannot be made prior to design. Therefore, for portions of the system which have not been designed, a stress level of 0.5 is assumed. The failure rate data to be used are based on an assumed ambient temperature of 35°C within the equipment enclosures. Where better temperature estimates can be made, the basic data are corrected by means of a temperature-vs.-failure-rate trade-off curve.

The process of combining part failure rates to obtain preliminary block failure rates or reliabilities; of adjusting block rates, or probabilities; and of computing system reliability is precisely the same for predesign and interim predictions as for final design predictions.

A Predesign Reliability-Prediction Procedure for Ground Electronic Systems†

The procedure outlined in this section is intended to fill the need for a "first look" at the potential reliability of a ground electronic system during

---

*See Chapter 6 for a discussion on feasibility studies.

†Derivation of the technique is given in H. R. Jeffers and H. S. Balaban, *System Reliability Prediction by Function*, ARINC Research Corporation, Publication No. 241–01–1–375 (May 27, 1963). (Prepared for Rome Air Development Center, Contract AF 30 [602]–2838.)

the early planning stages, allowing the planners to estimate the requirements that can reasonably be placed on the system and the degree of risk involved in meeting specific goals. The predictions are based on rather broad system information—the type of information that can only be estimated during the planning stages. Therefore, we can place less reliance on these predictions than on later ones based on more precise inputs. As an indication of the confidence that can be associated with the predesign predictions, the procedure described herein includes a means for estimating confidence intervals about the point predictions of system reliability.

BASIC CONSIDERATIONS

The equation for predesign reliability prediction is linear in $\ln \hat{\theta}$:*

$$. \ln \hat{\theta} = b_0 + b_1 X_1 + b_2 X_2 + \ldots + b_n X_n, \tag{9.8}$$

where $X_1, X_2, \ldots, X_n$ refer to specific equipment parameters; $b_1, b_2, \ldots, b_n$ are coefficients of these parameters; and $\hat{\theta}$ is the predicted *MTBF*. $\hat{\theta}$ may be obtained from $\ln \hat{\theta}$ by use of tables of natural logarithms.

This equation appears herein in four forms (designated I through IV). The equation form to select for a particular application is determined by the type of information available. Each equation is accompanied by the standard deviation of residuals, commonly known as the *standard error of estimate*, obtained from the regression analysis. Where a choice exists as to the prediction equation to be used, the equation with the minimum standard deviation should be selected. The use of the standard deviation in the computation of a confidence interval will be discussed under "Prediction of Confidence Interval," page 301.

The predicted *MTBF*, $\hat{\theta}$, is for a nonredundant system. In the numerical evaluation of the system parameters (X_i), only nonredundant portions of the system were considered. Suggestions for consideration of some forms of redundancy are presented in the original report.

An exponential distribution of failure times was assumed for the systems studied in this program. This assumption was confirmed by analysis of associated failure data.

Each of the parameters listed in Table 9.2 are used in one or more of the prediction equations. Upper and lower data limits noted in Table 9.2 are values observed for the systems of this study, and indicate the ranges over which the predictions are valid. For systems whose parameters lie outside these ranges, the predictions must be considered as extrapolations, and the confidence limits will not apply.

*ln represents logarithm to the base *e*.

PREDICTION EQUATIONS

The reliability of a system in the predesign stage can be predicted from the appropriate equation among the four given below. Step-by-step procedures and worksheets for the solution of the equations appear under "Procedures and Worksheets", on page 296.

Equation Number	Prediction Equation	Standard Error of Estimate(s)
9.8 I	$\ln \theta = 7.173 - 0.801X_1 - 0.136X_2$ $- 0.596X_3 + 0.513X_5$	0.378
9.8 II	$\ln \theta = 7.910 - 0.121X_8 - 0.740X_1$ $- 0.090X_2 - 0.359X_3$	0.382
9.8 III	$\ln \theta = 2.731 - 0.484X_7 - 1.457X_6$ $+ 2.206X_5 - 0.929X_9 - 0.139X_2$	0.626
9.8 IV	$\ln \theta = 2.172 - 0.277X_2 + 2.327X_5$ $- 1.075X_3 - 1.770X_6 + 0.192X_4$	0.698

$(\theta = \text{Predicted } MTBF \text{ in hours.})$

The equation with the lowest standard error of estimate will generally yield the most accurate prediction (minimum confidence interval).

The term "system" is used to connote a system, subsystem, or equipment. However, a distinction must be made among these levels in certain applications of the prediction equations. Large, complex systems should be broken down into subsystems or equipments and the equations applied separately—particularly Equations 9.8 III and 9.8 IV, which do not contain direct measures of complexity.

The data available for this program provided an insufficient basis on which to establish a general rule for subdividing systems for prediction purposes. Until more data are available, the best rule is to apply the prediction procedure to systems no larger than those considered in this study. (See Table 9.2.) Transceivers, radars, transmitters, receivers, and computers are classified as single units. However, a carrier telephone/radio relay terminal should be considered as two units: the radio system (transmitter and receiver) and the telephone multiplexing system. Should this terminal also include a teletype multiplexing system, this system would constitute a third unit. The prediction procedure would be applied separately to each unit.

PROCEDURES AND WORKSHEETS

Steps for predicting a system $MTBF$, θ, are given in this section. Worksheets are also included for facilitating the procedure. It is suggested that these worksheets be reproduced for the application of these equations, since

Table 9.2 Definitions of System Parameters

Parameter Designation (Ranges of Validity)	Definitions
$X_1 = \ln C - 2.303$	Natural logarithm of system complexity (C) minus 2.303. Complexity is measured in terms of an adjusted active element count. A worksheet is provided (Table 9.4) to aid in the numerical evaluation of X_1.

$C =$ 1.0 × No. of electron tubes (all types)

+ 0.3 × No. of analog transistors

+ 0.4 × No. of solid state power rectifiers

+ 0.006 × No. of solid state digital diodes

+ 0.02 × No. of digital transistors

+ 0.1 × No. of solid state analog diodes

+ 10^{-5} × No. of magnetic core devices (e.g., magnetic amplifiers, memory cores, etc.; passive devices, such as transformers, are not included)

Upper limit = 8.087 ($C = 32,488$)
Lower limit = -1.609 ($C = 2$)

The number of each type of active element in the systems of this study ranged from zero to the following maxima:

Tubes — 3,037
Analog transistors — 4,167
Solid state power rectifiers — 204
Digital diodes — 25,000
Digital transistors — 3,083
Analog diodes — 692
Magnetic cores — 16,200

An element is considered as digital in application if the function depends primarily on the existence of a pulse, rather than on the shape or amplitude of a signal.

X_2
Upper limit = 11
Lower limit = 0

Number of cathode-ray tubes in the system.

X_3
Upper limit = 2
Lower limit = 0

Number of transmitters in the system.

$X_4 = \ln F - 2.303$

Upper limit = 6.908 ($F = 10,000$ Mc)
Lower limit = -9.211 ($F = 0.001$ Mc)

$F =$ Frequency (megacycles)

Natural logarithm of highest frequency characteristic of the system, in megacycles per second, minus 2.303.

(In many cases, the highest frequency characteristically found in a system must be estimated on the basis of engineering judgment, as in the case of a radar repeater. The influence of this parameter on the result is such that an error in the estimate of frequency of a factor of two or three will be acceptable.)

In the case of tunable equipment, the upper limit of the tuning range will be the estimated highest characteristic frequency; e.g., in the case of a 225-Mc/400 Mc radio set, the frequency to be used is 400 Mc. The units of frequency in the equation are megacycles.   (continued)

Table 9.2 Definitions of System Parameters (continued)

Parameter Designation (Ranges of Validity)	Definitions
X_5 Upper limit = 2 Lower limit = 1	The value of X_5 is 1 for systems which are primarily digital in nature, and 2 for systems which are primarily analog in nature. In the case of mixed equipments, the choice is made according to the nature of the majority of the important functions. The distinction between digital and analog is the same as noted in the definition of X_1
X_6 Upper limit = 1 Lower limit = 0	The value of X_6 is 1 if the system has a steerable antenna. Otherwise, the value is zero.
X_7 Upper limit = 5.489 (242 kV) Lower limit = −4.200 (0.015 kV)	Natural logarithm of the maximum dc voltage in the system, in kilovolts.
X_8 Upper limit = 5.989 (399 kW) Lower limit = −3.412 (0.033 kW)	Natural logarithm of total rated power consumption, in kilowatts.
X_9 Upper limit = 0.984 Lower limit = 0	X_9 is the ratio of the tuning range of the system to the highest characteristic frequency: $$X_9 = \frac{\text{tuning range (Mc)}}{\text{frequency (Mc)}}.$$ For equipments with no tuning provision, X_9 has the value of zero.

they contain all fixed factors pertaining to the equations. As previously stated, the procedure outlined below pertains to nonredundant systems.

Step 1. Selection of Prediction Equation.

The system parameters which enter into the prediction equations are defined in Table 9.2. For each of these parameters, mark the appropriate blocks of Table 9.3 if it will be possible to estimate the parameter's value for the system under consideration. The first column in which all the blocks are marked indicates which of the four equations to use.

Step 2. Evaluation of Parameters.

Instructions are given below for the numerical evaluation of each of the nine parameters. In any specific case, only those parameters need be evaluated which are required for the selected prediction equation. The rest may be ignored.

Step 2a. Evaluation of X_1 (Use Table 9.4).

1. Enter in column b the number of each type of active element in the system.

2. Multiply each value in column b by its corresponding factor in column c and enter the product in column d.

3. Add the individual products of column d and enter the sum in the space marked TOTAL.

4. Compute and enter the values for the remaining blocks. The final entry is X_1.

Step 2*b*. Evaluation of Other Parameters.

The remainder of the parameters needed for the prediction equation are obtained as indicated in Table 9.5.

Table 9.3 SELECTION OF PREDICTION EQUATION

Parameter	Equation Number (see "Prediction Equations", page 296)			
	I	II	III	IV
X_1 (Number of active elements by type)			✓	✓
X_2 (Number of cathode-ray tubes)				
X_3 (Number of transmitters)			✓	
X_4 (Highest frequency)	✓	✓	✓	
X_5 (Analog or digital)		✓		
X_6 (Steerable antenna; yes/no)	✓	✓		
X_7 (Maximum dc volts)	✓	✓		✓
X_8 (Rated power)	✓		✓	✓
X_9 (Ratio of tuning range to highest frequency)	✓	✓		✓

Procedure

Mark the blocks for which information is available. The column in which all blocks are marked indicates which prediction equation to use. Where there is a choice between equations, use the one represented by the lowest numeral (I through IV).

Step 3. Solution of Prediction Equation for θ (Table 9.6).

Step 3*a*. For the prediction equation selected in Step 1, enter the values of X_i from Tables 9.4 and 9.5 into the appropriate spaces of column a, Table 9.6.

Step 3*b*. Multiply each value in column a by the corresponding factor in column b and enter the product in column c, including the algebraic sign of the product.

Step 3c. Add the products in column c, including the constant term, and enter the sum as ln $\hat{\theta}$.

Step 3d. Enter the antilog of ln $\hat{\theta}$ (i.e., $e^{\ln \hat{\theta}}$) in the final blank ($\hat{\theta}$).

Table 9.4 WORKSHEET FOR CALCULATION OF X_1

(a) *Active Element Type*	(b) *Quantity*	(c) *Multiplying Factor*	(d) *Product* (b) (c)
Electron Tube		1.0	
Analog Transistor		0.3	
Solid State Power Rectifier		0.4	
Digital Diode		0.006	
Digital Transistor*		0.02	
Analog Diode		0.1	
Magnetic Core		10^{-5}	

	Total (Equipment Complexity, C)	
	ln C	
	Subtract from ln C	2.303
$X_1 = \ln C - 2.303 =$		

Editor's Note: Although no experience was obtained with integral circuits used in digital applications in the course of this study, recent data now indicate that an integral circuit package should be as reliable as a discrete transistor. (See *Investigation of Factors Affecting Early Exploitation of Integral Solid Circuitry*, ARINC Research Corporation, Publication No. 234–02–3–362 [April 1, 1963].)

PREDICTION OF CONFIDENCE INTERVAL

The point estimation of an *MTBF* by the procedure outlined in the preceding section must be qualified to reflect the inherent difficulties in attempting to predict system reliability during the predesign stage. To assure that the prediction represents a valid measure of system reliability, the predicted *MTBF* is better represented by a range of values in which it is likely to fall. In the procedure outlined here, the range will be chosen such that we can state with 95 per cent confidence that the actual *MTBF* will fall within the upper and lower confidence limits.

Since $\ln \theta$ is the basic quantity predicted in the procedure outlined earlier, and since the distribution of this quantity is approximately normal, then the confidence interval is in terms of $\ln \theta + K$. Expressed as hours, this interval is from θe^{-K} to θe^{+K}. For example, if $e^{K} = 2$, the confidence interval is from $\theta/2$ to 2θ.

Table 9.5 WORKSHEET FOR CALCULATION OF X_2 THROUGH X_9

Parameter	Instructions	Parameter Value
X_2	Enter the number of cathode-ray tubes in the system.	$X_2 =$
X_3	Enter the number of transmitters in the system.	$X_3 =$
X_4	(a) Enter natural logarithm of highest frequency of system, in megacycles.	
	(b) Subtract constant value 2.303	
	(c) Enter the difference as the value of X_4.	$X_4 =$
X_5	Enter 1 for digital systems or 2 for analog systems.	$X_5 =$
X_6	Enter 1 if the system has a steerable antenna; otherwise enter 0.	$X_6 =$
X_7	Enter natural log of maximum dc voltage in kilovolts.	$X_7 =$
X_8	Enter natural log of rated power consumption in kilowatts.	$X_8 =$
X_9	Enter the ratio $\dfrac{\text{tuning range}}{\text{highest frequency}}$.	$X_9 =$

Table 9.6 WORKSHEET FOR CALCULATION OF $\hat{\theta}$

Equation 9.8 I		
(a) Parameter Value	(b) Multiplying Factor	(c) Product (a) (b)
$X_1 =$	−0.801	
$X_2 =$	−0.136	
$X_3 =$	−0.596	
$X_5 =$	+0.513	
Add Constant Term		7.173
ln $\hat{\theta}$ = Total		
$\hat{\theta}$(hr) =		

Equation 9.8 II		
(a) Parameter Value	(b) Multiplying Factor	(c) Product (a) (b)
$X_8 =$	−0.121	
$X_1 =$	−0.740	
$X_2 =$	−0.090	
$X_3 =$	−0.359	
Add Constant Term		7.910
ln $\hat{\theta}$ = Total		
$\hat{\theta}$(hr) =		

Equation 9.8 III		
(a) Parameter Value	(b) Multiplying Factor	(c) Product (a) (b)
$X_7 =$	−0.484	
$X_6 =$	−1.457	
$X_5 =$	+2.206	
$X_9 =$	−0.929	
$X_2 =$	−0.139	
Add Constant Term		2.731
ln $\hat{\theta}$ = Total		
$\hat{\theta}$ (hr) =		

Equation 9.8 IV		
(a) Parameter Value	(b) Multiplying Factor	(c) Product (a) (b)
$X_2 =$	−0.277	
$X_5 =$	+2.327	
X_3	−1.075	
$X_6 =$	−1.770	
X_4	+0.192	
Add Constant Term		2.172
ln $\hat{\theta}$ = Total		
$\hat{\theta}$ (hr) =		

A step-by-step procedure for computing the 95 per cent confidence interval around the estimated *MTBF* is presented in this section. A table for facilitating the computation is included for each of the four prediction equations (see "Prediction Equations," page 296). The step-by-step procedure outlined below applies to all the tables (9.7 through 9.10).

Step 1. Select the table corresponding to the chosen prediction equation.

Step 2. Enter the values of the independent variables into the first column (X_i).

Step 3. Subtract from these values the mean values of the independent variables ($\overline{X}_i$, second column), and enter the difference into the third column ($X_i - \overline{X}_i$). Care should be taken to enter the algebraic sign correctly.

Step 4. Enter the appropriate deviation values in columns a and b of the bottom portion of the table.

Step 5. Form the products of (factor 1) × (factor 2) × (multiplying factor) for each line and enter into the product column (d).

Step 6. Form the sum of all the products plus the constant term. This sum is labeled f.

Step 7. Compute the square root of f.

Step 8. Form the product $K = 2s\sqrt{f}$.* (The value of s is given at the bottom of the table.) The confidence interval is then $\ln \hat{\theta} \pm K$.

Step 9. To determine the confidence interval in hours, evaluate the function $e^K = g$. The confidence interval is from $\hat{\theta}/g$ to $\hat{\theta}g$.

Example. A receiving set operates in the 2-to-32 megacycle band. This set has 29 vacuum tubes but no other active elements. The equipment is analog in nature and operates from a separate fixed antenna, which is not included in the prediction. The maximum dc voltage of the receiver is 165 volts, and the rated power consumption for the entire set is 88.8 watts. Using Equation 9.8 I, the computed *MTBF* is 1550 hr and the 95 per cent confidence interval is from 710 to 3385 hr. (The observed *MTBF* of this set in field operation was 2063 hr with a 95 per cent confidence interval ranging from 1808 to 2359 hr.)

---

*The term $2s\sqrt{f}$ corresponds to the 2α limits of the normal curve, or a confidence interval of approximately 95 per cent. For other confidence levels, say (100) $(1-\alpha)$ per cent, use the factor $t_{\alpha/2} \, s\sqrt{f}$, where $t_{\alpha/2}$ represents the value corresponding to the upper (100) $(\alpha/2)$ per cent tail of the t distribution with $(m-r-1)$ degrees of freedom. (The quantity m is the number of observations; r is the number of independent variables.) The $t_{\alpha/2}$ values can be obtained from any standard text on statistics.

Table 9.7 WORKSHEET FOR COMPUTING CONFIDENCE INTERVAL
FOR EQUATION 9.8 I

Independent Variables X_i	Mean of Independent Variables $\bar{X}_i$	Deviation $(X_i - \bar{X}_i)$
$X_1 =$	2.971	
$X_2 =$	1.911	
$X_3 =$	0.551	
$X_5 =$	1.847	

(a) Factor 1	(b) Factor 2	(c) Multiplier	(d) Product (a) (b) (c)
$(X_1 - \bar{X}_1) =$	$(X_1 - \bar{X}_1) =$	0.026	
$(X_2 - \bar{X}_2) =$	$(X_2 - \bar{X}_2) =$	0.005	
$(X_3 - \bar{X}_3) =$	$(X_3 - \bar{X}_3) =$	0.098	
$(X_5 - \bar{X}_5) =$	$(X_5 - \bar{X}_5) =$	0.602	
$(X_1 - \bar{X}_1) =$	$(X_2 - \bar{X}_2) =$	−0.016	
$(X_1 - \bar{X}_1) =$	$(X_3 - \bar{X}_3) =$	−0.024	
$(X_1 - \bar{X}_1) =$	$(X_5 - \bar{X}_5) =$	+0.210	
$(X_2 - \bar{X}_2) =$	$(X_3 - \bar{X}_3) =$	−0.001	
$(X_2 - \bar{X}_2) =$	$(X_5 - \bar{X}_5) =$	−0.069	
$(X_3 - \bar{X}_3) =$	$(X_5 - \bar{X}_5) =$	−0.188	

Subtotal

Add   1.020
Constant Value

$f =$ Total

$\sqrt{f} =$

$s =$ 0.378

$K = 2(\sqrt{f})(s) =$

$g = e^K =$

Table 9.8 WORKSHEET FOR COMPUTING CONFIDENCE INTERVAL
FOR EQUATION 9.8 II

Independent Variables X_i	Mean of Independent Variables $\bar{X}_i$	Deviation $(X_i - \bar{X}_i)$
$X_8 =$	1.550	
$X_1 =$	2.971	
$X_2 =$	1.911	
$X_3 =$	0.551	

(a) Factor 1	(b) Factor 2	(c) Multiplier	(d) Product (a) (b) (c)
$(X_8 = \bar{X}_8) =$	$(X_8 - \bar{X}_8) =$	0.046	
$(X_1 - \bar{X}_1) =$	$(X_1 - \bar{X}_1) =$	0.080	
$(X_2 - \bar{X}_2) =$	$(X_2 - \bar{X}_2) =$	0.003	
$(X_3 - \bar{X}_3) =$	$(X_3 - \bar{X}_3) =$	0.160	
$(X_8 = \bar{X}_8) =$	$(X_1 - \bar{X}_1) =$	−0.115	
$(X_8 = \bar{X}_8) =$	$(X_2 - \bar{X}_2) =$	−0.013	
$(X_8 - \bar{X}_8) =$	$(X_3 - \bar{X}_3) =$	−0.119	
$(X_1 - \bar{X}_1) =$	$(X_2 - \bar{X}_2) =$	0.011	
$(X_1 - \bar{X}_1) =$	$(X_3 - \bar{X}_3) =$	0.158	
$(X_2 - \bar{X}_2) =$	$(X_3 - \bar{X}_3) =$	0.004	
		Subtotal	
		Add Constant Value	1.020
		$f = $ Total	
		$\sqrt{f} =$	
		$s =$	0.382
		$K = 2(\sqrt{f})(s) =$	
		$g = e^K =$	

Table 9.9 WORKSHEET FOR CONFIDENCE INTERVAL FOR
EQUATION 9.8 III

Independent Variables X_i	Means of Independent Variables $\overline{X}_i$	Deviation $(X_i - \overline{X}_i)$
$X_7 =$	1.529	
$X_6 =$	0.237	
$X_5 =$	1.847	
$X_9 =$	0.325	
$X_2 =$	1.911	

(a) Factor 1	(b) Factor 2	(c) Multiplier	(d) Product (a) (b) (c)
$(X_7 - \overline{X}_7) =$	$(X_7 - \overline{X}_7) =$	0.016	
$(X_6 - \overline{X}_6) =$	$(X_6 - \overline{X}_6) =$	0.364	
$(X_5 - \overline{X}_5) =$	$(X_5 - \overline{X}_5) =$	0.322	
$(X_9 - \overline{X}_9) =$	$(X_9 - \overline{X}_9) =$	0.248	
$(X_2 - \overline{X}_2) =$	$(X_2 - \overline{X}_2) =$	0.006	
$(X_7 - \overline{X}_7) =$	$(X_6 - \overline{X}_6) =$	−0.020	
$(X_7 - \overline{X}_7) =$	$(X_5 - \overline{X}_5) =$	0.082	
$(X_7 - \overline{X}_7) =$	$(X_9 - \overline{X}_9) =$	0.046	
$(X_7 - \overline{X}_7) =$	$(X_2 - \overline{X}_2) =$	−0.009	
$(X_6 - \overline{X}_6) =$	$(X_5 - \overline{X}_5) =$	−0.174	
$(X_6 - \overline{X}_6) =$	$(X_9 - \overline{X}_9) =$	0.143	
$(X_6 - \overline{X}_6) =$	$(X_2 - \overline{X}_2) =$	−0.058	
$(X_5 - \overline{X}_5) =$	$(X_9 - \overline{X}_9) =$	−0.103	
$(X_5 - \overline{X}_5) =$	$(X_2 - \overline{X}_2) =$	−0.028	
$(X_9 - \overline{X}_9) =$	$(X_2 - \overline{X}_2) =$	−0.008	
	Subtotal		
	Add Constant Value	1.020	
	$f =$ Total		
	$\sqrt{f} =$		
	$s =$	0.626	
$K = 2(\sqrt{f})(s) =$			
$g = e^K =$			

Table 9.10 WORKSHEET FOR COMPUTING CONFIDENCE INTERVAL
FOR EQUATION 9.8 IV

Independent Variables X_i	Means of Independent Variables X_i	Deviation $(X_i - \overline{X}_i)$
$X_2 =$	1.911	
$X_5 =$	1.847	
$X_3 =$	0.551	
$X_6 =$	0.237	
$X_4 =$	1.164	

(a) Factor 1	(b) Factor 2	(c) Multiplier	(d) Product (a) (b) (c)
$(X_2 - \overline{X}_2) =$	$(X_2 - \overline{X}_2) =$	0.017	
$(X_5 - \overline{X}_5) =$	$(X_5 - \overline{X}_5) =$	0.019	
$(X_3 - \overline{X}_3) =$	$(X_3 - \overline{X}_3) =$	0.129	
$(X_6 - \overline{X}_6) =$	$(X_6 - \overline{X}_6) =$	0.154	
$(X_4 - \overline{X}_4) =$	$(X_4 - \overline{X}_4) =$	0.081	
$(X_2 - \overline{X}_2) =$	$(X_5 - \overline{X}_5) =$	−0.004	
$(X_2 - \overline{X}_2) =$	$(X_3 - \overline{X}_3) =$	0.010	
$(X_2 - \overline{X}_2) =$	$(X_6 - \overline{X}_6) =$	−0.237	
$(X_2 - \overline{X}_2) =$	$(X_4 - \overline{X}_4) =$	0.004	
$(X_5 - \overline{X}_5) =$	$(X_3 - \overline{X}_3) =$	0.012	
$(X_5 - \overline{X}_5) =$	$(X_6 - \overline{X}_6) =$	0.123	
$(X_5 - \overline{X}_5) =$	$(X_4 - \overline{X}_4) =$	−0.061	
$(X_3 - \overline{X}_3) =$	$(X_6 - \overline{X}_6) =$	−0.161	
$(X_3 - \overline{X}_3) =$	$(X_4 - \overline{X}_4) =$	−0.084	
$(X_6 - \overline{X}_6) =$	$(X_4 - \overline{X}_4) =$	−0.284	
		Subtotal	
		Add Constant Value	1.020
		$f =$ Total	
		$\sqrt{f} =$	
		$s =$	0.698
		$K = 2(\sqrt{f})(s) =$	
		$g = e^K =$	

9.6 Failure Rate Data

Essential to the prediction process is the availability of dependable failure-rate data. Different sets of data are based on dissimilar failure criteria, and the documented rates can vary considerably. Table 9.11, for instance, lists a large number of basic failure rates from 14 sources, from which it is obvious that variations of three orders of magnitude for many types of parts are not uncommon. This section lists some of the major data sources, with comments on the related premises, correction factors, and methods of generation.

Sources of Failure Data

Table 9.12 lists several failure-rate data sources that have relatively wide circulation and use. It is instructive to compare the premises on which two sets of data were formulated. The data presented in MIL-HDBK-217 (Reference 2, Table 9.12) were developed by RCA on an Air Force contract; they were based on three classes of ground-based equipments—a long-range search radar, a communications radio set, and a radar identification set—and incorporate these major premises:

1. The only types of failure were random catastrophic failures and random failures resulting from rapid deterioration.
2. Part types were limited to electronic, electrical, and electromechanical.
3. Equipment design was considered to be entirely adequate.
4. Equipment maintenance was extremely sound.
5. Basic part failure rates are functions of stresses under ground-based or laboratory bench conditions.

The data presented in Table 9.13 (Reference 1, Table 9.12) were based on 200 million part-operating hours accumulated in nine different types of airborne electronic systems; they were obtained over a nine-month period ending in March 1962. The data and the associated prediction procedure incorporate the following premises:

1. The operator's opinion of the equipment is the criterion for determining satisfactory or unsatisfactory operation.
2. Equipment failure may be caused by deterioration mechanisms and time-dependent drift characteristics of the piece parts, although individual parts parameters may be within their respective specified values.
3. Equipment failure may occur as a result of the design.
4. Equipment failure may occur as the result of the quality of the maintenance performed during corrective and/or preventive maintenance repair actions.

Table 9.11 PART FAILURE RATE COMPARISONS (FAILURES PER MILLION HOURS)

Part Type	ARINC (1)	Collins (2)	Boeing (3)	Autonetics (4)	Autonetics (5)	G.E. (6)	Philco (7)	STL (8)	VITRO (9)	RCA (10)	IBM (11)	BTL (12)	RCA (13)	T-I (14)
Alternators			775.0											
Batteries (Secondary)			1429.0											
Capacitors (N.O.C.)	3.4			10.0	1.0					0.6		0.01		5. (Trimmer)
Capacitors (Air)		0.44										1.0		
Capacitors (Ceramic)		0.49				0.2		0.01	0.9	0.35	0.1		0.35	0.17
Capacitors (Electrolytic)		2.34								0.6			2.92	
Capacitors (Mica)		0.43	1.0			0.1		0.01	0.39	0.15	0.1		0.01	
Capacitors (Paper)		0.94	5.0					0.01	0.8	0.17	0.6		0.17	5. (Mylar)
Capacitors (Tantalum)						10.0		0.15					1.12	5.
Coil, RF		1.5		1.0	1.0	1.0		0.1					0.5	5.
Connector	1.5	0.7		10.0	2.0	0.5	2.0			2.0	0.1	0.01	0.3/ Pin	
Crystal (Quartz)				10.0	2.0						36.2		0.2	
Diode (N.O.C.)	17.2 Sig. 64.5 Pow.	4.5		0.3	0.3	3.5								
Diode (Germanium)						0.3				8.6		0.3	0.54	
Diode (Selenium)	50.0									7.15	6.0			
Diode (Silicon)	120.0 (RF Mixer)		3.0					0.15		3.85	172.0 (RF Mixer)	0.2	0.5	
Fuse			36.0	3.0	3.0								0.1	
Generator (DC)			160.0	20.0	10.0								21.0	
Gyros	394.0										82.0		50.0	
Heaters (Elec.)	57.0		100.0	5.0	2.0			0.15			7.5		1.0	
Inductors												0.02	0.5	
Lamp (Indicator)				3.0	3.0	0.1							1.0	
Motor, AC	88.0	23.0	556.0	10.0	5.0		10.0	0.50		11.5	15.0	0.15	16.0	
Motor (Synchro)	45.0		200.0	50.0	20.0					11.5	18.0		10.0	
Potentiometers	70.0	1.52		10.0	5.0	10.0		0.1			17.5	0.5	0.72	
Printed Ckt. (Beard)				20.0	1.0									
Photoconductors								0.2						
Relay	125.0	5.0	50.0	50.0	50.0	10.0	1.1	0.3		2.5	9.0	0.1- 0.5	5.17	
Regulators (Pressure)								0.15						
Resistor (N.O.C.)	3.4	0.09		2.0	1.0							0.01		
Resistor (Comp.)		0.04	1.0			0.5		0.06	0.6	0.66	0.5		0.2	0.15
Resistor (Film)		<0.5				0.2		0.02		0.29	1.0		0.52	0.2
Resistor (W.W.)		0.17	5.0					0.15	6.0	1.50 Prec. 1.89 Pow.	2.0		1.50 Prec. 1.30 Pow.	
Socket				60.0	3.0									
Solenoid	36.0			10.0							9.5			
Switch	25.0	7.0	33.0	20.0	5.0	0.4	0.3			1.5	10.0	0.03	0.58	
Transformer (Pulse)				20.0	5.0						500.0	0.1	0.4	
Transformer (Mag-Amp)				10.0	5.0						10.8		0.25	

(continued)

Table 9.11 PART FAILURE RATE COMPARISONS (FAILURES PER MILLION HOURS) (continued)

Part Type	ARINC (1)	Collins (2)	Boeing (3)	Autonetics (4)	Autonetics (5)	G.E. (6)	Philco (7)	STL (8)	VITRO (9)	RCA (10)	IBM (11)	BTL⁻ (12)	RCA (13)	T-I (14)
Transformer (Filament)				10.0	2.0						13.5	0.02	0.15	
Transformer (N.O.C.)	16.3	4.0				10.0		0.2		1.5	1.9		5.0	
Transistor (N.O.C.)	17.2 Sig. 64.5 Pow.	10.0	14.0	10.0	1.0	5.0-20.0	0.1 0.2						9.0	
Transistor (Germanium)											2.0	0.9	0.76	
Transistor (Silicon)								0.45			2.0		0.7	
Vacuum Tube (N.O.C.)	22.5	34.0		24.0	4.0				27.0		15.0	2.0	5.0-20.0	30.0
Valves								0.3						

(N.O.C.—not otherwise classified)

Data Sources for Part Failure Rate Comparisons

1. ARINC AFRAP Report
   Capacitors @ 50% Voltage
   Resistors @ 50% Rating
   Vacuum Tubes @ Class 4

2. Collins CTR 195—No Trade-off Figure

3. BAC DOC D6-2770—No Trade-off

4. Autonetics Bulletin 41—No Trade-off
   Commercial Grade Parts

5. Autonetics Bulletin 41—No Trade-off
   Autonetics Controlled Parts

6. G.E. Heavy Military Electronic Dept.
   No Trade-off. 5th R & Q.C. Symposium

7. Philco Observed Rates—Transac S-2000 Computer

8. STL Rates Used for "STEER"
   STL Rates Taken at 25% Rating, 30°C

9. Vitro Tech. Report #80
   Capacitors @ 50% Voltage
   Resistors @ 50% Rating

10. RCA TR-1100 (Old Edition)
    All Components @ 50% Rating, 40°C

11. IBM Extrapolated Failure Rates
    for Airborne Equipment

12. BTL Rates for Lab. Equipment. All Parts Operated
    at <25% of Rating in 15-25°C Environment. Rates
    to be Multiplied by Following Factors for Military
    Field Use:

Use	Vacuum Tube Equipment	Transistorized Equipment
Lab.	1.0	1.0
MIL Ground Station	1.1	1.1
Ship	10.0	7.7
Trailer	20.0	14.3
Aircraft	33.3	25.0
Missile	83.3	62.5

13. RCA TR-59416-1
    All Components @ 50% Rating, 40°C

14. Texas Instrument Prediction, Project
    Mercury Telemetry Tx.

Although the failure data listed in Table 9.13 are listed as *part* failure rates and *part* adjustment rates, they would be more accurately identified as rates of *system* repair by action to a specific part type. For simplicity, the terms "part failure rate" and "part adjustment rate" are retained and the sum of these two is referred to as the *estimated part malfunction rate*.

Development of Derating Curves

The data for the generation of derating curves come from several sources, such as life tests of component parts, system life tests, or field operation. As previously indicated, there are wide variations in the effects of factors such as criteria of failure and applied stresses, and in the quality of failure analysis and data analysis. A large number of derating curves for many types of parts are unavailable (see References 2 and 4, Table 9.12). Table 9.14 and Figure 9.6* demonstrate how such curves might be generated.

*D. A. Adams, "Component-Part Failure Rate Curve Considerations," *Proceedings Ninth National Symposium for Reliability and Quality Control* (January, 1963). Adapted with permission.

Generally, the information shown in Table 9.14 is not available for the vast majority of published curves. Curves of this type are based on the assumption that temperature is the primary stress and consequently that Arrhenius' Law, which defines the chemical reaction rate and temperature relationships, can be used to establish interpolated stress-failure-rate contours.

Correction Factors for Stress

As indicated earlier in this chapter, many reliability prediction procedures require that the basic part failure rates be modified, depending on anticipated environmental, electrical, and thermal stresses. The result usually requires an extensive stress analysis so that a reasonable point estimate of system reliability can be made. This work proves useful for pinpointing potential areas of unreliability, even though the predicted reliability may not be as accurate as desired.

The correction factors generally take the form of an equation similar to

$$\lambda_n = \lambda_0(K_1 K_2 K_3 K_4 K_5 K_6) \tag{9.9}$$

where

λ_n is the adjusted failure rate;

λ_0 is the basic, or standard, failure rate;

K_1 corrects for applied stresses;

K_2 relates the proportion of likely tolerance failures to random catastrophic failures;

Table 9.12 SOURCES OF FAILURE RATE DATA

1. *Prediction of Field Reliability for Airborne Electronic Systems*, ARINC Research Corporation, Publication No. 203–1–344 (December 31, 1962).

2. *Reliability Stress Analysis for Electronic Equipments*, MIL-HDBK-217 (December 31, 1961).

3. *Failure Rates*, D. R. Earles and M. F. Eddins, AVCO Corporation (April, 1962). [An updated version appears in *Proceedings*, Ninth National Symposium on Reliability and Quality Control (January, 1963).]

4. *Bureau of Naval Weapons Failure Rate Data Handbook*, U.S. Naval Ordnance Laboratory, Corona, Calif. (Available only to qualified contractors and government agencies.)

5. *Reliability Prediction and Measurement of Shipboard Electronic Equipments*, Vitro Corporation, Report No. 98 (April 15, 1957).

Table 9.13 PART RELIABILITY DATA FOR AIRBORNE ELECTRONIC SYSTEMS

1	2	Observed Data	Computed Estimates		6
Part Category	Part Type	Part Hours, T_i (Millions)	Failures per Hour (multiply by 10^{-6})	Required Adjustments per Hour (multiply by 10^{-6})	Note References
		3	4	5	1
			Normalized for Zero On-Off Cycles and Corrected for Clustering		
Attenuator	Variable	0.070			2
Capacitor (fixed)	Ceramic	10.936	0.11		6
	Electrolytic	1.495	2.21		6
	Glass	4.886			2, 6
	Mica	10.337			2, 6
	Paper	22.128	0.19		6
	Plastic Film	0.206	8.74		6
	Printed	0.210			2, 6
	Tantalum	1.708	4.39		6
Capacitor (variable)	All	3.263		0.74	2
Connector (plug or receptacle)	Coaxial	2.234	6.04		4
	All others	358.758	0.02		3, 4
(plug only)	All	5.726		1.15	4

Crystal	Oscillator	2.958	0.61		
Diode (solid state)	Germanium	1.430	2.94		6
	Magnesium-Cop.Sulph.	0.021			2, 6
	Selenium	1.300	2.54		6
	Silicon	2.616	33.4		6
Electron Tube	Cathode Ray	0.041	141.5		5, 6
	Klystron	0.052	238.5	28.8	5, 6
	Magnetron	0.035	557.1		5, 6
	TR	0.052	423.1		5, 6
	Miniature, Ampl. (a)	0.396			
	(b)	4.142	8.44		
	Miniature, Rect. & Gas (a)	0.409			
	(b)	0.708	20.1		

(continued)

Notes

1. Values in the normalized columns must be modified to include the effects of on-off cycling by multiplying the values by the term $1+8N$, where N is equal to the expected number of on-off cycles per operating hour.

2. No replacements or repairs were observed for this part type. As a conservative estimate of the failure rate, $\hat{\lambda}_i$, the upper 50% confidence limit on zero observations may be used. This value is computed by the equation

$$\hat{\lambda}_i = \frac{0.693}{\text{observed part hours}}.$$

The corresponding confidence statement is $P(0 \leq \lambda \leq \hat{\lambda}_i) = 0.50$.

3. The part hours shown on this line are pin and socket hours for the plugs and receptacles. Correspondingly, the value in column 9 represents the rate of plug and receptacle failure per pin and socket hour. To obtain the estimated failure rate for a plug, therefore, the number of pins is multiplied by 0.02; similarly, for a receptacle, the number of sockets is multiplied by 0.02.

4. The adjustment data for this category has been presented on a separate line since it pertains only to plugs. For the malfunction rate of plugs, the adjustment rate must be added to the failure rate; the malfunction rate of receptacles is equal to the failure rate alone.

5. Letter (a) refers to parts groups in the AN/ARC-34 and AN/ARC-65 equipments; letter (b) refers to parts groups in equipment types AN/ARA-25, AN/APX-25, AN/ARN-21, AN/APN-89, AN/APN-89A, AN/ASB-4, and AN/ASB-4A.

6. When abnormal stress conditions are expected, modification of basic failure rates may be required. (See page 6 of Reference 1, Table 9.3.)

Table 9.13 PART RELIABILITY DATA FOR AIRBORNE ELECTRONIC SYSTEMS (continued)

1	2	Observed Data 3	Computed Estimates 4	5	6
Part Category	Part Type	Part Hours, T_i (Millions)	Failures per Hour (multiply by 10^{-6})	Required Adjustments per Hour (multiply by 10^{-6})	Note References
			Normalized for Zero On-Off Cycles and Corrected for Clustering		
Electron Tube (continued)	Subminiature, Ampl. (a)	0.929	} 6.66		5, 6
	(b)	5.907			
	Subminiature, Rect. & Gas (a)	0.327	} 10.3		5, 6
	(b)	0.734			
	Other, Ampl. (a)	0.241	} 60.7		5, 6
	(b)	0.364			
	Other, Rect. & Gas (a)	0.086	} 11.2		5, 6
	(b)	0.191			
Filter	Band Pass	0.163			2
	Harmonic	0.038			2
	R.F. Interference	1.054			2
Heater	—	1.159	3.11		
Inductor (fixed & variable)	All	7.961	1.06		
Meter & Counter	—	0.284			2

Motor	Blower, AC	0.158	28.5	2
	Blower, DC	0.069	21.7	
	Motor & Pump, AC	0.079	11.4	
	Servo or Set, AC	0.813	15.5	
	Servo or Set, DC	0.378	1.59	
	Timing or Clock, AC	0.017	17.6	
	Timing or Clock, DC	0.038		
Motor Generator	—	0.705	43.4	
Relay	Rotary	0.138	41.3	
	Switching	4.474	21.6	
	Switching, Dry circuit	0.940	66.7	
	Time Delay	0.127	16.5	

(continued)

Notes

1. Values in the normalized columns must be modified to include the effects of on-off cycling by multiplying the values by the term $1 + 8N$, where N is equal to the expected number of on-off cycles per operating hour.

2. No replacements or repairs were observed for this part type. As a conservative estimate of the failure rate, $\hat{\lambda}_i$, the upper 50% confidence limit on zero observations may be used. This value is computed by the equation

$$\hat{\lambda}_i = \frac{0.693}{\text{observed part hours}}.$$

The corresponding confidence statement is $P(0 \leq \lambda \leq \hat{\lambda}_i) = 0.50$.

3. The part hours shown on this line are pin and socket hours for the plugs and receptacles. Correspondingly, the value in column 9 represents the rate of plug and receptacle failure per pin and socket hour. To obtain the estimated failure rate for a plug, therefore, the number of pins is multiplied by 0.02; similarly, for a receptacle, the number of sockets is multiplied by 0.02.

4. The adjustment data for this category has been presented on a separate line since it pertains only to plugs. For the malfunction rate of plugs, the adjustment rate must be added to the failure rate; the malfunction rate of receptacles is equal to the failure rate alone.

5. Letter (a) refers to parts groups in the AN/ARC-34 and AN/ARC-65 equipments; letter (b) refers to parts groups in equipment types AN/ARA-25, AN/APX-25, AN/ARN-21, AN/APN-89, AN/APN-89A, AN/ASB-4, and AN/ASB-4A.

6. When abnormal stress conditions are expected, modification of basic failure rates may be required. (See page 6 of Reference 1, Table 9.3.)

Table 9.13 PART RELIABILITY DATA FOR AIRBORNE ELECTRONIC SYSTEMS (continued)

Part Category	Part Type	Observed Data	Computed Estimates		Note References
1	2	Part Hours, T_i (Millions) 3	Failures per Hour (multiply by 10^{-6}) 4	Required Adjustments per Hour (multiply by 10^{-6}) 5	6
			Normalized for Zero On-Off Cycles and Corrected for Clustering		1
Resistor (fixed)	Carbon Composition	69.740	0.16		6
	Carbon, Deposited	1.582	0.57		6
	Carbon Film	18.393	0.11		6
	Wire	11.483	1.83		6
Resistor (variable) and Potentiometer	Carbon Composition	0.977	4.91	54.0	6
	Carbon Comp., Computing	0.473	6.98	54.0	6
	Carbon Film	0.202	53.5	54.0	6
	Wirewound	1.122	8.82	54.0	6
	W.W., Infinite Resol.	0.144	68.8	54.0	6
	Wirewound, Computing	0.884	19.0	54.0	6
Solenoid	Axial	0.532	6.20		
	Rotary	0.158	45.6		
Switch	Bomb Release	0.021	157.1		
	Cam	0.123	290.2		
	Commutator Type	0.021	314.3		
	Momentary Contact	0.030	10.0	11.5	
	Microswitch/S.A.	1.694	3.19		
	Push Button	0.240	11.3		

Category	Type				Notes
	Rotary	3.017	1.69		2
	Thermostat	1.165			2
	Toggle	0.836			
Synchro	Control Transformer	0.422	0.71	4.59	2
	Differential Generator	0.041		4.59	
	Synchro Generator	0.370	0.81	4.59	2
	Synchro Repeater	0.085		4.59	
	Synchro Resolver	0.521	16.1	4.59	
Transformer	High Voltage	0.111			2, 6
	Power and Filament	1.326	0.45		6
	Others (IF, Audio, etc.)	3.336	0.27		6
Transistor	All	0.201			2

Notes

1. Values in the normalized columns must be modified to include the effects of on-off cycling by multiplying the values by the term $1 + 8N$, where N is equal to the expected number of on-off cycles per operating hour.

2. No replacements or repairs were observed for this part type. As a conservative estimate of the failure rate, $\hat{\lambda}_i$, the upper 50% confidence limit on zero observations may be used. This value is computed by the equation

$$\hat{\lambda}_i = \frac{0.693}{\text{observed part hours}}.$$

The corresponding confidence statement is $P(0 \le \lambda \le \hat{\lambda}_i) = 0.50$.

3. The part hours shown on this line are pin and socket hours for the plugs and receptacles. Correspondingly, the value in column 9 represents the rate of plug and receptacle failure per pin and socket hour. To obtain the estimated failure rate for a plug, therefore, the number of pins is multiplied by 0.02; similarly, for a receptacle, the number of sockets is multiplied by 0.02.

4. The adjustment data for this category has been presented on a separate line since it pertains only to plugs. For the malfunction rate of plugs, the adjustment rate must be added to the failure rate; the malfunction rate of receptacles is equal to the failure rate alone.

5. Letter (a) refers to parts groups in the AN/ARC-34 and AN/ARC-65 equipments; letter (b) refers to parts groups in equipment types AN/ARA-25, AN/APX-25, AN/ARN-21, AN/APN-89, AN/APN-89A, AN/ASB-4, and AN/ASB-4A.

6. When abnormal stress conditions are expected, modification of basic failure rates may be required. (See page 6 of Reference 1, Table 9.3.)

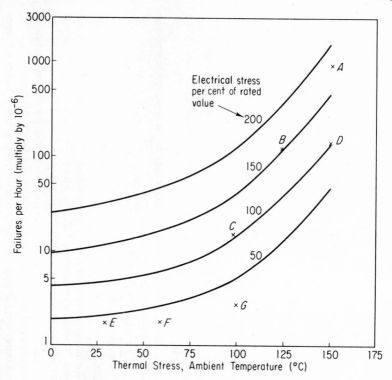

Fig. 9.6 Derating curves showing data curves from various sources.

K_3 adjusts for changes in external environments;

K_4 is a possible adjustment required to account for different maintenance practices which can have an effect on observed system failures;

K_5 denotes system complexity—the more complex the system, the greater will be its failure rate;

K_6 accounts for observed cycling effects.

Prediction techniques vary considerably in the degree that these factors are utilized or considered. For example, the AVCO method* generic failure rates (normalized to laboratory conditions) must be multiplied by application (derating) factors and then by factors representing the installation environment.

The procedure recommended by MIL-STD-756† uses the procedures of MIL-HDBK-217, wherein the predicted failure rate of a system is determined

---

*Reference 3, Table 9.12.

†*Procedures for Prediction and Reporting Prediction of Reliability of Weapon Systems* (October 3, 1961).

Table 9.14 DATA FOR FIGURE 9.6

| Data Source | Data Type | Stress | | Failure Criteria | Part Operating Hours | Observed Failures Per Hour (Multiply by 10^{-6}) |
		Electrical (Per cent)	Thermal (°C)			
A	Part Life Test	175	150	Parameter Change ±10%	10^5	720
B	Part Life Test	150	125	Parameter Change +5, −2%	10^5	140
C	Part Life Test	100	100	Parameter Change ± 1%	10^5	22
D	Part Life Test	100	150	Parameter Change ± 1%	10^5	158
E	System Field Operation	45	30	Any Change to Cause System Malfunction	10^6	2
F	System Field Operation	20	60	Any Change to Cause System Malfunction	10^4	2
G	Circuit Life Test	25	100	Any Change to Cause Circuit Malfunction	10^6	5

by considering basic failure rates of the parts suitably modified by anticipated electrical and thermal stress factors, then further modified by a suitable factor related to the anticipated system environment. In this case, the environmental factors are

Shipborne/Fixed Ground	1.0
Aircraft	6.5
Missiles	80.0
Satellite: Launch and Boost Phase	80.0
Satellite: Orbit Phase	1.0

Section 9.4, under the heading "Modify Preliminary Block Failure Rates or Reliabilities," page 292, recommends a variation of this method in which the adjustment factors are treated as exponents in the final reliability equation. In addition, separate factors are provided to account for the differences in reliability between analog and digital circuitry.

The data provided in column 4 of Table 9.13 are for *airborne electronic systems*; they generally will not require adjustments for stress conditions, since all important factors have been included. This approach has the advantage of providing a quick means of making a reliability prediction when a detailed stress analysis is not required. Reference 1 of Table 9.12 provides failure-rate correction procedures for use where the expected stress/operating conditions are deemed to be significantly abnormal. The data in column 5 may be added to that of column 4, if required, to compensate for maintenance adjustments. Another correction factor that can be applied is to multiply rates in column 4 by the factor $1 + 8N$, where N is the expected number of on-off cycles per operating hour.

9.7 Computing Confidence Interval about the Predicted Point Estimate of the System Malfunction Rate*

One of the deficiencies of most part reliability data is the absence of total part-operating hours associated with the stated failure rate under given environmental conditions. When this type of data is provided (and there are indications that this is now being done in some recent compilations— Table 9.15, for example) we can compute an approximate confidence

Table 9.15 CONSTANTS AND FORMULAS FOR OBTAINING $100 (1 - \alpha)$ PER CENT CONFIDENCE INTERVAL FOR SYSTEM MALFUNCTION RATE, λ_s

	Constants Confidence Level			Formulas*	
α	[$100(1-\alpha)$ per cent]	C_α	D_α	Two-Sided Interval	One-Sided Interval
0.01	99	2.576	2.326	$P[lcl \leq \lambda_s \leq ucl] =$ $1 - \alpha$	Lower-limit interval: $P[lcl \leq \lambda_s < \infty] =$ $1 - \alpha$
0.02	98	2.326	2.054		
0.05	95	1.960	1.645	$lcl = \hat{\lambda}_s - C_\alpha \sqrt{\sum_{i=1}^{n} Z_i}$	$lcl = \hat{\lambda}_s - D_\alpha \sqrt{\sum_{i=1}^{n} Z_i}$
0.10	90	1.645	1.282		
0.20	80	1.282	0.842	$ucl = \hat{\lambda}_s + C_\alpha \sqrt{\sum_{i=1}^{n} Z_i}$	Upper-limit interval: $P[0 \leq \lambda_s \leq ucl] =$ $1 - \alpha$
0.40	60	0.842	0.253		
0.50	50	0.674	0.000		$ucl = \hat{\lambda}_s + D_\alpha \sqrt{\sum_{i=1}^{n} Z_i}$

*lcl = lower confidence limit.
ucl = upper confidence limit.
n = number of part types in the system.

*H. S. Balaban and A. Drummond, *Prediction of Field Reliability of Airborne Electronic Systems*, ARINC Research Corporation, Publication No. 203-1-344. This technique is valid only for the exponential assumption and for series systems.

interval about the predicted point estimate of a constant system malfunction rate as follows:

1. For each part type in the system, determine the value Z_i from the equation

$$Z_i = \frac{M_i^2}{T_i} \hat{\lambda}' \qquad (9.10)$$

where

M_i = number of parts of the ith type in the system,

T_i = total observed operating hours on the ith part type, from data source, and

$\hat{\lambda}_i'$ = estimated malfunction rate for the ith part type.

2. *For a two-sided confidence interval*, use Table 9.15 to determine the value of C_α for the selected confidence interval.

3. Insert the value of Z_i and C_α into the appropriate equation in Table 9.15 and compute the upper and lower confidence limits.

4. *For a one-sided confidence interval* (upper or lower limit), use Table 9.15 to determine the value of D_α for the selected confidence level.

5. Insert the values of Z_i and D_α into the appropriate equation in Table 9.15 and compute the required confidence limit.

Example. Given a point estimate of system reliability, $\hat{\lambda}_s = 65 \times 10^{-6}$ failures per hour, compute upper (single) 90 per cent confidence limit for $\hat{\lambda}_s$ and lower (single) 90 per cent confidence limit for $\hat{\theta}_s$ based on the following parts data:

Part Type	No. of Parts in System M_i	Estimated Failure Rate for Use Condition— λ_i	Based on T_1 hr of Test Data	Based on T_2 hr of Test Data
Transistors	10	6×10^{-6}	10^4	10^8
Composition resistors	30	0.1×10^{-6}	10^5	10^{10}
Paper capacitors	20	0.1×10^{-6}	10^5	10^{10}

1. For failure rates based on T_1 hr:

$$Z_1 + Z_2 + Z_3 = 6.13 \times 10^{-8}.$$

One-sided upper confidence limit $= \hat{\lambda}s + 1.28 \sqrt{6.13 \times 10^{-8}}$

$$= 65 \times 10^{-6} + 3.175 \times 10^{-4}$$

$$= 383 \times 10^{-6}$$

$$(\hat{\theta}_s = \frac{1}{65 \times 10^{-6}} = 15{,}300 \text{ hr}; \ lcl = \frac{1}{383 \times 10^{-6}} = 2600 \text{ hr}).$$

2. For failure rates based on T_2 hr:

$$\text{One-sided upper confidence limit} = \hat{\lambda}s + 1.28 \sqrt{6.013 \times 10^{-12}}$$
$$= 65 \times 10^{-6} + 3.15 \times 10^{-6}$$
$$= 68.15 \times 10^{-6}$$

$$(\theta = \frac{1}{65 \times 10^{-6}} = 15,300 \text{ hr}; \, lcl = \frac{1}{68.15 \times 10^{-6}} = 14,600 \text{ hr}).$$

9.8 A Reliability Prediction Method for Propulsion Systems

The method used in predicting the reliability of propulsion systems is similar, but not identical, to the method used in predictions of reliabilities of electronic systems. A basic ingredient in each is the reliability block diagram, a pictorial representation of a failure effects analysis. One difference lies in the block identification. In electronic systems, the blocks in the diagram are usually identified as parts or components with only infrequent reference to mechanism or mode of failure. In propulsion systems, the blocks are identified both by the component or components involved and by the modes or mechanisms of failure. Thus, in propulsion system reliability predictions, we rarely speak abstractly about "failure rates of parts"; we must speak about "types of failures of parts in specific applications."

Definition of Failure

The discussion above emphasizes the importance of an accurate, precise definition of failure; the conclusions are critically dependent on such a definition. This statement applies not only to the definition of system failure, but also to the definition of subunit failure, where the subunits are those portions of the system for which failure data are available and are to be used in the system prediction. Now the tolerance of the system to subunit performance variation, including inoperability, is the controlling factor in determining the meaning of subunit failure. In propulsion systems this tolerance varies significantly with type and timing of subunit performance variation. A sticking valve may or may not affect system performance, depending on whether the valve sticks open or closed and also when it sticks. A LOX or fuel leak may or may not cause trouble, depending on amount, location, and timing of the leakage. Even thrust chamber burnout may not adversely affect system performance if safe shutdown is accomplished, guidance problems are not extreme, and timing is satisfactory. In other words, subunit failure in propulsion systems cannot often be defined except in the "context" of the system. The net effect of unusual subunit performance on the system in each individual case, as well as the series or parallel arrangement involved, must be considered in defining failure.

The preceding discussion of the meaning of failure implies a yes-no definition for use in reliability prediction. "Degree of failure" can be considered if we are studying performance degradation in the larger context of system effectiveness.

Types of Failures and Their Relationship to Reliability Prediction

In addition to defining failure for systems and subdivisions thereof, it has been found helpful to classify failures into four different types in propulsion system reliability analyses. The four categories overlap to some extent, but in the main they are reasonably well identified. The first two are predominantly human errors while the third and fourth are more closely associated with design characteristics. The four failure types are as follows:

Failure Type A. This type has been designated a human error which inflicts accidental damage to the hardware after it has passed the last point of inspection or checkout, the result being either erroneous operation of equipment or reduction of safety margins. Type A failures can be estimated or identified by a design review to uncover areas of obvious vulnerability which can then be redesigned. If historical data on similar systems exist, such data can be used for identifying failure mechanisms and areas which require special design attention.

Failure Type B. This type is a quality-control failure (often considered the result of a type of human error) which permits defects to remain undiscovered prior to use. (Design failures occur even though the prescribed and specified tolerances are apparently met.) The frequency of occurrence of Type B failures in a new design cannot be estimated. Recourse to historical data may give some indication of the efficiency of the quality control group, and this may or may not be of significance in a new design.

Failure Type C. This type is an "embryonic" design failure which results from unknown or unpredicted failure mechanisms. Such failures may be due to either a more severe operating environment or lower component resistance than was anticipated. In general, these failures are discovered early in the operational or flight test program and are, in fact, the reason for such a program. By definition, Type C failure modes cannot be predicted. The frequency of occurrence of such failures may be partly predictable, however, since there is apparently some relationship between their occurrence and the presence of components and subsystems which cannot be analyzed and tested under simulated flight environments.

Failure Type D. This type results from an acceptable design risk (that is, use of components and equipments which are not completely reliable) taken because of compromises, trade-offs, or constraints imposed by performance, schedules, technology, funds, production processes, weight, etc. Type D failures can be predicted if failure rates for the components and equipment used can be obtained. Such data, when they do exist, often include those Type A failures which are characteristic of the component design and those Type B failures which are characteristic of the agency or contractor performing the tests.

Failure Interactions

In addition to placing emphasis on the effect of malfunctions on system operation, analyses of multiple part-failure modes have a large and important role in propulsion-system reliability analysis. For example, consider a liquid oxygen tank of a rocket propulsion system with two vent valves. Engineering analysis may show that, if either of the two valves fails open, sufficient venting capacity exists to cause cavitation of the LOX pump and subsequent engine shutdown. However, if one of the valves fails closed, the remaining valve could handle the operational requirements. Thus, it is not enough to speak of vent valve failure—it is essential to speak of modes of failure and to have data available for frequency of failure by mode. We need to compute three different probabilities of system failure due to these vent valves:

1. Valve #1 fails open.
2. Valve #2 fails open.
3. Valves #1 and #2 both fail closed.

Active and Passive Components in Propulsion Systems

In discussing failure interactions, we have observed the importance of relating mode of failure to its effect on system performance, with special mention of the components with more than one functional mode of operation and the relationship between their failure mode probabilities and the probability of system failure. To identify these component types in propulsion systems, we use the term *passive component* to denote a component with a single failure mode, and the term *active component* to denote one with more than one failure mode. For example, a high-pressure gas bottle has only one mode of failure—inability to hold pressure when required. (Of course, it has many mechanisms of failure which can cause the single failure mode.) An active component can be illustrated by a valve which can operate when not needed or can fail to operate when needed—it can close when it should be open or remain open when it should close. (Of course, the valve too can have many mechanisms or causes of failure for either of the modes.)

Redundancy usually increases reliability, but, with active elements in propulsion systems, redundancy can actually lower reliability, depending on probability relationships for the different failure modes. This will be illustrated after we discuss the formula for redundancy in this case.

Consider two valves, A and B, each with two failure modes and associated probabilities as follows:

	Valve A	Valve B
Failure-free operation	a	b
Fail open	a_o	b_o
Fail closed	a_c	b_c

Then

$$a + a_o + a_c = b + b_o + b_c = 1.$$

Assume that system failure occurs if *either* valve fails open or if *both* valves fail closed. Probabilities of all possible combinations of valve operations can then be developed by expanding the product

$$(a + a_o + a_c)(b + b_o + b_c).$$

(Numerically, the product equals unity, but we are interested in the separate terms of the polynomial form of the product.) However, if we are concerned only with system success, we can simplify the algebra by dropping terms having either a_o or b_o, since they represent sure failure. Hence, all successful combinations, together with some failure combinations, are included in terms of the product

$$(a+a_c)(b+b_c) = ab + ab_c + a_c b + a_c b_c.$$

Indeed, only the last term represents a failure condition, so the successful system operation has probability

$$ab + ab_c + a_c b. \tag{9.11}$$

The block diagram representation of this combination of two valves is not easily drawn. The "fail open" situation suggests a series arrangement, since failure of either valve means system failure. The "fail closed" situation suggests a parallel arrangement, since both valves must fail to cause system failure. Hence we might be tempted to draw the block diagram representation shown in Figure 9.7. This block diagram will serve the purpose if it is

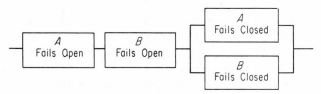

Fig. 9.7 Reliability diagram of a two-valve system.

properly interpreted. As each valve is represented twice, conditional probabilities must be used. The "fails closed" events are drawn after the "fails open" events. Hence, "fails closed" probabilities are conditional on the assumption of no "fails open" event if we are computing reliability.

Thus, the probability of "fails open" events not occurring is

$$(1-a_o)(1-b_o) = (a+a_c)(b+b_c).$$

Given the above—no "fails open"—the probability of no "fails closed" in the parallel configuration is

$$1 - \frac{a_c}{a+a_c}\frac{b_c}{b+b_c}.$$

Then the probability of no failure is the product

$$(a+a_c)(b+b_c)\left(1-\frac{a_c}{a+a_c}\frac{b_c}{b+b_c}\right) = (a+a_c)(b+b_c) - a_cb_c$$

$$= ab + ab_c + a_cb$$

as before.

In this example, suppose the two valves were of the same type, so that

$$a = b, a_c = b_c, a_o = b_o.$$

Then, the probability of successful operation of the redundant configuration would be

$$a^2 + 2aa_c.$$

If a single valve had been used instead of the two, then the probability of successful operation would have been a. Now consider the conditions under which the redundancy reduces the reliability, namely

$$a^2 + 2aa_c < a, \quad \text{or} \quad a < 1 - 2a_c.$$

For example, if $a_c = 0.1$, the nonredundant configuration would yield higher system reliability for

$$a < 1 - (0.2), \quad \text{or} \quad a > 0.8.$$

This implies $a_o < 0.1$ in this case.

Need for a Complete Malfunction-Effect Analysis

Despite the magnitude of the task, it is nevertheless necessary to begin a propulsion reliability prediction with a malfunction analysis. Great care must be exercised to be sure that all important failure modes and mechanisms are identified and described. For this purpose, "unimportant" failure modes and mechanisms are those for which, individually and collectively, the probabilities of occurrence are so small that they can all be disregarded.

This statement is not precise, and it cannot be made precise, since we are in an area of approximation. Ideally, all possible malfunctions should be included, but, in practice, this is rarely possible. The only compromise is to perform sufficient analyses to provide reasonable confidence that the neglected malfunctions are indeed insignificant contributors to the probability of system malfunction.

Identification of malfunction types in propulsion systems is, to a large extent, an art. By contrast, electronic component malfunctions are somewhat simpler—we can usually determine that the output of a part, black box, or component is not what it should be and that it cannot serve adequately as an input to the next item of equipment in a sequence. An analog in propulsion is not as simple—a pump may or may not cavitate due to a valve leak. Recognition of the potential for cavitation requires careful design analysis by a qualified engineer—one who looks for and recognizes chances for failure and the results which can be expected. Thus, prediction of propulsion system reliability is not yet a routine process requiring only limited engineering understanding.

Custom Construction in Propulsion Systems

Some of the parts used in propulsion systems are directly available from vendors. For such parts, it is sometimes possible to obtain information on performance capability and failure probability by failure mode in a form suitable for direct use in system reliability assessment. This situation is perhaps the exception, for many parts are specially designed and other portions of the system are custom-constructed from stock materials—tubing, bars, sheet metal, etc. It is a very difficult step from stock specifications to estimating custom-part failure-mode probabilities. Data for specially manufactured subunits must come from research and development test programs.

Stress Adjustments

In order to make available data applicable to new systems, it is necessary to adjust them for differences in stresses—those internal to the system and those due to the external environment in which the system operates. This data adjustment process is a major element in electronic system reliability prediction—it is almost nonexistent in propulsion system evaluations. In the case of boosters, the reason is that, in large part, the propulsion system failure data come from the test programs, which apparently correlate closely with flight test experience—according to studies made by rocket engine manufacturers and government agencies. In addition, many of the more critical stresses internal to the over-all system are generated by the propulsion subsystem itself. Of course the propulsion subsystem parts are qualified to stand these self-produced stresses, and the design capability is checked in both ground and flight tests.

Thus, in the present state of the art in propulsion system reliability assessment, the method is based on the assumption that stress adjustments are already included in the failure data. The validity of this assumption may be open to question in some propulsion systems—for example, those operated in hard vacuum. Later studies should give careful consideration to this problem.

Time Dependence

Propulsion system failures are certainly expected to be time-dependent— the longer the operating time, the greater the chance of failure. Furthermore, we might expect different failure risks during three important operational modes:

1. start
2. run
3. shutdown.

However, most of the available data are expressed as survival or failure probabilities for a specified operating time in each mode. Thus, failure probabilities may be known for a start cycle extending to 3 seconds after full thrust is obtained; they are not known for any other time. As a result, predictions of system failure or survival can be made directly for only the one time interval—the one for which data are obtainable. This situation contrasts markedly with electronic system predictions, which usually express estimated system reliability not for a fixed time, but as a function of time. The restriction, then, is one imposed by the data rather than one resulting from propulsion system characteristics.

One-shot devices present special problems in most reliability analyses. However, when computations are based on a fixed time span, as in most propulsion predictions, one-shot devices are treated in much the same manner as devices with time-dependent failure patterns. The main point is to exercise care in developing correct formulas for the prediction and in making certain that data are properly collected and perhaps adjusted.

Conditions of Operation and Operational Modes

The usual definition of reliability includes the identification of conditions of use—specifically, the handling of the system and the demands placed upon it. These factors (and others of similar nature) can vary from mission to mission and during the course of a particular mission. The varying modes of operation must be properly considered in the reliability prediction process. For instance, in some systems the start, run, and shutdown phases might represent three different reliability configurations. Some system components are required to operate only during one of the phases; others, only during two phases; and still others, during all three. Thus, component failures may or may not affect system failure, depending on phase and operational requirements at the time. Also, depending on the operational mode, components may be effectively in a series array in one phase and in a parallel array in another phase. Obviously, operational modes must be examined separately with respect to effects of component failure on potential system failure.

System Definition

For convenience, the discussion has been phrased in terms of a propulsion "system." Usually the prediction is for a "subsystem," and we must delineate the boundaries of the subsystem within the over-all system under study. More specifically, the interface between subsystems must be established precisely, to ensure that electronic, structural, and propulsion reliability assessments are all-inclusive and nonoverlapping. This is not a "cut and dried" division. Portions of the propulsion subsystem may have structural functions. Other propulsion components may have self-contained electronic controls. Thus, subsystem boundaries may have to be assigned somewhat arbitrarily, and it may be necessary to use data and methods in the manuals on electronics and structures to develop a complete propulsion-subsystem reliability prediction. The term "system" will be used even though the propulsion system is only a subsystem of the over-all system under study.

System Subdivision

Frequently the very size of the propulsion system requires that it be divided into a number of subdivisions just to make the work load manageable. Here again, the problem of subsystem boundary definition arises. It is desirable to select the subsystems on the basis of some logical criterion, such as producer function, independence of other subsystems, geometry or location in the system, or critical relation to mission success. For example, in a missile or booster, it is natural to treat the single engine as a subsystem.

In general, making a suitable breakdown of the system into subunits is not difficult. Since the task is primarily a matter of convenience, there is no "right" way. The "best" subdivision is the one which yields the simplest arithmetic and the most easily obtained interpretations—especially the latter. Of course, it must be possible to combine subsystem results to obtain the over-all propulsion system evaluation. This requirement is a strong argument for the use of function as a subsystem definition criterion. An example of a subsystem breakdown will be given later.

Summary of the Reliability Prediction Procedure for Propulsion Systems

The method used in predicting the reliability of a propulsion system is summarized in Figure 9.8. Input information is shown by the two blocks labeled "Vehicle Design" and "Hardware Experience." The system to be studied is defined by the schematics which describe vehicle design in detail. The input data come from hardware experience, which may include data from other programs as well as test and operational data from the program which is developing the system under study.

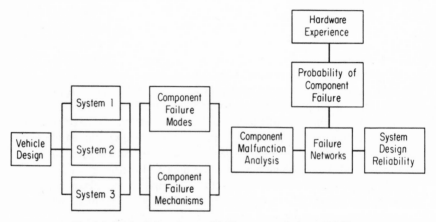

Fig. 9.8 Methodology for establishing system design reliability.

The steps of the method as indicated by the other blocks of Figure 9.8 are as follows:

1. Divide the over-all system into a convenient number of systems (or subsystems) for ease of handling.

2. Analyze schematics to identify significant failure modes, including all modes if possible.

3. Determine component failure mechanisms which would generate these failure modes.

4. From the results of steps 2 and 3, prepare a component malfunction analysis which relates component malfunctions to the system malfunctions.

5. From the results of step 4, prepare a failure network which is, in reality, a pictorial presentation of the malfunction effects analysis.

6. From hardware experience, generate component malfunction probabilities.

7. From results of steps 5 and 6, synthesize the system reliability prediction, using standard probability techniques.

The preceding discussion has indicated some of the logic behind the method for predicting propulsion system reliability and has given a number of hints as to the detail required. The summary of the method focuses attention on the kinds of activities required but does not provide the rules necessary for implementation. Essentially, we have seen that the method depends on system definition, subdivision into meaningful, manageable subsystems, a failure-effects analysis based on a study of failure modes and mechanisms, and, finally, substitution of probability estimates based on past experience into the system formula generated in the failure-effects analysis.

An Illustration of the Propulsion System Prediction Method

As an example, consider the prediction of the reliability of a propulsion system. This stage consists of a single tank with one compartment for liquid hydrogen and another for liquid oxygen, and is powered by a liquid hydrogen engine. It is unnecessary to comment about electronic and structural features, since the reliability prediction is used here solely for illustration.

Because the example is a second stage, no hold-down feature exists. Thus the study includes special consideration of engine ignition problems. Naturally it involves normal engine operation (the run mode) and the shut-down mode.

SYSTEM SUBDIVISION

It was decided to divide the propulsion system into five subdivisions:

1. Fuel system (includes tankage and associated plumbing required to deliver fuel to the engine system)

2. Oxidizer system (includes tankage and associated plumbing required to deliver oxidizer to the engine system)

3. Hydraulic control system (required for engine gimballing)

4. Ambient helium control system (supplies helium at ambient temperature for valve actuation)

5. Engine system.

It is apparent that this subdivision of the system is based essentially on functions. In general this is the best criterion to use. But it should be restated here that the critical element in system subdivision is reduction of

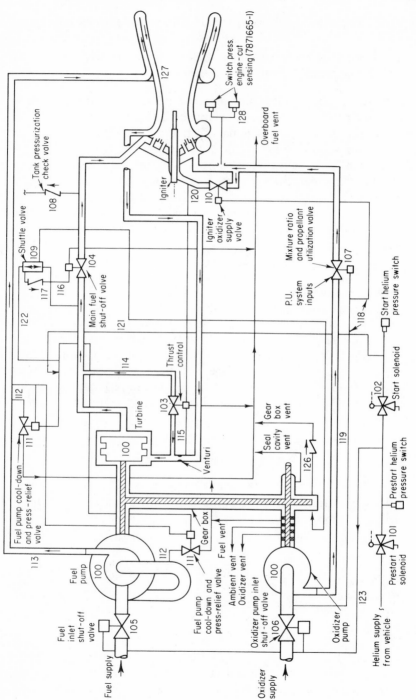

Fig. 9.9 Propellant flow schematic for a propulsion system.

analytical complexity, and good alternative methods of subdivision no doubt exist for almost any propulsion system. The selected method is not nearly as dominant in determining accuracy of the prediction as the way the subsequent analysis is carried out.

LINE DRAWINGS AND MALFUNCTION ANALYSIS

After the system subdivisions have been determined, it is necessary to obtain a clear understanding of their operational characteristics—the basic design information. For this purpose, schematics or line drawings are obtained for each subdivision. In this case it was convenient to use line drawings for only portions of subsystems which gave essentially a finer subdivision. This will no doubt be a common occurrence in the study of complex systems. As an example, Figure 9.9 shows the Propellant Flow Schematic. This is followed by Table 9.16, which shows the malfunction analysis of a portion of the propulsion system as required in the prediction process.

The reliability block diagram representation of the failure mode analysis is illustrated by Figure 9.10. This figure shows the diagram for the propulsion system, which matches the tabular form of the failure analysis in Table 9.16. Diagrams in Figure 9.10 for two operational modes—start and run—show the serial or parallel configurations of the failures and the probabilities that each occurs and causes system failure. These probabilities are the inputs needed in computing the required reliability estimate for the system.

FAILURE MODES AND MECHANISMS

The seven steps used in propulsion system reliability prediction emphasize the significance of the malfunction analysis based on the schematics and the engineering talents of the analysts. This process will be explained by an exposition of what the engineering analysis involves; it is not practical to specify a series of simple steps which can be followed in a routine fashion. The explanation will consist primarily of a description of the entries in the malfunction analysis shown in Table 9.16.

Each component in the schematic is examined to determine potential failure modes. The component is named in the first column, and the component reference number from the schematic is shown in the second column. The next five columns give the names of failure modes, list an identification number for each, and show associated probabilities by operational mode. (Sources of these probabilities will be discussed later.) Engineering analysis then develops possible mechanisms which could cause the identified failure mode, shown next in the table. The balance of the table concerns the effect of the failure. It identifies an immediate effect of the component failure and the ultimate effect on the system. This system effect is usually the consequence

Table 9.16 EXAMPLE OF MALFUNCTION ANALYSIS FOR A PROPULSION SYSTEM

Failure Types:

Out of specifications	Within specifications
A. Human Error	C. Embryonic Design
B. Quality Control	D. Calculable or Acceptable Design Risk

Component		Component Failure Mode					Component Failure Mechanisms		
Name	Find Number	Failure Mode	Number	Failure Probability* per Phase (times 10^{-4})			Possible Mechanisms	Most Probable Failure Type	Immediate Effect of Component Failure
				Start	Run	Shut-down			
Turbopump assembly	100	Mechanical failure or explosion	1	1	33	0	Impeller rubbing or gear failure		Loss of propellant pressure
Main fuel shutoff valve	104	Fail closed	8	8	3	3	Bellows rupture, jamming, contamination		No fuel reaches thrust chamber
		Fail open	9	3	0	6	Mechanical jamming, gate deformation		Fuel dribbles through thrust chamber during cool-down period
Fuel pump inlet shutoff valve (normally closed)	105	Fail closed	10	2	2	2	Mechanical jamming or failure of rack		No fuel to pump
		Fail open	11	0	0	0	Mechanical jamming or failure of rack		Fuel pump inlet pressure is maintained in fuel lines up to main shutoff valve
Line to turbo-pump gear box	125	Leakage or rupture	40	0	2	0	Fatigue, improper installation, etc.		Loss of gear box pressurization
Gear box vent and pressure relief valve	126	Fail closed	41	0	2	0	Mechanical jamming due to contamination or freezing can cause failure in either mode		Gear box pressure may exceed nominal 25 psi maximum
		Fail open	42	0	2	0	See above		Gear box pressure will drop below 18 psi minimum allowable pressure
Thrust chamber	127	Rough combustion, injector failure, burn through, etc.	43	2	28	0	Fatigue, contamination, pre-flight damage, etc.		Loss of thrust
Chamber pressure monitoring pressure switch (either of two)	128	Operate when not required	44	0	(2)6	0	Fatigue, contamination, pre-flight damage, etc.		Signal to shutdown engine
		Fail to operate when required	45				Fatigue, contamination, pre-flight damage, etc.		None

Table 9.16 Example of Malfunction Analysis for a Propulsion System (continued)

System Failure Modes:

0. No Failure	3. Guidance Failure
1. All Engines Fail	4. Catastrophic Explosion
2. Single Engine Failure	* Most Probable System Failure Mode

Ultimate Effect of Component Failure				
By Itself	*In Combination*	*System Failure Mode*	*System Failure Mechanisms*	*Comments, Remarks and/or Recommendations*
✓		2 or 4	Failure could be an immediate explosion or it could result in a serious leakage due to rupture of case or fittings.	Damage to other engines due to turbopump explosion could be eliminated through use of turbopump explosion containing shield, with prevalves required to prevent propellant loss.
✓		1, 2 or 4	Engine flames out if this occurs during run phase. Pump cavitates, housing may rupture. If bellows ruptures, helium will leak overboard until start solenoid is shut off.	
✓		1, 2	Fuel loss during prestart phase. May cause undesirable thrust chamber cool-down prior to ignition.	Fuel leakage (intentional or otherwise) past this valve during the cool-down period is mixed with cool-down LOX in the thrust chamber. The possibility of explosive mixtures forming in boattail region during cool-down should be investigated.
✓		2 or 4	Failure to start. Engine shuts down during run. Pump cavitates and may cause pump rupture. Possible thrust chamber burnout due to low coolant flow at high mixture ratios.	This valve (oxidizer inlet valve) could serve the same function as a prevalve (isolate engine in case of rupture upstream of the main fuel or lox valves) if it was moved upstream, away from the engine. In its present location any violent engine failure will also destroy the valve, negating any prevalve operation.
✓		2 or 4	Turbopump continues to operate. Main fuel shutoff stays open. Fuel rich burnout on shutdown.	
✓		2 or 4	Turbopump failure due to overheating of gear train. Air may leak into gear box.	
✓		2 or 4	Possible gear box rupture, or seal failure could cause loss of pump.	
✓		2 or 4	Possible gear train failure due to overheating. Air could leak into gear box.	
✓		2, 4	Safe shutdown initiated by thrust chamber pressure monitor, or adjacent engines damaged by burn-through.	
	✓(with other switch)	2	Safe engine shutdown of "good" engine.	Calculations of failure probability assume that two switches are used in a configuration requiring that both must send a malfunction signal in order to effect engine shutdown.
	✓(corresponding engine failure)	1, 4	Propellant depletion or fire and explosion.	

of the single failure. However, it may happen that the system effect is produced only by simultaneous occurrence of two or more component failures (e.g., the last two in the table, component 128, failure modes 44 and 45). The columns of the table which deal with these effects are self-explanatory and need not be discussed in detail.

BLOCK DIAGRAM OF FAILURE MODE ANALYSIS

The block diagram representation in Figure 9.10 is in reality the formula to be used in combining malfunction probabilities to obtain the system failure probability. The important element is the indication of the series or parallel arrangement of the various malfunctions: They are in series except for the cases in which two (it could be more than two) malfunctions are required to cause system malfunction. The various malfunctions shown represent a set of mutually exclusive events, assumed to be independent, when the parallel configuration is treated as one event. Hence, the probability of system failure is obtained by adding the probabilities of the events as shown, treating the parallel series by the proper formula. Thus, for the parallel set for component 44 in the run operational mode, the probability of system failures is $(0.0006)^2 = 36 \times 10^{-8}$, a negligible amount with the rounding level used here.

The probability of system survival for both modes is obtained as the product of the two reliabilities, each obtained by taking one minus the mode failure probability.

PROBABILITIES OF PROPULSION-SYSTEM COMPONENT FAILURE AND SURVIVAL

Failure probabilities in the reliability block diagram are shown as numbers because reliability and time-to-failure density functions are not available. The form in which propulsion data are collected forces the use of a number of compromises and assumptions. These warrant careful attention because they serve as guides for similar approximations in the future and because they emphasize the necessity for documenting all such assumptions in the reliability prediction reports. It will also be observed that the system operational requirements influence the effect of such approximations. For example, the short operating time involved in the booster prediction imposed less stringent requirements on data adjustment than one would encounter in analysis of propulsion systems of spacecraft for longer interplanetary missions. In fact, the data sources are restricted largely to missile and space booster programs. Nuclear propulsion systems, for example, will present data problems of serious magnitude.

Data on rocket engine failures show a sharply declining hazard rate with time in passing from the start mode to the run mode of operations. Hence,

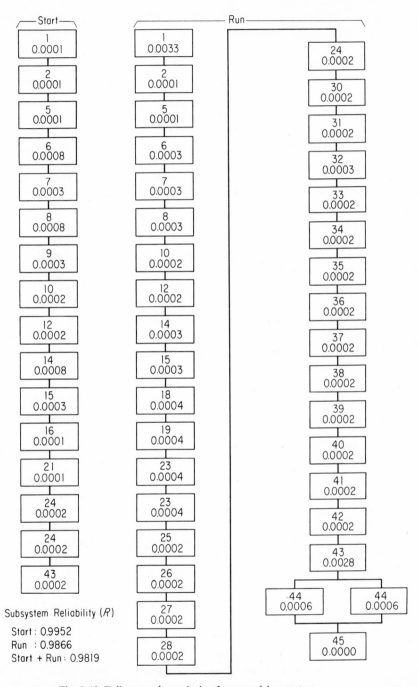

Fig. 9.10 Failure mode analysis of a propulsion system.

lengthening the run time would affect over-all reliability less critically than a change in operating environment, which might significantly raise or lower the hazard rate during both the start and run operating modes. Since operating environments assume considerable importance in propulsion system reliability, and since these environments are difficult to anticipate and hence to simulate with confidence, the best failure data are obtained from results of actual flight tests. However, such tests are few in number and some of the data are difficult to interpret. Nevertheless, the missile flight programs have furnished significant quantities of maintenance data, and these have been used to identify failure modes. Many of these data have been generated during checkout, and even the observed frequencies under checkout conditions cannot be directly extended to the flight environment.

Static test results are the major source of data usable in estimating malfunction probabilities. Failures may occur during a test activity before actual firing, and these stimulate corrective action before firing, with no premature shutdown being involved. Such failures probably do not represent flight failures nearly as well as those which occur during firing and which cause engine cutoff and premature shutdown. Hence, cutoff failures are used here as the basis for estimating probabilities. They are assumed to apply to well-developed components which have been properly inspected prior to use (the static tests are performed by well-trained personnel, and it is therefore assumed that the firing team is also well-trained).

To extend the engine data to make them applicable to propulsion components outside the engine system, failure probabilities were developed for generalized groupings of components such as solenoid valves and pressure switches, in an attempt to average out the effects of design and environment. It had to be assumed that such averaging was effective in producing relatively unbiased reliability predictions. Paucity of data forced the use of this technique.

The division of failure probabilities for the three operational modes—start, run, and shutdown—could be accomplished by looking at recorded time of occurrence. However, when a component had two failure modes, such as a valve failing open or failing closed, data were not available for estimating probabilities of failure in the separate modes. Hence it was assumed that the two modes were equally likely.

DATA EXTRAPOLATION TO OTHER OPERATING TIMES

It was observed that propulsion system data have been published in the form of single-operating-time, numerical survival probabilities. Future data collection should be planned to develop hazard rate-vs.-time curves, showing how hazard rate varies from ignition on. A curve should reflect premature shutdown, with a special analysis to show the failure pattern during planned shutdown with and without fuel depletion. Until such data become available, some method should be developed for estimating probabilities for various run times.

It is reasonable to use an approximation method based on the exponential. If the published probabilities apply for a specified time, t, then we compute a synthetic hazard rate, λ, from the equation

$$R(t) = e^{-\lambda t}$$

where $R(t)$ = the nominal run mode probability for time t.
Then

$$\lambda = \frac{\ln R(t)}{t}.$$

This synthetic failure rate can be used to estimate probabilities for other run times. The extension to cycle-dependent failure patterns is obvious.

ILLUSTRATION OF SPECIAL ANALYSES—THE ENGINE CLUSTER

In some systems, a cluster of several engines provides redundancy when the mission is such that it can be successfully accomplished if one or more engine failures occurs. This feature must be included in the reliability prediction analysis, since the system reliability is markedly influenced thereby. Such unusual problems frequently require special types of analyses to be used in prediction studies.

In some propulsion systems, one reliability problem is the risk of explosion. Even with some engine-out capability, a single engine malfunction which results in explosion is assumed to result in complete propulsion system failure. Hence, the design problem is to develop a method for sensing the engine malfunction and effecting a safe shutdown. We speak of this as "isolation of the failure," and the failure isolation factor, I, represents the fraction of single-engine failures which do not cause failure of the multiple-engine system. We can think of I as the product of two probabilities,

$$I = R_c R_s \tag{9.12}$$

where

R_c = the probability that an engine failure will not result in catastrophic explosion, and

R_s = the probability that the engine failure-sensing and shutdown systems will operate successfully when a nonexplosive engine failure occurs.

A cluster of n engines which can operate with up to k safe engine failures will have a cluster reliability, R, given approximately by the formula

$$R = \sum_{j=o}^{j=k} \frac{n!}{(n-j)!\,j!} (R_e F_s)^{n-j} (1 - R_e F_s)^j\, I^j \tag{9.13}$$

where

R_e = the single engine reliability, and

F_s = the probability that an individual engine failure-sensing and shut-down system will not transmit a false failure signal.

For a system with six engines and a single engine-out capability, this reduces to

$$R = (R_e F_s)^6 + 6I(R_e F_s)^5 (1 - R_e F_s).$$

The formula above is approximate unless time is the control on planned engine shutdown. A slightly different formula would apply if shutdown were controlled by fuel depletion.

In order to apply the theory above it was necessary to consider the probability that the system failure mode would be an engine-out—a mode which would result from the engine subsystem failures but could also result from other subsystem failures. For example, a failure of the Hydraulic System, Single Engine Control Subsystem, could force the engine into an unacceptable thrust angle, thus forcing shutdown. Hence, R_e must be computed from all failures which cause shutdown regardless of the subsystem involved.

COMBINING SUBSYSTEM PREDICTIONS TO GENERATE SYSTEM PREDICTION

The final step in the computation involves the combination of subsystem predictions to generate the system prediction. In selecting system subdivisions and the associate malfunction analyses, the subsystem probabilities were generated as independent in the probability sense. In other words, the subsystems form series elements in a block diagram representation of the system reliability. Hence the desired system reliability prediction is obtained as the product of the subsystem survival probabilities. In doing this, it is necessary to consider the correct subdivisions. If an engine cluster is involved, the single engine subsystem reliability is not used directly, but rather as an input in computing survival probability of the cluster.

In this final computation, as indeed in all others in reliability prediction, appropriate probability relationships must be developed from the engineering characteristics of the system under study. This can be accomplished only through the combination of competent engineering and statistical talents.

Summary of Reliability Prediction Method for Propulsion Systems

While we cannot hope to reduce the preceding discussion of the method of propulsion-system reliability prediction to the "cookbook" style of a sequence of routine steps, we can summarize the over-all procedure into a list of fairly specific tasks or instructional guides. These are intended to serve as a check list of items to be accomplished, and it is assumed that the process will become a bit easier as experience is gained. The tasks involved in reliability prediction for propulsion systems are described below.

Task 1. In order to reduce the over-all task to manageable size, the system must be divided into a number of subsystems. This should be done on a functional basis, for the process involves reliability prediction for each subsystem and later recombination of these predictions to give the over-all system prediction. Notice that the very statement of this task implies careful system and subsystem definition. Since the environmental stresses and even the system definition can change during a mission, the mission profile must be determined and the system and subdivision definitions must be changed during the analysis to reflect operational requirements and conditions.

Task 2. Schematic engineering drawings must be obtained for each subdivision and studied in detail to determine all of the significant failure modes. This task involves knowledge of the component mode of reaction to failure and the subdivision and system modes of reaction to failure. The analysis is best performed by sequencing the failure mode analysis by component; however, some modes result from simultaneous failures of more than one component, and these must be included. Implicit in this task is the definition of failure—an essential part of the analysis, but one which cannot be treated in general except to say that failure must mean operation not in conformity with some well-stated performance requirement. Notice further that this task can be accomplished only by the application of engineering talent. It requires both knowledge and ingenuity.

Task 3. The failure mechanisms are the basic causes of the failures; the failure modes are the reactions to the failure mechanisms. It is now necessary to determine all the component mechanisms which could give rise to each of the failure modes. Some modes may result from the occurrence of any one of a set of mechanisms or from the simultaneous occurrence of two or more particular mechanisms. All possible single and combined component mechanisms should be identified for each failure mode. Here, as in Task 2, the successful performance of the tasks is dependent on engineering capability.

Task 4. The failure modes and mechanisms determined in Tasks 2 and 3 are now tabulated and analyzed to show the relationships between component and system malfunctions. This task involves a summarization of all the reliability information derived from the design schematic drawings and analyzed through the engineering skills required for successful completion of the preceding steps. The remainder of the effort (Tasks 5, 6, and 7) requires predominantly statistical skills.

Task 5. The tabular form of the malfunction analysis contains all the inputs required to prepare the failure network. This is nothing more than a reliability block diagram which shows the series or parallel arrangement of the malfunctions identified in the previous steps. Independent component failure mechanisms generate series elements in the diagram, since the single failure generates system failure. The parallel elements in the diagram are generated when the system failure involves the simultaneous occurrence of more than one component failure mechanism.

Task 6. Numerical inputs are now needed. These take the form of probabilities of the occurrence of the specified failure mechanisms, which are represented by the blocks in the block diagram developed in Task 5. Experience with the components during developmental testing, or experience with similar components, provides the basis for computing (or estimating) these probabilities. Such data are regularly being accumulated, reduced, and recorded by firms and government agencies involved in the development and testing of propulsion systems.

Task 7. This final task consists of using the probabilities obtained in Task 6 with the block diagram obtained in Task 5 to synthesize the system reliability prediction. The formula for accomplishing this synthesis is developed from the diagram of Task 5 by applying ordinary theorems of probability. The basic task requires only a modest amount of statistical training; however, problems arising from uncommon combinations of failure mechanisms will require more capability in the area of mathematical theory. In some situations the block diagram symbolism needs to be supplemented by word descriptions to express these uncommon features. The varying nature of such problems precludes the writing of detailed steps required for their solution.

PROBLEMS

1. Refer to the flip flop circuit on the next page. (a) Determine the *MTBF* of this circuit for environments of 25°C and 85°C, using the hourly failure rates indicated. (b) What is the probability of survival for one year if the power source is 100 per cent reliable?

2. Using the failure data in Table 9.13, for the spectrum amplifier on the next page, determine (a) θ and (b) the lower 90 per cent (single-sided) confidence limit for θ if the heater voltage is at the rated value of 6.3V. Repeat (a) and (b) if the heater voltage is maintained at 6.9V. (Tube failure rates increase approximately 80 per cent for a 10 per cent increase in heater voltage.)

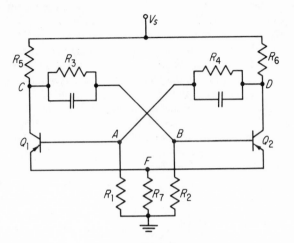

Hourly Failure Rates

	25°C Half Rating	85°C Full Rating
Resistors	0.25×10^{-6}	1×10^{-6}
Capacitors	0.05×10^{-6}	1×10^{-6}
Transistors	2×10^{-6}	4×10^{-6}

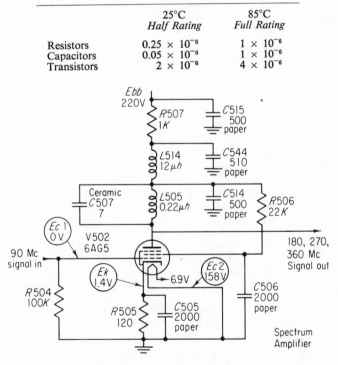

Notes. 1. All resistance values in ohms unless otherwise specified. (Carbon composition.)

2. All capacitor values in micromicrofarads.

3. All voltages shown are positive with respect to ground.

3. A block diagram of a three-phase inverter with one redundant single-phase inverter is shown below. The reliability of each inverter is 0.973 for a 100-hr mission. The failure detector and switching circuit as a unit have a reliability of 0.992 for a 100-hr mission. Determine the reliability of this system with and without the redundant configuration. *Hint*: Set up a reliability logic diagram showing all possible operating conditions. Assume that 1 million of these devices will be built, and calculate the probability of success of each mode of operation, using the method described in Section 2.2 of Chapter 2. A solution to this problem is given by W. E. Marshall in "Graphic Solution of Reliability Logic Equations," *Proceedings, Seventh Joint Military-Industry Guided Missile Symposium*, San Diego, California (1963).

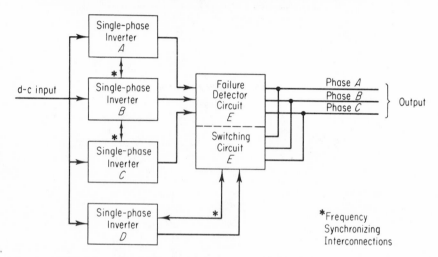

4. An airline maintains a fleet of 4-engine aircraft. Maintenance records show that, on the average, a typical engine fails 3 times in 10^4 operating hours in spite of the preventive maintenance program. What is the probability of 2 or more engines failing during a typical 8-hr period of utilization?

5. As shown in Figure 9.9, the fuel pump cool-down and pressure-relief valve actually consists of 2 valves, which are normally open. What are possible failure modes, mechanisms of failure, system failure modes, and system failure mechanisms?

6. Using the data given under the heading "Example" on page 303, calculate θ and the associated upper and lower 95 per cent confidence limits for each of the four prediction equations 9.8 I through 9.8 IV, page 296. Compare results with a feasibility estimate obtained from Figure 6.1 (Chapter 6).

7. What is the prime purpose of reliability prediction? Outline the differences in approach used for reliability prediction of electronic, propulsion, and structural systems. Relate prediction, stress analysis, and failure mode analysis.

Maintainability Analysis

10.1 Introduction

From time to time, statistics are generated which emphasize the costliness of maintenance actions. While estimates of actual dollar costs vary, they invariably reflect the immensity of maintenance expenditures. Maintenance expenditures of the Department of Defense have been estimated at more than $25 million per day—about $8 billion per year—representing more than 25 per cent of the total defense budget. According to one source, approximately 800,000 military and civilian technicians are directly concerned with maintenance.* Another source states that for a sample of four equipments in each of three classes—radar, communication, and navigation—the *yearly* support cost is 0.6, 12, and 6 times, respectively, the cost of the original equipment.† Such figures clearly indicate the need for continually improved maintenance techniques.

In addition to these cost considerations, maintainability has a significant effect on other system-effectiveness characteristics. For example, equipment availability is directly related to down time (an element of maintainability).

*Col. V. J. Bracha, *Analysis of Reliability Management in Defense Industries*, USAF Ballistic Systems Division (BSD-62-48).

†J. A. Cafaro and H. D. Voegtlen, "The Measurement and Specification of Product Abilities," *Quality Control*, Vol. XVIII, No. 9 (March, 1962).

Numerous systems, such as missiles and rockets, require an availability which approaches unity. Such requirements cannot be achieved without careful attention to maintenance factors during the design and development phases.

10.2   Measures of Maintainability

Maintainability is defined as the probability that a failed system is restored to operable condition in a specified down time when maintenance is performed under stated conditions. In the evaluation of any system, the measure of maintainability is quite important; how often the system fails (reliability) and how long it is down (maintainability) are vital considerations in determining its worth. In practice, the trade-off between these two concepts is dictated by cost, complexity, weight, and operational and other requirements.

The first step in measuring maintainability is to define its constituent elements. *Down time* is the interval during which the system is not in an acceptable operating condition (i.e., the time from the initiation of a complaint or most routine maintenance actions to the restoration of the system to satisfactory operating condition). Down time is divided into (1) active repair time, (2) logistic time, and (3) administrative time. *Active repair time* is the number of down-time hours during which one or more technicians actually work on a system to restore it to operable condition. *Logistic time* is the number of down-time hours consumed in awaiting parts or units needed to effect a repair.* *Administrative time* is that portion of down time not covered by active repair time or logistic time. Based on a 24-hr day, it includes overnight time, weekends, and normal administrative delays.

Active repair time is usually indicative of the complexity of the system, the nature of its design and installation, the adequacy of test facilities, and the skill of maintenance personnel.

Logistic time is generally a function of the supply methods associated with the operational activity, but it can be influenced by the design of the system. For example, if large numbers of nonconventional parts are used in a system, then the supply organization must handle greater quantities of special items; this situation could cause delays in the receipt of spares or replacement parts.

Administrative time is a function of the structure of the operational organization involved; it is influenced by work schedules and the assignment of nontechnical duties to maintenance personnel. Generally, this time can in no way be charged against the manufacturer of the system. In practice, it would be greatly reduced in an emergency or wartime situation.

---

*Logistic time as used here comprises only the time consumed when parts must be ordered. The time needed to obtain parts which are immediately available is included as an element of active repair time (see Figure 10.1).

Repair time can be reduced in most situations by the use of additional manpower. For this reason, records are maintained on the man-hours expended during a given maintenance action. *Man-hours* is defined as the sum of the times all technicians worked on the system during a given maintenance action. These data can be used effectively to determine the average maintenance support required to maintain a system, often expressed in terms of man-hours per 1000 flight-hours (Maintenance Support Index, or *MSI*).

Maintainability data are often presented in terms of the probability that a given maintenance action will be completed in some finite time or less. The observed data can frequently be described by a theoretical log-normal distribution in which the logarithms of the time values are normally distributed. (Techniques for computing maintainability and repairability functions are presented in Chapter 5.)

10.3   Maintainability Prediction*

Introduction

The best time to work toward minimizing maintenance requirements is during the system design and development phases. During this period, any unsatisfactory conditions indicated by a system maintainability analysis can be corrected economically. There is an obvious correlation between the complexity of a system and the time required to maintain it: The bigger and more complex the system, the longer the required maintenance time. Other factors related to hardware, such as accessibility, built-in measuring and metering devices, warning lights, and packaging, may also furnish clues about expected maintenance time. However, the system-hardware approach does not take into account all of the factors which influence maintenance time. Human factors, for instance, which are acknowledged to have a considerable effect on maintenance time, would be neglected in the system-hardware approach. Consequently, a different approach to prediction of maintenance time has been developed, based on

1. Determination of the probabilities of occurrence of certain fundamental maintenance activities
2. Establishment of the distribution of times required to perform these activities, and the statistical combination of these (and other) time distributions to obtain an ultimate prediction of system down time.

These principles are considered to have general application. However, the

*This section is based on the report, *Maintainability Prediction—Theoretical Basis and Practical Approach*, by A. Drummond, G. T. Harrison, and H. R. Leuba, ARINC Research Corporation, Publication No. 207–1–275 (February, 1962). Contract AF33(657)–7382.

specific prediction procedures developed thus far apply only to those electronic systems for which maintenance at the system level is accomplished predominantly by "black box" replacement.

Theoretical Concepts of Maintainability Prediction

In early studies of system maintainability, it was established that any maintenance action can be classified as falling within one of the following categories:

1. Preparation
2. Malfunction verification
3. Fault location
4. Part procurement
5. Repair
6. Final malfunction test.

The time required to perform each of these categories varies from zero to several hours, depending on numerous conditions associated with particular maintenance events. Weather, for example, causes great variations in the time required for preparation. Other variables include the skill level of maintenance technicians, their familiarity with the system under repair, and even the manner in which symptoms are reported to them. This variability in performance time would limit the accuracy of any maintenance-time predictions based on maintenance-category time distributions.

The first step toward a successful prediction approach was the recognition that data collectors use certain descriptive phrases repeatedly in their narrative recording of maintenance actions. These phrases represent elemental activities; for example, such phrases as "five minutes—turn on and warm up equipment" and "eight minutes—opening radome" appear frequently in the description of preparation time. It was also noted that the variation in time required to perform each elemental activity is much less than the variation observed for the entire maintenance category to which the activity belongs. If the narrative recording were made even more detailed and if the various activities were timed with a stop watch, the elemental activities could be broken down into even smaller and less complex actions. These actions would, in general, involve much shorter and more uniform execution times.

As an example of this last statement, consider the following ways in which a maintenance event involving replacement of spark plugs in an automobile engine can be described:

Method 1: "Fixing" the engine

Method 2: Replacing six spark plugs in the engine

Method 3:   Opening hood
                 Removing spark plug wires
                 Picking up and assembling plug wrench
                 Setting ratchet direction
                 Inserting socket over plug No. 1
                 Etc.

"Fixing" the engine (Method 1) could comprise a simple, quickly performed operation, such as adjusting the carburetor, or a more difficult, time-consuming operation, such as replacing the crankshaft. Consequently, the time duration for the repair operation could range from a few seconds to several hours. A distribution of such widely varying performance times would be of little use in predicting repair time for the engine. On the other hand, the frequency of occurrence of "fixing" the engine might be predicted fairly accurately from the failure rate of the engine.

A large number of different motions similar to those given as examples for Method 3 could occur during a complex maintenance event. Consider, for example, the repairing of an engine. Since the variation in time required to perform each motion would be small, engine repair time could be estimated from the times required for the individual motions. By their general nature, however, such motions as described by Method 3 cannot be associated—from the standpoint of frequency of occurrence—with any particular unit or part of the engine. For example, no correlation could be expected between any characteristic of the engine and the frequency of occurrence of the motion, "setting ratchet direction." It is true that a plug wrench is required for replacement of a spark plug and that spark plug replacement is a maintenance event which might be predicted from a system characteristic—namely, the failure rate of the plug. However, there is no assurance that the direction of the ratchet on the plug wrench was not properly set from a previous maintenance event, in which case that particular action would not be executed.

In summary, both Methods 1 and 3 of describing a maintenance event have limitations. The Method 1 description would permit a rather accurate estimation of the frequency of occurrence of the event, but *not* a determination, within tolerable limits, of the amount of time required to complete the event. The Method 3 description would permit an accurate prediction of the time required to complete each constituent motion of the event (and thus the time for the complete event), but *not* the establishment of all factors responsible for the occurrence of each motion.

Method 2 is a compromise: the frequency of occurrence of the described activity ("replacing six spark plugs in the engine") would be a function of a system characteristic (failure rate of spark plugs), and the distribution of time required to perform the activity would be narrow enough for prediction

Table 10.1 CATEGORIES AND ELEMENTAL ACTIVITIES OF ACTIVE REPAIR TIME

Category	Elemental Activity	Activity #
	System turn-on, warm-up, setting dials and counters as necessary	1
	Code 1 plus time awaiting particular component stablization	2
	Opening and closing radome	3
	Gaining access and reinstalling covers (other than radome)	4
	Obtaining test equipment and/or Tech Orders	5
Preparation	Checking maintenance records	6
	Conferences	7
	Procuring components in anticipation of need	8
	Replacing aircraft spares used for in-flight maintenance	9
	Setting up test equipment	10
	Leveling aircraft	11
	Observing indications only	1
	Using test equipment to verify malfunctions inherently not reproducible on ground	2
	Performing standard test problems or checks	3
Malfunction	Testing for pressure leaks	4
Verification	Attempting to observe elusive or non-existent symptom(s)	5
	Using standard test equipment	6
	Using special test equipment designed specifically for this equipment	7
	Making a visual integrity check	8
	Fault self-evident from symptom observation	1
	Interpreting symptoms by mental analysis only (from knowledge/ experience)	2
	Interpreting displays at different settings of controls	3
	Interpreting radar meter readings	4
	Removing unit(s)/subunit(s) and checking in shop (for flight-line maintenance only)	5
	Switching and/or substituting unit(s)/subunit(s)	6
Fault	Switching and/or substituting part(s)	7
Location	Removing and checking parts	8
	Making a visual integrity check	9
	Checking voltages, continuity, waveforms, and/or signal tracing	10
	Consulting Tech Orders	11
	Conferring with Tech Reps or other maintenance personnel	12
	Performing standard test problem(s)	13
	Isolating pressure leak	14
	Deciding whether maintenance is necessary	15
	Using special test equipment designed specifically for this equipment	16

Table 10.1 (continued)

Category	Elemental Activity	Activity #
	Obtaining replacement component from aircraft spares	1
	Obtaining replacement(s) from bench stock	2
	Obtaining replacement component(s) from pre-issue stock	3
Part	Obtaining replacement component(s) by cannibalization	4
Procurement	Attempting to obtain replacement component(s). Unavailable	5
	Obtaining replacement component(s) from tool box	6
	Obtaining replacement component(s) from shop	7
	Obtaining replacement component(s) from other sources	8
	Replacing unit(s)/subunit(s)	1
	Replacing parts	2
	Correcting improper installation or defective plug-in connection(s)	3
	Making adjustments in aircraft	4
Repair	Making adjustments in shop	5
	Baking magnetron	6
	Precautionary repair activity (includes so-called fault location, part procurement, and repair times spent when symptom not verified)	7
	Repairing wiring or connections	8
	Cleaning equipment	9
Final Malfunction Test	Function checkout following completion of repair	1

purposes. Accordingly, the maintenance-event description procedure of Method 2 is used as the basis of the maintainability prediction technique.

Characteristics of Elemental Maintenance Activities

The AN/ASB-4 bombing/navigation system in the B-52 aircraft was selected for the development of a maintainability prediction technique. It was found that 99 per cent of all observed maintenance actions on that system could be described by the 53 "elemental activities" listed in Table 10.1. The following four conditions were considered in the selection and phrasing of the activities:

1. The mean time required for performance of an elemental activity must be independent of system design and support facilities.
2. The frequency of occurrence of an elemental activity must, however, correlate to some factor of system design or of support facilities.

3. The elemental activities in any maintenance category must be mutually independent.

4. The total time required in any maintenance category must be completely accounted for by one or more of the elemental activities in the category.

Two characteristics of each elemental activity are of primary concern: the distribution of time required for completion of the activity, and the frequency of occurrence of the activity during maintenance. Since the time distribution is assumed to be independent of the type of system, it is only necessary to observe an activity being performed several times on one system to obtain a first approximation of the time distribution for that activity on any system. Observations of the performing of the 53 elemental activities on the ASB-4 system provided approximate time distributions which would apply to these same activities on all Air Force electronic systems. (The assumption that the distribution of performance time for an elemental activity is not dependent on the type of system is not so rash as it may appear. For instance, large differences would not be expected between the times required to open and close a radome on two different types of aircraft or between the times required to make a simple electrical adjustment, such as setting a potentiometer, in two different equipments.)

The times required to perform basic actions are characterized by a normal distribution. By fitting such a theoretical function to each set of observed times, the elemental activities associated with the maintenance of the ASB-4 system were characterized by the parameters given in Table 10.2. The data in this table can be used to predict the time required to perform these elemental activities in the maintenance of other Air Force electronic systems.

The frequency of occurrence of an elemental activity is expressed as a *probability value*. For some systems, the probabilities of certain activities will be zero, since these activities will not be performed. For other systems, the probabilities of some activities will be the same as those observed for the ASB-4 system.

For a majority of the elemental activities, the probability of occurrence, P, is obtained from the general expression

$$P = KS \tag{10.1}$$

where

$K =$ a proportionality constant, and

$S =$ a quantitative characteristic of the system.

The value of K for each activity was determined by substituting, in this expression, the values of P and S observed for the ASB-4 system. For example, the occurrence of the elemental activity, opening and closing radome, is assumed to be a function of the ratio of radome component

Table 10.2 PARAMETERS OF NORMAL DISTRIBUTIONS FOR USE IN
DESCRIBING TIMES REQUIRED TO PERFORM ELEMENTAL ACTIVITIES*

Category	Elemental Activity†	Mean (hours)	Standard Dev. (hours)
Preparation	1	0.102	0.068
	2	0.665	0.384
	3	0.235	0.135
	4	0.276	0.196
	5		
	6	0.140	0.104
	7	0.045	0.050
	8	0.070	0.041
	9		
	10	0.107	0.127
	11		
	1 & 3	0.283	0.109
	1 & 4	0.326	0.373
	1 & 5	0.328	0.277
	1 & 6	0.241	0.170
	1 & 8	0.256	0.100
	1, 5 & 6	0.412	0.299
Malfunction	1	0.105	0.263
Verification	2	0.593	0.416
	3	0.397	0.435
	4	0.329	0.605
	5	1.394	1.236
	6		
	7		
	8		
	2 & 3	0.986	0.858
	3 & 5	0.973	0.844
Fault Location	1	0.010	0.0
	2	0.019	0.040
	3	0.333	0.259
	4	0.141	0.172
	5	0.821	0.788
	6	0.324	0.346
	7	0.436	0.480
	8	0.181	0.191
	9	0.140	0.213
	10	0.807	1.000
	11		
	12	0.466	0.430
	13	0.582	0.541
	14	0.683	0.960
	15		
	16		
	1 & 16	0.534	1.047

Table **10.2** (continued)

Category	Elemental Activity†	Mean (hours)	Standard Dev. (hours)
Part Procurement	1	0.022	0.033
	2	0.316	0.326
	3	0.301	0.211
	4	0.315	0.171
	5	0.199	0.144
	6		
	7	0.102	0.202
	8		
	2 & 3	0.520	0.336
	3 & 5	0.402	0.256
	4 & 5	0.371	0.186
Repair	1	0.394	0.518
	2	0.380	0.603
	3	0.066	0.055
	4	0.415	0.541
	5	0.993	0.862
	6	1.416	0.701
	7	0.752	0.747
	8	0.810	1.114
	9	0.353	0.165
	1 & 2	0.646	0.546
	1 & 4	0.756	0.640
	2 & 4	1.478	1.272
Final Malfunction Test	1	0.358	0.350

*Times less than 0.01 hr randomly selected from the normal distributions described by these parameters are to be considered as 0.01 hr.

†Table 10.1 lists the Elemental Activities represented by the numbers in this column.

failure rates (λ_2) to the system failure rate (λ_1). For the ASB-4, this quantity, S, is 0.328. The observed probability of occurrence of the activity in that system is 0.055. Hence,

$$K = P/S = \frac{0.055}{0.328} = 0.168.$$

For any system, the probability of occurrence of the activity, opening and closing radome, may be calculated from the expression

$$P = KS = 0.168(\lambda_2/\lambda_1).$$

The K factor in this case takes account of those situations in which the opening of the radome occurs together with another elemental activity within the preparation-time category. Such a concurrence of two activities is not counted as individual occurrences of each, but rather as a single, combined action. For example, suppose that the radome is opened during 33 of every 100 maintenance actions. Suppose, further, that during 28 of these 33 occasions, the maintenance man performs another activity within the preparation-time category, such as turning on and warming up the system. The activity, opening and closing radome, will have occurred only 5 times, yielding a probability of occurrence of 0.05 for this activity. However, the ratio of λ_2 (failure rate of components in the radome) to λ_1 (system failure rate) would indicate a probability of occurrence of 0.33. Hence, a value of K of 5/33 (or 0.15) is necessary to derive the correct probability of the solitary occurrence of the activity, opening and closing radome.

The K factor also takes account of those situations in which the S-factor assumptions are inaccurate. For example, while λ_2/λ_1 may be a true measure of how frequently a radome component fails, there may be occasions when the maintenance man suspects that a problem is in the radome, opens it, and finds otherwise.

The K factor has a third function. The occurrence of any elemental activity is a function of a large number of variables, most of which have only a small effect on the occurrence probability. The major variable (S) is included, and the K factor is an adjustment to include all other variables.

The procedure for calculating the probability of occurrence of each of the elemental activities is outlined in Table 10.3. The probabilities are calculated from Equation 10.1, in which

$S = 1$ when there is no dependence between system characteristics and the observed probability of occurrence of the elemental activity. For this condition, therefore, the predicted probability is that observed for the ASB-4;

$S = 0$ when system characteristics preclude occurrence of the elemental activity.

Although the functional relationships embodied in most of the following equations are subjective to some degree, the reasoning upon which they are based becomes clear in most cases upon examination of the term S. Where deemed necessary, additional explanation has been included in the notes at the end of Table 10.3. (The steps so referenced are denoted by an asterisk. Symbols are explained in the legend to Table 10.3.)

Synthesis of Time Distributions

The synthesis of a predicted distribution of system down times from basic distributions of elemental-activity times is ideally a task for a computer. For this reason, the detailed step-by-step procedure for a laborious hand computation will not be discussed. Rather, the theory on which both the computer program and the hand process are based will be presented.

Step in Calculation Procedure	Elemental Activity Number	Instructions	Equation, $P = KS$		
			P	K	S
	Preparation				
1	1		0.485	0.485	1
2	2	Does system contain components which require an unusually long time to reach thermal or positional steady-state condition? (Time-delay relays and magnetron warm-up are not considered unusual.)			
		If yes —	0.025	0.025	1
		If no —	0	0.025	0
3	3			0.168	λ_2/λ_1
4	4			0.086	λ_{12}/λ_1
5	5		0	(Not observed in ASB-4 maintenance)	
6	6			1.2×10^{-4}	N_1
7	7			3.0×10^{-5}	N_1
8	8			3.48×10^{-4}	$N_2 - N_5$
9	9		0	(Not observed in ASB-4 maintenance)	
10	10	See Note 1		2.63	P_5
11	11		0	(Not observed in ASB-4 maintenance)	
12	1 & 3			3.0	$P_1 \times P_3$
13	1 & 4			1.08	$P_1 \times P_4$
14	1 & 5	See Note 1		56.5	$P_1 \times P_5$
15	1 & 6			19.8	$P_1 \times P_6$
16	1 & 8			1.3	$P_1 \times P_8$
17	1, 5, & 6	See Note 1		6,100	$P_1 \times P_5 \times P_6$
18		Normalize each of the values of P calculated for preparation by multiplying each by the reciprocal of their sum.			
	Verification				
19	1			7.84	$(N_3 - N_4)/N_1$
20	2	Is special test equipment used to reproduce any in-flight conditions not otherwise reproducible on the ground?			
		If yes —	0.0077	0.0077	1
		If no —	0	0.0077	0

Table 10.3 (continued)

Step in Calculation Procedure	Elemental Activity Number	Instructions	Equation, $P = KS$		
			P	K	S
21	3			0.5	N_4/N_3
22	4	Does the system contain any dynamically pressurized components?			
		If yes —		0.0065	N_6
		If no —	0	0.0065	
23	5		0.197	0.197	1
24	6		0	(Not observed in ASB-4 maintenance)	
25	7	Is test equipment, designed especially for this system, used for flight-line maintenance?			
		If no, enter	0		
		If yes, enter	0.146	0.146	1
26	8		0	(Not observed in ASB-4 maintenance)	
27	2 & 3		6.06	6.06	$P_{20} \times P_{21}$
28	3 & 5			1.9	$P_{21} \times P_{23}$
29		Normalize each of the values of P calculated for verification by multiplying each by the reciprocal of their sums.			
	Fault Location				
30	1		0.191	0.191	1
31	2	See Note 4		0.026	N_1/N_3
32	3			0.305	N_7/N_3
33	4			0.0766	N_8/N_3
34	5	Are there provisions such as mock-up, go-no-go tester in the shop for checking flight-line replaceable "black boxes"?			
		If yes —		0.1	$\dfrac{\lambda_1 - \lambda_3 + \lambda_5 - \lambda_4}{\lambda_1}$
		If no — See Note 2		0.1	0
35	6	See Note 3		1.875	$(\lambda_1 - \lambda_8)P_8 + \left(\dfrac{\lambda_3 - \lambda_5}{\lambda_1}\right)$
36	7			0.193	$\left(\dfrac{\lambda_8}{\lambda_1}\right)P_8 + \left(\dfrac{\lambda_9 - \lambda_{11}}{\lambda_1}\right)$

Table 10.3 (continued)

Step in Calculation Procedure	Elemental Activity Number	Instructions	Equation, $P = KS$		
			P	K	S
37	8	Are there provisions such as mock-up, go-no-go tester in the shop for checking flight-line replaceable parts?			
		If yes —		0.012	$(\lambda_8 - \lambda_{10})/\lambda_8$
		If no —	0	0.012	0
38	9		0.066	0.066	1
39	10			0.116	N_9/N_3
40	11		0	(Not observed in ASB-4 maintenance)	
41	12			25×10^{-6}	N_1
42	13			0.375	N_4/N_3
43	14			0.005	N_6
44	15		0	(Not observed in ASB-4 maintenance)	
45	16	Is test equipment, designed especially for this system, used for flight-line maintenance?			
		If no, enter	0		
		If yes, enter	0.122	0.122	1
46	1 through 16			0.23	$\sum\limits_{m=30}^{45} P_n$
47		Normalize each of the values of P calculated for fault location by multiplying each by the reciprocal of their sum.			
	Part Procurement				
48	1			0.47	N_5/N_2
49	2			0.95	λ_8/λ_1
50	3			0.62	$(\lambda_1 - \lambda_8)/\lambda_1$
51	4		0.0075	0.0075	1
52	5			0.1	P_L
53	6			0.45	$\dfrac{\lambda_3}{\lambda_1}$
54	7			0.835	P_{49}
55	8		0	(Not observed in ASB-4 maintenance)	
56	3 & 5			13	$P_{50} \times P_{52}$
57	4 & 5			204	$P_{51} \times P_{52}$

Table 10.3 (continued)

Step in Calculation Procedure	Elemental Activity Number	Instructions	Equation, $P = KS$		
			P	K	S
58	2 & 3			0.296	$P_{49} \times P_{50}$
59		Normalize each of the values of P calculated for part procurement by multiplying each by the reciprocal of their sum.			
	Repair				
60	1			0.74	$\dfrac{\lambda_1 - \lambda_7}{\lambda_1}$
61	2			1.80	$\dfrac{\lambda_8}{\lambda_1 - \lambda_8}$
62	3			3.3×10^{-4}	N_1
63	4			3.78	λ_8
64	5			0.1	P_{63}
65	6	Does system contain one or more magnetrons?			
		If yes —	0.006	0.006	1
		If no —	0	0.006	0
66	7		0.102	0.102	1
67	8			9×10^{-6}	N_1
68	9		0	(Not observed in ASB-4 maintenance)	
69	1 & 2			0.277	$P_{60} \times P_{61}$
70	1 & 4			0.338	$P_{60} \times P_{63}$
71	2 & 4			0.478	$P_{61} \times P_{63}$
72		Normalize each of the values of P calculated for repair by multiplying each by the reciprocal of their sum.			

LEGEND, **Table 10.3**
(Failure rates to be expressed on a per-hour basis.)

N_1 Number of flight-line-replaceable components in system

N_2 Number of different types of flight-line-replaceable components in system

N_3 Number of readouts in system, as determined by use of Table 10.6

N_4 Number of system readouts whose function is to evaluate a standard test problem

N_5 Number of different types of spares carried aboard the aircraft

N_6 Number of connectors (electrical or mechanical) which maintain dynamic pressure integrity

N_7 Number of CRT's in the system (excluding built-in test scopes)

N_8 Number of circuit parameters monitored by built-in meters

N_9 Number of test points in the system

P_L Probability of occurrence of logistic time

P_n Probability calculated in the nth step of Table 10.3

T_f Anticipated average flight length

λ_1 Failure rate of the system

λ_2 Summation of failure rates of flight-line-replaceable components located in the aircraft radome

λ_3 Summation of failure rates of flight-line-replaceable components for which spares are carried aboard the aircraft

λ_4 Summation of failure rates of flight-line-replaceable components whose operation is reflected by a readout

λ_5 Summation of failure rates of those flight-line-replaceable components whose operation is reflected by a readout and for which a spare is carried aboard the aircraft

λ_6 Summation of failure rates of flight-line-replaceable components which contain adjustment(s)

λ_7 Summation of failure rates of flight-line-replaceable components which contain either adjustment(s) or flight-line-replaceable parts, or both

λ_8 Summation of failure rates of flight-line-replaceable parts

λ_9 Summation of failure rates of flight-line-replaceable parts for which spares are carried aboard the aircraft

λ_{10} Summation of failure rates of flight-line-replaceable parts whose operation is reflected by a readout

λ_{11} Summation of failure rates of flight-line-replaceable parts whose operation is reflected by a readout, and for which a spare is carried aboard the aircraft

λ_{12} Summation of failure rates of components accessible through—or which contain—access covers (other than radome), removal of which may be necessary for flight-line maintenance.

NOTES, Table 10.3

Note	Step	
1	—	On only one occasion was elemental activity no. 5 of preparation employed for ASB-4 maintenance. Its probability of occurrence, based on this single use, is 0.002. Although no distribution of times for the activity can be determined if based on a sample of one, this observed occurrence probability is to be used as the value of P_5 as necessary for the calculation of other probabilities.

2	34	The numerator is the sum of the failure rates of those units which have neither a readout nor a spare in the aircraft, and which are therefore likely to be carried to the shop for checkout. The probability of occurrence of this elemental activity is considered to be a function of the ratio of this summation of failure rates to the failure rate of the system.
3	35	The first term is the probability of anticipatory procurement of either a flight-line-replaceable part of a flight-line-replaceable black box, multiplied by the probability that a black box (as opposed to a part) has failed. The first term, then, is a contributor to the probability of locating a fault by black-box substitution. The second term represents the probability that the component which has failed is one which has no readout but for which an aircraft spare may be substituted. This term also contributes to the probability of fault location by black-box substitution.
4	31	If $N_4/N_3 = \infty$, assume $N_4/N_3 = 38.4$ so that $P = 0.026 \times 38.4 = 1.00$.

MAINTENANCE CATEGORIES

If the occurrence of a maintenance category (see Figure 10.1) involves all of its elemental activities, the predicted distribution of times required can be easily synthesized. A value could be selected at random from each of the time distributions for the various activities, and their sum taken as one value for the maintenance-category distribution. This process is repeated until sufficient points have been obtained for plotting the maintenance-category function.

However, the process of adding equally-sampled variates does not usually apply. The probabilities of occurrence of the various elemental activities within a maintenance category are not likely to be equal, and must be predicted by methods previously discussed herein. Hence, the over-all distribution is synthesized by the adding of variates unequally sampled (according to their probabilities of occurrence) from the distributions of the elemental activities.*

An exception to the foregoing procedure is the category of final malfunction test time. This category includes only one elemental activity, and thus the time distribution for the category is the same as for the activity.

MALFUNCTION ACTIVE REPAIR

As indicated in Figure 10.1, the synthesized distributions for the maintenance categories are in turn used to synthesize the predicted distribution of

*The sampling is performed by straightforward statistical procedures employing Monte Carlo techniques.

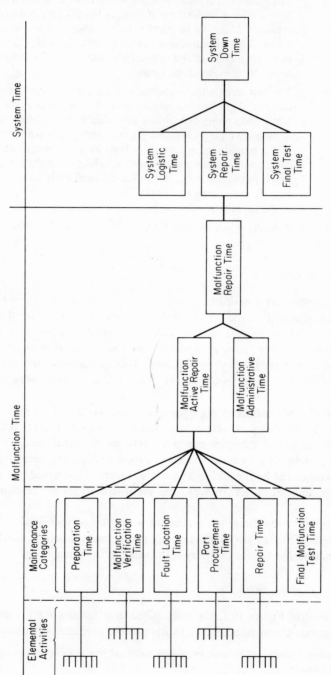

Fig. 10.1 Structure of time elements in "fix" of malfunction.

times required for malfunction active repair. The probabilities of occurrence in any one repair action are not equal for the six maintenance categories, and consequently the synthesizing process is again one of adding unequally sampled variates. In this case, the variates are sampled from the six category distributions according to an occurrence matrix based on observed data from ten different electronic systems. Contingency tests have indicated that this matrix, shown in Table 10.4, represents the occurrence pattern of the six maintenance categories during repair actions associated with the ten observed systems. It is therefore reasonable to assume that the matrix is applicable to the repair actions of all Air Force electronic systems.

ADMINISTRATIVE TIME

Analysis of ASB-4 repair actions reveals no correlation between the physical characteristics of a system and the length of administrative time. This time category is affected to a significant extent only by the following factors associated with repair:

1. Active repair time
2. Number of units involved
3. Number of men involved
4. Skill level of men involved.

The observed relationships between these four characteristics and administrative time are shown in Figures 10.2 through 10.5. The observed results are not unreasonable or unexpected when such factors as the following are considered:

1. As active repair time increases, the probability of interruption for meals, recreation, sleep, etc., increases.
2. As the number of units involved increases, the probability of a requirement arising for transporting a replacement unit—with its attendant delays—also increases.
3. As the number of personnel involved increases, the time spent at lunch or "coffee break" tends to increase.
4. The confidence and job-familiarity of personnel at the higher skill levels are reflected in a more liberal attitude toward recreation breaks.

Because of the strong correlation among the four significant characteristics and the difficulties foreseen in predicting the last two, only the dependence on active repair time and on the number of units involved is pursued in the method used for predicting administrative time.

Table 10.4 MODIFIED COMBINING MATRIX FOR ACTIVE REPAIR TIME SYNTHESIS (x = OCCURRENCE)

Instructions:

1. Select $10n$ random samples from the time distributions of each category (n = total number of x's in any column).

2. For each row, sum the samples of those maintenance categories denoted by an "x." Multiply each sum by the occurrence probability of that row. Repeat this procedure 10 times for each row. The result is 280 samples of malfunction active repair time.

Row Number	Preparation	Malfunction Verification	Fault Location	Part Procurement	Repair	Malfunction Test	Occurrence Probability of Row
1	x	x	x	x	x	x	0.162
2		x	x	x	x	x	0.021
3	x		x	x	x	x	0.021
4	x	x	x		x	x	0.061
5	x	x	x	x	x		0.165
6			x	x	x	x	0.018
7	x	x			x	x	0.011
8	x	x	x			x	0.007
9		x	x	x	x		0.028
10	x		x		x	x	0.006
11	x	x	x		x		0.073
12	x		x	x	x		0.023
13		x	x		x	x	0.010
14	x	x				x	0.008
15	x	x	x				0.010
16			x	x	x		0.048
17		x	x		x		0.017
18	x	x			x		0.030
19	x		x		x		0.010
20		x			x		0.007
21		x				x	0.008
22	x				x		0.007
23			x		x		0.006
24	x		x				0.004
25		x	x				0.007
26	x	x					0.165
27			x				0.003
28		x					0.062

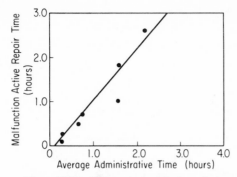

Fig. 10.2 Observed correlation between administrative time and malfunction active repair time.

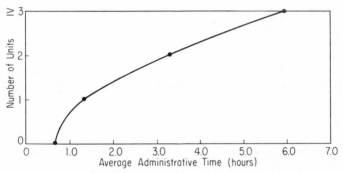

Fig. 10.3 Observed correlation between administrative time and number of units.

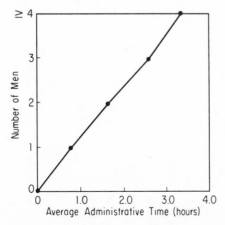

Fig. 10.4 Observed correlation between administrative time and number of personnel.

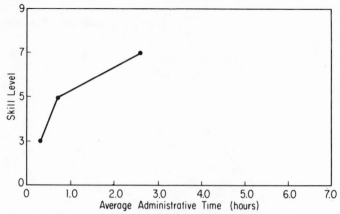

Fig. 10.5 Observed correlation between administrative time and skill level of personnel.

The cumulative distribution function of administrative times observed for the ASB-4 system during 1,378 occasions of malfunction repair is shown in Figure 10.6. This distribution is described by the Weibull function,

$$P_t = 1 - \exp\left(-\frac{t^\beta}{\alpha}\right). \tag{10.2}$$

Regression analysis of the data provides equations for predicting the parameters of the distribution of administrative times for a given value of malfunction active repair time. The variable, number of units involved, does not appear in the equations since the analysis shows that it may be set at a constant value of one without unduly prejudicing the accuracy of the α and β parameters. The equations derived from regression analysis are as follows:

$$\beta = 0.027X^3 - 0.233X^2 + 0.521X + 0.230 \tag{10.3a}$$

$$\alpha = \frac{\beta}{0.310 - 0.064X} \tag{10.3b}$$

where

α and β = parameters of the Weibull function for length of administrative time; and

X = malfunction active time, where $0 < X < 4.84$.

The probability that administrative time will be t hr or less for a given value of malfunction active repair time may be calculated by combining Equations 10.2, 10.3a and b, with the Weibull function modified as follows:*

$$P_t = 1 - \exp\left(\frac{-10t^\beta}{\alpha}\right).$$

*In the process of obtaining the equations for the α and β parameters, the observed values of administrative time were multiplied by 10 to attain greater sensitivity in the use of the Weibull graph paper. Since the α and β terms were obtained under this circumstance, t must also be multiplied by 10.

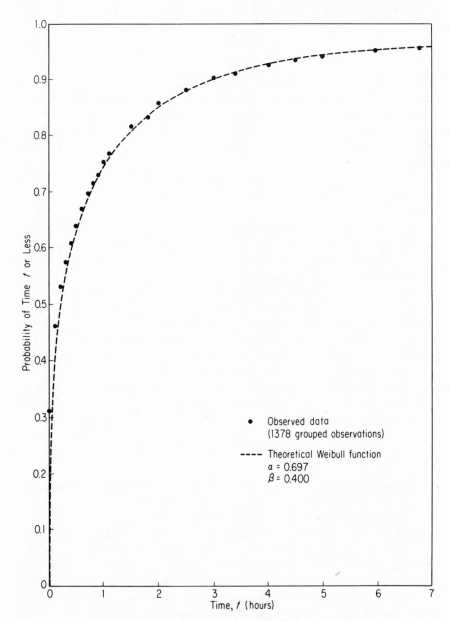

Fig. 10.6 Distributions of malfunction administrative time.

The modified Weibull equation may be rewritten as

$$t = \frac{[-\alpha \ln (1 - P_t)]^{1/\beta}}{10}, \tag{10.4}$$

from which a predicted value of administrative time t for a given value of malfunction active repair time and a given probability of occurrence of time t may be computed. This computation is used in the following procedure for combining the predictions for malfunction active repair time and administrative time to generate the predicted distribution of malfunction repair times.

MALFUNCTION REPAIR TIME

The sum of a given value of malfunction active repair time and a value randomly selected from its dependent distribution of administrative times is called *malfunction repair time*. A distribution of malfunction repair times is generated from a number of such sums, each originating from a value selected by an appropriate random process from the predicted distribution of malfunction active repair times. The value for the corresponding administrative time is computed from Equations 10.3a, 10.3b, and 10.4, with the required probability input selected by an appropriate random process. The number of such solutions should be no less than 200 to provide reasonable confidence in the predicted distribution.

In the form of a cumulative probability function, this distribution of malfunction repair times is commonly referred to as the *repairability function* for the system.

SYSTEM REPAIR TIME

The conversion of malfunction repair time to system repair time is dependent upon the number of malfunctions which must be corrected. Under ideal maintenance circumstances there would be only one malfunction, since corrective action would begin the moment a system exhibits a failure symptom. However, for a complex system having many functions and different modes of operation, a malfunction may disable only an occasionally used operating mode and simply degrade—rather than discontinue—system performance. In such a case, the system's continued operation affords the possibility for other malfunctions to occur. Similarly, during system "on" time for the correction of a malfunction, other malfunctions may appear.

Following a flight which provokes a complaint, the total number of malfunctions is the sum of those occurring during both flight and post-flight maintenance. The average number of malfunctions resulting from the flight itself can be predicted from the expression for the Poisson distribution truncated at zero (i.e., no zeros will occur).* The expression is

$$v \doteq \frac{\lambda t}{1 - e^{-\lambda t}} \tag{10.5}$$

where

v = the mean of a conditional Poisson distribution (the condition being that at least one malfunction has occurred),

λ = the system failure rate (predicted by a standard reliability prediction technique), and

t = the anticipated average flight length.

The average number of malfunctions occurring during post-flight maintenance may be estimated by multiplying the failure rate of the system by the predicted maintenance operating time. Assuming that the system is energized during the whole period of maintenance, the maintenance operating time, T_m, may be predicted by the expression,

$$T_m = v \bar{t}_{ar}, \tag{10.6}$$

where

v = the average number of in-flight malfunctions as predicted in Equation 10.5, and

$\bar{t}_{ar}$ = the median of the predicted distribution of malfunction active repair times.

The slight inaccuracy introduced by disregarding the time spent on repair of the malfunctions which occur during maintenance does not warrant the employment of the complicated calculations necessary to account for such time. Furthermore, this inaccuracy is counterbalanced by the small error related to the assumption that the system is energized throughout active repair time.

The expression for the total average number of malfunctions following a flight which provokes a complaint is, therefore,

$$N = \frac{\lambda t}{1 - e^{-\lambda t}} + \lambda \left(\frac{\lambda t}{1 - e^{-\lambda t}} \right) \bar{t}_{ar} = \frac{\lambda t}{1 - e^{-\lambda t}} (1 + \lambda \bar{t}_{ar}) \tag{10.7}$$

*This type of equation is used—rather than one expressing the product of flight time and failure rate—because only those flights in which at least one malfunction occurs are of interest in this discussion. For derivation of the expression, see Paul R. Rider, "Truncated Poisson Distributions," *Journal of the American Statistical Association*, Vol. 48 (December, 1953), 826–830.

where

λ = the predicted failure rate of the system,

t = the anticipated average flight length, and

t_{ar} = the median of the predicted distribution of malfunction active repair time.

For further accuracy, the value of N obtained by the foregoing equation must be adjusted to account for overlapping or concurrent maintenance activity. An overlap time of approximately 5 per cent has been observed among the malfunctions of a maintenance event. This overlap is assumed to be constant for all maintenance events. Therefore, an effective average number of malfunctions is $0.95N$. This number, when multiplied by the distribution of malfunction repair times, yields the predicted distribution of system repair times.

Discussion of the prediction process has now evolved to a point where consideration can be given to the synthesis of system down time from its three constituents: system repair time, logistic time, and final-test time.

SYSTEM LOGISTIC TIME

In this prediction technique, the parameters associated with the time distribution of logistic delay for the ASB-4 system (see Table 10.5) are assumed to be the same for all Air Force electronic systems. However, the probability that a logistic requirement will arise has been observed to be a function of the characteristics of a particular system. The simple but gross correlation presented in Figure 10.7 has been observed between the number of flight-line replaceable components contained in a system and the probability of a logistic requirement. This correlation can be expected to yield the approximate probability figure for logistic occurrence in systems of the type observed.

SYSTEM FINAL-TEST TIME

The probability of occurrence of most of the elemental activities can be calculated because of their close relationship with those system characteristics responsible for their occurrence. No such relationship has been observed for system final-test time. Until other systems can be studied, the probability of occurrence of system final-test time during a maintenance event is assumed to be 0.5, the value observed for the ASB-4.

The parameters of the distribution of system final-test time vary as functions of the number and type of system readouts which can reflect the

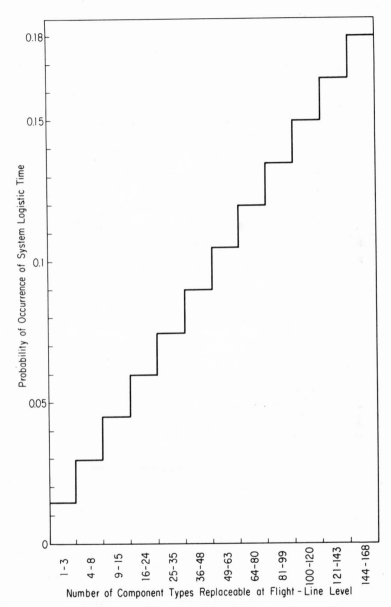

Fig. 10.7 Probability of occurrence of system logistic time.

<table>
<tr><td colspan="2" align="center">**Table 10.5** OBSERVED
DISTRIBUTIONS OF SYSTEM
LOGISTIC TIME</td></tr>
<tr><td>Time (t)
in Hours</td><td>Probability of
Time (t) or Less</td></tr>
</table>

Time (t) in Hours	Probability of Time (t) or Less
0.01	
0.014	
0.02	
0.028	
0.04	
0.056	
0.08	
0.112	
0.16	
0.224	
0.32	0.01
0.448	0.01
0.64	0.02
0.896	0.08
1.28	0.14
1.79	0.20
2.56	0.24
3.58	0.30
5.12	0.38
7.19	0.42
10.2	0.43
14.3	0.53
20.4	0.79
28.7	0.81
40.8	0.82
57.4	0.94
81.6	0.94
114.8	0.98
163.2	1.00

Table 10.6 OBSERVED DISTRIBUTIONS OF SYSTEM FINAL-TEST TIME

Time (t) in Hours	Probability of Time (t) or Less
0.01	
0.014	
0.02	
0.028	
0.04	
0.056	
0.08	0.02
0.112	0.02
0.16	0.02
0.224	0.08
0.32	0.23
0.448	0.39
0.64	0.61
0.896	0.75
1.28	0.92
1.79	0.97
2.56	0.99
3.58	1.00

operational integrity of the system. The distribution for any system is calculated as the product of the distribution observed for the ASB-4 and a *readout factor*. This factor is determined by multiplying the number of each of several types of system readouts by a weighting factor representative of the relative time required to interpret that type of readout. The sum of these products (a value of 1 for the ASB-4) is the quantity by which the observed distribution, presented in Table 10.6, is multiplied. The format for computing the readout factor is shown in Table 10.7.

Table 10.7 SYSTEM READOUTS

Readouts which can reflect operation of system	(A) Quantity	(B) Normalized Weighting Factor	(C) Product (A) (B)
I. Visual			
A. *Information Devices*	—	0.05	—
(i.e., position, velocity, angle, indicating instruments)			
B. *Circuit Monitors*			
(i.e., voltage current, power pressure, frequency indicators)	$n* =$ $n =$ $n =$ $n =$	$0.03 +$ $0.003(n-1)^\dagger$	— — — —
C. *Cathode Ray Displays*	—	0.14	—
D. *Optics*	—	0.05	—
II. Auditory			
A. *Audio*	—	0.03	—
B. *Transmitter Sidetone*	—	0.03	—
Total = (Number of System Readouts)		Total = (Readout Factor)	

* n is the number of different parameters which can be read on each monitor by a switching arrangement.

† This weighting factor must be multiplied by *each* of the circuit monitors listed in column A.

SYSTEM DOWN TIME

The distribution of system down times is generated by combining the predicted time distributions for system repair, logistic delay, and final system test. This procedure is a straightforward statistical exercise—the adding of variates unequally sampled from the three distributions in accordance with their respective probabilities of occurrence, using a Monte Carlo technique. The probability for repair time is 1.0, the probability for system final test is 0.5, and the probability for logistic delay is computed as discussed above under "System Logistic Time."

Application of Prediction Procedure

The accuracy obtainable from this prediction procedure is evidenced in Figures 10.8 and 10.9. These figures present the predicted and observed distributions of system down times for the APN-89 and the MD-1 systems in the B-52 aircraft.

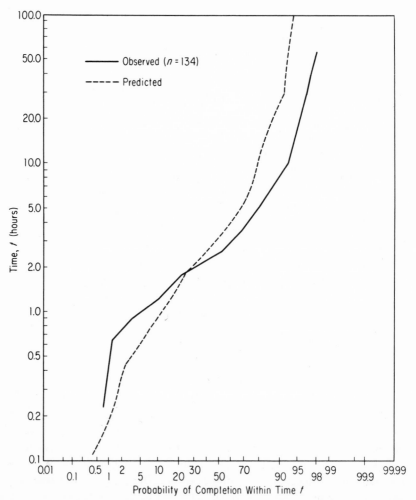

Fig. 10.8 Distribution of down time for AN/APN-89 system (B-52 configuration).

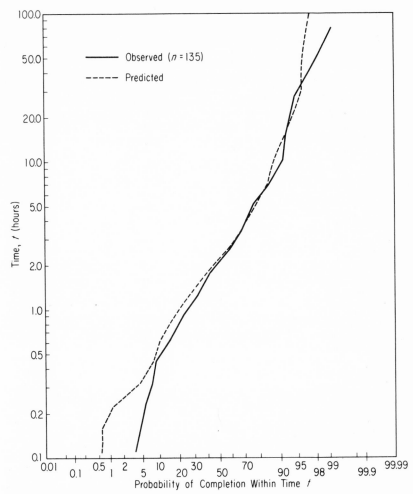

Fig. 10.9 Distribution of down time for MD-1 system (B-52 configuration).

10.4 Other Approaches to Maintainability Measurement and Prediction

American Institute for Research

An index for measuring electronic equipment maintainability has been developed by the Human Factors Office of the American Institute for Research (AIR), Pittsburgh, Pennsylvania. This work was carried out under contract to the U.S. Army Signal Equipment Support Agency, Fort Monmouth, New Jersey. The development and use of the method is described in detail in the Institute's report AIR-275-59-Fr-207, and two other associated documents.*

The maintainability index developed in this study is based on the presence or absence of 241 maintenance design features for which maintenance effects or consequences have been determined. The design features are organized into 9 major design-factor categories:

1. Displays and controls
2. External accessibility
3. Test points
4. Cables and connectors
5. Internal accessibility
6. Cases
7. Lubricants and tools
8. Manuals
9. Test equipment.

Each design feature is weighted according to its relative importance in each of 4 "maintenance consequence" areas. These weights were determined from data gathered from experienced field maintenance personnel. The consequence areas are

1. Maintenance time
2. Logistic requirements
3. Equipment damage
4. Personnel injury.

To determine a score for a specific equipment, the evaluator must indicate whether or not each of the 241 design features is present in the equipment. From these answers, scores between zero and 100 are derived for each of the 4 consequence areas and for each of the 9 design-factor areas. These scores are then plotted to give a score profile for the equipment.

In conjunction with the maintainability index, a set of maintainability standards for each consequence and design-factor score was developed. The input data for computing these standards were obtained from command-level users of various equipment types—i.e., the standards are based on the expressed requirements of command-level personnel. The standards provide a method for determining whether the maintainability of an equipment, as measured by the index, meets the requirements of the users of that equipment.

*The ASTIA document numbers and titles are: AD-219987, *An Index of Maintainability Evaluation Booklet*; AD-219988, *Development of an Index of Electronic Maintainability*; and AD-219989, *An Index of Maintainability Instruction Manual*.

The profile sheets on which the evaluator plots his scores are preprinted with the standards, so that a direct comparison between the equipment scores and the standards is easily made. The standards are shown as percentile ranges rather than as single values. For example, if the score equals or exceeds the standard value for the 75th percentile, this means that the requirements of 75 per cent of the users will be met in that particular design or consequence area. For the 50th percentile, the requirements of 50 per cent of the users will be met, etc.

The development of this procedure appears to be a significant step in the qualitative evaluation of equipment maintainability. The method seems best suited for evaluation when at least a prototype of the equipment is available. However, it does not seem that the method would be applicable in the initial design stages. The scores are of value principally in pointing out areas of the equipment which will benefit from redesign.

The evaluation procedure requires from 2 to 4 hours for the evaluator to answer the questionnaire, the time depending on the complexity of the equipment and the evaluator's familiarity with it. The computations necessary to derive the scores require another 3 to 4 hours using a desk calculator.

The AIR evaluation technique does not consider the actual length of equipment down time for maintenance, nor the frequency with which maintenance may be required—factors which must enter into a maintainability prediction. It is felt, however, that the AIR technique can be of considerable value as a supplement to other methods of maintainability evaluation.

Federal Electric Corporation

The Federal Electric Corporation (FEC) has performed extensive work for the Bureau of Ships under Contract NObsr-75376 to determine the feasibility of describing numerically the maintenance activities associated with shipboard electronic equipments. The primary objective of the work was to provide a design tool for the quantitative prediction of electronic equipment and system maintainability early in the system design stages. The work under this contract is described in detail in two reports published by FEC and delivered to the Bureau of Ships in July 1960. The first of these, *A Maintainability Study of Shipboard Electronic Equipment*, describes the derivation and development of the procedures. The second, *A Maintainability Prediction Procedure for Designers of Shipboard Electronic Equipment and Systems*, describes the step-by-step procedure for predicting a maintainability index and also provides work sheets and the necessary tabulated values of maintenance task times.

Since the FEC methods are well documented in the reports listed above, they will be discussed only briefly here. The basis of the FEC procedure is the computation of a maintainability index (*MI*) using a formula first proposed by the Naval Electronics Laboratory, San Diego, California. This formula is

$$MI = \left(1 - \frac{M_{c91} + M_{s91}}{2184}\right) \times 100 \qquad (10.8)$$

where

M_{c91} = total down time, measured in hours, for corrective maintenance tasks over a 91-day calendar interval;

M_{s91} = total down time, measured in hours, for scheduled or preventive maintenance tasks over a 91-day calendar interval;

2184 = 24 hr per day for 91 days, which is a hypothetical continuous cruise period of three calendar months anticipated under combat conditions.

The index above is essentially a measure of equipment up time in terms of the percentage of time during a 91-day period that the equipment is available to perform its intended function. This measure has also been termed *operational availability.*

Measured values of M_{c91} and M_{s91} may be substituted directly in Equation 10.8 to arrive at a value of *MI* for a specific equipment already in use. The process of estimating *MI* from information available early in equipment design stages is considerably more complex.

FEC has assumed that the effect of M_{s91} on the predicted *MI* will be negligible, and this assumption has been substantiated by field measurements on equipments aboard destroyer-class vessels.* This assumption simplifies Equation 10.8 to

$$MI = \left[1 - \frac{M_{c91}}{2184}\right] \times 100.$$

The prediction process then requires an estimate of M_{c91} for insertion in the formula. FEC's method involved predicting M_c, the equipment down time resulting from 2184 hr of equipment operation, and, since a 91-day calendar period does not necessarily correspond to 2184 hr of equipment operation, correcting M_c to give M_{c91} by

$$M_{c91} = \frac{2184 \, M_c}{(2184 + M_c)}.$$

The predicted value of M_c is arrived at by the use of reliability data (frequency with which maintenance is required), and maintenance-time data (average time required to perform a specific task). M_e is given as the sum-

---

*Data collected by ARINC Research Corporation on many equipment types also support this assumption.

mation of the products of failure rates and task times for individual parts, modules, subassemblies, or other applicable levels of maintenance.

FEC'S computations of MI values using this method show excellent agreement between the predicted values and the values computed from actual field data.

The M_c predicted by the FEC method is the *active repair time* for 2184 hr of equipment operation and does not include any of the delay times which may be associated with equipment maintenance. M_c, or active repair time, is the only element of total equipment down time which is controllable to any extent by the design engineer. Delay time, which is composed of administrative and logistic time, is not "design controllable"; however, in actual operational situations, delay time may contribute appreciably to total equipment down time. The FEC reports point out this situation, and an attempt was made to determine a correlation factor between delay time and active repair time. Such a factor could not be determined with any degree of confidence in its general applicability, and it was concluded that delay time was not predictable.

The fact remains, however, that in practical maintenance situations, delay time has been observed to range from 1/2 to 10-or-more times active repair time, and would therefore have a significant effect on equipment down time or unavailable time. The MI is therefore not a true measure of equipment "percentage available time" or operational availability, *unless* some consideration has been given to administrative and logistic down time in arriving at the value of M_c to be substituted in the MI formula. Where M_c represents only active repair time, the resultant MI represents a measure of an ideal availability or intrinsic availability.

While delay time will undoubtedly be minimized to the greatest possible extent for critical systems in combat situations, it probably will continue to exist in most maintenance actions. For these reasons, use of the MI as a measure of equipment *operational availability* is not adequate where M_c represents only active repair time. The MI is adequate as a measure of *intrinsic availability*.

Radio Corporation of America

The Government Services Section of the RCA Service Company, a Division of Radio Corporation of America, is developing a method for measuring and predicting the maintainability of Air Force ground electronic equipments. This work is being performed for the Rome Air Development Center, Air Research and Development Command, under Contract AF 30(602)-2057. The general plan and some preliminary results are described in detail in a report prepared for RADC by RCA (RADC-TN-60-5, January 6, 1960).

Before considering maintainability as a separate attribute, the RCA report discusses the concept of system effectiveness—i.e., a measure of the net worth, or value, of a system to the user. *System effectiveness* is a function of system performance capability, system dependability, and system cost. *Performance capability* includes the capacity to meet specified requirements, such as range, power output, sensitivity, accuracy, and the like. *Dependability* is a measure of the degree of consistency of performance and is essentially the same as operational availability. *Availability* is, in turn, a function of reliability and maintainability. *System cost* must include the total amount for development, production, and service-life support of the equipment.

Maintainability, then, is only one part—although a very important part—of the measurement of over-all system worth. This conceptual framework proposed by RCA is very similar to the approach of ARINC Research Corporation. The RCA report defines two kinds of availability:

1. Intrinsic availability, A_i, expressed as

$$A_i = \frac{\bar{t}_f}{\bar{t}_f + \bar{t}_r} \qquad (10.9)$$

where

$\bar{t}_f$ = mean time between failures, and

$\bar{t}_r$ = mean time to repair (active repair time only);

2. Operational availability, A_o, expressed as

$$A_o = \frac{\bar{t}_f}{\bar{t}_f + \bar{t}_r'} \qquad (10.10)$$

where

$\bar{t}_r'$ = mean time to repair measured on a calendar time basis (time includes all maintenance delay times).

For equipments which are required to operate continuously, operational availability may be expressed as

$$A_o = \frac{\text{(total operate time)}}{\text{(total operate time)} + \text{(total down time)}}. \qquad (10.11)$$

In Section 2.7.1 of the RCA report, entitled "Discussion of Definitions," the Department of Defense definition of maintainability is quoted:

"Maintainability is a quality of the combined features and characteristics of equipment design which permits or enhances the accomplishment of maintenance by personnel of average skills, under the natural and environmental conditions in which it will operate."

The definition above is highly qualitative and is not subject to quantification without further specification. The report states further that "—The search for a single definition that encompasses all the attributes of maintainability in a quantitatively measurable term is, for the present, unrewarding. It is first necessary to identify and measure the most relevant factors that make up this end measurement. It is likely that no single final measurement will adequately serve all purposes."

In line with this reasoning, RCA suggests several possible indices which may be useful in the quantitative description of maintenance activity. Among these are

1. Ratio of satisfactory operation time to total required time

2. Average down time per unit of calendar time (or other stated time base)

3. Mean time to repair

4. Man-hour requirements per unit of operating time

5. Total man-hour requirements per unit of calendar time

6. Waiting time per unit of time (calendar or other stated time)

7. Material requirements per unit of time

8. Cost of support per unit of calendar time.

It is probable that any or all of the indices above may be needed in one situation or another, plus, perhaps, other special indices.

For purposes of quantitative prediction, RCA has elected to express maintainability in terms of time required to perform a maintenance action. It is assumed that this time is a function of such governing factors as equipment design, supply and logistics, test equipment, training, technical orders, operational circumstances, malfunction criticality, and personnel requirements. This concept may be stated symbolically as:

$$M_t = f(X_1, X_2, \ldots, X_n),$$

where

M_t = maintenance time, and

$X, \ldots, X_n$ = values which quantitatively express the n governing factors described above.

The X_i are quantified by a numerical scoring procedure, and the following linear relationship is assumed:

$$M_t = a + b_1 X_1 + b_2 X_2 + \ldots + b_n X_n.$$

The constants, $a, b, b_1, \ldots, b_n$, are determined by regression analysis techniques based on a large number of observed values of M_t and the associated values of $X, \ldots, X_n$. Once the regression constants are known, the equation can then be used to estimate M_t for other situations where the X's can be measured or estimated.

Coefficients for this equation have been determined and are documented in an RCA report.* An empirical procedure for predicting maintenance down time, based on this report, has been incorporated into Air Force Maintainability Specification MIL-M-26512C, described in Section 10.5.

Republic Aviation Corporation

The Republic Aviation Corporation is performing research and development work in the area of maintainability and supportability evaluation techniques for Air Force weapons systems. This work was initiated by Wright Air Development Division under Air Force Contract No. AF 33(616)-6495. The long-range project consists of three successive phases, with objectives as follows:

Phase I. Establish maintainability and supportability (M&S) evaluator formulae which will express the degree of capability of a given type Air Force weapon system to meet the specified operational support parameter in time and cost terms.

Phase II. Apply the M&S formulae to contractor-selected portions of a representative system requirement or development design area to determine adequacy.

Phase III. Apply the M&S formulae to development designs in system areas selected by the Air Force.

A report has been prepared covering the results of Phase I of the project. This report, *Maintainability and Supportability Evaluation Techniques*, was distributed as WADD Technical Note 60-82, dated March 1960. This report discusses work in the following four areas:

1. Definition of maintainability
2. Specification of maintainability
3. Contractor accomplishment of maintainability
4. Procuring agency evaluation of maintainability.

The Republic Aviation report states that: "Maintainability is a function of time and resource parameters expressed quantitatively in terms of manpower, support equipment, facilities, consumables, and the monetary values thereof." This statement lead to the following definition:

*Radio Corporation of America, *Maintainability Prediction Technique*, Publication No. RADC-TR-62-156 (prepared for Rome Air Development Center) (March 15, 1962).

"Maintainability is the degree with which a product meets its required, or specified, operational parameters in a given maintenance environment, within the time specifications, while being maintained by the resources programmed for the task. The degree may be expressed as a ratio or percentage comparison between the actual resource requirements and the programmed resources."

The definition must then be translated into a maintainability specification to control product design and operational function. The following basic factors must be covered in the specification:

1. Operational parameters

2. Time specifications

3. Maintenance environments

4. Programmed resources.

In addition, the procuring activity must provide the contractor with the following operational information:

1. Number of units to be considered as a planning entity

2. Required in-commission or operable percentage or number

3. Operational plan

4. Environments

5. Life cycle

6. Allowances for major duty-time factors.

The report states further that it is essential for both the contractor and the procuring agency to have personnel capable of assuming the responsibility of maintainability evaluation—the contractor, in order to assure design for the achievement of maintainability in the system; and the procuring agency, in order to assure itself that this maintainability has in fact been achieved in the end product.

The report then presents a number of specific formulae for calculation of specific maintainability parameters. It is not practicable to present the details of the analysis in this brief summary, nor is that the purpose of this survey. The general plan establishes a framework within which the Air Force can specify, control, and assure achievement of required maintainability. The plan is directed toward, and is consistent with, the current Air Force concept of evaluation of the over-all weapons system, rather than the individual sub-systems or equipments.

General Electric Company

The Technical Military Planning Operation (TEMPO) of the General Electric Company has developed a method for analyzing and estimating the effectiveness or operational availability of a multimodal system. The technique has been used in predicting the operational availability of the Polaris Fire Control System, and reports on this analysis are available to authorized personnel. However, the method is described in an unclassified paper by Bovaird and Zagor.* Because of the increasing use of multimodal complex systems, the method will be discussed briefly.

The technique is referred to by TEMPO as "modal analysis," and is applicable to those complex systems capable of operating in various states or modes with different degrees of task capability for each mode. The analysis uses basic reliability and maintainability information on subsystems to arrive at a predicted operational availability for each mode of operation. The various modes are ordered according to decreasing operational effectiveness or degree of task capability, with the last mode being unacceptable (complete system failure). Other modes may be unacceptable for mission performance under certain conditions.

The multimodal system is first divided into two or more functional equipment groups which can fail independently. The groups which do not fail are still capable of providing some degree of task capability less than that of the primary mode. The operational availability of a particular mode is the fraction of mission time (or the probability) that the configuration of failed and nonfailed groups of the mode exists.

The probability that a given group has not failed at a random point in mission time is given by

$$A = \frac{t_f}{t_f + t_d} \qquad (10.12)$$

where

A = availability of the group,

t_f = mean time to failure (in mission time), and

t_d = mean down time per failure (in mission time).

The availability of each operational mode is then obtained from the group availabilities by application of probability theory to the groups comprising each mode. The system availability or probability of being in an acceptable mode is then the sum of the availabilities of the acceptable modes. The

*Electronic Maintainability, Vol. 3, Third Electronic Industries Association Conference on Maintainability of Electronic Equipment (Elizabeth, N.J.: Engineering Publishers), 132.

system reliability is then

$$R = \frac{1}{A} \sum A_i R_i \qquad (10.13)$$

where

A = probability that system is in an acceptable mode at time use is required,

A_i = probability that system is in mode i at time of demand, and

R_i = probability that system remains in an acceptable mode for the duration of demand time, given that it is initially in mode i and that mode i is acceptable.

The results of this type of analysis are of value in two different situations: (1) to aid the tactical commander in making a decision whether to use the system in a substandard mode, or wait for repair to return it to a primary mode; and (2) to aid the system designer in making a decision as to the addition of other functional groups to increase the system's effectiveness. In each of these situations the decision is based on certain trade-offs which are expressed quantitatively by the analysis. The tactical commander must weigh the reduced effectiveness of a substandard mode against the penalty of time lost awaiting system repair. The system designer must weigh the advantages of increased effectiveness against the disadvantages of higher cost, increased weight and complexity, and other constraints which may be placed on the system.

This type of analysis, when properly applied and interpreted, can be a valuable tool in the design and operation of multimodal systems for maximum operational effectiveness.

10.5 Summary of Military Maintainability Specification Requirements

The importance of specifying maintainability requirements has been recognized by the military services, as exemplified by the following specifications:

1. MIL-M-23313 (Bureau of Ships)
2. WS-3099 (Bureau of Naval Weapons)
3. MIL-M-26512C (USAF)
4. AR-705-26 (Army).

These specifications emphasize the principles of maintainability, namely,

1. Minimize the complexity of maintenance tasks
2. Provide for rapid recognition of failures or marginal performance

3. Provide optimum accessibility
4. Minimize types and quantities of tools and test equipment
5. Provide maximum safety for the product and personnel
6. Provide adequate service manuals
7. Provide optimum fault location, fault isolation, disassembly, interchange, reassembly, alignment, and checkout features
8. Provide optimum ease of maintenance action under field or depot maintenance environments.

The contractor's maintainability program requirements are defined in each specification. They generally involve design reviews; consideration of subcontractor and vendor selection and performance, to assure compliance with the maintainability requirements of the specification; and a maintainability analysis. This analysis includes prediction, determination of potential trouble areas, trade-offs, and a demonstration to assure that the equipment meets quantitative maintainability objectives, permitting the achievement of operational requirements.

Some differences will be noticed in the three specifications cited, with regard to maintainability prediction and demonstration. The Bureau of Ships specification, based on work performed by Federal Electric Corporation, requires that a maintainability prediction be made for the specified equipment. Reliability prediction data (NAVSHIPS 93820) are combined with standardized time data for fault location, isolation, disassembly, reassembly, alignment, and checkout activities. The specified procedure permits the calculation of the geometric mean time to repair ($MTTR_G$). This value is not to exceed the specified equipment repair time (ERT). The specification suggests that the specification writer first establish a maximum value for ERT, based on operational requirements which cannot be accepted more than 10 per cent of the time. The following relationship is used (and is derived in the specification):

$$ERT = 0.37 \, ERT \, \text{max.} \qquad (10.14)$$

A maintainability (repairability) demonstration test is then performed by a qualified technician. Twenty defective parts, selected in proportion to the over-all failure rate contribution to the system (by part class), are introduced one at a time, and appropriate repair time measurements are made. Acceptance is accomplished when the measured geometric mean time to repair ($MTTR_G$) and the standard deviation (S)* produce the following result:

$$\log MTTR_G \leq \log ERT + 0.397S. \qquad (10.15)$$

Experience has shown that to predict corrective maintenance time with use

---

*A formula for calculating S is given in the specification.

of this technique requires 2 hours for each active element group (*AEG*). One hour of this estimate is used for making a reliability prediction.*

The Bureau of Naval Weapons' specification requires mean-time predictions for both corrective and preventive maintenance actions. Required also is a calculation of the mean maintenance time ($\bar{M}$)—which includes both preventive and corrective maintenance, and a calculation of the Maintainability Index (*MI*), for which a special formula is given in the specification. The intent of the specification is that the mission requirements, such as availability (intrinsic availability as defined in Chapter 1) or permissible $\bar{M}$ of the product per specified time period shall be the basic criteria for the maintainability philosophy.

A maintainability demonstration test is required to show that the measured mean maintenance time, $\bar{M}$, meets specification requirements within the appropriate confidence limits.

The Air Force specification provides a means for determining the number of corrective and preventive maintenance actions to be included in the maintainability demonstration test. The mean corrective and preventive maintenance times are determined from this test and are combined to give the mean system down time for a period of 5000 operating hours.

According to the Air Force specification, maintainability must be quantitatively demonstrated and evaluated. The specification presents a demonstration technique, but the contractor may propose alternatives. The technique outlined in MIL-M-26512C provides a means for determining the number of corrective and preventive maintenance actions to be included in the demonstration test. The following relationships are used for establishing specification requirements and analyzing test data:

1. Mean corrective maintenance down time, $\bar{M}_{CT}$, is given by:

$$\bar{M}_{CT} = \frac{\sum_{1}^{N_C} M_{CT}}{N_C} \tag{10.16}$$

where

M_{CT} = active maintenance down time per corrective maintenance task.

N_C = number of simulated corrective maintenance tasks.

2. Mean preventive maintenance down time, $\bar{M}_{PT}$, is given by:

$$\bar{M}_{PT} = \frac{\sum_{1}^{N_P} M_{PT}}{N_P} \tag{10.17}$$

*G. Margulies and J. Sacks, "Bureau of Ships Maintainability Specification," *Proceedings Ninth National Symposium on Reliability and Quality Control*, San Francisco, California (January, 1963).

where

M_{PT} = active maintenance down time per preventive maintenance task.

N_P = number of simulated preventive maintenance tasks.

3. Total active-maintenance down time, D_T, is given by:

$$D_T = \bar{M}_{CT} f_C + \bar{M}_{PT} f_P \qquad (10.18)$$

where

f_C = number of expected corrective maintenance tasks in the specified operating time period.

f_P = number of expected preventive maintenance tasks in the specified operating time period.
(For ground equipments, the operating time period is at least 5000 hr; for other equipments, shorter times may be required.)

4. Mean down time, $\bar{M}$, is calculated as

$$\bar{M} = \frac{MTBF\,(1 - A_i)}{A_i} \qquad (10.19)$$

where

A_i = inherent availability (which excludes administrative and logistic down time).

5. Mean down time, $\bar{M}$, can also be calculated as

$$\bar{M} = \frac{\bar{M}_{CT} f_C + \bar{M}_{PT} f_P}{f_C + f_P}. \qquad (10.20)$$

6. $\text{Log } M_{\max} \approx \overline{\log M_{CT}} + 1.645 \sqrt{\dfrac{\displaystyle\sum_1^{N_C} (\log M_{CT})^2 - \dfrac{\left(\displaystyle\sum_1^{N_C} \log M_{CT}\right)^2}{N_C}}{N_C - 1}}$

$$(10.21)$$

where

$M_{\max}$ = maximum corrective maintenance down time.

(Ninety-five per cent of all repair time intervals are restricted to less than the value given by $M_{\max}$.)

10.6   Preventive Maintenance Scheduling

Preventive maintenance is sometimes considered as a procedure intended primarily for the improvement of maintenance effectiveness. However, it is more proper to describe preventive maintenance as a particular category of maintenance, designed to optimize the related concepts of reliability and availability.

The value of preventive maintenance has long been a subject of debate. Some strongly believe that a satisfactorily operating equipment should be left alone. Others advocate strict adherence to a preventive maintenance schedule. In either event, it is apparent that established procedure must be founded on an analytical evaluation of the equipment involved. Such matters as failure-rate patterns of the system or its piece parts must be considered.

Preventive maintenance is advantageous for systems and parts whose failure rates increase with time. Kamins and McCall* show that cost savings accrue for preventive maintenance (planned replacement) only if the parts under consideration exhibit increasing failure rates. Many types of electron tubes, batteries, lamps, motors, relays, and switches fall within this category. Most semiconductor devices and certain types of capacitors exhibit decreasing failure rates, while complex systems generally have constant failure rates. In the latter case, certain classes of parts within the systems display increasing failure rates; consequently, the effectiveness of a preventive maintenance program depends on how well it detects these deteriorating parts. Frequently, a decrease in reliability is noted after a large-scale overhaul or preventive maintenance action. For example, in one situation involving electronic equipments overhauled after each 1000 hr of operating time, 30 per cent of all reported failures occurred within 24 hr after each overhaul.

Reduction of operational failures is the real purpose of scheduled or preventive maintenance. To achieve a balance between reliability and maintenance costs for any equipment, several factors must be weighed simultaneously and a suitable trade-off point selected. The various factors to be considered are: (1) the reliability index and time duration desired; (2) the cost of an in-service failure; (3) the cost of replacement before failure; (4) the most economical point in the equipment life to effect this replacement; and (5) the predictability of the failure pattern of the equipment under consideration. The ideal procedure would be to replace a unit just prior to failure, and then

*M. Kamins and J. J. McCall, Jr., *Rules for Planned Replacement of Aircraft and Missile Parts*, The Rand Corporation, RM-2810-PR (November, 1961). See also, E. L. Welker and C. L. Bradley, *The Dollar Value of Improved Reliability*, ARINC Research Corporation, Publication No. 101-31-175 (August 15, 1960).

realize the maximum of trouble-free life. The relationship used here was developed by W. E. Weissbaum* and gives the average hourly costs in terms of two costs, K_1 and K_2, and the failure probability distribution of the item. The model is as follows:

$$A(\tau) = \frac{K_1 - (K_1 - K_2) - G(\tau)}{\displaystyle\int_0^\tau G(t)dt} \qquad (10.22)$$

where

$A(\tau) = $ the average hourly cost,

$K_1 = $ the total cost of an in-service failure,

$K_2 = $ the total cost of a scheduled replacement,

$G(\tau) = $ the probability that a new unit will last at least τ units of time before failure, and

$\tau = $ the time to replacement after the last replacement.

Application of this technique enables the optimum replacement interval to be determined if the failure distribution is known. If the replacement interval is too short, considerable loss of useful equipment life would result and the average hourly cost would be high. However, if the replacement interval is too long, then the costs of an in-service failure, in terms of mission aborts and manpower, are quite intolerable. The ratio of K_1 (the cost of in-service failure) to K_2 (the cost of scheduled replacement) is the critical factor in arriving at a decision regarding scheduled replacement policy. As the ratio increases, the lowest average hourly cost is realized by replacing the part after a shorter life, as shown in Figure 10.10.

In the model, the family of curves is plotted for various ratios of K_1 to K_2 and is denoted as K. When $K = 1$, there is no advantage to scheduled replacement, and the equipment should be allowed to run to failure. When $K > 1$, there is an advantage to scheduled replacement. If, for example, the cost of in-service failure was 10 times the cost of a scheduled replacement, then the $K = 10$ curve shows that replacement should be scheduled at approximately 1/3 of the *MTBF*, or at 80 hr. The cost would be the least at this point, and the probability of the unit surviving for this 80-hr period is in excess of 0.90. The y-axis scale is determinable only after the K_1 and K_2 costs have been assigned actual dollar values. This model requires information only on the

*W. E. Weissbaum, "Probability-Theoretic Solution of Some Maintenance Problems," Fourth Signal Maintenance Symposium (1960). See also M. Kamins and J. J. McCall, Jr., "Rules for Planned Replacement of Aircraft and Missile Parts," Memorandum RM-2810-PR, The Rand Corporation (November, 1961); and E. L. Welker, *Relationship Between Equipment Reliability, Preventive Maintenance Policy, and Operating Costs*, ARINC Research Corporation, Publication No. 101-9-135 (February, 1959).

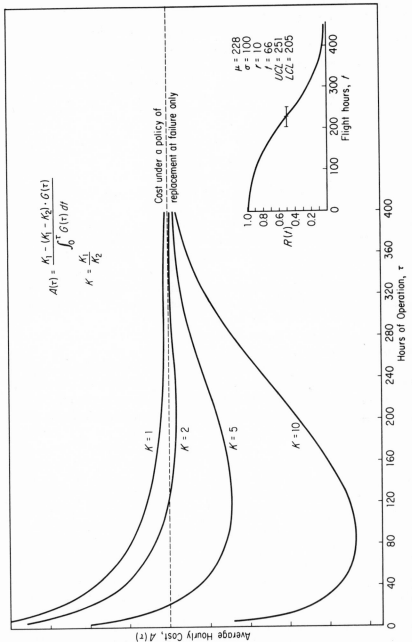

Fig. 10.10 Average hourly cost of scheduled replacement after τ hours of operation for a component exhibiting normal wear-out.

failure distribution and the values of K_1 and K_2, and is easily adaptable for computer programming.

The example shown in Figure 10.10 is based on an expensive mechanical component used in military aircraft. The theoretical normal function is shown; the mean (μ) is 228 flight hr and the standard deviation (σ) is 100 hr. The family of curves based on different K values was derived by using a K_2 of $2000 (estimated overhaul cost). K_1 was derived from the formula

$$K = \frac{K_1}{K_2},$$

from which

$$K_1 = \frac{K}{K_2}.$$

This method of deriving K was used because the cost of an in-service failure (K) is difficult to ascertain in terms of dollars.

This technique is directly applicable to any system component that has an increasing failure rate with time. It can be used to determine the optimum overhaul point in the unit's life, taking cost into consideration. However, it must be remembered that this technique cannot be used on equipment that exhibits a constant failure rate with time. This type of equipment would be most economically replaced only at the time of failure.

10.7 Prediction of Spare Parts Requirements

Ebel and Lang* have developed a rather simple procedure for determining spare parts requirements. The procedure assumes a constant failure rate for the parts under consideration; the probability, then, of having r or fewer failures in time t is given by

$$P(r) = \sum_{n=0}^{r} \frac{(\lambda t)^n}{n!} e^{-\lambda t} \tag{10.23}$$

where

$n =$ number of failures,

$\lambda =$ failure rate of parts being considered,

$t =$ operating time.

The number of spares to stock for a given period, to back up a given number of items having a known failure rate, can be found from Figure 10.11.

*G. Ebel and A. Lang, "Reliability Approach to the Spare Parts Problem," *Proceedings, Ninth National Symposium on Reliability and Quality Control*, San Francisco, California (January, 1963).

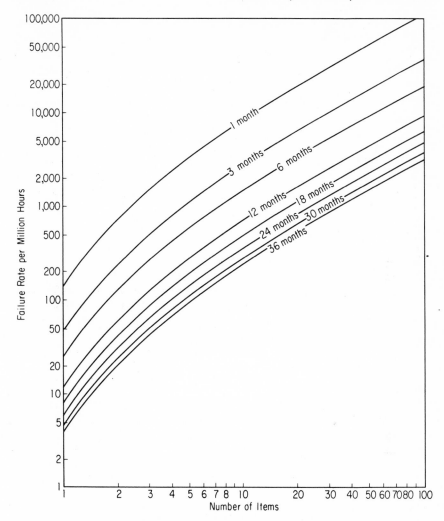

Fig. 10.11 Minimum spare requirements for various operating periods (90 per cent confidence level). [SOURCE: G. Ebel and S. Lang, "Reliability Approach to the Spare Parts Problem," *Proceedings, Ninth National Symposium on Reliability and Quality Control*, San Francisco, California (January, 1963).]

This figure is used as follows:

1. Determine the failure rate (λ) of the part or assembly.
2. Determine the number of times the item is used in the equipment to be serviced by the supply point, i.e., the number of times the item appears in a system times the number of systems serviced by the supply point.

3. Determine the total operating time for the item; e.g., a part operating 8 hr per day for one year has a total operating time of 4 months.
4. Enter the left side of the chart at the appropriate failure-rate level. Move up the sloped line until it intersects the vertical line representing the number of items used in the system.
5. Move horizontally to the heavy curve representing operating time, as determined in Step 3. The number of spares required is read on the abscissa.

Example. An equipment uses two transmitting tubes, each of which has an expected hourly failure rate of 500×10^{-6}. The equipment is to be used 8 hr per day. How many spares are required for a 3-calendar-year period? Answer: 14.

PROBLEMS

1. Following is an ordered listing of active repair times for an aircraft electric system. Plot the observed data and the corresponding theoretical log-normal function. Show the 95 per cent confidence limits for the median time to repair for the theoretical function.

Active Repair Time, Hours	Number of Observations
0.5	2
1.0	24
1.5	5
1.7	1
2.0	46
2.2	1
2.3	2
2.5	3
2.6	1
3.0	13
3.2	1
3.5	1
4.0	16
4.4	1
4.5	1
5.0	7
6.0	6
7.0	1
7.5	2
8.0	2
9.0	1

10.0	6
10.5	1
15.0	1
16.0	1
18.0	1
20.0	2

2. The system discussed in Problem 1 accumulated a total of 6000 operating hours. Listed below are total maintenance man-hours associated with the repair of the system. (a) Plot the observed data and the theoretical function. (b) Why are the time values greater than those of Problem 1? (c) Calculate the Maintenance Support Index.

Man-Hours	Number of Observations
1.0	24
1.5	5
1.7	1
2.0	35
2.2	1
2.5	2
2.6	1
3.0	15
3.2	1
3.5	1
4.0	21
4.3	2
4.4	1
5.0	4
5.5	1
6.0	1
7.5	2
8.0	9
10.0	4
10.5	1
11.0	2
12.0	4
13.0	1
15.0	2
16.0	1
18.0	1
20.0	1
25.0	1
28.5	1
30.0	2
38.0	1

3. Discuss problems associated with scheduling preventive maintenance for (a) military systems and (b) industrial or commercial systems.

4. A certain type of battery has an expected hourly failure rate of 10×10^{-3}. This battery is used 24 hr per day. How many spares are required for a 3-calendar-month period?

5. A system has an *MTBF* requirement of 100 hr, an operational-availability requirement of 95 per cent, and an inherent-availability requirement of 97 per cent. Solve the following problems, using the method of MIL-M-26512C. (a) Determine the system's mean down time. (b) From test data, it has been found that the mean corrective-maintenance time is 72 minutes and the mean preventive-maintenance time is 13.4 minutes. The number of corrective maintenance actions expected over the service life of the equipment is 150, and the equivalent number of preventive maintenance tasks is 50. Calculate the system's total down time. (c) Determine the mean system down time based on test data. What is the relationship between this value and that found in (a)? (d) Using the relationship, $\log M_{max} = \log \bar{M}_{CT} + 0.5$, determine the value of M_{max} which should be specified.

Quantification of System Effectiveness*

11.1 Introduction

Previous chapters discussed and illustrated methods by which certain basic information on system operating time, malfunctions, and maintenance time could be used to estimate reliability and maintainability functions. These functions describe in a quantitative manner the way in which a system operates, fails, and is restored to operable condition. In this chapter, the analysis is carried one step further to show how the same basic reliability and maintainability data can be combined, together with necessary information regarding the adequacy of the system to accomplish its mission, to express quantitatively the over-all effectiveness of the system.

The first portion of the chapter gives a general discussion of the equations and techniques for estimating effectiveness. It also illustrates the application of the techniques by the use of actual operational data on an airborne search radar system. The latter part of the chapter introduces the application of system effectiveness analysis to multimodal operations.

*Much of the material presented in this chapter was developed under Contract AF33 (657) -10594, *Development of Quantitative Analysis Techniques*, Systems Command, Aeronautical Systems Division, Wright-Patterson Air Force Base, Ohio.

11.2 System Effectiveness

The probabilities and time elements involved in the determination of system effectiveness were defined in Chapter 1. The definitions are summarized and the interrelationships are shown graphically in Figure 1.2 of that chapter. The following symbols will be used in the development of the equations for effectiveness:

$$P_{SE} = \text{System Effectiveness}$$
$$P_R \ \ = \text{Mission Reliability}$$
$$P_{OR} = \text{Operational Readiness}$$
$$P_{DA} = \text{Design Adequacy.}$$

Then,

$$P_{SE} = P_{OR} \times P_R \times P_{DA}. \tag{11.1}$$

Equation 11.1 states that the effectiveness of the system is the product of three probabilities: (1) the probability that the system is operating satisfactorily or is ready to be placed in operation, multiplied by (2) the probability that the system will continue to operate satisfactorily for the period of time required for the mission, multiplied by (3) the probability that the system will successfully accomplish its mission given that it is operating within design limits.

P_{OR} in Equation 11.1 may be computed from the equation

$$P_{OR} = P_A(1-P_S) + k_1 P_S \tag{11.2}$$

where

$P_S =$ the probability that the system is not being used or is in storage at any point in time, given that it was operating satisfactorily when last used or when placed in storage.

$(1-P_S) =$ the probability that the system either is in operation or is undergoing maintenance at any point in time.

$k_1 =$ the probability that the system does not fail during storage or nonuse.

$P_A =$ availability, computed as discussed below.

System availability may be computed as the following ratio:

$$P_A = \frac{(\text{total operating time})}{(\text{total operating time}) + k_2\,(\text{total down time})} \tag{11.3}$$

and

if $0 < k_2 < 1$, free time exists;

if $\quad k_2 = 1$, no free time exists.

The case in which no free time exists represents a situation where there is no possibility of performing maintenance during periods when operation of the system is not required. The factor k_2 would equal unity only if the system were required to operate continuously (i.e., 24 hr per day). However, if free time exists, and if it is possible to schedule an appreciable portion of maintenance activity during periods when use of the system is not required, the net effect is to increase availability. The basic reasoning here is that the portion of down time during which the system is not needed should not degrade availability. This suggests a method for estimating k_2. A first approximation of k_2 may be obtained from the ratio of the total time during which system operation is required (demand time) to total calendar time. This value of k_2 is almost certain to result in an optimistic value for P_A—i.e., too low a value, since it is never possible, in an operational situation, to perform all maintenance during free time.

Considerable judgment must enter into closer estimation of k_2. The use of any detailed information on the system use cycle (logistic delays) and the times during which maintenance crews are available for work will permit a better estimate of the factor. In the ideal situation where the hours available for maintenance during system free time are known, these hours may be subtracted from total down time, and Equation 11.4 may be written directly without incorporating the factor k_2*:

$$P_A = \frac{\text{(total operating time)}}{\text{(total operating time)} + \text{(total down time)} - \text{(total maintenance time)}}. \quad (11.4)$$

P_R in Equation 11.1 may be obtained directly from $R(t)$, the system reliability function. P_R is the point on the reliability curve corresponding to the period of time required for completion of the mission. Stated in equation form, this is:

$$P_R = R(t_o), \quad (11.5)$$

where t_o = length of the mission.

P_{DA} in Equation 11.1 must be estimated from knowledge of the system's performance when it is operating within specifications, and from knowledge of the performance required to achieve success in the mission for which effectiveness is being evaluated.

To clarify this statement, assume an airborne communication system with a mission length of 6 hr. Further, consider that for 10 per cent of the mission time the aircraft is operating at a range beyond the performance capabilities of the particular system, even though the system is operating at optimum

---

*It would also be possible to use an observed system maintainability function to arrive at an estimate of k_2 if the system were required to operate for a definite period of time during each day. This could be done by calculating the probability that a repair would be completed in, say, X hr.

performance levels—i.e., within design specifications. For this situation, $P_{DA} = 0.90$, or the system is adequate for the mission only 90 per cent of the time.

Another example of the design adequacy factor is shown in Figure 11.1 as a plot of the probability of detecting a target of specified size as a function of range.

To illustrate the effect of P_{DA} on system effectiveness, assume that examination of the reliability function indicates a 95 per cent probability that the system will operate within specifications for a period of 6 hr. An estimate of P_{OR}, based on operating time, down time, free time, and available spares, shows this probability to be 98 per cent. Then, from Equation 11.1,

$$P_{SE} = (0.98)\,(0.95)\,(0.90)$$
$$= 0.84 \text{ or } 84 \text{ per cent.}$$

If the mission requirements had been perfectly compatible with the system design, then P_{DA} would equal unity, and

$$P_{SE} = (0.98)\,(0.95)\,(1.0)$$
$$= 0.93 \text{ or } 93 \text{ per cent.}$$

This computation simply points out that the effectiveness of a system is degraded if the system is used for a task which is beyond its design performance capabilities, even though the reliability and maintainability of the system may approach ideal values. (See Section 11.6 for additional considerations of the relationship between design adequacy and system effectiveness.)

11.3   Intrinsic Availability

In Chapter 1, Section 1.3, page 10, intrinsic availability was defined so as to exclude all but the effects of the built-in reliability and repairability of the system. This characteristic, therefore, depends only on operating time and active repair time. The numerical value of intrinsic availability, P_{Ai}, may be estimated from the equation

$$P_{Ai} = \frac{\text{(total operating time)}}{\text{(total operating time)}+\text{(total active repair time)}} . \tag{11.6}$$

If the operating times are exponentially distributed and the repair times are log-normally distributed (as they usually are), a better estimate of P_{Ai} may be given by

$$P_{Ai} = \frac{\text{(mean time between malfunctions)}}{\text{(mean time between malfunctions)}+\text{(mean active repair time)}} . \tag{11.7}$$

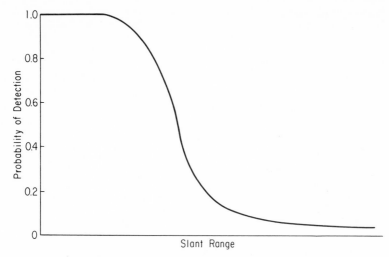

Fig. 11.1 Probability of detection vs. slant range for a radar system.

If improvement in intrinsic availability is indicated, responsibility can properly be assigned to design and production engineers.

Although intrinsic availability is influenced to some degree by the ability of maintenance personnel and the availability of adequate test equipment and technical manuals, it is generally a satisfactory descriptor of the merit of the physical system. Therefore, specification of both reliability and active repair time leads to the specification of intrinsic availability for new systems. As indicated in Chapter 1, trade-offs between reliability, maintainability, and cost of alternative system configurations are necessary for the optimization of system design. In the same manner, evaluation of reliability, maintainability, and cost factors can lead to an optimized solution for proposed field modifications.

Figures 11.2 and 11.3 are useful for evaluating the effects on availability of work accomplished during free time and the effect of operable nonuse time on operational readiness. Note that Figure 11.2 was developed under the assumption that the probability (k_1) that the system does not fail during storage or nonuse is equal to 0.95. Similar graphs for values of k_1 that differ from 0.95 can be developed through the use of Equation 11.3.

11.4 Illustration of Computational Procedures

This section presents the computations for system effectiveness, using actual field data for an operational military system—the AN/APS-20E radar. The data are used solely for the purpose of illustrating, by a concrete numerical example, the computations of the various probabilities and other

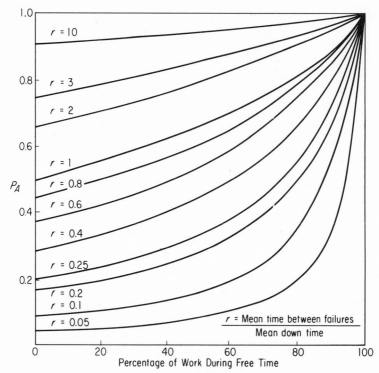

Fig. 11.2 Effect of work accomplished during free time on availability.

factors. The numerical results obtained from the analysis should not be used as the bases for any firm conclusions regarding the effectiveness of this particular radar system, because the surveillance was not conducted (and the data were not collected) in a manner that would permit illustration of all the concepts discussed in Chapter 1. Therefore, certain assumptions must be made in order to fill the gaps in the information.

The AN/APS-20E radar system is a relatively high-powered, airborne search radar which has been used by the Air Force and the Navy for a number of years. The primary mission of the system when it was under surveillance was crew training and patrol work for the purpose of submarine detection. Although the system is reasonably satisfactory for this mission, it was not designed specifically for this work. Therefore, it was to be expected that the design adequacy of the system, as previously defined, would be somewhat less than unity.

The data presented here were collected over an 8-month period of surveillance of 24 systems in 2 squadrons of operational aircraft. Figure 11.4 shows the apportionment of the accumulated time for the 24 systems for the 8-month period. The data were collected by individual system, but the chart provides

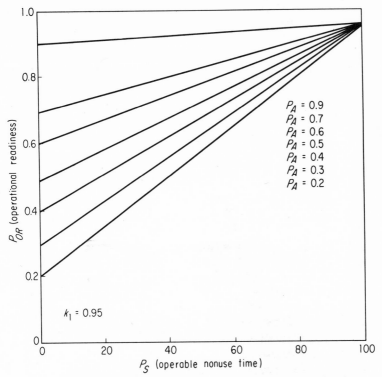

Fig. 11.3 Effect of operable nonuse time on operational readiness.

totals in system hours as needed for present purposes. Note also that Total Flight Time has been divided according to either of two alternative criteria (Section 1.4).

Table 11.1 presents supplementary data on the number of missions and the number of system malfunctions.

Table 11.1 MALFUNCTION AND FLIGHT DATA

Number of flights or missions	2068
Number of missions on which radar was used or its use was attempted	1067
Mean length of mission, hours	4.70
Number of malfunctions detected in flight by radar operator	96
Number of malfunctions detected on ground by maintenance technician	86
Total number of system malfunctions, all types	182
Total number of repair actions	135

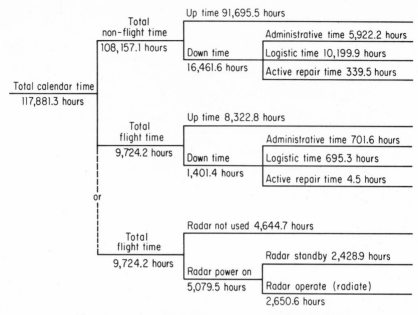

Fig. 11.4 Time breakdown for 24 AN/APS-20E radar systems for a period of eight months. (All time values are given in system hours.)

Intrinsic Availability

First, it will be of interest to compute the intrinsic availability, P_{Ai}, to provide an idea of the optimum reliability and repairability of the system. Equations 11.6 and 11.7 may both be used. Substituting first in Equation 11.6, we find,

$$P_{Ai} = \frac{2650.6}{2650.6 + 344.0} = 0.89.$$

This result assumes that both the operating times and the repair times are exponentially distributed.

The operating times were found to fit an exponential distribution. The mean operating time between malfunctions ($MTBM$), based on actual radiate time, is

$$MTBM = \frac{2650.6}{182} = 14.6 \text{ hr.}$$

The active repair times, on the other hand, were better described by a log-normal distribution, as is indicated in Figure 11.5. The note under the figure shows that the mean repair time is about 2.5 hr.

Using the foregoing information in Equation 11.7, we find the intrinsic availability to be

$$P_{Ai} = \frac{14.6}{14.6 + 1.0} = 0.94 \text{ if the median repair time is used,}$$

or

$$P_{Ai} = \frac{14.6}{14.6 + 2.5} = 0.85 \text{ if the mean repair time is used.}$$

The repairability function presented in Figure 11.5 shows that 75 per cent of the system repairs required 2.5 hr or less (the mean value), and that 50 per cent required 1.0 hr or less (the median value). It must be remembered that P_{Ai} is an index that reflects only the reliable operating time and the *minimum*

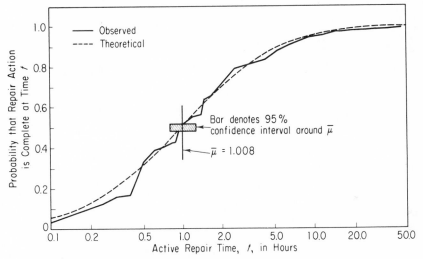

Fig. 11.5 Observed and theoretical repairability functions for AN/APS-20E radar systems.

Theoretical repairability function where

$$M_R(t) = \frac{2}{\sqrt{2\pi}\,\sigma} \int_{-\infty}^{x} \exp\left[-\frac{(x-\xi)^2}{2\,\sigma^2}\right] dx$$

$\bar{\mu} = 1.008$
$\xi = 0.0037$

Where

$\sigma = 0.5891$

$x = \log_{10} t$

$n = 135$ repair actions

$\xi = $ mean value of x

$\sigma = $ standard deviation of x

$\bar{t} = \dfrac{\Sigma t_i}{n}$ 2.52 hr

$\bar{\mu} = $ antilog ξ

95 per cent confidence interval around $\bar{\mu} = $ antilog $(\xi \pm 1.96\ \sigma/\sqrt{n})$
where $n = $ number of observations.

system down time once failure occurs. The actual observed availability would seldom approach this value in field operation because of various delay times. For specification purposes, the mean value of repair time should be used since most military maintainability specifications require this measure.

Availability

The first step in the computation of availability from Equation 11.3 is the estimation of the factor k_2 in the equation. AN/APS-20E data lead one to believe that only a negligible amount of in-flight free time would be usable for repairs and that only non-flight time (or a portion thereof) would be available for maintenance work. Active work time during non-flight periods constitutes only 0.29 per cent of calendar time, while administrative and logistic delays constitute 5.02 per cent and 8.65 per cent, respectively. These percentages add up to 13.96 per cent (16,461.6 hr) of the total non-flight down time. Clearly, some of the remaining time is free time and should not degrade availability. In the absence of better information, k_2 will be computed as described in Section 11.2. The estimate of k_2 then becomes

$$k_2 = \frac{\text{demand time}}{\text{total calendar time}} = \frac{2650.6}{117,881.3} = 0.0225.$$

Substituting the appropriate values, Equation 11.3 becomes

$$P_A = \frac{2650.6}{2650.6 + 0.0225\,(16,461.6 + 1401.4)} = 0.87.$$

Operational Readiness

Strictly speaking, for this application of the AN/APS-20E system, there is no such thing as storage time since the operating squadrons are not provided with spare systems. However, when a system has not flown or been checked for three days to a week or more, it is usually given a pre-flight check; and there is a small probability that a discrepancy will be discovered, even though the system performance was satisfactory the last time it was used or checked. This situation is somewhat similar to storage of the system and will be used to illustrate the effect of k_1 on P_{OR}, using Equation 11.2. Information on pre-flight checks shows that about one in ten checks discloses some malfunction or discrepancy. Therefore, the probability of nonfailure during nonuse is

$$k_1 = 0.9.$$

Examination of Figure 11.4 shows that the system non-flight up time is 91,695.5 hr, which is 77.79 per cent of calendar time. Therefore $P_S = 0.78$, $P_A = 0.87$, and $k_1 = 0.90$, from the previous computations. Substituting these values in Equation 11.2 gives

$$P_{OR} = (0.87)\,(0.22) + (0.90)\,(0.78)$$
$$= 0.89.$$

By way of further illustration, assume that the probability of failure during nonuse is negligible—i.e., $k_1 = 1.0$. Then

$$P_{OR} = (0.87)\,(0.22) + (1.0)\,(0.78)$$
$$= 0.97.$$

Thus it is seen that operational readiness is enhanced by periods of nonuse or storage time, provided that the probability of failure during nonuse is low. This means simply that operational readiness is increased by the provision of spares if the conditions of storage are such that the spares do not fail in storage.

Mission Reliability

The average mission length for this particular application of the AN/APS-20E is 4.7 hr of flight time. On the average, the radar is used (accumulating time) approximately 25 per cent of this time, or 1.2 hr. The mean time between malfunctions is determined on the basis of the operating time and the number of malfunctions. The reliability function based on an assumed exponential distribution of times is shown in Figure 11.6. Examination of this function shows that the probability of satisfactory operation for a period of 1.2 hr is 0.921. Of course, this could be computed from

$$e^{-t/\theta} = e^{-1.2/14.6} \simeq 0.921$$

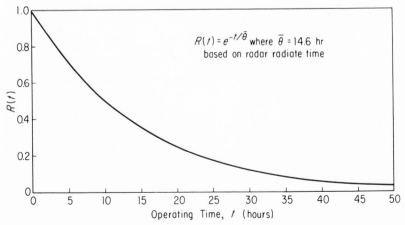

Fig. 11.6 Theoretical reliability function for AN/APS-20E radar system.

if the function had not been plotted. Therefore, the mission reliability is

$$P_R = 0.921 \cong 0.92.$$

Design Adequacy

As discussed in Section 11.4, the AN/APS-20E radar was not specifically designed for low-altitude, over-water search as required in antisubmarine missions. In this particular application, even when the radar performance characteristics are well within design limits, the probability of detecting a submarine varies over a wide range, depending on numerous factors such as sea state, range, and altitude of search. The operators frequently reduce the transmitted power in an attempt to minimize the effects of sea clutter. For purposes of this example, the probability of successfully accomplishing the mission (detecting the submarine), given that the radar is operating within design limits, will be estimated, somewhat optimistically, as falling between 80 per cent and 90 per cent. Therefore, the estimate of design adequacy is

$$P_{DA} = 0.85.$$

Effectiveness

Substituting the computed values of P_{OR}, P_R, and P_{DA} in Equation 11.1, the effectiveness of the system is

$$P_{SE} = (0.89)\,(0.92)\,(0.85)$$
$$= 0.696 \approx 0.70.$$

This, according to the definition of P_{SE}, states that there is a 0.70 probability that the AN/APS-20E system will successfully meet its operational demand on an antisubmarine mission of approximately 5 hr under the average operating conditions and with the operating and maintenance personnel encountered in the type of squadron from which the data were taken.

Certainly, 70 per cent seems to be a somewhat less than desirable or even acceptable effectiveness. Examination of the steps taken in the computation of P_{SE} permits a quick determination of the factors or areas which contribute most to this low value.

For example, the first and most obvious remedial step would be to improve the design adequacy. This might be impractical for economic or other reasons, but even if the performance of the system were perfectly compatible with the mission (i.e., if $P_{DA} = 1.0$) the effectiveness would be only

$$P_{SE} = (0.89)\,(0.92)\,(1.0)$$
$$= 0.82,$$

which may still be considered low.

Consider next the possible improvement in availability, which is a major factor in determining operational readiness. Reduction in administrative delays and in delays awaiting parts can increase P_A so that, ideally, it may approach P_{Ai} and intrinsic availability. The administrative and logistic times total 13.67 per cent of calendar time. This is roughly 50 times the percentage of calendar time devoted to active repair. Obviously, reduction of these "non-working" delays would effect considerably more improvement in P_A than a reduction in active repair time. Also, active repair time is probably already at a minimum for the existing system, repaired by maintenance personnel with the present level of skill.

Let it be assumed that these delays could be halved (from 13.67 per cent to 6.84 per cent). Conceivably this could be done by rescheduling of the working time of maintenance crews and by provisioning of spare parts more efficiently. Then the non-flight down time would become 8400.6 hr, and

$$P_A = \frac{2650.6}{2650.6 + 0.0225\,(8400.6)}$$

$$= 0.93,$$

using the same value of $k_2 = 0.0225$ as before. With no other changes from the preceding computation,

$$P_{SE} = 0.84.$$

Figure 11.7 presents a plot of the equation

$$P_{SE} = P_{OR} \times P_R \times P_{DA},$$

for four different values of P_{OR} and several values of P_{DA}. The purpose of Figure 11.7 is to permit a quick evaluation of the effect on P_{SE} of changing any of the three variables.

For example, assume (a) that the AN/APS-20E system is used on a less demanding mission, so that $P_{DA} = 0.95$, and (b) that other factors are improved, so that $P_{OR} = 0.90$. Assume that P_R remains 0.92 as in the previous example. Using the second plot in Figure 11.7 ($P_{OR} = 0.9$), we read $P_{SE} = 0.78$ for $P_R = 0.92$.

The plots in Figure 11.7 are intended not for reading of P_{SE} but simply to illustrate in a general way the effect of the three factors; Operational Readiness, Design Adequacy, and Mission Reliability. Computation of P_{SE} directly from the equation is very simple and is as accurate as the values substituted in the equation.

11.5 Effectiveness of Multimoded Systems

In Chapter 7, it was assumed that redundant elements or modes of operation are equally as effective as the primary or nonredundant elements. When

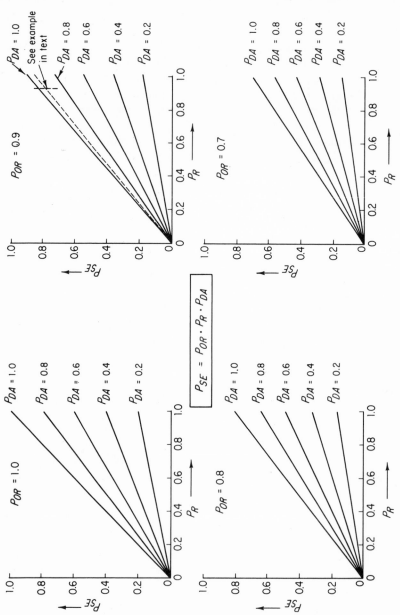

Fig. 11.7 System effectiveness *vs.* operational readiness, design adequacy, and mission reliability.

redundancy is introduced at the part or circuit level, this is usually true. However, at equipment or subsystem levels, redundant modes of operation may be less effective than nonredundant ones with respect to meeting all the criteria of mission success.

The internal antenna of a television set is a simple example of a redundant mode which, when utilized, may lead to reduced system effectiveness. If the roof antenna fails, TV reception can still be obtained with the redundant (internal) antenna, although it will probably be much poorer—i.e., the system will be less effective.

In a military mission, modal effectiveness is demonstrated in the bombing/navigation system located within a plane. Bombing by means of the radar subsystem is the primary mode; if the radar fails, optical bombing is an alternative or secondary mode. System effectiveness in the redundant mode (optical) is lower than that in the primary mode (radar).

Mission reliability refers to the ability of all components and the complete set to remain within design specification limits—i.e., to give a TV picture. A TV set that is operationally ready is one that is available for use and operates reliably when turned on. If 5 per cent of the time the set is being repaired when the user would like to see a particular program, $P(OR) = 0.95$. Design adequacy refers to the ability of the set to satisfy the viewer by having a desirable picture for a particular program if the set is operationally ready and operates reliably for the duration of the program. A program originating 75 miles away will not always result in a satisfactory picture even though the set is available for use, is operable when turned on, and remains operable for the duration of the program. If satisfactory reception of this particular program can be achieved about 10 per cent of the time, then $P(DA) = 0.10$ *for this particular "mission"*.

Effectiveness is the product of (a) the probability that the set is available for use and will operate when turned on; (b) the probability that if available and initially operating within design specifications, it will remain so for the duration of the program; and (c) the probability that if it remains within design specifications it will satisfy the viewer for a particular set of use conditions. Therefore, effectiveness is essentially the probability that, given a particular program or set of use conditions, the viewer will be satisfied, in that the operational demand of good program reception will be maintained throughout the period of interest.

Equation 11.1 for P_{SE} applies to a nonredundant system—i.e., there is only one mode of operation. If there are m possible modes, the equation for the redundant or multimoded system effectiveness becomes

$$P_{SE} = \sum_{i=1}^{m} P_i(M) P_{E_i}, \tag{11.8}$$

where $P_i(M)$ is the probability that the system requires operation in the ith mode (the probability that there will be a switching to mode i), and P_{E_i} is the effectiveness of the ith mode. It is assumed that the system is initially operated in the most effective mode and, upon modal failure, switches sequentially to redundant modes in order of their design adequacy.

Replacement of P_{E_i} by its equivalent gives

$$P_{SE} = \sum_{i=1}^{m} P_i(M)\, P_i(R)\, P_i(OR)\, P_i(DA). \tag{11.9}$$

If $P(OR)$ and $P(DA)$ are the same for all modes, then let

$$P(OD) = P(OR)\, P(DA).$$

This can replace $P_i(OR)\, P_i(DA)$ for all i in the equation for P_{SE}. P_{SE} then becomes

$$P_{SE} = P(OD) \sum_{i=1}^{m} P_i(M)\, P_i(R). \tag{11.10}$$

It will now be shown that $\displaystyle\sum_{i=1}^{m} P_i(M)\, P_i(R)$ is exactly equivalent to the

equations for R given in previous sections. In these equations it was implicitly assumed that both operational readiness and design adequacy were the same for all modes and, in fact, were equal to one [$P(OD) = 1.0$ for all redundant paths].

A Simple Example

In the simple case of two modes (elements) which can be represented as a parallel configuration, as shown in the following diagram,

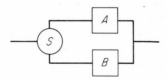

A is the primary mode—that is, the mode which operates initially—and B is the secondary or redundant mode, which is switched in by S if A fails. Assuming that S is failure-free* and that $P_a(OR) \times P_a(DA) = P_a(OD)$, $P_b(OR) \times P_b(DA) = P_b(OD)$, effectiveness is given by

$$P_{SE} = P_a(M)\, P_a(R)\, P_a(OD) + P_b(M)\, P_b(R)\, P_b(OD), \tag{11.11}$$

---

*The previous discussion of switching failures (Chapter 7) is applicable to this section, and the results should be applied if the assumption of failure-free switching is not valid. For most situations where the redundant modes involve equipments or subsystems, the switching is performed manually and therefore can be considered to be highly reliable.

where $P_i(M)$ ($i = a, b$) is the probability that mode i is required, and $P_i(R)$ is the reliability of mode i or, in the notation of previous sections, is equal to $P(A)$ for $i = a$ or $P(B)$ for $i = b$.

$P_a(M) = 1.0$ since the mission is always initiated in A, the primary mode.

$P_b(M) = [1-P(A)]$ since mode B is required only if mode A fails.

Therefore,

$$P_{SE} = 1.0\,P(A)\,P_a(OD) + [1-P(A)]\,P(B)\,P_b(OD)$$
$$= P_a(OD)\,P(A) + P_b(OD)\,[P(B)-P(A)\,P(B)]. \tag{11.12}$$

If $P_a(OD) = P_b(OD) = P(OD)$,

$$P_{SE} = P(OD)\,[P(A) + P(B) - P(A)\,P(B)]$$
$$= P(OD)\,R.$$

Since it has been shown that for equal $P_i(OD)$,

$$P_{SE} = P(OD) \sum_{i=1}^{m} P_i(M)\,P_i(R),$$

the last two equations above give

$$R = \sum_{i=1}^{m} P_i(M)\,P_i(R). \tag{11.13}$$

For this type of analysis, R represents the probability of system operation in at least one mode or, if modal operation is mutually exclusive, the probability of operation in a mode.

A Complex Example

An effectiveness analysis of multimoded systems is somewhat more complicated than a straightforward reliability analysis, since the probability of requiring each mode, $P_i(M)$, has to be individually calculated if the $P(OD)$'s differ. Some modes may use common as well as unique elements, with the combinations of common and unique elements differing for each mode. The following system block diagram is an illustrative example of this type of arrangement:

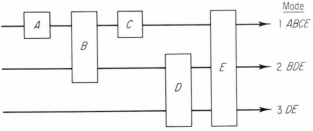

In order to determine $P_i(M)$, the failure events that will require the ith mode must be described. These events will be denoted by (M_i). For example, Mode 1 is always initially used. Therefore, $P_i(M) = 1.0$. Mode 2 will be required if M_1 fails, but only if (1) A fails or (2) C fails, since failure of B will require M_3 and failure of E results in system failure. Therefore,

$$(M_2) = \bar{A} \text{ or } \bar{C}.$$

(A bar above a letter represents failure of the corresponding element.) Mode 3 is necessary if (1) A or C fails in M_1 and B fails in M_2, or (2) B fails in M_1. (Failure of D in Mode 2 results in system failure.) Hence,

$$(M_3) = \bar{A}\bar{B} \text{ or } \bar{B}\bar{C} \text{ or } \bar{B}.$$

A tabular method can be used for schematic representation of these events (see Table 11.2).

Table 11.2 TABULAR METHOD FOR OBTAINING THE EVENTS (M_i)*

Operating Mode	Element Failure†				New Mode
	$\bar{A}$	$\bar{B}$	$\bar{C}$	$\bar{D}$	
M_1	1				M_2
		1			M_3
			1		M_2
M_2	1	2			M_3
		2	1		M_3

*(Entries in the table refer to failures of the element in the operating mode indicated. All elements are assumed to be energized even though they may not be in the operating mode.)

†Element E is not included since its operation is required for all three modes.

The 1 under the $\bar{A}$ column $(\bar{A}_1)$ represents failure in M_1, the original mode. The block diagram shows that this failure results in a switch to M_2. Similarly $\bar{B}_1$ is the failure of B in M_1, which in turn will require a new mode, M_3 (M_2 is skipped since it requires B). This procedure is followed for each of the possible element failures. ($\bar{D}_1$ is not shown, since a failure of D will not require switching, and $\bar{D}_1$ with any other element failure will result in system failure.)

For the next mode, M_2, the entries in M_1 that led to a switching-in of M_2 are duplicated. With each of these entries ($\bar{A}_1$ and $\bar{C}_1$) one repeats the process of entering all element failures in M_2 that will lead to a switching-in of higher modes. For example, the first row of M_2 ($\bar{A}_1\bar{B}_2$) represents the failure of A in M_1 which caused a switch to M_2, and the failure of B in M_2, which will require a new mode, M_3.

After entries have been made for the first $(m-1)$ modes, the events (M_i) can be determined by writing down each of the corresponding row entries for the ith new mode. In the example above, this gives (omitting subscripts)

$$M_2 = \bar{A} \text{ or } \bar{C}$$

$$M_3 = \bar{B} \text{ or } \bar{A}\bar{B} \text{ or } \bar{B}\bar{C},$$

which is the same as the result found by the block diagram method. Before calculating the $P_i(M)$'s, some Boolean manipulations are required, since some entries may include others. For example, X includes XY as shown in the following diagram:

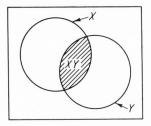

Therefore, for (M_3)

$$\bar{B} \text{ or } \bar{A}\bar{B} \text{ or } \bar{B}\bar{C} = \bar{B} \text{ or } \bar{B}\bar{C}$$
$$= \bar{B}.$$

Since it is assumed that all elements are energized, $\bar{A}$, $\bar{B}$, etc., are not mutually exclusive*; therefore

$$P[(M_1)] = P_1(M) = 1.0$$

$$P[(M_2)] = P_2(M) = P(\bar{A}) + P(\bar{C}) - P(\bar{A}\bar{C})$$

$$= 1 - P(A)P(C)$$

and

$$P[(M_3)] = P_3(M) = P(\bar{B}) = 1 - P(B).$$

To illustrate the complete computation, the following probabilities will be assumed:

Element Reliabilities:

$$P(A) = 0.7, P(B) = P(C) = 0.8, P(D) = P(E) = 0.9$$

---

*To avoid the complications of time-to-failure densities, only constant probabilities are being considered. Therefore, this assumption is necessary.

Modal Reliabilities:

$$P(R) = P(A)\,P(B)\,P(C)\,P(E)$$
$$= 0.40$$
$$P(R) = P(B)\,P(D)\,P(E)$$
$$= 0.65$$
$$P(R) = P(D)\,P(E)$$
$$= 0.81$$

Modal P(OD)'s:

$$P(OD) = 1.0$$
$$P(OD) = 0.8$$
$$P(OD) = 0.5$$

Modal $P_i(M)$'s:

$$P(M) = 1.0$$
$$P(M) = 1 - P(A)\,P(C)$$
$$= 0.44$$
$$P(M) = 1 - P(B)$$
$$= 0.20.$$

A Generalized Method for Determining System Effectiveness of Multimoded Systems

If the system is provided with alternative modes of operation so that the mission can be accomplished in several different ways, the effectiveness of all possible modes must be considered. The basic equation for system effectiveness (Equation 11.1) must be modified to incorporate all success possibilities, and additional terms are required to measure the full system potential. A framework for constructing such a model is described below. Let

$$\underline{V} = [v_1, v_2, \ldots, v_s]$$ where v_i = probability that the system is in state i at time 0.   (11.14)

$$\underline{W} = \begin{bmatrix} w_1 & & & 0 \\ & w_2 & & \\ 0 & & \ddots & \\ & & & w_s \end{bmatrix}$$ where w_i = probability that the system will be used for the mission, given state i at time 0.   (11.15)

$$\underline{R}(t) = \begin{bmatrix} R_{11}(t), R_{12}(t), \ldots, R_{1s}(t) \\ R_{21}(t), R_{22}(t), \ldots, R_{2s}(t) \\ \ldots \\ R_{s1}(t), R_{s2}(t), \ldots, R_{ss}(t) \end{bmatrix}$$ where $R_{ij}(t)$ = probability that the system is in state j at time t, given state i initially.   (11.16)

$T_{mk} = t'_{mk} - t_{mk}$ equals the mth required performance time interval for the kth subfunction, where t_{mk} and t'_{mk} are the beginning and end of this interval, respectively.

$$\underline{A}(T_k) = \begin{bmatrix} a_1(T_k) & & & 0 \\ & a_2(T_k) & & \\ & & \cdot & \\ 0 & & \cdot & \\ & & & a_2(T_k) \end{bmatrix}$$

where $a_i(T_k) =$ probability that, given system state i at time t_k, no state transition occurs before time t'_k.   (11.17)

$$\underline{D}(T_k) = \begin{bmatrix} d_1(T_k) \\ d_2(T_k) \\ \cdot \\ \cdot \\ \cdot \\ d_s(T_k) \end{bmatrix}$$

where $d_i(T_k) =$ probability that ith system state will successfully perform kth subfunction during interval T_k.   (11.18)

The model framework includes the necessary system characteristics for optimization, which are described below:

V matrix—state readiness: reliability on previous missions and during ground maintenance; maintenance diagnosis of the state of system efficiency and restoration to a more satisfactory state through repair.

W matrix—mission readiness: diagnosis of system state (alert phase of mission-time profile), mission requirements, operational policy, system flexibility, and system backup.

R matrix—state transition: } equipment reliability and in-flight repair
A matrix—state continuance: } capability.

D vector—design adequacy: mission requirements, performance capabilities, and human factors.

By expressing the probability elements of these vectors and matrices as functions of the trade-off parameters, the framework for trade-off is established.

The probability that the kth subfunction will be successfully performed in a performance interval T_{mk}, assuming no allowable state transitions, is then

$$SE_k(T_{mk}) = \underline{V}\,\underline{W}\underline{R}(t_{mk})\,\underline{A}(T_{mk})\,\underline{D}(T_{mk}), \qquad (11.19)$$

which is a scalar of the form

$$\sum_{j=1}^{s} \sum_{i=1}^{s} v_i w_i R_{ij}(t_{mk}) a_j(T_{mk}) d_j(T_{mk}). \tag{11.20}$$

Thus, the general term of the equation is

$P(\text{state } i \text{ at } t = 0) \times P(\text{use system state } i \text{ at } t = 0)$

$\times P(\text{transition from } i \text{ to } j \text{ at } t_k) \times P(\text{remains in } j \text{ until } t_k)$

$\times P(k\text{th subfunction successful state } j \text{ for } T_k).$

To illustrate the application of this model, consider the block diagram of a very simplified computer system:

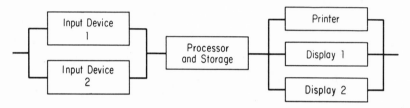

If we assume that each equipment of this simple system can be in only one of two possible states, success or failure, there are $2^6 = 64$ possible system states. However, from an analysis of operational requirements and a subsequent functional analysis, this number can be reduced. For example, the processor cannot be in a failed state; at least one of the two input devices must operate. These considerations, in effect, represent the process of identifying those states for which the corresponding elements in the design-adequacy vector equal zero.

Table 11.3 POSSIBLE SYSTEM STATES IN A SIMPLIFIED COMPUTER SYSTEM

System State	Design Adequacy	I_1	I_2	Processor	Printer	D_1	D_2
1	0.95	S^*	S	S	S	S	S
2	0.90	$F^\dagger$	S	S	S	S	S
3	0.50	F	S	S	F	S	S
4	0.70	F	S	S	S	F	S
5	0.70	F	S	S	S	S	F
6	0.90	S	F	S	S	S	S
7	0.50	S	F	S	F	S	S
8	0.70	S	F	S	S	F	S
9	0.70	S	F	S	S	S	F

*S = success
†F = failure

From an analysis of operational requirements and functional diagrams, assume that the states listed in Table 11.3 are those for which design adequacy is greater than zero. The example will be further simplified by considering only states 1, 2, and 3 to represent the system effectiveness equations numerically. From a consideration of ground and airborne maintenance factors, operational reliability, and mission requirements, the following data inputs are assumed to hold:

$$V = [0.85 \ \ 0.10 \ \ 0.05] \ \ W = \begin{bmatrix} 1.0 & 0 & 0 \\ 0 & 0.8 & 0 \\ 0 & 0 & 0.5 \end{bmatrix} \ \ R(t) = \begin{bmatrix} 0.92 & 0.06 & 0.02 \\ 0.15 & 0.80 & 0.05 \\ 0 & 0.10 & 0.90 \end{bmatrix},$$

and

$$A(T) = \begin{bmatrix} 0.99 & 0 & 0 \\ 0 & 0.96 & 0 \\ 0 & 0 & 0.95 \end{bmatrix} \ \ D = \begin{bmatrix} 0.95 \\ 0.90 \\ 0.50 \end{bmatrix}.$$

Then,

$$\begin{aligned} SE(T_1) = & \ (0.85 \times 1.0 \times 0.92 + 0.10 \times 0.80 \times 0.15) \ (0.99) \ (0.95) \\ & + (0.85 \times 1.0 \times 0.06 + 0.10 \times 0.80 \times 0.80 + 0.05 \times 0.50 \\ & \times 0.10) \ (0.96) \ (0.90) + (0.85 \times 1.0 \times 0.02 + 0.10 \times 0.80 \times 0.05 \\ & + 0.05 \times 0.50 \times 0.90) \ (0.95) \ (0.50); \end{aligned}$$

$$SE(T_1) = 0.869.$$

The over-all system effectiveness for a mission time, T, is then some multiplicative function of the time and subfunction values. The actual expression can be quite complex since consideration must be given to such situations as: (1) failure of an equipment when it is not needed, but which is repaired before a performance time interval, or (2) system state transition during the performance of a subfunction.

It is quite possible that a description of over-all system effectiveness by a single analytic expression may not be feasible for certain systems. However, through detailed examination of the expected mission profile (or a set of weighted mission profiles), we can formulate a set of "rules" for describing successful system operation that, in conjunction with the model framework described above, can be programmed for computer simulation.

For the over-all performance criterion, we can consider a loss vector:

$$L_k(t) = \begin{bmatrix} 1_{1k}(t) \\ 1_{2k}(t) \\ \cdot \\ \cdot \\ \cdot \\ 1_{sk}(t) \end{bmatrix}, \quad \begin{array}{l} \text{where } 1_{ik}(t) \text{ represents the fractional loss in perform-} \\ \text{ance for the } k\text{th subfunction if the system is} \\ \text{in state } i \text{ at time } t. \end{array}$$

$$(11.21)$$

The expected loss for the kth subfunction (assumed here to be needed continuously) is then

$$L_k(t) = \underline{V}\ \underline{W}\ \underline{R}\ (tk)\ \underline{L}_k(t),$$
(11.22)

and the expected fractional loss for a mission of length T is

$$E[L_k(T)] = \frac{1}{T}\int_0^t L_k(t)dt.$$
(11.23)

Appropriate modifications should be made to these equations to account for actual performance time intervals for different subfunctions. Assuming some additive function of subfunction loss, the system loss is

$$E[L(T)] = \sum_k^* [E\,L_k\,(T)],$$
(11.24)

where $\sum_k^*$ represents an appropriate combinational operator on individual subfunction losses.

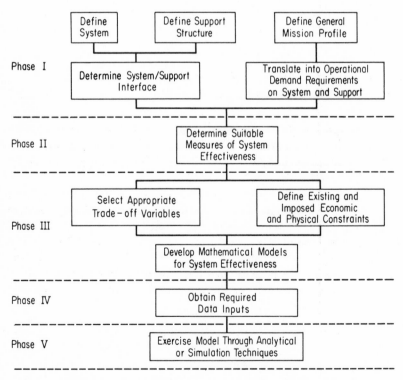

Fig. 11.8 Flow diagram for a typical system effectiveness analysis.

11.6 System Effectiveness Analysis

The preceding sections have developed the basis for quantification of system effectiveness. In this section, we will examine additional factors that may have to be considered in the analysis of alternative design configurations or in the field evaluation of systems in use. Figure 11.8 is a flow diagram for a typical program.

The analysis is performed at several system and support levels. The first application is at the major equipment component level (e.g., data processor, display console, and input/output devices), and at the associated major support functions. Application of the model at this stage will yield early decisions at these levels and provide a basis for further analysis at lower hardware and support levels. (Detailed descriptions of the phases of the analysis appear in following sections.)

Until the final design is frozen, the results of each analysis on model application are fed back into the "analytical loop" to generate a continual cycle of model refinement and improved design-decision criteria.

The timing of the analysis is important if the results are to provide management with a tool to guide the design. Figure 11.9 shows a timing chart of the work involved in the analysis, the information produced, and the use of the information in the design process. As is apparent from this figure, the hierarchical structure of the system is important.

The analysis proceeds to translate system requirements and constraints into requirements and constraints on the parameters of progressively smaller parts of the system. Concurrently, the design groups roughly define these smaller blocks. When the design group begins to implement the rough outlines, it can draw on qualitative information as to *how* different parameters of smaller systems affect the over-all system performance.

During the implementation of design, the analysis quantifies the relationships between the different component parameters and between the system effectiveness and system restraints.

At the end of this preliminary phase, the design will have been reduced to a relatively small number of alternatives from which the analyst can choose the best.

Phase I: Definition of System, Support, and Mission

This phase involves a detailed study of those aspects of system, support, and mission that will eventually form the basis for the effectiveness model and the establishment of design-decision criteria.

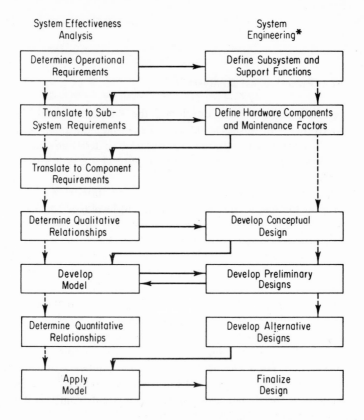

*An excellent reference on system engineering concepts is A. D. Hall, *A Methodology for Systems Engineering*, (Princeton, N.J.: D. Van Nostrand Company, Inc., 1962).

Fig. 11.9 Timing chart of system effectiveness analysis and system design.

The system must be clearly defined with respect to its boundaries, its major subsystems, and its functions. This definition is accomplished through hardware and functional block diagrams. Study should be made of the interface with the operational user, the support functions, the larger system (e.g., the related communication equipment, and the power sources), and the environment.

In the early stages of system development, detailed block diagrams should be used as a vehicle for performing analyses on multimodal capabilities, failure effects, and redundancy utilization. Analyses should also be made of other reliability and maintainability design-improvement methods such as cooling, minimum stress, serviceability features, and integral electronics.

The support structure should be examined with regard to such aspects as logistics, repair facilities, ground checkout, preventive maintenance, and ground and airborne corrective maintenance. Various approaches for achieving high maintainability and readiness must be formulated for further study through the effectiveness model.

The third task in this phase should involve translating the general system objectives into quantitative operational requirements. The specific system concept naturally involves a multitude of factors that have direct bearing on the mission profile. While the over-all purpose of the system may be simply stated as command and control after enemy attack, we must consider, in evaluating a proposed system, such factors as the type of enemy attack, the environment, the information available, the information required, the real-time requirements, and the system backup.

These considerations may lead to numerous possible mission types and, thus, numerous sets of operational requirements. The number of sets of operational requirements should be reduced to a manageable size and, then, possibly weighted by importance and the probability of occurrence. An average set of requirements may evolve that will yield an effective system for the primary or most likely mission types.

The major purpose of Phase I is to define the scope of the analysis with respect to the system, support structure, and mission. An important secondary purpose is to define the areas of effectiveness responsibility and control. Major avenues of approach should be formulated, leading to a basis for an effectiveness model and the establishment of design-decision criteria.

Phase II: Measures of Effectiveness

The definitions of system, support, and operational requirements that were developed in Phase I are used as the basis for formulating measures of effectiveness. It is during this phase that we must consider two major system operational demand requirements. The first is associated with *time* requirements. The system effectiveness concept is based on meeting an operational demand within a specified time constraint. Usually the type of function performed will indicate the nature of this time constraint, but, in some cases, there are several possible choices. Three types of time requirements are given in Table 11.4.

Table 11.4 Time Requirements Related to System Operational Demands

Type of Requirement	Description	Example
Pointwise	System must meet the demand at a given instant of time for a negligibly small interval of time.	Retro-rocket package for re-entry of orbiting satellite.
Continuous Interval	System must meet the demand continuously for a given interval of time.	Plane on a bombing mission.
Fractional Interval	System must meet the demand for a specified fraction of a given interval of time.	Reconnaissance (picture-taking) satellite.

A somewhat artificial but illustrative example of one system to which any one of the three types of time requirements may apply is the bomb-release mechanism of a plane. If the plane carries only one bomb, the pointwise requirement applies. If the plane is to drop "X" bombs on "X" different targets, the continuous interval requirement applies. If the plane carries more bombs than are necessary to destroy a designated target, the fractional interval requirement applies.

The second major operational demand requirement of a system is related to performance requirements. The effectiveness of a system cannot be quantified without a clear definition of how performance is to be measured and of the criterion for evaluating such measures. In this discussion we shall be concerned only with criteria of evaluation.

There are two basic types of performance criteria:

1. Minimum Performance Criterion

The minimum performance criterion (*MPC*) is one for which quantitative bounds on the output of the system are prescribed. These bounds define the range of acceptable performance; there is no assessment of the degree of acceptability within these bounds. Thus, this performance criterion leads to the commonly used dichotomous description of system performance—success or failure. As an example, a criterion which states that a bomb must be dropped within d miles from the center of a target is a minimum performance criterion.

2. Over-All Performance Criterion

The over-all performance criterion (*OPC*) is one for which the complete distribution of output is considered either in terms of the actual probability distribution or by a related statistical measure such as the expected value. This type of criterion may be used when the degree of conformance to mission or functional requirements is of interest and where a quantitative requirement in terms of an *MPC* is too artificial. As an example, the criterion for measuring the performance of a bomber may be to estimate or measure the probability distribution of miss distances and perhaps calculate the mean and variance of miss distance.

The choice of which of the two criteria to use is not always clear. Several of the more important factors that are associated with this choice will now be discussed.

The system function and the associated output to accomplish the function sometimes dictate the evaluation criterion. If the functional output is dichotomous (e.g., for a detection system, the output might be detection or nondetection), the choice would be the *MPC*. For a dichotomous or binomial output, the probability of meeting the minimum performance condition and its complement is, of course, the *probability distribution.*

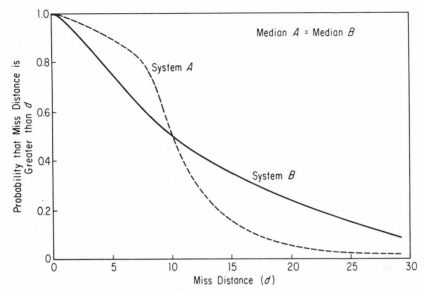

Fig. 11.10 Evaluation of bomber systems A and B: observed miss-distance distributions.

For a multi- or continuous-system output spectrum, we may use the *OPC* if success boundaries would be highly artificial, or if interest centers on the mean and variance of miss distance.

The choice between the *MPC* and the *OPC* can be critical. Suppose that, in the example of the bomber (extremely simplified by neglecting such factors as cost, range, and payload), two bombers are being evaluated. The observed miss-distance distributions for a given mission are shown in Figure 11.10.

Inspection of the two curves reveals that for a critical miss distance of less than 10, system B is more effective since it has the lower probability of having a miss distance (d) that is greater than a critical value in this range. However, the converse is true for a critical miss distance that is greater than 10. Thus, if the mission were such that the target would receive little damage if the miss distance were less than, say, 5, system B would be the choice. If the target damage was significant for any miss distance less than, say, 25, consideration would have to be given to the expected amount of damage (the value of the output) as determined from the miss-distance distributions and the target damage curves.

Table 11.5 shows an assumed target damage assessment and the miss-distance probabilities of bomber systems A and B. The expected percentage of target destroyed by system A is $\sum_d P_A(d)T(d) = 62.3$, while for system B the percentage is $\sum_d P_B(d)T(d) = 54.7$. This type of evaluation leads to a conclusion that system A is more effective.

Table 11.5 BOMBER SYSTEMS A AND B: ASSESSMENT OF ASSUMED TARGET
DAMAGE AND PROBABILITIES OF MISS DISTANCE

Miss Distance, d	Average Percentage of Target Destroyed, $T(d)$	Probability of Miss Distance for System A, $P_A(d)$	Probability of Miss Distance for System B, $P_B(d)$
0 to < 5	95%	0.18	0.30
5 to < 10	80%	0.32	0.20
10 to < 15	50%	0.35	0.15
15 to < 20	20%	0.10	0.11
20 to < 25	5%	0.02	0.10

Of course, there are other evaluation approaches, some of which embrace a minimum performance criterion, others the over-all performance criterion. The choice of the evaluation criterion must ultimately depend on the purpose of the evaluation. If the purpose is for information, the *OPC* is naturally more desirable since it includes the *MPC* as a special case. However, this type of evaluation usually requires a great deal more input information than does an evaluation to decide the acceptance of a design concept or to choose between several competing designs. In this latter evaluation, the criterion could very well be the *MPC*, provided that critical performance boundaries can be specified realistically. The criterion should be chosen before the analysis since the choice depends on system objectives and should not be influenced by numerical results.

Clearly, then, an effectiveness measure based on a *minimum performance criterion* is one for which specified bounds on the system output define the region of acceptable performance and lead to the classifications of success or failure. The other means of quantifying effectiveness is one based on an *over-all performance criterion*. In this case, the complete distribution of output is considered and is usually quantified by an appropriate statistical measure such as the mean output or the expected amount of information loss.

It is likely that some types of system output can be considered to fall within the minimum performance criterion. However, outputs for which partial or degraded information return may be of some value could possibly be measured under the over-all performance criterion. The associated time requirements of alert readiness and mission performance will also be studied and defined on an appropriate basis.

Therefore, the conceptual vector of system output and associated effectiveness measures form the basis for evaluating the effectiveness of proposed designs. This evaluation is performed through the mathematical models that are developed to represent these effectiveness measures and the cost or burden involved in their achievement.

Phase III: Mathematical Model

Three tasks are to be performed in this phase. They are

1. Selection of appropriate "trade-off" variables
2. Definition of existing and imposed economic constraints
3. Development of the mathematical relationships between the variables to express the effectiveness measures.

The trade-off variables are the system and support parameters that will appear in the model. Within the defined limitations on these parameters and the limitations of the model, the proper selection of the values of these parameters will yield an optimum system. These variables, therefore, are those that have an important influence on the reliability, maintainability, and performance characteristics of the system; they are variables that can be traded off against one another on the basis of some common denominator which may be called *burden*.

Some of the more obvious variables are: complexity, the number of redundant units and the type of redundancy, the number and level of replaceable modules, and the type and schedule of preventive maintenance. As the system analysis-loop progresses, more "detailed variables" must be considered; e.g., stresses on parts and components and the numbers and types of maintenance readouts and check points.

The associated burden of the various alternatives will also be determined. Thus, the increased weight, cost, and complexity of redundancy as a reliability improvement procedure is compared to, say, the cost and feasibility of developing and procuring ultrahigh part reliability.

The second task is to list existing and imposed physical and economic limitations. The imposed physical limitations are those which are either explicitly required (e.g., available space) or implied by the operational requirements; the existing constraints are those defined by the current state of the art, such as achievable failure rate levels, maintenance repair rates, and data printout speeds. Also to be listed are such economic limitations as total cost, development time, maintenance manpower, and skill levels.

These factors are suitably introduced into the mathematical model to insure that the produced system will meet its operational requirements within the required burden limitations.

The third task is to develop the mathematical effectiveness model. The model is the vehicle for: (1) estimating the effectiveness of proposed systems, (2) evaluating the effectiveness potential of various alternatives, (3) trading off these alternatives, and (4) determining reliability and maintainability requirements at equipment and unit levels.

The most general form of the model to be used can be described in terms of system states that are defined by the states of major system elements. By assessing the performance of these elements with respect to each subfunction, the effect of design and support decisions on over-all system effectiveness can be determined.*

A framework for constructing such a model is described in Section 11.5.

At this point, a cost/effectiveness study should be introduced. By building in high reliability and maintainability, initial investment costs obviously increase. However, this addition should lower the support costs over the system service life. This type of trade-off is one which should be investigated at major system and support levels to narrow the choice of alternatives early in the development stage.

Cost models should be developed and applied to the analysis of costs with respect to the reliability and maintenance characteristics (Chapter 15). Reliability or, actually, unreliability, triggers the support system and thus determines *how often* a particular equipment will involve the support system and consume a corresponding amount of funds. The expenditure in terms of manpower and maintenance time is a function of the maintainability characteristics.

The framework of the cost model should provide the following information:

1. Comparative cost data on the operation of major subordinate organizations

2. The complete cost of the present support of units to ascertain which units might repay engineering changes

3. Factors which measure the elasticity of support costs against the frequency of unit failures and the direct labor time required for repair (These factors can be used to evaluate the savings to be obtained from projected engineering changes.)

4. Information needed to compare the costs of maintenance at different echelons; e.g., which repairs are made cheaper at depot and which at base level

---

*It was shown in Section 11.2 that design adequacy can be determined by field evaluation of the system. At the design stage, prediction of design adequacy should be a system engineering function, which will furnish essential data inputs for the evaluation of system effectiveness, and, ultimately, system worth. The required data take the form of probabilistic performance indices. As an example, when a value is established for the cumulative probability of detection at some preselected range, a true measure of radar performance is then available. An excellent discussion of this problem as applied to search radar performance is given in H. G. Fridell and H. G. Jacks, "System Operational Effectiveness (Reliability, Performance, Maintainability)," *Research and Development Reliability*, American Society for Quality Control (February, 1961).

5. Information needed to trade off possible support savings against increases in fixed investment; e.g., the introduction of new test equipment

6. A simulation tool with which suggested changes in the support organization can be evaluated before they are initiated.

Phase IV: Data Inputs

The required data inputs to the model will consist of such items as: part or component reliability and maintainability parameters, cost inputs, weight and space estimates, and other necessary data on pertinent physical, engineering, or economic factors.

Initially, such inputs may be obtained from past experience and from appropriate estimation techniques (Chapters 6 through 10). As early design approaches become more definite and component and unit development progresses, these inputs should be refined for the iterative application of the model to reflect actual observation as recorded on development tests.

Phase V: Model Implementation

The basis of the approach is essentially one of: (1) "designing" many systems that satisfy the constraints; (2) computing the resultant values of effectiveness and burden; (3) comparing these values with requirements; and (4) making generalizations concerning appropriate combinations of design and support factors which are then re-fed into the model as part of the feedback loop.

Schematically, this process might be as shown in Figure 11.11; only those designs which meet physical and economic limitations are actually evaluated by the model. The range of designs, however, is naturally restricted by the customer's requirements. It is through the use of the model that this basic approach is translated into actual hardware configurations and support planning, which, within reasonable bounds, should be of an optimal nature.

Undoubtedly, the actual process for model implementation will require extensive computer utilization. Those portions of the model which defy analytical representation can be handled by Monte Carlo simulation. Such techniques as linear and dynamic programming, queuing theory, renewal theory, and network synthesis can be utilized whenever required and programmed into the model for effective decision making. This process can be used to define reasonable requirements at system sublevels. In conjunction with the system effectiveness and cost models, we can then make appropriate decisions on optimization and also define contractual requirements.

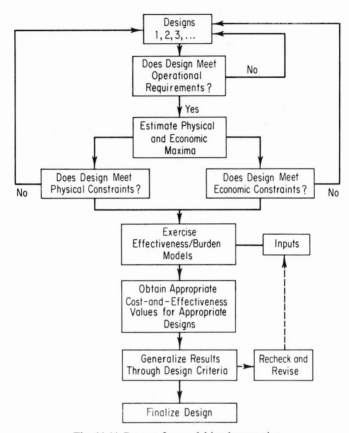

Fig. 11.11 Process for model implementation.

PROBLEMS

1. The data presented below apply to a radar system that is used in shipboard air-search operations. The data were collected over a period of approximately seven months, during which time the system was required to operate nearly continuously.

Total Calendar Time	5448.0 hr
Total Power-On Time	4869.0 hr
Total Operate Time	1381.0 hr
Total Down Time	137.5 hr
Total Active Repair Time	66.0 hr
Total Administrative Time	77.9 hr
Total Free Time	436.6 hr
Number of Malfunctions	25

(a) A distribution of the active repair time spent in correcting the 25 mal-functions, cited at the beginning of this problem, is shown below:

Active Repair Time, in Hours	Number of Repair Actions
0.1	2
0.3	2
0.4	1
0.5	4
0.7	1
1.0	1
1.1	1
1.5	1
1.8	1
1.9	1
2.0	1
2.1	1
3.0	2
3.3	1
5.0	1
6.6	1
8.0	1
9.3	1
11.0	1

Compute the observed and theoretical log-normal repairability functions. Plot both functions. Find P_{Ai} using mean and median active repair times.

(b) k_2 is assumed to be unity since the radar system is required to provide air search and surveillance on a continuous basis. The total power-on time is assumed to be up time. Find P_A.

(c) Technically speaking, there is no storage time on this equipment. However, there are periods during which all power is off except as needed for repairs. These nonuse periods may be likened to storage time. Experience with this equipment shows that whenever the equipment has been secured for a period of time, the initial checks after application of power disclose a malfunction or a discrepancy at a rate of 1 for every 14 times. Therefore, k_1, the probability of nonfailure during nonuse, is 0.93. Examination of the table shows that there were 436.6 hr of free or up time when the equipment was not in use. Using the information above, find P_{OR}.

(d) The reliability of this radar system is based on an exponential distribution of times to failure with an $MTBF$ of 194.8 hr. If mission length is assumed to be 24 hr, or one day's continuous operation, what is P_R for this time period?

(e) The design adequacy of the radar is assumed to 1.0. It is theorized that, when operating, the system is capable of aircraft detection to the limit of the designed range. Find P_{SE}.

2.* A radar system has been on a two-shift operation for a period of one year. During this period, field data show that, for several reasons (such as fewer and less-experienced men on duty during the night shift, or the main stock room being locked up so that certain types of spare parts are not readily accessible), the repair time for the night crew is greater than the repair time for the day crew; the result is greater down time for the night shift. Assume that the reliability of the system is 0.9 for the time period for which the radar can detect the target before it penetrates to within 50 miles, and that the operational readiness for each shift is:

$$OR \text{ day shift}   = 0.90$$
$$OR \text{ night shift} = 0.80.$$

The probability that the system will detect a target of specified size at a distance of 50 miles or more under ideal weather conditions is 0.9.

(a) Estimate system effectiveness for the foregoing conditions.

(b) It is found that 12 per cent of the targets employ noise jamming, which the radar has no capability to counter. What is the resulting level of system effectiveness?

(c) Independent of noise jamming, 20 per cent of the targets use lock-on jamming for which the radar has frequency stagger and jitter capability. Nevertheless, the radar allows 25 per cent of these targets to escape detection. What is the level of system effectiveness considering the results found so far?

---

*Based on a problem developed by L. R. Doyan, Wayland Laboratories, Raytheon Company. Used by permission.

Designing for Reliability and Maintainability

12.1 Introduction

An equipment whose design is not basically sound can extract severe maintenance penalties and necessitate considerable corrective action. Approximately 40 per cent of the maintenance actions performed on a typical electronic equipment during its first year of operation are traced to deficiencies in its electrical or mechanical design.* Furthermore, more than 50 per cent of the maintenance actions on a typical equipment are attributed to only 5 per cent of its parts.† The need for a sound basic design is clearly indicated. Means for making the inherent reliability and maintainability of an equipment as high as possible are discussed in this chapter.

12.2 Circuit Selection

Design reliability can often be improved by the utilization of (1) preferred circuits, (2) redundancy,‡ and (3) techniques protecting against the side-

---

*According to *NEL Reliability Design Handbook*, U.S. Naval Electronics Laboratory, San Diego, California.

†See *Electron Tube Requirements in Guided Missile Applications*, ARINC Research Corporation, Publication No. 101-3-125.

‡See Chapters 7 and 8.

effects of component failures. These factors are discussed in the following sections.

Preferred Circuits

In many cases, a circuit much like the required one has already been designed. This "preferred circuit" can often be adapted to meet specific requirements. Many design groups maintain a file of preferred circuits of proven performance and reliability.*

Redundancy

Redundancy can be defined as the provision of more than one means for accomplishing a given function, such that all means must fail before causing a system failure. Redundancy, therefore, permits a system to operate even though certain components have failed, thus increasing its reliability and availability.

Redundant configurations can be classified as either *active* or *standby*. Elements in active redundancy operate simultaneously in performing the same function. Two such elements will increase the mean time between failures (*MTBF*) of a single function by a factor of 1.5. Elements in standby redundancy are designed so that an inactive one will be switched into service when an active element fails. The *MTBF* of the associated function is increased by a factor equal to the number of standby elements (assuming, optimistically, that the sensing and switching devices of the redundant configuration are working perfectly).

A designer may often find that redundancy is

1. The quickest solution, if time is of prime importance;
2. The easiest solution, if the component is already designed;
3. The cheapest solution, if the component is economical in comparison with the cost of re-design;
4. The only solution, if the reliability requirement is beyond the state of the art.

On the other hand, in weighing its disadvantages, the designer may find that redundancy will

1. Prove too expensive, if the components are costly;
2. Exceed the limitations on size and weight, particularly in missiles and satellites;

*Numerous tube and transistor circuits are described in *Handbook of Preferred Circuits for Naval Aeronautical Electronic Equipment*, NAVWEPS 16-519 (Washington, D.C.: Government Printing Office).

3. Exceed the power limitations, particularly in active redundancy;

4. Attenuate the input signal, requiring additional amplifiers (which increase complexity);

5. Require sensing and switching circuitry so complex as to offset the advantage of redundancy.

Obviously, components of low or critical reliability are the ones to which redundancy can be the most beneficially applied.

Protective Techniques

It is generally desirable to include means in a design for preventing a failed component from causing further damage. Other components in the system, including power supplies, should also be protected from possible secondary effects of a component failure.

For these purposes, fuses or circuit breakers are used to sense excessive current drain and disconnect power from the failed component. Fuses within the circuits safeguard parts against excessive power dissipation and protect power supplies from shorted parts.

In some instances, means can be provided for preventing a failed component from completely disabling the system. A fuse or circuit breaker can disconnect a failed component from a system in such a way that it is possible to permit partial operation of the system after a component failure, in preference to total system failure.

By the same reasoning, degraded performance of a system after failure of a component is often preferable to complete stoppage. An example would be the shutting down of a failed circuit whose design function is to provide precise trimming adjustment within a deadband of another control system. Acceptable performance may thus be permitted, perhaps under emergency conditions, with the deadband control system alone.

Fuses, as well as holding or overload relays, may be used in active redundancy configurations (such as the parallel-series type). Where the reliability of the protective circuit is important, this type of redundancy should be considered—it can effectively reduce the possibility that the protective circuit will open prematurely or fail to open on over-currents.*

Thermostats may be used to sense over-temperature conditions and shut down the system, or a component of the system, until the temperature returns to normal.

In some systems, self-checking circuitry can be incorporated to sense abnormal conditions and operate adjusting means to restore normal conditions, or activate switching means to compensate for the malfunction.

*L. E. Dostert and T. T. Jackson, *Fusing Design Criteria for the Advent Satellite*, ARINC Research Corporation, Publication No. 176-6-304 (July 31, 1962).

Sometimes the physical removal of a component from a system can harm or cause failure of another component of the system, by removing load, drive, bias, or control circuits. In these cases, the first component should be equipped with some form of interlock to shut down or otherwise protect the second component.

The foregoing methods are referred to as *self-healing* techniques, in that they effect changes automatically to permit continued operation after a failure. More design effort is needed in devising additional automatic techniques for self-checking and self-adjusting. The ultimate design, in addition to its ability to act after a failure, would be capable of sensing and adjusting for drifts to avert out-of-tolerance failures.

In most of the foregoing examples of protective techniques, the basic procedure was to take some form of action, after an initial failure or malfunction, to prevent additional or secondary failures. In reducing the number of failures, such techniques can be considered as enhancing system reliability; they also increase availability, and thereby improve system effectiveness.

12.3   Parts Application Problems

The misapplication of electronic parts is one of the most serious problems relating to the reliability of electronic systems. Discussed in this section are the types of failures most frequently encountered, design techniques for accommodating part tolerances and drifting of electrical parameters, and some of the more important environmental considerations.

Failure Classifications

Failures can be classified into four categories: catastrophic, intermittent, out-of-tolerance, and maladjustment.

A *catastrophic failure* occurs when a part becomes completely inoperative or exhibits a gross change in characteristics. Examples are a shorted vacuum tube, an open or shorted resistor or capacitor, a leaky valve, a stuck relay, or a broken switch. The design of an equipment can sometimes be responsible for such failures; however, for parts of a given quality, and operated well within their ratings, some catastrophic failures are considered to be spontaneous or random in nature. Thus, the designer cannot be expected to eliminate all such failures.

Intermittent failures are also unpredictable, and the designer can do little to reduce them. These periodic failures usually occur within the piece parts themselves, and hence must be corrected in the design of the part, rather than in the design of the equipment or system. The designer should not only safeguard against the effects of intermittent failures, but also avoid creating possible modes of such failures (e.g., conditions permitting part-to-part or part-to-chassis voltage breakdowns).

Out-of-tolerance failures result from degradation, deterioration, drift, and wearout. Examples are the degradation of vacuum tube emission and transconductance, the drifting of resistor and capacitor values, and the wearing out of relays and solenoid valves. The changes come as a result of time, temperature, humidity, altitude, etc. When these gradual changes, considered collectively, reach the point where system performance is below acceptable limits, we say that the system has failed. Some steps that the designer can take to counteract this type of failure are discussed below, under "Part Tolerance and Drift Analysis."

Maladjustment failures are often due to human error. Such failures arise from improper adjustment of equipments, as well as abuse of adjusting devices due to lack of understanding of the adjustments and the capabilities of the equipments. Such failures are hard to evaluate and difficult to avoid, but must be considered if the reliability of the operational system is to be sustained.

Part Tolerance and Drift Analysis

If parts must be carefully selected in manufacture or subsequent field maintenance to meet performance requirements, the designer has not completed his job. A properly designed equipment will operate satisfactorily as long as the parameters of its parts are anywhere within their tolerances.

The design should also be capable of operating satisfactorily with parts which drift or change with time, temperature, humidity, altitude, etc. To guard against out-of-tolerance failures, therefore, the designer should consider the combined effects of (1) tolerances on parts to be used in manufacture, (2) subsequent changes due to the range of expected environmental conditions, and (3) drifts due to aging over the period of time specified in the reliability requirement. The component, therefore, should be designed to operate satisfactorily at the extremes of these parameter ranges.

Where high reliability is required, allowable ranges of these parts characteristics should be included in the procurement specification.

Three methods of dealing with part parameter variations are statistical analysis, worst-case analysis, and marginal testing. These methods are discussed in the following section.

STATISTICAL DESIGN ANALYSIS

In statistical design analysis, a functional relationship is established between the output of a circuit and the parameters of one or more of its parts.

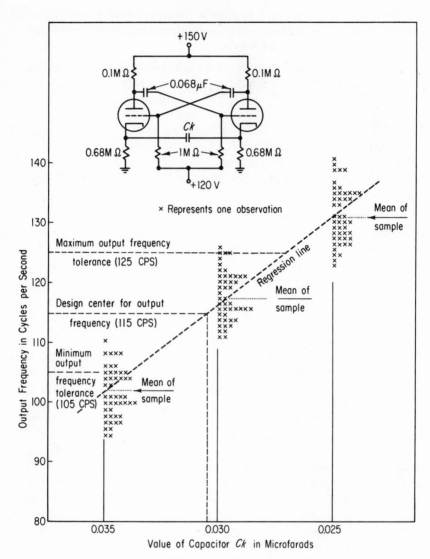

Fig. 12.1 Multivibrator circuit design improvement.

Figure 12.1 is an example of such a procedure; for an actual multivibrator circuit, the distribution of capacitances (Ck) of one of its capacitors is plotted against the output frequency of the circuit. A schematic diagram of the circuit appears in Figure 12.1. As indicated in the figure, a value of 0.030 microfarad for this particular capacitor yields maximum assurance that this circuit will perform within specifications.

When several parameters are considered, this method of analysis becomes quite complex and may require the use of a computer.*

WORST-CASE ANALYSIS

In worst-case analysis, the effect that a part has on circuit output can be evaluated on the basis of the part's end-of-life performance values. These degraded values will yield a conservative estimate of circuit output. Such an analysis can be performed for a part, a group of parts, or all of the parts in a circuit or component. However, the more parts considered, the more tedious the analysis becomes.

Worst-case design analysis can be accomplished in the following manner:

1. Write the circuit equation.

2. Determine which design criteria must be met and which parameters are fixed.

3. Determine how variations in fixed parameters will affect the design criteria.

4. Solve for the parameters that will meet the design criteria under the worst combination of fixed parameters.

An example of this procedure would be the designing, for use in a digital system, of a diode AND gate for negative-going signals (Figure 12.2). The design criteria are established so that when both inputs drop to $-V$, the output voltage, V_o, must drop to -8 volts within 1 microsecond, assuming

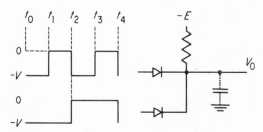

Fig. 12.2 Diode AND gate.

*Statistical design analysis techniques are discussed in greater detail in the following sources:

G. J. Blakemore, "Statistical Analysis of the Circuit Design of a Transistorized Multivibrator," a paper appearing as Appendix A in the ARINC Research Corporation publication, *Reliability of Semiconductor Devices*, by W. H. von Alven and G. J. Blakemore; Publication No. 144-6-270 (December 22, 1961). A. P. Lechler, D. G. Mark, and H. S. Sheffler, *Applying Statistical Techniques to the Analysis of Electronic Networks*, Battelle Memorial Institute, Columbus, Ohio (1962).

(1) infinite diode back resistance, (2) lead capacitance ranging from 50 to 100 pf, (3) 10 per cent voltage variations, (4) 10 per cent part tolerances, and (5) an input voltage, V, of -15 volts.

As long as one of the inputs is at ground, that diode is forward-biased and V_o is held at ground. When the last input drops to -15 volts, all of the diodes will be reverse-biased (assuming $-E > -V$), so the circuit can be shown as a series $R - C$ circuit with a negative unit step function as the applied voltage (Figure 12.3).

For a circuit as simple as shown in Figure 12.3, the voltage across the capacitor could be written directly. However, to illustrate the use of more general procedures, the Kirchoff voltage law is written in Laplace transform notation; this results in

$$-\frac{E}{S} = I\left(R + \frac{1}{SC}\right), \tag{12.1}$$

since there is no initial charge on the capacitor.

The output voltage is

$$V_o(s) = I \cdot \frac{1}{SC} \tag{12.2}$$

or

$$V_o(s) = E\,\frac{-1/RC}{S[S - (-1/RC)]},$$

and the inverse transform yields

$$V_o(t) = E\,(1 - e^{-t/RC}). \tag{12.3}$$

Solving for R,

$$R = -\frac{t}{C}\,\frac{1}{\ln\left(1 - \dfrac{V_o}{E}\right)}. \tag{12.4}$$

The output voltage is required to be -8 volts. The supply voltage, E, can be chosen to be -12 volts. Then, using overscored and underlined variables to represent maximum and minimum values, respectively, the maximum and

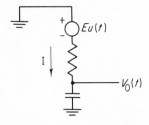

Fig. 12.3 Equivalent circuit.

minimum values of R can be determined. The equation for R maximum is

$$\bar{R} = -\frac{t}{\underset{\sim}{C}}\frac{1}{\ln\left(1-\dfrac{V_o}{\bar{E}}\right)}. \tag{12.5}$$

Noting that $t = 10^{-6}$ seconds, and

$\underset{\sim}{C} = 50 \, pf,$

$\bar{C} = 100 \, pf,$

$E = -12 \, v,$

$\underset{\sim}{E} = -10.8 \, v,$

$\bar{E} = -13.2 \, v,$

$V_o = -8 \, v,$

$\underset{\sim}{V}_o = -7.2 \, v,$

$\bar{V}_o = -8.8 \, v,$

then the maximum value of R is

$$\bar{R} = -\frac{10^{-6}}{5 \times 10^{-11}}\frac{1}{\ln\left(1-\dfrac{-7.2}{-13.2}\right)} = -20.10^3\frac{1}{\ln(1-0.5454)}$$

$$\bar{R} = 25.4 \, K\Omega.$$

The minimum value for R is

$$\underset{\sim}{R} = -\frac{t}{\bar{C}}\frac{1}{\ln\left(1-\dfrac{\bar{V}_o}{\underset{\sim}{E}}\right)} = \frac{10^{-6}}{10^{-10}}\frac{1}{\ln\left(1-\dfrac{-8.8}{-10.8}\right)}$$

$$= 10^4\frac{1}{\ln 0.1852} = \frac{10 \times 10^3}{0.685}$$

$$\underset{\sim}{R} = 14.6 \, K\Omega.$$

The value to choose is a standard 20 $K\Omega$. Then, within the 10 per cent tolerance, the value will range between 18 $K\Omega$ and 22$K\Omega$, well within the calculated limits.

For complex circuits, computer techniques can be employed for worst-case analysis.*

*See, for example, *Mandex—A Worst-Case Circuit Analysis Computer Program*, by H. S. Sheffler, J. J. Duffy, and B. C. Spradlin, Battelle Memorial Institute, Columbus, Ohio (1962).

The effects of aging on the performance of parts can be studied by simulating the normal aging process. The heater voltage on vacuum tubes can be lowered to simulate cathode deterioration and the corresponding change in the transfer function. Changes in gain of vacuum tubes due to aging can be simulated by lowering the plate voltage of triodes, or the screen voltage of pentodes. Increased forward resistance of diodes can be simulated by adding series resistance, and decreased reverse resistance can be simulated by adding shunt resistance.

In many switching applications, changes in transistor parameters (particularly H_{FE} and I_{CBO}) cause a number of circuit problems. Transistors having limit values of these characteristics can be used in the breadboard circuit to determine limit variations of circuit performance. Often, however, transistors with limit values are not available. In these cases, reduced values of H_{FE} can be simulated by placing a variable resistor in series with a forward-biased switching diode from base to emitter.*

Virtually all out-of-tolerance failures (see page 437) can be eliminated by careful attention during design to part parameter variations, after which part selection will usually not be necessary in manufacture or in subsequent maintenance.

Derating

Derating—a well known and commonly practiced procedure—is one of the most powerful reliability tools available to the designer.

The principle to keep in mind is that piece-part ratings (for voltage, current, temperature, power dissipation, etc.) are not distinct boundaries, above which immediate failure can occur and below which the part will operate indefinitely. Instead, the life of most parts increases in a continuous manner as the stress level is decreased below the rated value. The rating represents only a judgment of that stress level at which the part life is considered to be satisfactory. Above this value the rate of decline of part life is accelerated.

In general, the more a part is derated the longer it will last. Practically, however, there is a minimum stress level below which increased circuit complexity required to step up performance will offset the gain in reliability.†

---

*See "Analysis Testing for Improved Circuit Reliability," by B. J. Grinnell, *Eighth National Symposium for Reliability and Quality Control*, Washington, D.C. (January 9-12, 1962).

†Chapter 9 contains derating charts for several types of electronic parts. See Chapter 13 (pages 499-500) and Chapter 14 (page 528) for discussion of rating concepts.

12.4 Environmental Considerations

The designer of modern electronic systems must be well-versed in heat transfer, mechanical design, and the principles of radiation, shock, and vibration, in addition to electronic technology.* The following discussions highlight some of the more important environmental factors affecting the reliability of an equipment.

Shock and Vibration Protection

Shock and vibration are probably the most controversial areas in the environmental testing of electronic equipments, tubes, and other parts. Such tests should simulate the actual environmental stresses that the item will encounter in field use. The tests should also be reproducible, which requires that the essential characteristics of the test equipment be stable. Tables 12.1, 12.2, and 12.3 summarize shock and vibration requirements for a variety of military equipments.

The relative dynamic resistances of several common types of electronic parts are shown in Figure 12.4. The most frequent vibration-induced failures are

1. Flexing of electrical leads which support resistors and capacitors.

2. The dislodging or damaging of vacuum tubes. Monitored shock and vibration tests are often used to detect presence of foreign particles in transistors and diodes, as well as to determine their structural rigidity.

Equipments are sometimes specially mounted to counter the destructive effects of shock and vibration. Shock mounts often serve this purpose, but no effective means have been established for attenuating shock and vibration simultaneously. Isolation of an equipment against the effects of vibration requires that the natural frequency of the equipment be substantially lower than the undesired frequency of vibration.

Three basic kinds of isolators are available:

1. Elastomers made of natural or synthetic rubber, used in a shear mode or in a diaphragm to damp the induced shock or vibration.

2. Metallic isolators, which include springs, metal meshes, or wire rope. Springs lack good damping qualities, but meshes and rope provide smooth friction damping.

*Another difficult problem found by designers of electronic equipment is that of radio frequency interference. A consolidated source of information on the nature, design procedures, and specification requirements is the 4-volume series, *Handbook on Radio-Frequency Interference*, Frederick Research Corporation, Wheaton, Maryland.

Table 12.1 SHOCK REQUIREMENTS OF SOME MILITARY SPECIFICATIONS

Specification	Condition Simulated	Nature of Shock	Machine Commonly Used
MIL-S-901B (Navy), April 1954 Navy Dept., Spec 66S3* (Sept. 1945) MIL-T-17113 (Ships), July 1952 BuShips Spec 40T9* (Dec. 1946) MIL-E-005272B (USAF)† (June 1957) Procedure III	Navy shipboard shock. May also be used to simulate shocks initiated by impacts and explosions and transmitted through strong structures (ships, tanks, trucks, trains, buildings, etc.) to equipment mounted within.	Complex vibratory transient. Acceleration peaks about 1000 g for 1000 cps low pass filters. Velocity change 10 ft/sec. Displacement 1.5 to 3 in. Shock initiated by steel hammer-anvil impact.	Navy high impact shock machine for lightweight equipment.[1]† Navy high impact shock machine for medium-weight equipment.[1]
MIL-S-4456 MIL-STD-202A, Method 205 (Aug. 1958) MIL-E-005272B (USAF)† (June 1957) Procedure II MIL-E-5422D (ASG) (Nov. 1957) MIL-E-5400B (ASG) (May 1957)	Aircraft shock. May also be used to simulate shock for certain missile types and for rough handling.	Acceleration pulse amplitude of 15 or 30 g with 11 millisec duration is common specification. Machine used can provide up to 100 g and durations between 6 and 30 millisec. Velocity change 5 to 15 ft/sec. Velocity attained by free fall and stopped by impact on sand.	Sand-drop table or variable duration medium impact shock testing for lightweight equipment.[1,2]
MIL-STD-202A, Method 202A (Oct. 1956) JAN-S-44 (May 1944) MIL-E-005272B (USAF)† (June 1957) Procedure I MIL-M-3823 (May 1953) MIL-M-6B (May 1957) MIL-M-17275A (May 1955)	Idealized shock pulse for specimens weighing less than 4 lb. such as dial-indicating instruments. Provides a simple shock pulse for test of general ruggedness to non-oscillatory shock.	Half-sine acceleration pulse. 50 g amplitude of about 7 millisec duration is common specification. Amplitude is adjustable to about 100 g. Velocity change to about 10 ft/sec. Velocity attained by free fall and stopped by impact on a steel string.	Shock testing mechanism electrical indicating instruments.[3]

Table 12.1 (continued)

Specification	Condition Simulated	Nature of Shock	Machine Commonly Used
MIL-E-1C (Oct. 1956) (Drawing 180-JAN)	Shock excitation received by tubes in chassis.	Complex wave with a predominant initial acceleration pulse. Acceleration may extend to 1000g with 1000 cps low-pass filter. First pulse duration about $\frac{3}{4}$ millisec. Velocity change up to 10 ft/sec. Displacement about 4 in. Shock initiated by hammer-anvil impact.	Shock testing mechanism for electronic devices (flyweight machine).
MIL-STD-202A, Method 203 (Oct. 1956)	Repeated impacts due to handling and shipping.	Tumbling test. Specimen is mounted in a steel sleeve which is placed within a steel cage having baffles. Cage is rotated and specimen tumbles from baffles to lower areas.	Random drop test machine.

spectrum are required to be at least 100 g magnitude between 100 and 70 cps. The specifications state that such a shock can be obtained by a sawtooth pulse which rises linearly to 100 g in about 6 millisec and drops abruptly to zero. A discussion of this approach has been given by Lowe and Cavanaugh, *Environmental Engineering*, Vol. 1 (February, 1959), 24. The sawtooth pulse can be obtained by the plastic pellet type shock machine.

SOURCE: "The Fundamental Nature of Shock and Vibration," by Vigness, I., *Electrical Manufacturing* (June, 1959). Reprinted by permission.

*Obsolete.

†Supersedes MIL-E-5272.

[1] "Shock Testing Machines and Procedures," Kennard and Vigness, *Shock and Vibration Instrumentation* (June, 1956).

[2] American Standards Association Publication, S2.1/1.

[3] "Shock Testing Mechanism for Electrical Indicating Instruments," American Standards Association Publication, C39.3-1948.

NOTE: An interesting new approach to shock specification has been made by Ramo-Wooldridge (now Space Technology Laboratory) Specification GM 43.5-40. The shock is specified in terms of its shock spectrum. Both the positive and negative values of the

Table 12.2 VIBRATION REQUIREMENTS OF SOME MILITARY SPECIFICATIONS

Specification	Condition Simulated	Nature of Vibration and Test*
MIL-STD-202A, Method 204 (Oct. 1956)	Vibration on aircraft, missiles and tanks.	(1) Sweep between 10 to 500 cps with amplitude of 0.03 in. or 10 g, whichever is less. (2) Sweep between 10 and 2000 cps with amplitude of 0.03 in. or 15 g, whichever is less. Alternate procedures are also specified.
MIL-STD-202A, Method 201A (Oct. 1956)	Fundamental vibrations excited by reciprocating, rotary and propeller-type mechanisms.	Sweep between 10 and 55 cps with an amplitude of 0.03 in. Sweep cycle is 1 min. Test time is 2 hr in each coordinate direction.
MIL-STD-167 (Ships)† (Dec. 1954)	Navy shipboard vibration.	Equipment shall be vibrated for 2 hr at resonant frequencies below 33 cps that might cause damage. If no resonances are observed, the equipment shall be vibrated for 2 hr at 33 cps. Vibration amplitudes are 0.03, 0.02, and 0.01 in. for the respective frequency ranges of 5 to 15, 16 to 25, and 26 to 33.
MIL-E-005272B (USAF) (June 1957) Supersedes MIL-E-5272A	Aircraft vibrations. Different procedures are given according to where the equipment is located and whether or not isolators are employed.	A great variety of tests is prescribed, extending from 0.25 in. amplitude at low frequencies (5 to 8 cps) to 20 g for 80 to 500 cps. In addition, some procedures required circular motion for frequencies between 5 and 50 cps. Tests may be required to be run at 65 deg F and 160 F as well as at room temperature. Torsional (rotational) tests may be required.
MIL-E-5400B (ASG) (May 1957)	Aircraft vibrations.	(1) Vibration 0.04 in. amplitude between 5 and 10 cps, 0.03 in. between 10 and 55 cps, and 10 g amplitude between 55 and 500 cps. (2) Equipment normally with isolators shall be vibrated without isolators at 0.005 in. between 5 and 60 cps and at 2 g between 60 and 500 cps.
MIL-E-5422D (ASG) (Nov. 1957)	Aircraft vibrations.	Except for minor differences in amplitude the requirements are the same as MIL-E-5400A. However, procedures for testing are given in detail in 5422 whereas they are not given in 5400.
MIL-E-1C (Oct. 1955)	General field conditions encountered by vacuum tubes (low frequency components).	(1) Vibrations at 0.04 in. amplitude at 25 or 50 cps for about $\frac{1}{2}$ min. to determine change of tube output, or if shorts occur in tube. (2) Cycling tests over this frequency range are also specified, with a steady test of 1 min at the frequency having greatest effect on the tube output. (3) Fatigue tests are runs of 96 hr at 2.5 g amplitude (frequency of 25 cps and 0.04 in. amplitude suggested).

*Only the highlights of the tests are given; alternate procedures, exceptions and additional tests may be included in the original specification.

†This specification supersedes the vibration section of MIL-T-17113 (Ships) and 40T9 (Ships).

SOURCE: Vigness, I., op. cit., p. 445. Reprinted by permission.

Table 12.3 SUMMARY OF SELECTED VIBRATION AND SHOCK
SPECIFICATIONS FOR MISSILE ELECTRONIC COMPONENTS

1. A General Electric specification* for light electronic missile components requires that
   the equipment, while operating, shall be subjected to rms sinusoidal vibrations of 2 g
   from 15 to 2000 cps superimposed upon a Gaussian random vibration of 0.05 g^2/cps
   from 15 to 2000 cps in each of the three perpendicular planes. The vibration shall be
   applied for 2 min, and during this period the sinusoidal vibrations shall be swept once
   at a constant sweep rate over the frequency range. At low frequency the sinusoidal
   vibration shall be limited to an amplitude of 0.02 in. The instantaneous acceleration peaks
   shall be limited to 3σ (σ equals the rms value of the random vibration, which is about
   10 g).

2. A Boeing specification† requires acceleration densities for random vibration ranging
   from 0.002 to 0.02 g^2/cps from 5 to 10 cps; 0.02 g^2/cps from 10 to 35 cps; 0.02 to 0.22
   g^2/cps from 35 to 70 cps; 0.22 g^2/cps from 70 to 1100 cps; and 0.22 to 0.004 g^2/cps from
   1100 to 2000 cps. Many different procedures and amplitude ranges are given in other
   Boeing specifications. A Bomarc specification‡ requires frequency components up to
   4000 cps with an acceleration density of 0.3 g^2/cps above 250 cps.

3. The Space Technology Laboratory§ requires (for third-stage equipment) that sealed
   components, while operating, shall be pressurized to 15 psig and that the equipment
   shall be vibrated with Gaussian random vibrations having an acceleration of 0.1 g^2/cps
   over 20 to 2000 cps. Superimposed on this is a sinusoidal vibration having amplitudes
   of 2, 6, and 12 g rms in the respective ranges of 5 to 25, 25 to 700, and 700 to 2000 cps. A
   constant octave sweep rate shall be employed and the vibrations shall be for 15 min in
   each of the three coordinate directions.

4. A preliminary specification** recognizes some of the practical test problems in the procedures extracted here:

   a. Apply 10 g rms random broadband vibration (approximately flat 20 to 2000 cps)
      plus a superimposed sliding sine wave at 5 g rms swept at constant logarithmic rate
      from 2000 to 50 cps in time interval of 1 min. Vibrate once in each of three orthogonal
      axes for total vibration time of 3 min.

   b. Apply a 25 g rms sine wave vibration swept at constant logarithmic rate from 2000 to
      1000 cps in 30 sec. Vibrate once in each of three orthogonal axes for total vibration
      time of $1\frac{1}{2}$ min.

   c. The random vibration spectrum is specified only as approximately flat. It is
      suggested that this be obtained by adjusting suitable peak and notch-filters for table
      resonances only, using a rigid mass load of the same weight as the guidance subsystem.
      Adjust filters to about 3 db (2 to 1 in g^2/cps). Also adjust over-all spectrum so that
      the predominant peak, if any, occurs at about 1000 cps. A sharp cut-off should be
      employed above 2000 cps. Finally, set over-all level of 10 g rms.

---

*Vibration Shock and Acceleration Portion of Environment Test Specification for
Missile Guidance Electronic Equipment SK-52647-355-40. General Electric Co.,
Utica, New York (April, 1958).
†Boeing Co. Spec. D14999 (Jan., 1957).
‡Boeing Co. Spec. D80330 (Bomarc) (May, 1956).
§Environmental Type Test Specification for Airborne Electronic Equipment, D6003,
Space Technology Laboratory (May, 1958).
**Because of the preliminary nature of this specification the company name cannot be
divulged.

SOURCE: Vigness, I., op. cit., p. 445. Reprinted by permission.

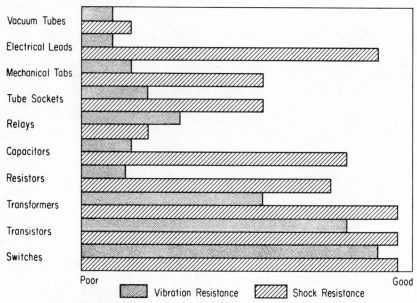

Fig. 12.4 Relative dynamic resistances of selected electronic components. [SOURCE: H. C. Lawrence, "Shock and Vibration Design," *Space/Aeronautics* (December, 1961). Reproduced by permission.]

3. Viscous dampers (similar to the type used on automobiles) which are velocity-sensitive and tend to become ineffective under high-frequency vibration and consequently may have limited application.* Resilient mounts must be used with caution. If improperly placed they can amplify the intensity of shock and vibration. The ideal goal is to design equipment to be resistant to shock and vibrations, rather than to isolate it from these forces.

Another protective technique is potting or encapsulation of small electronic assemblies. The choice of the material to be used is often critical. Some materials polymerize exothermically, and the self-generated heat may cause cracking of the casting or damage to such heat-sensitive parts as tantalum capacitors. Heat sinks or powdered fillers are often used to increase heat conduction. Another problem is shrinkage of the potting material—up to 16 per cent by volume depending upon the material used—during a change in state from liquid to solid. In some cases, molds are made oversized to compensate for shrinkage. Polyurethane materials are becoming more widely applied. They may be rigid or flexible; they have high abrasion resistance, low heat conduction and high mechanical shock resistance, and

---

*H. C. Lawrence, "Shock and Vibration Design," *Space/Aeronautics* (December, 1961).

good adhesive qualities. Other materials commonly used include epoxy resins, silicone elastomers, filled polyesterstyrene resins, and plastisols.*

Heat Transfer

Poor heat transfer is a major reliability problem of field electronic equipments. The fault is not always with the design; air intakes or outlets could be blocked because of faulty installation, or maintenance personnel might fail to replace air filters. Whatever the reason, poor heat transfer due to I^2R losses, hysteresis, or eddy current losses often result in physical damage or accelerated chemical reaction rates. The latter occurrence, affecting certain types of materials, can cause general degradation and catastrophic or intermittent failures.

Common methods of heat transfer in electronic equipments include (1) free convection, (2) forced air cooling, (3) liquid cooling, (4) conduction, (5) radiation, and (6) vaporization cooling. These methods are discussed briefly in the following paragraphs.

Free convection is relatively slow, even when adequate air space is provided. This technique is generally sufficient when heat dissipation is less than 0.25 watts per square inch under normal atmospheric conditions. Temperature-sensitive parts should be either isolated from heat sources or protected with polished radiation shields. Louvers should be provided in the enclosure to assure free circulation of air.

Forced air cooling permits heat dissipations of up to 2 watts per square inch. This technique requires fans and dust filters, which introduce a measure of unreliability into the system. In this regard, brushless a-c fan motors are preferred to brush-type a-c or d-c motors; these motors should be protected against overheating. In forced convection, heat transfer depends on air density; consequently, a larger volume of air has to be moved at higher altitudes than at sea level for the same amount of heat transfer.

Liquid cooling is often utilized in heat exchange systems for large electronic equipments, particularly those in which the heat to be dissipated exceeds 2-4 watts per square inch.

Heat conduction can be employed in electronic equipments when the heat source is not directly exposed to the coolant. This method is desirable in dense electronic packages, where radiation and convection are not effective cooling means. In such packages, cooling fins are generally the most effective method of heat conduction. Equipment designed for high-altitude operation is often sealed in an inert atmosphere to assure adequate heat conduction. The choice of gas to be used must be carefully considered. Molecules of helium, for example, may pass through the glass walls of electron tubes and

*C. Glenn Clark, "Potting, Embedment and Encapsulation," *Space/Aeronautics* (December, 1961).

become ionized; generally, nitrogen has been found to be the best compromise between effective heat conduction and containability.

Radiation is an effective method of heat transfer. The rate of radiant heat dissipation is proportional to the emissivity of the heat source. Nonmetallic materials (or coatings) have high values of emissivity; polished metal surfaces have very low values. Consequently, when thermal radiation is required from a metallic surface, the surface should be coated with a nonmetallic material.

Vaporization cooling is frequently used in electronic systems where heat dissipation is greater than 7 watts per square inch. One method employs an expendable evaporative coolant and has application in some missile systems. In this method the heat sources are submerged in the liquid; heated surfaces are either sprayed or wick-fed with the liquid. A second method of vaporization cooling involves continuous vaporization—the means for conventional refrigeration. Table 12.4 lists temperature precautions that should be taken for several types of conventional parts in the design of an equipment. Temperature-sensitive parts (such as frequency-control devices) which have to operate in a wide range of ambient temperatures can be mounted in a thermostatically controlled oven.

Table 12.4 PRECAUTIONS TO PREVENT OVERHEATING OF ELECTRONIC PARTS

Capacitors:
(Electrolytic and wax-impregnated types are particularly heat sensitive.)
1. Mount away from hot parts.
2. Use radiation shields if required.

Electron Tubes:
1. Use heat shields designed to aid heat radiation.
2. Measure bulb temperatures.

Resistors:
1. Clamp resistors on heat sinks, if practicable.
2. Use short leads if they go to cooler parts.
3. Mount power resistors vertically.
4. Prevent power resistors from radiating to temperature-sensitive parts.

Semiconductor Devices:
1. Minimize thermal resistance to chassis.
2. Use heat-radiating fins for power devices.
3. Locate away from heat-sensitive parts.
4. Consider the use of circuit breakers or thermostats for power devices.

Transformers:
1. Minimize thermal contact resistance to chassis.
2. Provide heat-radiating fins where appropriate.
3. Protect IF and RF transformers from radiated heat.

Temperature measurements are commonly made with thermocouples or temperature-sensitive lacquers. However, a new infrared scanning technique is being perfected, which will permit the development of an infrared profile of individual parts, printed circuit boards, or blocks of integral circuits.* This technique should prove to be valuable in both development and production activities.

Control of Corrosion and Biological Growths

In 1954, an estimate was made that various forms of deterioration caused material losses (excluding foodstuffs) of more than $12 billion per year in the United States. Losses due to corrosion alone were estimated at $6 billion.† The earth's environment contains numerous deteriorators: oxygen, carbon dioxide, nitrogen, snow, ice, sand, dust, water salt spray, chemicals, etc. To cope with this serious problem, equipment specifications often require demonstration that the equipment will withstand specified levels of temperature, humidity, altitude, salt spray, fungus, sunshine, rain, sand, and dust.

In addition to the deterioration problems associated with the external environments to which equipments are subjected, the design engineer must also contend with deterioration of parts in their normal operating modes. Numerous types of parts—including vacuum tubes, batteries, and certain types of capacitors—are susceptible to chemical actions and biological growths.

CHEMICAL ACTIONS

A material can undergo change in a number of ways; among them are chemical interactions with other materials and modifications in the material itself (recrystallization, phase changes, or changes induced by irradiation).

Corrosion-resistant materials should be used as much as possible. Although aluminum is excellent in this respect due to the thin oxide coating which forms when it is exposed to the atmosphere, it tends to pit in moist atmospheres if dust particles are present. Furthermore, aluminum may corrode seriously in a salt-laden marine atmosphere. Either anodization or priming with zinc chromate can provide additional protection.

Magnesium is sometimes used to minimize the weight of electronic equipments, but must be protected by surface additives against corrosion and electrolytic reaction. A coating of zinc chromate serves both purposes. Because magnesium is the most reactive metal normally used for structural purposes,

---

*E. R. Jervis, *Investigation of Factors Affecting Early Exploitation of Integrated Solid Circuitry*, ARINC Research Corporation, Publication No. 234-2-342 (January 1, 1963).

†G. A. Greathouse and C. J. Wessel, *Deterioration of Materials* (New York: Reinhold Publishing Corp., 1954).

the grounding of the magnesium structure requires careful selection of another metal for making the connection. Zinc- or cadmium-plated steel are the more commonly chosen connective materials.

Copper, when pure, is quite resistant to corrosion. However, under certain conditions, copper-made components will become chemically unstable. IF or RF transformers have occasionally failed due to the presence of d-c currents, impurities in the insulation, and moisture. These conditions favor electrolytic reaction, which causes the copper to dissolve and eventually results in an open circuit.

Iron and steel are used in many components because of their good magnetic and structural properties. However, only certain stainless steels are reasonably resistant to corrosion. Various types of surface platings are often added, but since thin coatings are quite porous, undercoatings of another metal such as copper are often used as a moisture barrier. Cadmium and zinc platings in some marine environments corrode readily, and often grow "whiskers" which can cause short circuits.

Materials widely separated in the electromotive series are subject to galvanic action, which occurs when two dissimilar metals are in contact in a liquid medium. The most active metal dissolves, hydrogen is released, and an electric current flows from one metal to the other. Coatings of zinc are often applied to iron so that the zinc, which is more active, will dissolve and protect the iron (a process known as *galvanization*). Galvanic action is known to occur within the same piece of metal, if one portion of the metal is under stress and has a higher free-energy level than the other. The part under stress will dissolve if a suitable liquid medium is present. Stress-corrosion cracking occurs in certain magnesium alloys, stainless steels, brass, and aluminum alloys. It has also been found that a given metal will corrode much more rapidly under conditions of repeated stress than when no stress is applied.

Proper design of an equipment therefore compels one to (1) select corrosion-resistant materials, (2) specify protective coatings if required, (3) avoid use of dissimilar metallic contacts, (4) control metallurgical factors to prevent undue internal stress levels, (5) prevent water entrapment, (6) use ceramic coatings in high-temperature environments, and (7) control the environment through dehydration, rust inhibition, and electrolytic and galvanic protective techniques.

BIOLOGICAL GROWTHS

High humidity and warm temperatures often favor the growth of fungus on all types of equipment. As a result, most insulating materials are affected in humid environments by a decrease in electrical and mechanical properties. Equipments designed for use under tropical conditions should receive a moisture- and fungus-proofing treatment, which consists of applying a moisture-resistant varnish containing a fungicide. Air-drying varnishes generally are not as moisture-proof as are the baked-on types. This treatment is usually quite effective, but the useful life of most fungicides is less than that of the varnish, and retreatments may be necessary.

Designing for the Space Environment

Table 12.5 summarizes the effects of space environment, along with protective measures that can be taken. Two of the more serious problems are the effects of high vacuum and radiation on materials and electronic parts used in space systems.*

HIGH VACUUM

In a high vacuum, materials with a high vapor pressure will sublimate or evaporate rapidly, particularly at elevated temperatures. In some plastics, the loss of plasticizers by evaporation will cause cracking, shrinking, or increased brittleness. Metals such as magnesium, which would normally evaporate rapidly (1 gram/ cm^2/year at 250°C), can be protected by inorganic coatings with low vapor pressures.

In a high vacuum, adjoining solid surfaces can become cold-welded after losing absorbed gases. Some form of lubrication is therefore necessary. Conventional oils and greases evaporate quickly. Graphite becomes unsatisfactory (actually an abrasive) because of the loss of absorbed water. However, thin films of soft metals, such as lead, silver, or gold, are effective lubricants in a high vacuum. Thin films of molybdenum disulfide are often sprayed over chrome or nickel plating, forming easily sheared layers. The film also releases sulfur at the interfaces during sliding, performing the same function as water vapor does for graphite.

RADIATION EFFECTS

The radiation environment in space near the earth is composed, primarily, of the Van Allen, auroral, solar flare, and cosmic phenomena. Others of lesser importance are solar wind, thermal energy atoms in space, neutrons, albedo protons, plasma bremsstrahlung, and man-made nuclear sources. In the electromagnetic spectrum are gamma rays, x-rays, ultraviolet and Lyman-alpha radiation.

*For a detailed discussion of these problems, see *Equipment Design Considerations for Space Environment*, by S. N. Lehr, L. J. Martire, and V. J. Tronolone, Space Technology Laboratories, Inc. (February, 1962).

Table 12.5* Summary of Space Environment Design Factors

Environment	Effects	Design Factors
Temperature	Thermal energy within vehicle produced by solar radiation, earthshine, earth radiation, and internal heating. No convection heating outside earth atmosphere.	Design for temperature control by means of absorptive and reflective surfaced (α/ϵ ratio) with heat control servo circuits. Isolate internal equipment thermally for temperatures between 0° and 60°C depending on requirements. For heat transfer use radiation and conduction heat sinks. Spin space vehicle, where feasible, to eliminate temperature gradients around surface.
Vibration	Unimportant in space (except for launch environment). No acoustic, frictional, or combustion vibration problems except for special applications	Any vibration, acceleration, or shock levels in space which may occur would be very small compared with those during the boost phase. The equipment must be designed to withstand the levels during boost and therefore the lower levels encountered in space should pose no problems.
Acceleration	Unimportant in space except for special applications.	
Shock	Unimportant in space except for meteorite and micrometeorite impacts.	
High vacuum	Sublimation and evaporation of materials occur in high vacuum.	Use materials with low sublimation rates. Allow sufficient thickness for sublimation and evaporation over expected operating life.
	Chemical atmosphere produced by outgassing and sublimation may have corrosive, plating, or chemical effects.	Select material with care to avoid hazardous conditions.
	Electrical arc-over or corona discharge.	Provide adequate insulation material and insulation paths.
	Cold welding of contact, sliding or rolling surfaces.	Provide solid state or low vapor pressure lubricants to avoid clean surfaces, hermetic or labyrinth seals.
Magnetic fields	No effects except for fine instrumentation.	Avoid use of instruments not shielded against variations outside earth's magnetic fields.
Gravitational fields	No effect on materials or parts.	None for materials or parts. For manned vehicles, physiological considerations are involved.

Table 12.5 (continued)

Environment	Effects	Design Factors
Meteorites and micrometeorites	Collisions with particles of varying sizes occur.	Statistically calculated risk is involved. Use pre-roughened surfaces or oxide finishes to minimize changes in α/ϵ ratio. Use sufficient outer skin thickness or secondary outer shell to provide protection against small particles.
Ultraviolet light	Increases sublimation rates in high vacuum, may produce bond precision in organic surface coatings or exposed plastic parts.	Minimize sublimation by selection of materials and by providing sufficient material thickness allowances or radiation resistant substances.
X-rays and gamma rays	Ionization of material occurs, possibly causing atomic displacements which produce changes in material characteristics or composition.	Intensity of primary radiation is negligible but shielding with heavy material may be considered for secondary ionizing radiation effects.
Neutrons	Intensity too low to require consideration of atomic displacement effects.	No special precautions necessary because of low intensity.
Trapped electrons	Ionizing radiation occurs primarily in Van Allen belts, possibly causing atomic displacements which produce changes in material characteristics or composition.	Protection required for externally mounted equipment such as solar cells. Space vehicle shell normally provides protection for internal equipment.
Trapped protons	Ionizing radiation occurs primarily in Van Allen belts, possibly causing atomic displacements which produce changes in material characteristics or composition.	For low energy protons in outer Van Allen belts, protection requirements are similar to those for trapped electrons. For high energy protons in inner Van Allen belts there is no known adequate protection.
Biological organisms.	Contamination of space environment.	Sterilization of vehicle and contents now required by international agreement to prevent possible biological contamination of lunar or planetary environment.

*Source: Lehr et al., op. cit., p. 453.

In general, metals are quite resistant to radiation damage in the space environment. Semiconductor devices may be affected by gamma rays which increase leakage currents. The lattice structure of semiconductors can be damaged by high-energy electrons, protons, and fast neutrons, which cause permanent effects through atomic displacement and damage to the lattice structure. Organic materials are particularly susceptible to drastic physical changes in cross-linking and scission of molecular bonds. Formation of gas, decreased elasticity, and changes in hardness and elongation are some of the predominant changes in plastics which have been subjected to radiation of the type encountered in the space environment. Electronic circuits are affected by a lowering of input and output impedances. Solar cells particularly are in need of shielding. (In the case of the Telstar satellite, complete protection against electron damage to the solar cells in the Van Allen belt for a two-year period was obtained by the use of 0.3 gram/cm^2 of sapphire shielding. Complete shielding against protons would require an intolerable amount of shielding material. Reliability considerations in this instance involved a trade-off between the weight penalty of radiation protection and the increase in the number of solar cells to allow for decreasing power output per cell because of the radiation damage as a function of time.*)

12.5   Designing for Operability

Operability is the ability of an equipment to be operated in its intended manner. This concept recognizes that an equipment's full capability can sometimes be compromised because of shortcomings of the personnel who operate it. Failures reported by operating personnel are frequently unconfirmed in subsequent maintenance. The incidence rates of unconfirmed failures are usually the highest in those systems requiring the greatest degree of operator skill. Unconfirmed failures have as adverse an effect on system effectiveness as confirmed ones, since the system must be taken out of service for maintenance in all cases.

Human engineering, therefore, must be considered in designing for operability, which is mainly a packaging problem. The complexion of the system which is presented to the operator—the array of components, including ancillary equipment, cable connectors, patch panels, switches, knobs, controls, panel markings, meters, oscilloscopes, horns, bells, panel lights, etc., is of concern in designing for operability.†

*P. S. Darnell, *Some Reliability Considerations for Telstar*, Bell Telephone Laboratories, Whippany, New Jersey (September, 1962).

†A procedure for the quantitative evaluation of electronic equipment operability has been developed by the American Institute for Research (AIR), Pittsburgh, Pennsylvania. This work was carried out for the U.S. Army Electronics Proving Ground, Ft. Huachuca,

The various components or modules comprising the system should be clearly labeled for easy identification. Those items which must be manipulated in order to connect the system (cable connectors, patch panels, switches) should be arranged in a simple and logical order, and plainly marked.

Controls should be oriented for operation in a logical order. The order might be (1) the desired sequence of adjustments, (2) the nature of their functions (pitch-up, yaw left, etc.), (3) the proximity of the controls to displays associated with their functions, or some other plan whose meaning is obvious to the operator. In any case, the controls should be within convenient reach of the operator, and easily manipulated.

Displays in the form of meters, lights, alarms, cathode-ray tubes, etc., should be provided to indicate that the various components of the system are operating properly. Displays of the "all-or-none" type are the most effective—panel lights, flag indicators, or alarm horns leave no doubt in the operator's mind that something has happened. The significance or interpretation of signal lights, abnormal meter readings, and oscilloscope patterns should be indicated on the equipment as ready reminders; the operator should not have to rely on the availability of an instruction manual. Function selector positions should be labeled not "1, 2, 3," etc., but in terms of their functions.

Meters are also required for setting, checking, and monitoring parameters; for adjusting nulls; for tracking; and for making quantitative readings. Moving pointer meters are the most satisfactory when located near the eye-level of the operator in his normal position. Scale divisions should not be finer than necessary for the required reading accuracy and should be omitted in favor of tolerance bands where applicable. Protection should be provided against apparently valid null readings of center-null meters after power or signal failure. Biasing the meter or including a light or other signal will suffice.

A blown fuse should be replaced during an operational period. Fuseholders which indicate with a signal light when a fuse has blown are time savers, especially when many fuses are required. Proper labeling of fuse functions is also important, but often neglected.

A common design mistake is the arranging of operational controls in such a way that the operator, under pressure, may reach for the wrong control or be unsure which meter he should observe during the adjustment. The operator may forget the correct adjustment setting or forget that a function or range selector must be changed in order to make the adjustment. Some

---

Arizona, on Contract DA-36-039-SC-80555, and resulted in the report, *An Index of Equipment Operability*. The procedure provides estimates of the time and accuracy of operator performance, identification of design features which degrade operator performance, and guidance for selection and training of operators. Twenty basic equipment characteristics are used as the basis for the AIR procedure.

combination of control settings, made in error, may have deleterious effects on equipment operation. Placards cautioning the operator are not usually satisfactory.

Operator problems are also created by the omission of displays or lack of built-in test equipment in the system. The complexity of a checkout system often makes it difficult to determine whether a fault is in the unit under test or in the checkout system. In such cases, it is important that the operator should have the benefit of some display means to indicate whether the related and pertinent functions of the checkout system are being performed in a normal manner. Displays can be provided in some cases to monitor such functions, or if the insertion of test equipment is necessary to make a determination, then the test equipment can often be incorporated as a part of the system. For example, in testing a guidance system whose digital computer is intended to respond to pulse commands, a discrete number of short pulses would be delivered by the checkout equipment. A failure of the system to respond might be due to several causes. However, the cause can be confirmed as a failure within the guidance system if an oscilloscope and digital counter built into the checkout equipment indicate that proper pulses in the correct number are delivered at the input to the guidance system.

Built-in test equipment and self-checking techniques, in addition to aiding the operator, are of great benefit in expediting maintenance. In fact, most of those design features which simplify the operator's job also aid the maintenance technician.

12.6 Designing for Maintainability

No operational system can be perfectly reliable, in spite of the designer's best efforts. Maintenance, therefore, also becomes an important consideration in the long-term performance of a system.

Maintainability is the probability that, when a specified maintenance action is initiated, a failed system (or product) will be restored to operable condition in a specified down time. According to this definition, design features which will expedite maintenance will enhance maintainability. Just as designing for operability is designing with the operator in mind, so designing for maintainability means inclusion in the design of those features which can be conceived to assist the maintenance technician.

In checking out an equipment, the maintenance technician must also operate it. Therefore, the design considerations relating to operability are, in general, also applicable to maintainability. In addition, many design considerations relating to maintenance are more complex than those relating to operation.

The designer, in acknowledging that his design can fail, should be aware of failure modes and should be able to identify the marginal or critical areas. By concentrating on these areas, he should try to provide means for facilitating removal of the failed unit, trouble shooting, access to the failed unit, and repair or replacement. Such means, in addition to improving maintainability, will also improve reliability through the averting of subseqent failures due to human errors during maintenance.

Table 12.6 DESIGNING FOR MAINTAINABILITY

1. Design for minimum maintenance skills. Some technicians are neither well-trained nor well-motivated.

2. Design for minimum tools. Special tools and laboratory test equipment may not always be available.

3. Design for minimum adjusting. Adjustment for shift, drift, and degradation should not be necessary in most cases.

4. Use standard interchangeable parts wherever possible; special parts create problems.

5. Group subsystems (e.g., power supply components) so they can be easily located and identified. (Use colors.)

6. Provide for visual inspection. Burnt resistors, diodes, and broken terminals can be located quickly if they are visible. Visible tube filaments and heaters are helpful.

7. Provide trouble-shooting techniques—panel lights, tell-tale indicators, built-in test panels, etc.

8. Provide test points. Use plain marking and adequate spacing and accessibility.

9. Label units. Labels on top of components should agree with instruction manuals to aid in locating suspected components.

10. Use color-coding. (Choose colors carefully; about 6 per cent of the people are color-blind.) Use different colored wires and tracers to facilitate trouble-shooting wiring harnesses.

11. Use plug-in rather than solder-in modules. Ease in replacement avoids errors and mutilation of harnesses.

12. Orient sockets all in same direction to avoid need for looking at each key position. (Especially tube sockets.)

13. Use captive-type chassis fasteners which can be manipulated without tools and which cannot be lost.

14. Avoid large cable connectors where possible. Label connectors. *Key* connectors so they cannot be inserted wrongly. Provide adequate separation between connectors. Wherever possible, follow same letter connector-to-connector (e.g., A-A-A-A-). Use "P" for plate, "C" for collector, etc.

15. Provide handles or hand-holds on heavy components for easy handling.

16. Use BAD-GOOD meters, red-lined meters, or tolerance bands. Where possible, avoid using meters which must be read and evaluated from a document or table. Arrange controls for metered values adjacent to meters. (Again use colors.)

17. Use overload indicators, alarms, and lighted fuses to assist in locating a failure.

18. Design for safety; use interlocks, safety covers, and guarded switches, and do not permit exposed high voltages.

Many design details, often overlooked, are important to the maintenance technician. A list of "Do's" and "Don't's" in designing for maintainability is presented in Table 12.6.

Design mistakes are often the bane of the maintenance technician. The designer is not always made aware of these mistakes; Table 12.7 lists design errors which he should avoid.

Table 12.7 WARNING TO DESIGNERS; SOME COMMON DESIGN ERRORS
AFFECTING MAINTAINABILITY

1. Don't place tubes or transistors where they cannot be removed without removal of the whole unit from its case or without first removing other parts.

2. Don't put an adjustment out of arm's reach.

3. Don't put an adjusting screw in such a place that the technician has difficulty in locating it.

4. Don't install a low-reliability part under other parts so that the maintenance technician has to disassemble many components to get at the suspect part.

5. Don't screw subassemblies together in such a way that the maintenance technician cannot tell which screw holds what; *do* mount inside parts on the same screws.

6. Don't neglect to provide enough room for the technician to get his gloved hand into the unit to make an adjustment.

7. Don't use chassis and cover plates that fall when the last screw is removed.

8. Don't make sockets and connectors for modules of the same configuration so that the wrong unit can be installed.

9. Don't provide access doors with numerous small screws.

10. Don't use air filters which must be replaced frequently; don't place these filters so that it is necessary to shut down power and disassemble the equipment to reach them.

11. Don't put a screwdriver adjustment underneath the module so that the technician has great difficulty in reaching it.

12. Don't omit the guide from a screwdriver adjustment, so that while the technician is adjusting and watching a meter, the screwdriver slips out of the slot.

13. Don't locate adjusting screws beside a hot tube or an exposed power supply terminal.

14. Don't omit handles and don't place fragile components just inside the bottom edge of the chassis where the technician will be sure to put his hand.

15. Don't provide unreliable, built-in test equipment which will cause the operator to report false failures of the system.

In summary, designing for maintainability is designing with a view toward aiding the maintenance technician—specifically, in recognizing the weak or critical areas and making repairs as easily as possible.*

*Several recommended guides for maintenance designing are:
J. D. Folley and J. W. Altman, *Designing Electronic Equipment for Maintainability*, Penton Publishing Company. (Available from ARINC Research Corporation.)

12.7   Design Reviews and Reliability Controls

During the entire period of design development and test of a component, reliability should be continuously monitored by the reliability group. In addition, a program of periodic design reviews should be established to

1. Review progress of the design

2. Monitor reliability growth

3. Assure the reliability requirements will be met

4. Provide information feedback to all concerned with the program.

The membership of the design review group and the technical matters requiring their attention are shown in Table 12.8.

Table 12.8 DESIGN REVIEW BOARD

Members	1. Project engineer 2. Electrical design engineer 3. Mechanical design engineer 4. System engineer 5. Reliability engineer 6. Packaging engineer 7. Consultants as required 8. Management representatives
Technical Considerations	1. Parts lists 2. Part application 3. Tolerance studies 4. Environmental effects 5. Drift, aging, and end-of-life parameters 6. Regression or worst-case analysis 7. Redundancy 8. Reliability analysis 9. Reliability predictions 10. Trade-off studies (if any) 11. Maintenance factors 12. Test data

Maintenance Engineering Guide for Ordnance Design, ORD P 20-134, U.S. Army Ordnance Corps (December, 1961).

Maintainability Design Criteria Handbook (NAVSHIPS 94324), Fleet Effectiveness Branch, Bureau of Ships, Department of the Navy (April, 1962).

Maintainability Design, Signal Corps Technical Requirement, SCL-4301B (March 29, 1962).

The design review meetings scheduled for any design program should include at least the following:

1. Preliminary design review

2. Detailed analysis review

3. Final design review

4. Test correlation review.

The purposes and details of these design review meetings are described in the following paragraphs.

Preliminary Design Review

Several alternate designs for meeting performance requirements are generally available. The primary purpose of the Preliminary Design Review is to make a choice among alternative design approaches. The choice should be one of the following, in order of preference: (1) the simplest design which meets the reliability requirement, (2) the design which has the highest reliability, or (3) the design which shows the greatest promise of meeting the reliability requirement.

In order to reach such a decision, reliability predictions must first be made for each design approach. These predictions need only be rough estimates based on parts counts and failure rates. In general, the accuracy of the failure rate data is not important—provided the data are the same and consistent for all predictions—since the predictions need be only for comparison. It is important, however, to exercise care in dealing with parts peculiar to a particular design. In such cases, an erroneous failure rate could influence the decision.

The outputs of this first design review should include an understanding of the weak areas in the chosen design and the areas in need of special attention in order to improve reliability. Evolving from the discussions should be a reliability block diagram of the chosen design, showing the series and parallel elements.

The Preliminary Design Review should also reveal any lack of data or need for more design information, such as

1. more complete part specifications,

2. more information on part drift or degradation, and

3. application data (e.g., new ratings).

Detailed Analysis Review

By the time the design has progressed from the preliminary stage to the breadboard, parts have been selected and part applications have been determined. At this point, with only one design to consider, a careful reliability analysis should be performed.

To estimate if the design will meet the reliability requirement, a reliability prediction must be made. This prediction, again based upon a parts count and part failure rates, must be more accurate than the previous comparative estimate, and should be based upon realistic failure rates. For this prediction the assistance of a reliability engineer is recommended.

If the prediction indicates that the reliability requirement will not be met, then—in a Detailed Analysis Review meeting—a management decision should be made regarding abandonment of the present design and starting anew, or concentrating effort on improving the design.

If improvement is needed, areas should be indicated in which more attention should be focused. This is the point where design decisions may be required as to (1) redundancy versus further derating, or (2) redesigning weak areas versus a search for high-reliability parts.

Planning should precede the meeting to ensure that the design review is patterned to the design. Any misapplications should be shown up in the meeting and noted. Any areas of overstressing or areas where severe environmental conditions appear to be troublesome, even tolerance problems, should become evident. Some problems may be indicated which should be earmarked for subsequent attention under the category of designing for maintainability.

In this connection, the list of needed test points may be started. Ideas may develop concerning easy-access packaging. The incorporation of sockets for replaceable modules or card-type circuit boards may be suggested. The physical design is still flexible in the breadboard stage, and such changes may be incorporated.

Management should determine, in analyzing the results of this design review, whether decisions made in the previous design review were valid, and how to plan the continuation of the design phase.

Final Design Review

After incorporation of changes indicated in the previous design review, the product has matured into the final stage. The purpose of the Final Design Review is to ascertain that all of the requirements have been met.

No one individual should be held responsible for remembering all of the detailed information accumulated to this point in a particular design, or for remembering which details are necessary for consideration in conducting the Final Design Review. The most common errors evolving from such a review are errors of omission. Therefore, the most useful tool in such a review is a detailed check list. Each design will require its special check list, which should be carefully prepared in a joint effort by design and reliability personnel. To illustrate the format, a typical check list is included herewith for guidance. (Table 12.9.)

Meeting design requirements is the prime consideration in the Final Design Review. For the reliability requirement, another reliability prediction must be performed. This prediction should be as precise as possible; therefore, failure rates should be as realistic as can be attained. Consideration must be given to the conditions of the tests which were used to establish the failure rates employed. The items of concern are the stress levels, environmental conditions, and the definition of failure. The failure rates must be modified in accordance with the stress levels of the design.

A reliability engineer should select the failure rates and make the reliability prediction. It need only be said that an optimistic reliability prediction at this point amounts to whistling in the dark and will inevitably be shown up as a design error when the product encounters the life testing phase.

Good collaboration is recommended between design and reliability personnel throughout the whole design phase. Design reviews should be collaborative efforts, and just as the design reviews should be patterned after the design, so also the designer must strive to meet the objectives of the design review.

Test Correlation Review

A test program is usually started upon completion of the design program, and is one of the most important activities in assuring that performance and reliability objectives have been achieved. The objectives of the Test Correlation Review are to evaluate the test results and compare them to the design requirements and to the theoretical predictions, in order to accomplish a final program evaluation.

Differences between test results and theoretical predictions will become apparent in such a comparison.

Life test results will either prove the integrity of the design or reveal errors in the reliability analysis and predictions. If the required reliability is demonstrated, the project—as far as reliability is concerned—is completed.

If the reliability predictions were carefully made and the life tests properly conducted, the disagreement, if any, should not be great. If life test results indicate that the reliability requirement has not been met, the designer and the reliability engineer should review all of the data in an attempt to determine if and where errors were made. Or, perhaps, the failure mode encountered in testing represents a critical area which was overlooked during the design. In any case, the designer should be in a position to propose minor design revisions to correct the problem.

Table 12.9 DESIGN RELIABILITY CHECK LIST

Item	*Yes*	*No*

1. Are the requirements for performance, signals, environment, life, and reliability established for each component?

2. Are the specific design criteria for reliability and safety of each component specified?

3. Are maintenance requirements and maintainability goals established?

4. Have acceptance, qualification, sampling, and reliability assurance tests been established?

5. Are standardized high-reliability parts being used?

6. Are unreliable parts identified?

7. Has the required failure rate for each part or part class been established?

8. Are component parts selected to meet reliability requirements?

9. Have "below-state-of-the-art" parts or problems been identified?

10. Has shelf life of parts chosen for final design been determined?

11. Have limited-life parts been identified, and inspection and replacement requirements specified?

12. Have critical parts which require special procurement, testing, and handling been identified?

13. Are derating factors being used in the application of parts? (Indicate level.)

14. Are safety factors being used in the application of parts?

15. Are circuit safety margins ample?

16. Are standard and proven circuits utilized wherever possible?

17. Has the need for the selection of parts (matching) been eliminated?

18. Have circuit studies been made considering variability and degradation of electrical parameters of parts?

19. Have adjustments been minimized?

20. Have all vital adjustments been classified as to factory, preoperational, or operator types?

21. Have stability requirements of all circuits or components associated with each adjustment been established?

22. Have specification limits for each adjustment—such as "no adjustment," "readjustment to nominal," or "classify as failure"—been established?

(continued)

Table 12.9 (continued)

Item	*Yes*	*No*

23. Is feedback being used wherever required to maintain constant gain of circuits?

24. Are regulated power supplies used for critical circuits?

25. Are similar plugs and connectors adequately protected against insertion in wrong sockets?

26. Are solid-state devices being used where practicable?

27. Are malfunction-indicating circuits or devices being used extensively?

28. Are self-monitoring or self-calibration devices installed in major systems where possible?

29. Are mechanical support structures adequate?

30. Are heat transfer devices or designs efficient?

31. Is there a concentrated effort to make the developmental model as near to the production model as possible?

32. Has equipment been designed for the elimination of shock mounts and vibration isolators where possible?

33. Are elapsed time indicators installed in major components?

34. Are fungus-inert materials and fungus-proofing techniques being used?

35. Has packaging and mechanical layout been designed to facilitate maintenance?

36. Has the theoretical reliability or *MTBF* of the component based on the actual application of the parts been determined?

    (a) Comparison made with reliability goal

    (b) Provision for necessary design adjustments

37. Are the best available methods for reducing the adverse effects of operational environments on critical parts being utilized?

38. Has provision been made for the use of electronic failure prediction techniques, including marginal testing?

39. Does the basic design provide for the system to meet serviceability goals established for field test operations?

40. Is there provision for improvements to eliminate any design inadequacies observed in engineering tests?

41. Have normal modes of failure and the magnitude of each mode for each component or critical part been identified?

(continued)

Table 12.9 (continued)

Item	Yes	No

42. In the application of failure rates of components to reliability equations, have the following effects been considered?

     (a) External effects on the system or branch in which the component is located

     (b) Internal effects on the component

     (c) Common effects, or direct effect of one component on another component, because of mechanical or electromechanical linkage

43. Has redundancy been provided where needed to meet reliability goals?

PROBLEMS

1. A voltage divider consists of two resistors in series with the nominal values, $R_1 = 15,000$ ohms and $R_2 = 5000$ ohms. The divider is connected across a voltage source of 24 volts, $\pm$ 5 per cent. Assuming no load, determine the worst case voltage levels across R_2, when (a) both resistors have a tolerance of 10 per cent, and (b) when both resistors have a tolerance of 1 per cent.

2. An airborne communication installation has a reliability requirement of 0.99 for 8 hr. The best available set has a demonstrated $MTBF$ of 300 hr. How could the mission requirement be met?

3. Redundant protection against spurious fuse blowing is required for a satellite circuit. It has been decided to use a four-ampere fuse (0.05 ohm) in series with a 0.40-ohm resistor. This combination is placed in parallel with another four-ampere fuse (0.05 ohm). The effect is to insure that 90 per cent of the current will pass through the fuse not in series with the resistor (Fuse A). Should that fuse blow, the current would be carried by the fuse (Fuse B) and resistor line (with a voltage drop of 1.0575 volts at the nominal 2.35 amperes, which the circuit can tolerate). The resistor thus may be considered as a failure-sensing and switching device. Determine the probability of failure when (a) in the absence of an over-current, Fuse A blows and then Fuse B blows, and (b) in the presence of an over-current, Fuse A does not blow, or Fuse A blows and Fuse B does not blow. (Fuse A must blow before Fuse B can blow since Fuse A is carrying a current load nine times greater than that of Fuse B.)

Sampling Procedures

13.1 Introduction

If all items in a given lot were identical, inspection (or testing) of any one item would suffice to describe the entire lot. In most cases, however, each item will differ from every other in some degree, even though they are manufactured by the same process. This does not suggest that *all* items in a given lot should be inspected. Inspection on a 100 per cent basis would be expensive, and could provide only the illusion of complete protection. Usually, it is far more economical and effective to examine a relatively few items (a sample) and to draw inferences from this limited investigation concerning the quality of the entire lot. Statistical methods for drawing such inferences are well developed. These methods provide a means for measuring the risk or degree of uncertainty associated with an inference, and for expressing the risk in terms of probability statements.

13.2 Test Classifications

The manner in which data are collected influences design of the reliability test procedure. For example, the most common procedures are

1. *Testing by attributes*, in which a device is classified quantitatively—that is, as a success or failure; or

2. *Testing by variables*, in which the output of a device (e.g., amplifier gain) is measured along a continuous numerical scale and is described in terms of its position along that scale.

Life tests can therefore be designed in which data (either attributes or variables) are collected in an ordered way along a continuous time scale for the specified test period.

13.3 Hypothesis Testing

Hypothesis testing forms the statistical basis of a reliability acceptance plan. In reliability testing, the hypothesis under test (or null hypothesis) is that the submitted lot conforms to the reliability requirement. In this discussion, H_0 will represent the null hypothesis and H_1 the alternative hypothesis that the reliability of the lot is at some unacceptable level. Table 13.1 shows the interrelationships among the consumer's and producer's risks,* the test decisions, and the true situation.

Table 13.1 RELATIONSHIPS BETWEEN TEST DECISION AND TRUE SITUATION

Test Decision	Actual Situation			
	H_0 True		H_1 True	
	Decision	Probability of Making Decision	Decision	Probability of Making Decision
Accept H_0	Right	$1 - \alpha$	Wrong	β (Consumer's risk)
Reject H_0	Wrong	α (Producer's risk)	Right	$1 - \beta$

Suppose, for example, that out of a population of 100 transistors, a random sample† of 20 is picked for estimation of the quality of the total population. The values of one property will determine the acceptability of the product. Several results are possible:

1. All 20 of the sample transistors may test good, yet all of the remaining 980 in the population may be defective;

*These terms are defined on pages 471-72.
†A random sample is one made up of items from a population in which every item has an equal probability of selection.

2. All 20 of the sample may test defective, yet all of the remaining 980 may be good;

3. Some of the sample may be defective and some of the population may be defective.

Although possibilities 1 and 2 could conceivably occur, the usual situation will be 3, in which an estimate of the probable quality of the entire population must be based on the known proportion of defective items in the sample.

Procedures based on attributes inspection have received most of the attention given the various techniques for acceptance sampling. Before some of these procedures are presented herein, the subject of sampling errors and the calculation of their associated risks will be briefly discussed.

A single-sampling attributes inspection plan is defined by a sample size (n) and an acceptance number (c). For example, where $n = 20$ and $c = 1$, a random sample of 20 items is drawn from each lot, and the lot is accepted if the number of defectives in the sample is less than or equal to one. If the number of defectives is greater than one, the lot is rejected.

Statistical methods are available for determining the probability that a lot having a given percentage of defective items will be accepted, depending on the values of n and c for a given plan. A plot of these probabilities for lots having percentages of defectives ranging from zero to one hundred yields the operating characteristic (O.C.) curve.

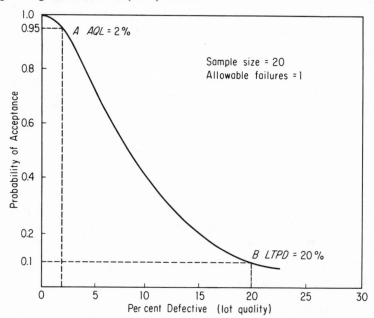

Fig. 13.1 Operating characteristic curve for a hypothetical sampling plan.

In this discussion, sampling will be assumed to be from large lots, unless stated otherwise. Accordingly, the defective items in the sample can be assumed to be binomially distributed. When the sample and lot sizes are small, the binomial approximation becomes invalid. In the latter case, the probabilities of acceptance are best computed by hypergeometric distribution methods, which take into account the actual value of lot size.

Figure 13.1 is an O.C. curve for a transistor sampling plan in which only 1 failure in a random sample of 20 is permitted. The abscissa shows lot quality in terms of the percentage of defective items produced. The ordinate shows the probability that the lot will be accepted, a condition requiring that fewer than 2 of the items in the sample of 20 be defective.

Figure 13.1 shows that if a lot has no defectives, no samples drawn from it will be defective, and the probability of acceptance as indicated by the O.C. curve is 1.0. If, at the other extreme, a lot has 100 per cent defectives, samples drawn from it will contain nothing but defectives, and its probability of acceptance is zero. Generally, the proportion of defective units inherent in a lot will be somewhere between 0 per cent and 100 per cent. In this case, the probability of acceptance of a sample drawn from this lot will be somewhere between 0 and 1. If the lot has excellent quality—a small percentage of defectives—the probability of acceptance will be high. If the lot quality is poor, its probability of acceptance based on sampling will be correspondingly low.

These probabilities of acceptance of the O.C. curve represent the proportions of lots that will be accepted in the long run. Two points on the curve of Figure 13.1 are of particular interest. *Point A* is referred to as the acceptable quality level (*AQL*) of the lot—2 per cent in this case. The *AQL* is usually interpreted to be that lot quality for which the probability of acceptance is 0.95. The manufacturer's process average is frequently chosen as this quality level. The 0.95 value of *AQL* means that if the manufacturer is producing lots exactly 2 per cent defective, then an acceptance test where $n = 20$ and $c = 1$ will yield an average of 19 out of 20 acceptable lots (95 per cent). The producer's risk (α) of having good lots rejected is therefore 5 per cent in this case.

If lots of lower quality are submitted, fewer will be accepted, and the producer will be forced to screen the rejected lots and either scrap them or sell them elsewhere. If lots of higher quality than necessary are submitted, the manufacturing costs may also be higher than necessary. Thus, prospective suppliers need to know the *AQL* of the O.C. curve in order to plan production and to set prices in bidding for contracts.

In order to assess the risk of accepting bad lots, it is customary to specify the quality of lots for which there is a 10 per cent probability of acceptance. This quality level (*Point B* on the curve of Figure 13.1) is referred to as the lot quality protection (*LQP*) or lot tolerance per cent defective (*LTPD*). The sampling plan shown in the figure allows the customer a 10 per cent risk (β) of accepting lots containing exactly 20 per cent defectives.

13.4 Selection of O.C. Curve

The producer's chief concern is to make sure that only a small proportion of good lots are rejected. The consumer's prime interest is in making certain that only a small proportion of bad lots are accepted. Thus, the producer is instrumental in setting the value of AQL for a given rejection risk, usually 0.05—a probability of acceptance of 0.95. Any sampling plan whose O.C. curve passes through this point would therefore protect the producer's interests, since any lot he submits containing quality better than this will be accepted at least 95 per cent of the time. Similarly, the consumer is the main force behind the specifying of the value for the $LTPD$, which is usually taken to be 0.10. Any sampling plan that has an O.C. curve passing through this point will satisfy the consumer, since any lot having more per cent defectives than the chosen value of $LTPD$ will be rejected at least 90 per cent of the time.

To satisfy both producer and consumer, the O.C. curve of a sampling plan must pass through both of the foregoing points. Many tables are available from which sampling plans corresponding to various values of AQL, $LTPD$, or both can be derived; such tables will be discussed later. As previously stated, the producer's risk is usually taken as 0.05 and the consumer's risk as 0.10. Although these values may be lowered either singly or collectively for added assurance, the consequence is an increase in the sample size. However, a producer's risk of 0.1 is often specified for the establishing of sampling procedures for expensive components or systems.

The O.C. curve (acceptance plan) that is best for one application may not be best for another. The basic problem is to choose the inspection plan in such a way as to minimize its total cost. The cost of an inspection plan is a function of the amount of discrimination provided by the plan, the amount of inspection required, the extent to which the plan encourages the producer to submit high-quality lots, and the ease with which the plan can be administered. Associated with the plan are requirements for quality and the relative amount of inspection. Where the inspection cost is low and the acceptance of a large percentage of defective items would be a serious disadvantage, more inspection is indicated than in a situation where the inspection cost is high and the acceptance of some defective items does not pose a critical situation. Perhaps the most difficult situation is that in which the cost of inspection is high and the acceptance of defective items is a serious problem.

Thus, in the selection of a plan, much depends on the nature of the intended application. Figure 13.2, based on information taken from a military specification, shows how plans will vary depending on the property under inspection.

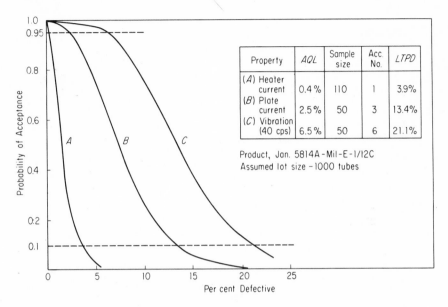

Product, Jan. 5814A - Mil - E - 1/12C
Assumed lot size - 1000 tubes

Property	AQL	Sample size	Acc. No.	LTPD
(A) Heater current	0.4%	110	1	3.9%
(B) Plate current	2.5%	50	3	13.4%
(C) Vibration (40 cps)	6.5%	50	6	21.1%

Fig. 13.2 Typical examples of O.C. curves.

13.5   Single Sampling

In single sampling, a decision is made to accept or reject a collection of devices on the basis of tests conducted on one sampling from that collection. The lot is accepted if a random sample of items, tested for a specified number of hours, produces fewer than a fixed number of failures. The unknown proportion, $R(t)$, of successful items in the lot is denoted by

$$R(t) = S/N \tag{13.1}$$

where

S = the actual, but unknown number of successful items in the lot;

N = the total lot size.

Based on the hypergeometric distribution, the probability of acceptance, $L[R(t)]$, of the lot is

$$L[R(t)] = P\{x \le c\} = \sum_{x=0}^{c} \frac{\binom{N-S}{x} \binom{S}{n-x}}{\binom{N}{n}} \tag{13.2}$$

for

$$0 \leq x \leq n, N > 0, 0 < n \leq N, 0 < S \leq N,*$$

where $P\{x<c\}$ = the probability of c or fewer failures in a random sample of n items; x = the actual number of failures.

The O.C. curve for this sampling plan can be established by assuming different values of S in Equation 13.2 and solving for $L[R(t)]$; the latter quantity is plotted as a function of S.

> **Example 1.** A lot of 10 items is submitted for acceptance tests. Five of the items are randomly selected for testing, with the requirement that none of them fail during the test period. It is assumed that 9 items of this particular lot will meet test requirements. What is the probability that the lot will be accepted?
>
> The given values are $N = 10$, $n = 5$, $c = 0$, and $S = 9$. Substituting these values in Equation 13.2,
>
> $$L[R(t)] = P\{x \leq 0\} = \sum_{x=0}^{0} \frac{\binom{10-9}{0}\binom{9}{5-0}}{\binom{10}{5}}$$
>
> $$= \frac{1!}{0!1!} \times \frac{(9!/5!4!)}{(10!/5!5!)}$$
>
> $$L[R(t)] = 0.50.$$

Tables 13.2 through 13.4 list representative percentages of the AQL and $LTPD$ for various combinations of (N, n, c). In these tables, the AQL is that percentage of the lot which can be defective while still yielding a probability of 0.95 that the lot will be accepted. The $LTPD$ is the percentage defective which will lead to a probability of 0.10 that the lot will be accepted. For example, if $N = 10$, $n = 2$, and $c = 0$, a lot 2.2 per cent defective will be accepted 95 per cent of the time; a lot 65 per cent defective will be accepted 10 per cent of the time.

The values for the AQL and $LTPD$ in Tables 13.2 through 13.4 can be converted to a reliability figure through the equation given in the tables. $R(t)$ in this case represents the proportion of the lot which survives a test of t hr.

> **Example 2.** A lot of 100 items is submitted for acceptance tests, with all other conditions the same as for Example 1 ($n = 5$, $c = 0$, $S = 0$). What is the reliability associated with a 0.95 probability of accepting the lot?

*An expression $\binom{N}{k}$ is equivalent to $\dfrac{N!}{k!(N-k)!}$

Table 13.2 Hypergeometric Sampling Plans for Small Lots (c = 0)

(N = Lot Size, n = Sample Size, c = Acceptance Number)

n	10 AQL	10 LTPD	20 AQL	20 LTPD	30 AQL	30 LTPD	40 AQL	40 LTPD	50 AQL	50 LTPD	60 AQL	60 LTPD	80 AQL	80 LTPD	100 AQL	100 LTPD	120 AQL	120 LTPD	150 AQL	150 LTPD	160 AQL	160 LTPD	200 AQL	200 LTPD	∞ AQL	∞ LTPD
2	2.2	65	2.5	66	2.5	67	2.5	67	2.5	67	2.5	68	2.5	68	2.5	68	2.5	68	2.5	68	2.5	68	2.5	68	2.5	68
4	1.2	36	1.2	40	1.2	42	1.2	42	1.3	42	1.3	43	1.3	43	1.3	43	1.3	43	1.3	43	1.3	44	1.3	44	1.3	44
5	1.0	29	1.0	33	1.0	34	1.0	35	1.0	35	1.0	35	1.0	36	1.0	36	1.0	37	1.0	37	1.0	37	1.0	37	1.0	37
8	0.5	15	0.6	20	0.6	22	0.6	23	0.6	23	0.6	23	0.6	24	0.7	24	0.7	24	0.7	24	0.7	24	0.7	25	0.7	25
10			0.4	15	0.5	17	0.5	19	0.5	19	0.5	19	0.5	20	0.5	20	0.5	20	0.5	20	0.5	20	0.5	20	0.5	21
16			0.2	6.9	0.25	10	0.25	11	0.3	12	0.3	12	0.3	12	0.3	13	0.3	13	0.3	13	0.3	13	0.3	13	0.3	13
20					0.2	6.8	0.2	8.0	0.25	9.0	0.25	9.0	0.25	9.4	0.25	10	0.25	10	0.25	10	0.25	10	0.25	11	0.3	11
25					0.15	4.3	0.15	5.7	0.2	6.9	0.2	6.9	0.2	7.4	0.2	7.5	0.2	7.6	0.2	7.7	0.2	7.8	0.2	7.9	0.2	8.8
32							0.1	3.7	0.1	5.0	0.1	5.0	0.1	5.5	0.1	5.4	0.15	6.0	0.15	6.2	0.15	6.3	0.15	6.3	0.15	6.9
40									0.1	3.4	0.1	3.4	0.1	4.0	0.1	4.5	0.1	4.6	0.1	4.9	0.1	5.0	0.15	5.0	0.15	5.6
50											0.1	2.3	0.1	2.9	0.10	3.3	0.1	3.5	0.10	3.7	0.1	3.7	0.10	3.9	0.10	4.6
64													0.08	1.7	0.08	2.2	0.08	2.5	0.08	2.7	0.08	2.8	0.08	2.9	0.08	3.6
80															0.07	1.5	0.07	1.7	0.07	2.0	0.07	2.1	0.06	2.2	0.06	2.9
100																	0.05	1.1	0.05	1.5	0.05	1.5	0.05	1.7	0.05	2.3
125																			0.04	0.8	0.04	0.9	0.04	1.2	0.04	1.8
128																			0.04	0.8	0.04	0.9	0.04	1.1	0.04	1.8
160																							0.03	0.7	0.03	1.4

These values can be converted to a reliability figure by the relationship

$$R(t) = 1 - \frac{AQL}{100}$$

SOURCE: *Acceptance Sampling Procedures for Small Lots*, JEDEC JT-11.3 Committee, Electronic Industries Association, New York, N.Y. (1963).

Table 13.3 Hypergeometric Sampling Plans for Small Lots (c = 1)

(N = Lot Size, n = Sample Size, c = Acceptance Number)

N →	10		20		30		40		50		60		80		100		120		150		160		200		∞	
n	AQL	LTPD	AQL	LTPD	AQL	LTPD	AQL	LTPD	AQL	LTPD	AQL	LTPD	AQL	LTPD	AQL	LTPD	AQL	LTPD	AQL	LTPD	AQL	LTPD	AQL	LTPD	AQL	LTPD
2	27	95	24	95	24	95	23	95	23	95	23	95	23	95	23	95	23	95	22	95	22	95	22	95	22	95
4	15	62	12	66	12	66	11	67	11	67	10	67	10	67	10	67	10	67	9.8	67	9.7	68	9.7	68	10.0	70
5	13	51	10	55	8.8	56	8.5	57	8.4	57	8.1	58	7.9	58	7.6	58	7.5	58	7.5	58	7.5	58	7.4	58	7.8	58
8	11	28	7.2	35	6.2	38	5.8	38	5.6	39	5.0	39	4.9	39	4.5	39	4.3	39	4.3	40	4.2	40	4.2	40	4.2	42
10			6.2	30	5.0	30	4.6	31	4.2	32	4.2	32	4.2	32	3.9	33	3.5	33	3.3	33	3.3	33	3.3	33	3.3	34
16			5.6	15	4.2	18	3.8	18	3.4	20	3.0	20	2.9	21	2.6	21	2.5	21	2.3	21	2.3	22	2.2	22	2.2	22.5
20					4.0	13	3.2	15	2.8	16	2.5	16	2.4	16	2.3	16	2.1	17	2.0	17	2.0	17	2.0	18	2.0	18
25					3.8	9.2	3.1	11	2.5	12	2.2	13	2.0	13	1.8	13	1.7	13	1.6	14	1.6	14	1.6	14	1.6	15
32							3.1	7.4	2.4	8.2	2.1	9.0	1.8	9.9	1.6	10	1.5	10.5	1.3	11	1.3	11	1.3	11	1.3	12
40									2.4	5.9	2.1	6.8	1.6	7.6	1.4	7.8	1.3	8.2	1.3	8.3	1.2	8.4	1.2	8.6	1.2	9.2
50											1.7	4.6	1.4	5.6	1.2	6.1	1.2	6.4	1.0	6.5	0.9	6.7	0.9	6.7	0.71	7.8
64													1.3	3.8	1.1	4.4	1.0	4.7	0.8	5.0	0.8	5.0	0.7	5.2	0.56	6.1
80															1.1	3.0	1.0	3.4	0.8	3.7	0.7	3.8	0.6	4.0	0.44	4.9
100																	0.9	2.5	0.7	2.8	0.7	2.8	0.6	3.0	0.36	3.9
125																			0.7	1.9	0.7	2.0	0.5	2.2	0.28	3.1
128																			0.7	1.7	0.7	1.9	0.5	2.2	0.28	3.0
160																							0.5	1.5	0.22	2.4

These values can be converted to a reliability figure by the relationship

$$R(t) = 1 - \frac{AQL}{100}$$

SOURCE: *Acceptance Sampling Procedures for Small Lots*, JEDEC JT-11.3 Committee, Electronic Industries Association, New York, N.Y. (1963).

Table 13.4 HYPERGEOMETRIC SAMPLING PLANS FOR SMALL LOTS ($c = 2$)

(N = Lot Size, n = Sample Size, c = Acceptance Number)

n	10		20		30		40		50		60		80		100		120		150		160		200		∞	
	AQL	LTPD	AQL	LTPD	AQL	LTPD	AQL	LTPD	AQL	LTPD	AQL	LTPD	AQL	LTPD	AQL	LTPD	AQL	LTPD	AQL	LTPD	AQL	LTPD	AQL	LTPD	AQL	LTPD
4	33	82	28	83	27	84	27	85	27	85	26	85	26	85	26	86	26	86	25	86	25	86	25	86	25	86
5	27	69	23	73	21	74	20	74	20	74	20	75	20	75	19	75	19	75	19	75	19	75	19	75	19	75
8	22	42	15	49	14	49	13	52	13	52	13	52	12	53	12	53	12	53	11	53	11	53	11	53	11	53
10			13	39	13	42	11	42	10	43	10	43	9.6	43	9.2	44	9.1	44	8.9	44	8.9	44	8.7	44	8.3	45
16			11	22	8.6	25	6.9	27	6.8	27	6.4	27	6.0	28	6.0	29	5.9	29	5.9	29	5.7	29	5.5	30	5.2	30
20					7.7	19	6.2	21	5.9	22	5.6	22	5.1	23	4.8	23	4.8	23	4.6	23	4.5	24	4.5	24	4.1	25
25					7.4	13	6.0	16	4.9	17	4.5	17	4.3	18	4.1	18	3.9	18	3.7	18	3.7	19	3.7	19	3.5	20
32							5.5	11	4.8	12	4.3	13	3.6	14	3.4	14	3.2	14	3.0	14.5	3.0	15	2.9	15	2.5	16
40									4.6	8.9	3.9	9.8	3.1	11	2.8	12	2.6	12	2.4	12	2.4	12	2.3	12	2.2	11.5
50											3.5	6.9	2.8	8.1	2.4	8.4	2.3	8.6	2.1	9.0	2.1	9.3	2.0	9.5	1.6	11
64													2.6	5.7	2.2	6.2	2.0	6.6	1.8	7.1	1.7	7.1	1.6	7.4	1.3	8.3
80															2.1	4.5	1.8	4.9	1.6	5.4	1.5	5.4	1.4	5.6	1.0	6.6
100																	1.8	3.5	1.4	3.9	1.4	4.0	1.2	4.4	0.82	5.3
125																			1.4	2.8	1.3	2.9	1.1	3.3	0.65	4.3
128																			1.4	2.6	1.3	2.9	1.1	3.2	0.64	4.1
160																							1.1	2.3	0.51	3.3

These values can be converted to a reliability figure by the relationship

$$R(t) = 1 - \frac{AQL}{100}$$

SOURCE: *Acceptance Sampling Procedures for Small Lots*, JEDEC JT-11.3 Committee, Electronic Industries Association, New York, N.Y. (1963).

From Table 13.2, $AQL = 1.0$. This value can be converted to a reliability figure according to the equation accompanying Table 13.2.

$$R(t) = 1 - \frac{AQL}{100} = 1 - \frac{1}{100} = 0.99.$$

The value of 0.95 for $L[R(t)]$ can now be confirmed. From Equation 13.1

$$R(t) = S/N; \; S = N \cdot R(t) = (100)(0.99) = 99.$$

The values $N = 100$, $S = 99$, and $n = 5$ are then substituted into Equation 13.2 to yield $L[R(t)] = 0.95$.

As the lot size increases ($N \to \infty$), and for fixed values of S/N, the hypergeometric distribution of Equation 13.2 converges to the binomial, which can be used as an approximation when the sample constitutes less than 10 per cent of the population; i.e., when $n/N < 0.1$. Then,

$$L[R(t)] = P\{x \leq c\} \approx \sum_{x=0}^{c} \binom{n}{x} [1 - R(t)]^x [R(t)]^{n-x}, \tag{13.3}$$

for

$$\frac{n}{N} < 0.1, \; 0 \leq x \leq n, \; 0 > n < N, \; 0 \leq R(t) \leq 1.$$

The O.C. curve can now be generated by assuming different values of S/N and solving for $L[R(t)]$, or through use of Table 13.5* pages 478-481.

Example 3. A lot of 200 items is submitted for acceptance tests. Five of the items are randomly selected for testing, with the requirement that none of them fail during the test period. It is assumed that $R(t)$, or S/N, equals 0.9. What is the probability that the lot of 200 items will be accepted?

Substituting the given values $N = 200$, $n = 5$, $c = 0$, and $R(t) = 0.9$ into Equation 13.3,

$$L[R(t)] = P\{x \leq 0\} \approx \sum_{x=0}^{0} \binom{5}{0} [1 - 0.9]^0 \, [0.9]^{5-0}$$

$$= \frac{5!}{0!5!} (0.1)^0 \, (0.9)^5$$

$$L[R(t)] = 0.59.$$

The ∞ columns of Tables 13.2 through 13.4 list the 0.95 and 0.10 proba-

---

Reliability Analysis Data for Systems and Component Design Engineers, General Electric Company, TRA-873-74 (July 15, 1961).

Table 13.5 PROBABILITY OF SURVIVAL OF AT LEAST S REQUIRED ELEMENTS FROM A
SET OF N REDUNDANT ELEMENTS OF RELIABILITY P

$$P(s,n) = \sum_{x=s}^{n} \binom{n}{x} p^x (1-p)^{n-x} \quad \text{where} \quad \binom{n}{x} = \frac{n!}{(n-x)!x!}.$$

n	p = s	0.99	0.98	0.97	0.96	0.95	0.94	0.93	0.92	0.91	0.90
2	1	0.999900	0.999600	0.999100	0.998400	0.997500	0.996400	0.995100	0.993600	0.991900	0.990000
	2	0.980100	0.960400	0.940900	0.921600	0.902500	0.883600	0.864900	0.846400	0.828100	0.810000
3	1	0.999999	0.999992	0.999973	0.999936	0.999875	0.999784	0.999657	0.999488	0.999271	0.999000
	2	0.999702	0.998816	0.997354	0.995328	0.992750	0.989632	0.985986	0.981824	0.977158	0.972000
	3	0.970299	0.941192	0.912673	0.884736	0.857375	0.830584	0.804357	0.778688	0.753571	0.729000
4	1	1.0	1.0	0.999999	0.999998	0.999994	0.999987	0.999976	0.999959	0.999935	0.999900
	2	0.999996	0.999969	0.999895	0.999752	0.999519	0.999175	0.998700	0.998075	0.997281	0.996300
	3	0.999408	0.997664	0.994814	0.990905	0.985981	0.980089	0.973272	0.965573	0.957035	0.947700
	4	0.960596	0.922368	0.885293	0.849347	0.814506	0.780749	0.748052	0.716393	0.685750	0.656100
5	1	1.0	1.0	1.0	1.0	1.0	0.999999	0.999998	0.999997	0.999994	0.999990
	2	1.0	1.0	0.999996	0.999988	0.999970	0.999939	0.999887	0.999809	0.999696	0.999540
	3	0.999991	0.999923	0.999742	0.999398	0.998842	0.998030	0.996920	0.995475	0.993659	0.991440
	4	0.999020	0.996158	0.991528	0.985242	0.977408	0.968129	0.957507	0.945639	0.932620	0.918540
	5	0.950990	0.903921	0.858734	0.815373	0.773781	0.733904	0.695688	0.659082	0.624032	0.590490
6	1	1.0	1.0	1.0	1.0	1.0	1.0	1.0	1.0	1.0	0.999999
	2	1.0	1.0	1.0	1.0	0.999999	0.999996	0.999991	0.999982	0.999968	0.999945
	3	1.0	0.999998	0.999989	0.999965	0.999914	0.999824	0.999679	0.999462	0.999153	0.998730
	4	0.999981	0.999848	0.999496	0.998832	0.997770	0.996236	0.994161	0.991488	0.988165	0.984150
	5	0.998540	0.994313	0.987544	0.978448	0.967226	0.954075	0.939180	0.922714	0.904874	0.885735
	6	0.941480	0.885843	0.832972	0.782758	0.735092	0.689870	0.646990	0.606355	0.567869	0.531441
7	1	1.0	1.0	1.0	1.0	1.0	1.0	1.0	1.0	1.0	1.0
	2	1.0	1.0	1.0	1.0	1.0	1.0	0.999999	0.999999	0.999997	0.999994
	3	1.0	1.0	1.0	0.999998	0.999994	0.999985	0.999969	0.999940	0.999894	0.999824
	4	1.0	0.999995	0.999974	0.999919	0.999807	0.999609	0.999293	0.998824	0.998164	0.997272
	5	0.999967	0.999737	0.999137	0.998017	0.996243	0.993706	0.990313	0.985986	0.980667	0.974309
	6	0.997979	0.992144	0.982907	0.970620	0.955619	0.938223	0.918726	0.897405	0.874519	0.850306
	7	0.932065	0.868125	0.807983	0.751447	0.698337	0.648477	0.601701	0.557847	0.516761	0.478297
8	1	1.0	1.0	1.0	1.0	1.0	1.0	1.0	1.0	1.0	1.0
	2	1.0	1.0	1.0	1.0	1.0	1.0	1.0	1.0	1.0	1.0
	3	1.0	1.0	1.0	1.0	1.0	0.999999	0.999997	0.999994	0.999988	0.999977
	4	1.0	1.0	0.999999	0.999995	0.999985	0.999963	0.999922	0.999851	0.999738	0.999569
	5	1.0	0.999990	0.999949	0.999843	0.999628	0.999254	0.998664	0.997797	0.996589	0.994976
	6	0.999946	0.999585	0.998650	0.996921	0.994212	0.990377	0.985301	0.978900	0.971113	0.961908
	7	0.997310	0.989663	0.977659	0.961853	0.942755	0.920838	0.896534	0.870241	0.842320	0.813105
	8	0.922745	0.850763	0.783743	0.721390	0.663420	0.609569	0.559582	0.513219	0.470252	0.430467

SOURCE: General Electric Company, *Reliability Analysis
Data for Systems and Component Design Engineers*
TRA-873-74 (July 15, 1961). Reprinted by permission.

(continued)

Table **13.5** (continued)

n	p = s	0.99	0.98	0.97	0.96	0.95	0.94	0.93	0.92	0.91	0.90
9	1	1.0	1.0	1.0	1.0	1.0	1.0	1.0	1.0	1.0	1.0
	2	1.0	1.0	1.0	1.0	1.0	1.0	1.0	1.0	1.0	1.0
	3	1.0	1.0	1.0	1.0	1.0	1.0	1.0	1.0	0.999999	0.999997
	4	1.0	1.0	1.0	1.0	0.999999	0.999997	0.999992	0.999982	0.999965	0.999936
	5	1.0	1.0	0.999997	0.999989	0.999967	0.999920	0.999834	0.999687	0.999455	0.999109
	6	0.999999	0.999982	0.999910	0.999726	0.999358	0.998722	0.997729	0.996285	0.994296	0.991669
	7	0.999920	0.999386	0.998021	0.995518	0.991639	0.986205	0.979088	0.970207	0.959522	0.947028
	8	0.996564	0.986885	0.971842	0.952234	0.928789	0.902162	0.872948	0.841679	0.808834	0.774841
	9	0.913517	0.833748	0.760231	0.692534	0.630249	0.572995	0.520411	0.472161	0.427930	0.387420
10	1	1.0	1.0	1.0	1.0	1.0	1.0	1.0	1.0	1.0	1.0
	2	1.0	1.0	1.0	1.0	1.0	1.0	1.0	1.0	1.0	1.0
	3	1.0	1.0	1.0	1.0	1.0	1.0	1.0	1.0	1.0	1.0
	4	1.0	1.0	1.0	1.0	1.0	1.0	0.999999	0.999998	0.999996	0.999991
	5	1.0	1.0	1.0	1.0	0.999997	0.999992	0.999981	0.999959	0.999919	0.999853
	6	1.0	1.0	0.999995	0.999979	0.999937	0.999848	0.999686	0.999415	0.998991	0.998365
	7	0.999998	0.999970	0.999853	0.999558	0.998972	0.997971	0.996424	0.994199	0.991166	0.987205
	8	0.999886	0.999136	0.997235	0.993787	0.988497	0.981162	0.971658	0.959925	0.945960	0.929809
	9	0.995734	0.983822	0.965494	0.941846	0.913862	0.882412	0.848270	0.812118	0.774553	0.736099
	10	0.904382	0.817073	0.737424	0.664833	0.598737	0.538615	0.483982	0.434388	0.389416	0.348678

n	p = s	0.88	0.86	0.84	0.82	0.80	0.75	0.70	0.65	0.60	0.55	0.50
2	1	0.985600	0.980400	0.974400	0.967600	0.960000	0.937500	0.910000	0.877500	0.840000	0.797500	0.750000
	2	0.774400	0.739600	0.705600	0.672400	0.640000	0.562500	0.490000	0.422500	0.360000	0.302500	0.250000
3	1	0.998272	0.997256	0.995904	0.994168	0.992000	0.984375	0.973000	0.957125	0.936000	0.908875	0.875000
	2	0.960256	0.946688	0.931392	0.914464	0.896000	0.843750	0.784000	0.718250	0.648000	0.574750	0.500000
	3	0.681472	0.636056	0.592704	0.551368	0.512000	0.421875	0.343000	0.274625	0.216000	0.166375	0.125000
4	1	0.999793	0.999616	0.999345	0.998950	0.998400	0.996094	0.991900	0.984994	0.974400	0.958994	0.937500
	2	0.993710	0.990176	0.985582	0.979821	0.972800	0.949219	0.916300	0.873519	0.820800	0.758519	0.687500
	3	0.926802	0.903199	0.877202	0.849107	0.819200	0.738281	0.651700	0.562981	0.475200	0.390981	0.312500
	4	0.599695	0.547008	0.497871	0.452122	0.409600	0.316406	0.240100	0.178506	0.129600	0.091506	0.062500
5	1	0.999975	0.999946	0.999895	0.999811	0.999680	0.999023	0.997570	0.994748	0.989760	0.981547	0.968750
	2	0.999063	0.998294	0.997143	0.995507	0.993280	0.984375	0.969220	0.945978	0.912960	0.868780	0.812500
	3	0.985681	0.978000	0.968241	0.956293	0.942080	0.896484	0.836920	0.764831	0.682560	0.593127	0.500000
	4	0.887549	0.853333	0.816509	0.777649	0.737280	0.632812	0.528220	0.428415	0.336960	0.256217	0.187500
	5	0.527732	0.470427	0.418212	0.370740	0.327680	0.237305	0.168070	0.116029	0.077760	0.050328	0.031250

(continued)

Table 13.5 (continued)

n	p= s	0.88	0.86	0.84	0.82	0.80	0.75	0.70	0.65	0.60]	0.55	0.50
6	1	0.999997	0.999993	0.999983	0.999966	0.999936	0.999756	0.999271	0.998162	0.995904	0.991696	0.984375
	2	0.999866	0.999715	0.999455	0.999037	0.998400	0.995361	0.989065	0.987678	0.959040	0.930802	0.890625
	3	0.997457	0.995453	0.992519	0.988449	0.983040	0.962402	0.929530	0.882576	0.820800	0.744736	0.656250
	4	0.973905	0.960546	0.943964	0.924137	0.901120	0.830566	0.744310	0.647085	0.544320	0.441518	0.343750
	5	0.844371	0.799726	0.752781	0.704406	0.655360	0.533935	0.420175	0.319080	0.233280	0.163567	0.109375
	6	0.464404	0.404567	0.351298	0.304007	0.262144	0.177978	0.117649	0.075419	0.046656	0.027681	0.015625
7	1	1.0	0.999999	0.999998	0.999994	0.999987	0.999939	0.999781	0.999357	0.998362	0.996263	0.992188
	2	0.999981	0.999954	0.999899	0.999799	0.999629	0.998657	0.996209	0.990993	0.981158	0.964294	0.937500
	3	0.999577	0.999118	0.998345	0.997131	0.995328	0.987122	0.971205	0.944393	0.903744	0.847072	0.773438
	4	0.994631	0.990566	0.984750	0.976872	0.966656	0.929443	0.873964	0.800154	0.710208	0.608288	0.500000
	5	0.958361	0.938031	0.913375	0.884585	0.851968	0.756409	0.647069	0.532283	0.419904	0.316440	0.226562
	6	0.798775	0.744404	0.688544	0.632334	0.576717	0.444946	0.329417	0.233798	0.158630	0.102418	0.062500
	7	0.408676	0.347928	0.295090	0.249285	0.209715	0.133484	0.082354	0.049022	0.027994	0.015224	0.007812
8	1	1.0	1.0	1.0	0.999999	0.999998	0.999985	0.999934	0.999775	0.999345	0.998319	0.996094
	2	0.999998	0.999993	0.999982	0.999959	0.999916	0.999619	0.998710	0.996429	0.991480	0.981877	0.964844
	3	0.999933	0.999837	0.999650	0.999319	0.998769	0.995773	0.988708	0.974682	0.950193	0.911544	0.855469
	4	0.998983	0.997921	0.996170	0.993484	0.989594	0.972702	0.942032	0.893909	0.826329	0.739619	0.636719
	5	0.990279	0.983211	0.973330	0.960261	0.943719	0.886185	0.805896	0.706399	0.594086	0.476956	0.363281
	6	0.939211	0.910923	0.877402	0.839180	0.796918	0.678543	0.551774	0.427814	0.315394	0.220130	0.144531
	7	0.751963	0.688897	0.625591	0.563385	0.503316	0.367081	0.255298	0.169127	0.106376	0.063181	0.035156
	8	0.359634	0.299218	0.247876	0.204414	0.167772	0.100113	0.057648	0.031864	0.016796	0.008373	0.003906
9	1	1.0	1.0	1.0	1.0	1.0	0.999996	0.999981	0.999921	0.999738	0.999243	0.998047
	2	1.0	0.999999	0.999997	0.999992	0.999981	0.999893	0.999567	0.998604	0.996199	0.990920	0.980469
	3	0.999990	0.999971	0.999929	0.999844	0.999686	0.998657	0.995709	0.988818	0.974965	0.950227	0.910156
	4	0.999819	0.999569	0.999093	0.998268	0.996934	0.990006	0.974705	0.946412	0.900647	0.834178	0.746094
	5	0.997939	0.995862	0.992515	0.987504	0.980419	0.951073	0.901191	0.828281	0.733432	0.621421	0.500000
	6	0.984150	0.973091	0.957981	0.938466	0.914358	0.834274	0.729659	0.608894	0.482610	0.361385	0.253906
	7	0.916741	0.879840	0.837112	0.789537	0.738198	0.600677	0.462831	0.337273	0.231787	0.149503	0.089844
	8	0.704884	0.634342	0.565157	0.498770	0.436208	0.300339	0.196003	0.121085	0.070544	0.038518	0.019531
	9	0.316478	0.257327	0.208216	0.167619	0.134218	0.075085	0.040354	0.020712	0.010078	0.004605	0.001953
10	1	1.0	1.0	1.0	1.0	1.0	0.999999	0.999994	0.999973	0.999895	0.999660	0.999024
	2	1.0	1.0	1.0	0.999999	0.999996	0.999971	0.999856	0.999460	0.998322	0.995498	0.989258
	3	0.999999	0.999995	0.999986	0.999965	0.999922	0.999584	0.998410	0.995179	0.987706	0.972608	0.945313
	4	0.999969	0.999915	0.999795	0.999560	0.999136	0.996495	0.989408	0.973976	0.945238	0.898005	0.828125
	5	0.999593	0.999050	0.998041	0.996331	0.993631	0.980273	0.952651	0.905066	0.833761	0.738437	0.623047
	6	0.996284	0.992674	0.986990	0.978677	0.967207	0.921873	0.849732	0.751495	0.633103	0.504405	0.376953
	7	0.976061	0.960036	0.938642	0.911659	0.879126	0.775875	0.649611	0.513827	0.382281	0.266038	0.171875
	8	0.891318	0.845470	0.793599	0.737199	0.677800	0.525593	0.382783	0.261607	0.167290	0.099560	0.054687
	9	0.658275	0.581560	0.508046	0.439163	0.375810	0.244025	0.149308	0.085954	0.046357	0.023257	0.010742
	10	0.278501	0.221301	0.174901	0.137448	0.107374	0.056313	0.028248	0.013463	0.006047	0.002533	0.000977

bilities of acceptance for values of $1 - S/N$, or $1 - R(t)$, for various combinations of (n, c).

If $R(t)$ is ≥ 0.9 and the sample contains less than 10 per cent of the lot, the hypergeometric may be replaced by the Poisson distribution with sufficient accuracy:

$$L[R(t)] = P\{x \leq c\} \approx e^{-n[1-R(t)]} \sum_{x=0}^{c} \frac{(n[1-R(t)]^n}{x!}, \tag{13.4}$$

for

$$R(t) \geq 0.9 \quad \text{and} \quad \frac{n}{N} < 0.1.$$

In this case the O.C. curve is generated by assuming various values of n $(1 - S/N)$ or $n[1 - R(t)]$, and solving for $L[R(t)]$; or by using Figure 13.3 or tables of the Poisson distribution.*

Example 4. Given that $N = 200$, $n = 5$, and $c = 0$, and assuming a value for $n[1 - R(t)]$ of 0.5, solution of Equation 13.4 yields a probability of 0.61 that the lot will be accepted.

MIL-STD-105†

Military Standard 105, whose sampling plans are indexed by AQL only, has been extensively used by the Armed Services in procurement operations. This document is also used by industry for controlling its own purchases and for in-process and end-item inspection of its manufactured products. This standard provides tables which list the sample size and acceptance number for corresponding lot sizes and selected values of AQL. The O.C. curves corresponding to each plan are included in the document.

MIL-STD-105, however, has met with some recent criticism, due chiefly to the facts that (1) the AQL is not specifically tied to a probability of acceptance of 0.95, and may vary from 0.85 to 0.995; and (2) the $LTPD$, for a given AQL, varies as the sample size varies.

As emphasis on reliability increased, the Department of Defense found it necessary to provide additional information in MIL-STD-105 for isolated lot protection in the form of a table giving $LTPD$ values associated with each sampling plan. Table 13.6 lists some frequently used MIL-STD-105 sampling plans with their equivalent $LTPD$ values.

*One source of these tables is *Poisson's Exponential Binomial Limit*, Molina, E.C., Bell Telephone Series, D. Van Nostrand Company, Inc. (1942).

†The latest issue in this series is Military Standard 105D, *Sampling Procedures and Tables for Inspection by Attributes.*

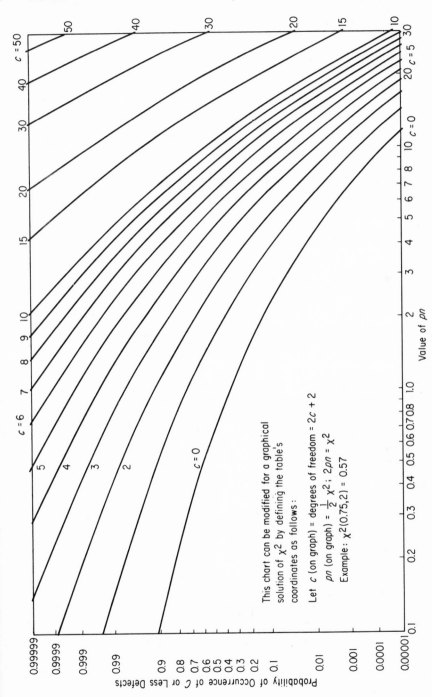

This chart can be modified for a graphical
solution of χ^2 by defining the table's
coordinates as follows:

Let c (on graph) = degrees of freedom = $2c + 2$

pn (on graph) = $\frac{1}{2}\chi^2$; $2pn = \chi^2$

Example: $\chi^2(0.75, 2) = 0.57$

Fig. 13.3 Cumulative probability curves—Poisson exponential. (SOURCE: H. F. Dodge and H. G. Romig, *Sampling Inspection Tables.*
New York: John Wiley & Sons, Inc., 1944. Reprinted by permission.)

Table 13.6 Some Frequently Used MIL-STD-105 Sampling Plans with Equivalent *LTPD* Values

Inspection Level	Lot Size	Sample Size Code Letter	Sample Size	AQL = 0.4 LTPD	Ac. No.	AQL = 0.65 LTPD	Ac. No.	AQL = 1.0 LTPD	Ac. No.	AQL = 2.5 LTPD	Ac. No.	AQL = 4.0 LTPD	Ac. No.	AQL = 6.5 LTPD	Ac. No.
L4	100	(A)	2	7	0	12	0	17	0	30	0	40	0	65	0
	1,000	(D)	7	6	0	11	0	15	0	27	0	32	0	37	1
	20,000	(G)	25	5	0	9	0	12	1	14	1	20	2	26	3
L6	100	(B)	3	7	0	11	0	15	0	28	0	35	0	56	0
	1,000	(F)	15	6	0	9	0	13	0	16	1	25	1	32	2
	20,000	(I)	50	5	0	6	1	7	1	13	3	16	4	23	6
L8	100	(D)	7	6	0	11	0	15	0	27	0	32	0	37	1
	1,000	(H)	35	6	0	7	0	9	1	13	2	18	3	25	5
	20,000	(K)	110	3	1	4	2	6	3	9	6	12	8	17	12
I	100	(D)	7	6	0	11	0	15	0	27	0	32	0	37	1
	1,000	(I)	50	5	0	6	1	7	1	13	3	16	4	23	6
	20,000	(N)	300	2	4	3	5	3	7	6	14	9	20	13	32
II	100	(F)	15	6	0	9	0	13	0	16	1	25	1	32	2
	1,000	(K)	110	3	1	4	2	6	3	9	6	12	8	17	12
	20,000	(N)	300	2	4	3	5	3	7	6	14	9	20	13	32
III	100	(H)	35	6	0	7	0	9	1	13	2	18	3	25	5
	1,000	(L)	150	3	2	4	3	5	4	8	8	11	11	16	17
	20,000	(O)	450	2	5	2	7	3	10	6	20	8	29	12	43

Note: Rejection Number equals Acceptance Number (Ac. No.) plus one.

Example 5. Relay specification MIL-R-5757D requires a 2.5 per cent *AQL* for certain Group B tests at an inspection level of L8. (The "inspection level" designates the relative amount of protection or discrimination selected for the property to be tested, and is obtained by striking a favorable balance between the cost of inspection and the cost of accepting defective product; see MIL-STD-105.) What is the equivalent *LTPD* for lot sizes of 100, 1000, and 20,000 relays?

MIL-STD-105 gives the respective sample sizes and acceptance numbers for the various lot sizes. The salient features of this specification are summarized in Table 13.7. The *LTPD* values can be obtained from Table 13.6 or from Table VI-A in MIL-STD-105D.

Table 13.7 CHARACTERISTICS OF A RELAY SAMPLING PLAN

Lot Size (Assumed)	AQL (Given)	Inspection Level (Given)	Sample Size Code Letter MIL-STD-105B	Sample Size Given by MIL-STD-105B	Accept. No. Given by MIL-STD-105B	LTPD
100	2.5%	L-8	D	7	0	27%
1,000	2.5%	L-8	H	35	2	13%
20,000	2.5%	L-8	K	110	6	9.3%

An *LTPD*-Oriented Sampling Plan

Sobel and Tischendorf developed plans indexed by the *LTPD* for the exponential distribution.† Their format was to present, for given acceptance numbers, minimum single-sample sizes to assure a customer that his maximum risk of accepting quality of mean life as poor as that specified is $(1 - P^*)$, where P^* is the consumer's risk.

Figures 13.4 and 13.5 are based on Sobel and Tischendorf's work, illustrating the relationship between sample size, number of defective items allowed in the sample, and the probability of acceptance of lots having various percentages of defective items for a producer's risk of 0.05 and a consumer's risk of 0.10. On the basis of *AQL* and *LTPD* values determined from these figures, it is possible to approximate the general shape of the true O.C. curve, and thus to estimate the discrimination and protection afforded by the test plan. Table 13.8 gives specific sample-size and acceptance-number requirements for several discrete levels of *LTPD* values.

†M. Sobel and J. Tischendorf, "Acceptance Sampling With New Life Test Objectives," *Proceedings, Fifth National Symposium on Reliability and Quality Control*, Philadelphia, Pennsylvania (January 12-14, 1959).

Example 6. In Air Force capacitor specification MIL-C-26244, the Group C inspection requirements for tests of such factors as vibration, corrosion, temperature, and immersion cycling specify a sample size of 48 and a maximum of 2 failures. The nomographs show that the equivalent *AQL* is 1.5 per cent and the equivalent *LTPD* is slightly more than 10 per cent.

As a matter of interest, newer military specifications for semiconductor devices are indexed by *LTPD* rather than *AQL*; however, a maximum *AQL* can be specified through use of a "minimum rejection number."

Example 7. The military specification for the type 2N1016B power transistor has an *LTPD* requirement of 10 per cent for a number of eiectrical parameters, and a minimum rejection number of 4 transistors. From Figures 13.4 and 13.5, the maximum allowable *AQL* is approximately 2 per cent, and 65 test samples would be required. (Table 13.8 shows that the exact number of test samples would be 65.) Note that the rejection number equals the acceptance number plus one.

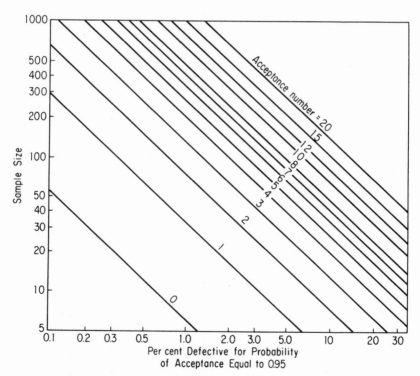

Fig. 13.4 Relationship between sample size, acceptance number, and lot quality, at the 95 per cent acceptance level on the O.C. curves.

Table 13.8 MINIMUM SIZE OF SAMPLES TO BE TESTED TO ASSURE, WITH A 90 PER CENT CONFIDENCE, A LOT TOLERANCE PER CENT DEFECTIVE NO GREATER THAN THE *LTPD* SPECIFIED

(The minimum quality required to accept, on the average, 19 of 20 lots is shown under the "Minimum Sample Sizes," for information.)

Each cell shows the minimum sample size (top) and the corresponding value λ (bottom).

Acceptance Number (A) ($R = A+1$)	\ Maximum Per Cent Defective (LTPD) (1): 0.1	0.2	0.3	0.5	0.7	1	1.5	2	3	5	7	10	15	20
0	2303 / 0.002	1152 / 0.005	770 / 0.007	461 / 0.01	238 / 0.02	231 / 0.02	153 / 0.03	116 / 0.04	76 / 0.07	45 / 0.11	32 / 0.16	22 / 0.23	15 / 0.34	11 / 0.46
1	3891 / 0.009	1946 / 0.018	1298 / 0.027	778 / 0.045	555 / 0.06	390 / 0.09	258 / 0.14	195 / 0.18	129 / 0.28	77 / 0.46	55 / 0.65	38 / 0.94	25 / 1.4	18 / 2.0
2	5323 / 0.015	2662 / 0.031	1777 / 0.046	1065 / 0.080	759 / 0.11	533 / 0.15	354 / 0.23	266 / 0.31	176 / 0.47	105 / 0.78	75 / 1.1	52 / 1.6	34 / 2.4	25 / 3.4
3	6681 / 0.018	3341 / 0.041	2228 / 0.061	1337 / 0.10	953 / 0.14	668 / 0.20	444 / 0.31	333 / 0.41	221 / 0.62	132 / 1.0	94 / 1.5	65 / 2.1	43 / 3.2	32 / 4.4
4	7994 / 0.025	3997 / 0.049	2667 / 0.074	1599 / 0.12	1140 / 0.17	798 / 0.25	531 / 0.37	398 / 0.50	265 / 0.75	158 / 1.3	113 / 1.8	78 / 2.6	52 / 3.9	38 / 5.3
5	9275 / 0.028	4638 / 0.056	3099 / 0.084	1855 / 0.14	1323 / 0.20	927 / 0.28	617 / 0.42	462 / 0.57	308 / 0.85	184 / 1.4	131 / 2.0	91 / 2.9	60 / 4.4	45 / 6.0
6	10533 / 0.031	5267 / 0.062	3515 / 0.093	2107 / 0.155	1503 / 0.22	1054 / 0.31	700 / 0.47	528 / 0.62	349 / 0.94	209 / 1.6	149 / 2.2	104 / 3.2	68 / 4.5	51 / 6.6
7	11771 / 0.034	5886 / 0.067	3931 / 0.101	2355 / 0.17	1680 / 0.24	1178 / 0.34	783 / 0.51	589 / 0.67	390 / 1.0	234 / 1.7	166 / 2.4	116 / 3.5	77 / 5.3	57 / 7.2
8	13995 / 0.036	6498 / 0.072	4334 / 0.108	2599 / 0.18	· 1854 / 0.25	1300 / 0.36	864 / 0.54	648 / 0.72	431 / 1.1	258 / 1.8	184 / 2.6	128 / 3.7	85 / 5.6	63 / 7.7
9	14206 / 0.038	7103 / 0.077	4739 / 0.114	2842 / 0.19	2027 / 0.27	1421 / 0.38	945 / 0.58	709 / 0.77	471 / 1.2	282 / 1.9	201 / 2.7	140 / 3.9	93 / 6.0	69 / 8.1
10	15407 / 0.04	7704 / 0.08	5147 / 0.120	3082 / 0.20	2199 / 0.28	1541 / 0.40	1025 / 0.60	770 / 0.80	511 / 1.2	306 / 2.0	218 / 2.9	152 / 4.1	100 / 6.3	75 / 8.4

(1) This table shall also be used for life testing. The life test failure rate lambda (λ) shall be defined as the *LTPD* per 1000 hr.

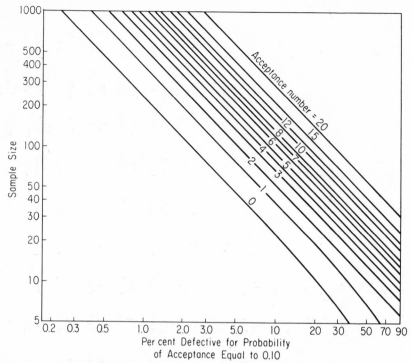

Fig. 13.5 Relationship between sample size, acceptance number, and lot quality, at the 10 per cent acceptance level on the O.C. curves.

13.6 Multiple Sampling

In multiple sampling, more than one sample may have to be tested before a decision to accept or reject the lot is reached, but the maximum number of samples and thus the maximum number of items to be tested is known. The conditions for applying the proper distribution for generating the O.C. curve are the same as for single sampling.

The most common form of multiple sampling is the double sampling plan. In this plan, a first sample, n_1, is tested. If c_1 or fewer failures occur, the lot is accepted. If the number of failures exceeds c_2, the lot is rejected. If the number of failures in sample n_1 exceeds c_1 but is not greater than c_2, a second sample, n_2, is tested. If the total number of failures in both samples is c_2 or fewer, the lot is accepted; otherwise it is rejected.

Denoting the number of failed items in the first and second samples by x_1 and x_2, respectively, the probability of acceptance, $L[R(t)]$, may be found as:

$$L[R(t)] = P\{x_1 \leq c_1 \quad \text{or} \quad x_1 + x_2 \leq c_2\}$$

$$= P\{x_1 \leq c_1\} + \sum_{x_1 = c_1 \pm 1}^{c_2} P\{x_1\} \, P\{x_2 \leq c_2 - x_1\} \tag{13.5}$$

where the probabilities $P\{x_1\}$ and $P\{x_2\}$ are given as hypergeometric distributions.* The conditions for applying the binomial or Poisson distributions are the same for generating the operating characteristic curve as for single sampling.

Multiple sampling, in comparison with single sampling, permits a saving in the average number of items tested while giving the same reliability assurance. The amount of saving depends on the reliability of the product submitted. For lots of very low or very high reliability, the average saving is larger than for those of intermediate reliability. However, this saving should be assessed in terms of the greater administrative difficulties (variable test loads, personnel requirements, etc.) in applying the multiple sampling plan.

13.7 Sequential Sampling

Sequential sampling is an extension of multiple sampling in that decisions to accept, reject, or continue sampling can be made after each item is tested. The decision is reached as shown in Figure 13.6: as sampling progresses, the number of failed items is plotted against the number of items tested; testing is continued until the plotted step function crosses one of the two decision lines.

For this plan, the expected sample size or number of observations, E_R, before a decision is reached for (1) an assumed reliability equal to 1, and (2) the same values for reliability R_0, associated with the producer's risk, and R_1, associated with the consumer's risk, are given by

$$E_{R=1}(n) = \frac{a_1}{b_2}. \tag{13.6}$$

$$E_{R=R_0}(n) = \frac{(1-\alpha)a_1 + \alpha a_2}{(1-R_0)b_1 + R_0 b_2}. \tag{13.7}$$

*The first term, $P\{x_1 \leq c_1\}$, is the probability of acceptance based on the first sample; that is, the number of actual failed items (x_1) occurring in the first sample (n_1) equals or is less than the allowable number of failed items (c_1) for the first sample. The second term represents the probability of acceptance on the basis of the second sample, when required; i.e., it is the sum of the joint occurrences of $(x_1 > c_1 + 1)$ failed items in the first sample and no more than $c_2 - x_1$ failures in the second sample.

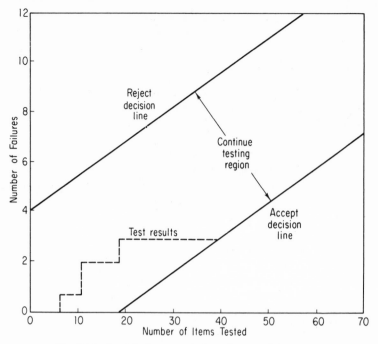

Fig. 13.6 Graphical representation of sequential acceptance test.

$$E_{R=R_1}(n) = \frac{\beta(a_1) + (1-\beta)a_2}{(1-R_1)b_1 + R_1 b_2} \tag{13.8}$$

where

$$a_1 = \ln \frac{\beta}{1-\alpha}, \; a_2 = \ln \frac{1-\beta}{\alpha}$$

$$b_1 = \ln \frac{1-R_1}{1-R_0}, \; b_2 = \ln \frac{R_1}{R_0}.$$

The decision lines are obtained as follows:

$$\text{Accept line} = h_0 + sn = \frac{a_1}{b_1 - b_2} + \frac{(-b_2)n}{b_1 - b_2} \tag{13.9}$$

$$\text{Reject line} = h_1 + sn = \frac{a_2}{b_1 - b_2} + \frac{(-b_2)n}{b_1 - b_2}. \tag{13.10}$$

After n items are tested and r_n failures are observed:

1. Accept the lot if $r_n \leq h_0 + sn$;
2. Reject the lot if $r_n \geq h_1 + sn$;
3. Otherwise, continue testing ($h_0 + sn < r_n < h_1 + sn$).

Generally, sequential sampling requires testing of fewer items than multiple or single sampling. However, in the sequential plan, an item is not tested until the previous item has been tested for a certain number of hours or cycles. This procedure has been applied mainly to one-shot devices, where the sample size and not the duration of the test is a factor in selecting a particular sampling plan.

13.8   Comparison of Attributes Sampling Plans

The single, multiple, and sequential sampling plans are compared in Table 13.9 with respect to such testing characteristics as sample size, basis of decision, etc.

Table 13.10 compares the variables associated with the three types of testing for given values of reliability, producer's risk, and consumer's risk.

13.9   Testing by Variables

Sampling plans based on the method of variables are not used to any great extent in reliability acceptance tests. However, in view of the high cost of demonstrating reliability through sampling by attributes, the economical features of variables sampling are well worth considering when this plan is applicable (see Table 13.11). The limiting factors are (1) variables testing requires that the parameters under test follow a normal density function, or be transformable into the normal distribution; and (2) considerably more record-keeping and calculations are needed than are normally required by attributes plans.

Following is a comparison of sampling sizes for single sampling by attributes and variables, with the plans having similar O.C. curves:

Attributes	Variables
10	10
25	15
50	25
150	40
1500	200

The technique of testing by variables is described in MIL-STD-414, *Sampling Procedures and Tables for Testing by Variables.*

Table 13.9 Comparison of Single, Multiple, and Sequential Sampling Plans

Characteristic	Single	Multiple	Sequential
Sample Size	Known.	Average can be computed for various quality levels. Generally less than single.	Average can be computed for various incoming quality levels. Generally less than single and multiple.
Decision Choices	Accept or reject.	Accept, reject, or take another sample until final sample is selected.	Accept, reject, or test another item.
Predetermined Characteristics	Two of the three quantities n, α, or β.	Same as single.	Fix α and β; n is a random variable.
Statistical Considerations	Must know distribution of sample statistic.	Same as single.	Compare sample against two constants. No need to know distribution of sample statistic.
Personnel Training	Requires least training.	More trained people required than for single.	Requires most training.
Derived Estimates	Unbiased.	Unbiased for only first sample.	Biased.
Ease of Administration	Easiest. Scheduling can be fairly precise and test-cost estimates can be made.	More difficult than single since the exact number of tests is unknown. Only average test costs can be estimated.	Most difficult in terms of testing, scheduling, and over-all administration. Most time-consuming.
Miscellaneous	Best used for testing situations where ease of administration is most important and cost of testing is relatively unimportant.	Has psychological advantage in that supplier is given a "second chance" by taking further samples if first sample results indicate a marginal lot.	Most efficient test in terms of required sample size. Will require approximately 50 per cent of sample size of single sampling plans. Best to use when test costs are most important.

Table 13.10 COMPARISON OF SINGLE, MULTIPLE, AND SEQUENTIAL TESTS

R_0	R_1	$1-R_1/1-R_0$	Single		Multiple						Sequential Expected Sample Size		
			n	c	n_1	n_2	c_1	c_2	ASN_{R_0}	ASN_{R_1}	$E_{1.0}(n)$	$E_{R_0}(n)$	$E_{R_1}(n)$
					$\alpha = 0.10$	$\beta = 0.10$							
0.99	0.98	2	950	13	500	1000	6	21	737	1372	199	437	582
	0.97	3	320	5	200	400	3	7	256	440	104	175	142
	0.95	5	110	2	75	150	1	2	95	95	53	69	43
	0.90	10	37	1	25	50	0	1	34	35	23	25	12
0.95	0.90	2	190	13	100	300	6	18	1155	359	40	103	87
	0.85	3	60	5	40	80	3	7	51	87	20	35	25
	0.75	5	20	2	15	30	1	2	19	19	9.3	12	7.8
0.90	0.80	2	80	11	50	100	6	16	74	137	19	48	40
	0.70	3	25	4	20	40	3	7	26	44	8.8	15	11
					$\alpha = 0.05$	$\beta = 0.10$							
0.99	0.98	2	1245	18	540	540	5	16	783	1057	222	639	619
	0.97	3	400	7	205	205	2	7	271	342	110	216	184
	0.95	5	140	3	70	70	0	3	105	112	55	81	59
	0.90	10	37	1	28	14	0	1	30	33	24	28	17
0.95	0.90	2	250	18	110	110	5	16	160	210	42	119	117
	0.85	3	80	7	45	45	2	7	62	74	20	39	35
	0.75	5	27	3	15	15	0	3	23	22	9.5	14	11
0.90	0.80	2	125	18	55	55	5	16	81	105	19	54	55
	0.70	3	40	7	20	20	2	7	26	34	9.0	17	16

Table 13.11 COMPARISONS OF ATTRIBUTES AND VARIABLES

Factor	*Attributes*	*Variables*
Use of Item	Single operation	Single operation
Reliability Goal	Per cent defective or probability of survival over a fixed time period.	Output tolerance limits which define success or failure possibly applying after a fixed period of operation.
Type of Sampling	Lot-by-lot, continuous, chain sampling	Lot-by-lot
Type of Information Yielded	Number of per cent of sample that failed to meet specified quality characteristics at a given point in time.	Distribution of some quantitative output at a given point in time. Provides most information for quality improvement.
Sample Size for Given Protection	Highest	Lower than attributes test for corresponding plan.
Truncation	Truncation or nontruncation	Nontruncation
Ease of Inspection	Requires relatively simple test equipment and less qualified personnel.	More complex test equipment and better trained people required than for attributes tests.
Simplicity of Application	Data recording and analysis is fairly simple. Single set of attributes criteria applies to all quality characteristics.	More clerical costs than attribute plans. Variables criteria needed for each quality characteristic.
Robustness	No assumptions on failure distribution required. Binomial distribution applies for most cases.	Requires a parametric assumption on the distribution of the characteristic considered.
Replacement	Nonreplacement	Replacement
Stress	Standard or accelerated	Standard or accelerated
Available Tables	Extensive tables are available.	Tables for the normal distribution are available.

13.10 Application of Sampling Procedures in Specifications for Electronic Parts

The Darnell Report* recently focused attention on the means by which guarantees of high reliability might be incorporated in military specifications for electronic parts. The results of the study showed that reliability guarantees for most electronic parts are either nonexistent or practically meaningless, and that greatly improved procedures must be adopted if the required reliability levels for military and space systems are to be achieved.

Parts Specification Management for Reliability, Vols. I and II, May 1960. U.S. Government Printing Office. (Price $1.50 for both volumes.)

Table **13.12** COMPARISON OF FIELD REMOVAL RATES AND SPECIFICATION
FAILURE-RATE GUARANTEES

Device Class	Typical Field Removal Rate, in removals per hr (multiply by 10^{-5})	Specification Failure-Rate Guarantees, in failures per hr, based on 90 per cent confidence (multiply by 10^{-5})	
		Conventional Military Specifications	High-Reliability Specifications
Receiving tubes	0.1 – 10	46	1 – 10
Transmitting tubes, < 100 watts	2 – 10	120	NA
Transmitting tubes, 100-1000 watts	10 – 20	160	NA
Magnetrons, L-S band (Megawatts peak power)	15 – 200	225	NA
Magnetrons, X band	100 – 700	225	NA
Semiconductor diodes			
Digital	<0.001 – 0.3	30 – 50	0.5 – 10
Rectifier-regulator	0.03 – 30	30 – 50	5 – 20
Transistors			
Digital	0.001 – 1.0	30 – 50	0.5 – 10
Analog	0.03 – 30	30 – 50	5 – 20
Capacitors			
Paper	0.01 – 0.1	100	1.0
Resistors			
Composition	0.01 – 1.0	20	1.0
Quartz crystals	0.1 – 20	60	NA
Relays	0.05 – 20	The prototype specification for relays in the Darnell Report recommended a basic failure-rate level of 1 per cent per 10,000 operations at 90 per cent confidence.	

Table 13.12 illustrates the relationship between typical field removal rates
for several classes of electronic parts and the failure rate "guarantees"
derived from an examination of typical military specifications. One of the
first serious efforts in improvement of parts specifications was instituted by
the Bureau of Ships (Navy Department), working with the JS-11 and JS-12

Committees of the Joint Electron Device Engineering Council (JEDEC) and the Semiconductor Device Panel of the Aerospace Industries Association, in regard to specifications for semiconductor devices. The Bureau of Ships, the Air Force Logistics Command, and the Army Electronics Material Command have prepared a large number of high-reliability specifications for all classes of electronic parts. The recently constituted Defense Electronics Supply Center (DESC), which has jurisdiction over tri-service procurement specifications, is now revising many older specifications to include improved reliability-assurance requirements, as well as coordinating new specifications.

However, with regard to parts used in the development of new systems, it has been estimated that the vast majority are procured to specifications prepared by the system manufacturer.* Those specifications developed for the Minuteman program have now been converted to a military format. [See MIL-M-38100 (USAF), *Reliability and Quality Assurance Requirements for Established Reliability Parts*, 15 April 1963.] Essentially this document requires the manufacturer to maintain a first-rate quality-control organization, with emphasis on the analysis, correction, and reporting of failures.

The Problem of Economics in Life Testing

To prove a failure rate of 0.001 per cent per 1000 hr† (with 90 per cent confidence) would require the testing of 532,300 devices for 1000 hr, with 2 failures. This would be equivalent to more than 12 years of testing if a more modest sample size of 5000 devices were used and 2 failures were allowed. It becomes obvious that life testing for acceptance purposes can be very expensive if use-condition failure rates are to be proved. The problem becomes particularly expensive for high-power components such as transmitting tubes, magnetrons, rectifiers, and transformers, for which test-facility and test operating costs can be prohibitive.

To circumvent this problem, several approaches have been used or suggested. One, which has found its way into some specifications, involves the lowering of the confidence level. Figure 13.7, taken from the Air Force specification, MIL-R-38100, illustrates the relationship between total test time, failure rates, and confidence level. For example, to demonstrate a failure rate of 0.01 per cent per 1000 hr, assuming no failures, would require

---

*In 1962, for example, it was estimated that from 70 to 90 per cent of all semiconductor devices procured for new military systems were in this category. See William H. von Alven, Ch. 1, *Semiconductor Reliability*, Vol. 2, Engineering Publishers, Elizabeth, N. J., 1962.

†For reliability prediction purposes, failure rates are often expressed in failures per hour; for specification and life-testing purposes, failure rates are generally expressed as X per cent per 1000 hr. Example: A failure rate of 10 per cent per 1000 hr is equivalent to 10×10^{-5} failures per hour.

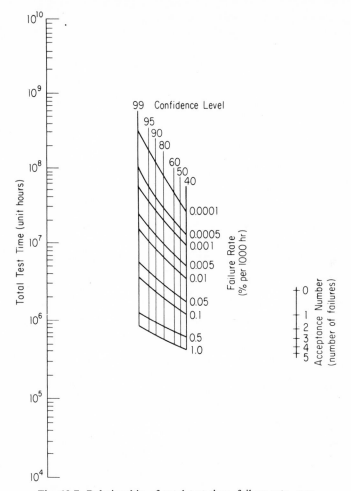

Fig. 13.7 Relationship of total test time, failure rate, confidence level, and acceptance number. [Source: Mil-R-38100 (USAF) 15 April 1963.]

a total test time of 2.3×10^7 unit hours at the 90 per cent level of confidence and only 10^7 unit hours at the 60 per cent level of confidence. (The equivalent failure rate for 10^7 unit hours of test time at the 90 per cent confidence level is approximately 0.02 per cent per 1000 hr.) It is important, therefore, to know the confidence level associated with a stated failure rate.

The Darnell Report proposed a data-accumulation plan based on the sequential life testing technique discussed in Chapter 14. Life test data taken from consecutive lots are accumulated and plotted on a sequential chart. This technique provides a means for determining the qualification of a product to much lower failure rate levels than would be possible by use of

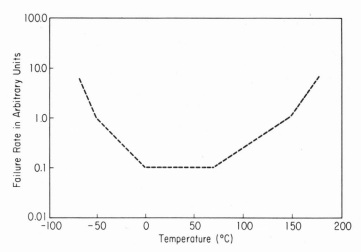

Fig. 13.8 A generalized pattern of failure rate vs. temperature on life tests (> 1000 hr) for silicon devices.

limited data from only one production lot. Care should be taken to assure that the data used come from production lots which are reasonably homogeneous and in which no major product-design or production-process changes have been made. For many types of devices, this restriction means that data from only a few, say 5 or 10, consecutive lots should be utilized. Frequently, moving average techniques are employed to avoid serious homogeneity problems.

The use of accelerated tests is sometimes proposed to reduce testing time and testing costs. The procedure is to determine failure rates under high-stress conditions and extrapolate the results to give an estimate of anticipated failure rates under use conditions. Experience has shown that it is not possible to generate *precise* acceleration factors for most electronic parts. Studies on paper capacitors, for example, show that life tests run at 140 per cent of maximum rated voltage result in acceleration factors that vary from 2 to 11 when failure rates are compared to those determined at maximum ratings. For the development of useful acceleration factors, there must be some assurance that they are based on some dominant failure mechanism. As shown in Figure 13.8, this situation may exist over only a limited range of stress. This figure illustrates the influence of temperature on silicon transistors manufactured in 1960. At very high temperatures, the failure rate increases when internal soldering begins to degenerate or the alloying or diffusion processes are allowed to continue. At high temperatures, the reverse leakage current exhibits a single failure mechanism; this relationship is depicted by a straight line plot of log of life vs $1/T$ (where T is in $°K$). At low temperatures, the failure rate increases due to thermal stresses on the hermetic seal. At still lower temperatures, the failure rate may increase further as the dew point of the internal atmosphere is approached.

Experience has shown that it is very difficult to determine failure rates less than 0.1 per cent per 1000 hr because of errors made in the recording and analysis of the vast quantities of data that are involved. As will be shown in the following discussion, the general approach now evolving is to require rating verification by means of a life test employing an economically feasible life-test sample together with appropriate screening procedures.

Influence of Ratings in Establishing Sampling Criteria

Before realistic ratings can be established, exploratory tests are required. Perhaps one of the most important test methods is that in which a stress is progressively increased, either in steps or in a continuous manner, with periodic or continuous measurements made of the performance characteristics. A properly designed test will yield considerable information regarding performance stability and failure rates, such as the stress points at which different failure modes occur. An example of this is shown in Figure 13.8.

In general, two types of failure characteristics are encountered:

1. Threshold types of failures for characteristics such as voltage, shock, high-acceleration vibration, and lead bends. These failure distributions usually plot as straight lines on normal probability paper.

2. Time-dependent types of failures for characteristics such as temperature, power, and vibration fatigue. (Failure distributions on parts such as semiconductor devices frequently plot as straight lines on log-normal probability paper.*)

The over-all objectives of exploratory testing should be to derive information as efficiently as possible. Often, statistically designed tests can be used for this purpose. A discussion of these techniques is beyond the scope of this text; however, they are described in standard texts on mathematical statistics.†

Rather than require proof of use-condition failure rates (which are generally under derated conditions), many parts specifications call for life tests conducted within 10 per cent of maximum ratings, and therefore can be considered as rating verification tests. To assist the circuit designer, some specifications provide application data indicating anticipated failure rates under various stress conditions, similar to the data shown in Figure 9.6 (Chapter 9).

---

*D. S. Peck, "Uses of Semiconductor Life Distributions," Chapter 2, and M. L. Embree, "Semiconductor Device Ratings," Chapter 20, *Semiconductor Reliability*, Vol. II., Engineering Publishers, Elizabeth, N. J., 1962.

†See, for example, D. K. Lloyd and M. Lipow, *Reliability: Management, Methods and Mathematics*, Prentice-Hall, Inc., 1962.

Absolute-maximum ratings are frequently used in military specifications and are considered to be limiting values above which the serviceability of any individual device may be impaired. It follows that a combination of absolute-maximum ratings cannot normally be attained simultaneously. The equipment designer has the responsibility of determining, for each rating, an average design value which is below the absolute value of that rating by some safety factor, so that the absolute values will never be exceeded under any unusual conditions of supply-voltage variation, load variation, or manufacturing variations in the equipment itself.*

It has been suggested that a standard failure percentile for rating determination be established for all parts specifications.† For semiconductor devices, the recommended stress condition is that producing 1 per cent failures for a 1000 hr life test; this would be considered an absolute-maximum rating. A realistic failure rate guarantee for such a test would be in the region of 5 to 10 per cent per 1000 hr at 90 per cent confidence. (See Table 13.8.)

Screening Procedures (Nondestructive Testing)

Aside from classification of electrical or mechanical properties, screening procedures are often used for classifying items which are potentially unreliable. Past experience has shown that screening procedures must be carefully evaluated before being instituted on a 100 per cent basis. A great number of screening methods for electronic parts were adopted under "panic" conditions resulting from the pressure of production commitments, without prior evaluation. All too often, an effective solution involving the redesign of circuits or parts has been unnecessarily delayed because an elaborate screening program has been substituted for an adequate engineering solution of the real cause of trouble. A classic example is given in Figure 13.9, which dramatically shows that certain types of incoming inspection for electron tubes proved to be ineffective and extremely expensive in terms of inspection time and the large quantities of good items that were rejected.

In many cases, however, nondestructive testing can be a useful tool. Some commonly used tests for materials and parts include:

1. Visual inspection	6. Leak detection
2. X-ray or fluoroscope	7. Monitored shock and vibration
3. Ultrasonic probing	8. Electrical noise
4. Magnetic-particle inspection	9. "Burn-in" or aging
5. Eddy-current testing	10. Infrared scanning.

*Military Specification, *Semiconductor Devices, General Specification For*, MIL-S-19500.
†Embree, op. cit.

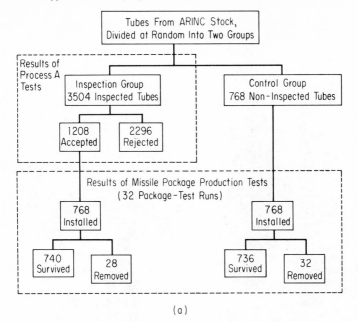

(a)

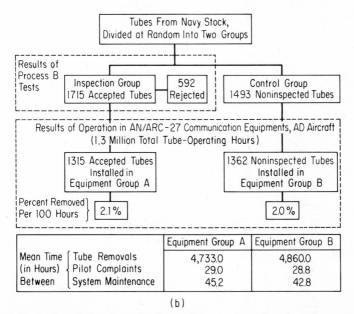

(b)

Fig. 13.9 Evaluation of three incoming-inspection procedures, based on the comparative performance of inspected and noninspected tubes in various environments. (a) Incoming-inspection process A (visual-microscopic and vibration tests). (b) Incoming-inspection process B (visual-microscopic, X-ray, polariscopic, and vibration-noise tests) (tubes operated in AN/ARC-27 communication equipments).

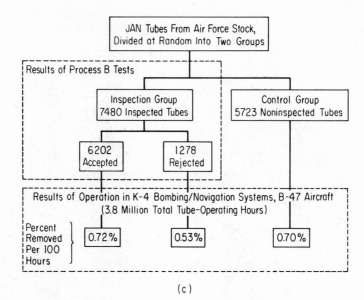

(c)

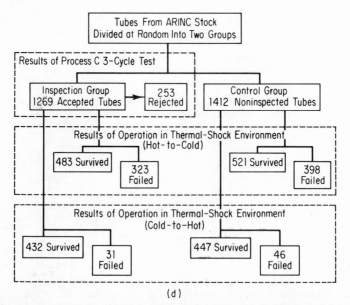

(d)

Fig. 13.9 (*continued*) (c) Incoming-inspection process B (tubes operated in K-4 bombing/navigation systems). (d) Incoming-inspection process C (thermal-shock tests).

Only burn-in and infrared scanning will be discussed here.* Burn-in or aging processes require that a product or material be subjected to a stress for some period of time, then examined for excessive drift in some characteristic or for catastrophic failures. The technique is particularly applicable to devices exhibiting decreasing failure rates with time, such as those illustrated in Figure 13.10. It should be noted that the presence of a decreasing failure rate for a given product is indicative of poor control of the causes of early failures. Frequently, these failures are the result of poor control of workmanship or processes—poor welds, cold solder joints, etc.

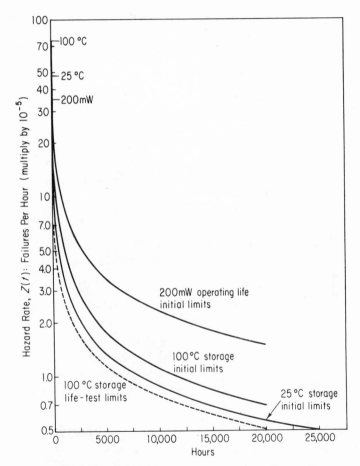

Fig. 13.10 Type 2N525 family, source Q: Hazard rates for three life-test conditions.

*Many of the tests are discussed in a text by W. J. McGonnagle, *Nondestructive Testing*, McGraw-Hill Publishing Company, New York, N.Y., 1961.

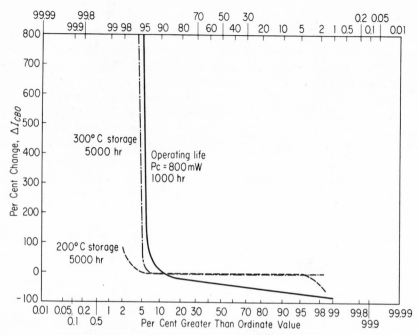

Fig. 13.11 Type 2N1613, source Y: distributions of $\triangle I_{CBO}$ in 1000-hr operating and 5000-hr storage life tests.

Aging can be used to identify devices having good electrical stability. An example is shown in Figure 13.11 for the type 2N1613 transistor. In this case, approximately 95 per cent of the devices tested proved to be extremely stable in regard to I_{CBO}.

Recently developed infrared scanning techniques should prove to be valuable in engineering design evaluation and quality control at the part and assembly levels.* These procedures do not affect the loading or upset the electrical and thermal equilibrium of the circuitry. They provide a means for disclosing the thermal characteristics of parts and assemblies, making it easy to detect poor or inadequate design, short-life components, heat damage, poor connections, and drifting characteristics of components.

Environmental Test Requirements

The Darnell Report also shows that military specifications for electronic parts have fallen far behind present environmental requirements for equipments. As a result, extensive effort is under way within the military services

---

*R. Vanzetti, "Infrared Techniques Enhance Electronic Reliability," *Proceedings, Ninth National Symposium on Reliability and Quality Control*, San Francisco, January, 1963.

to assure that the majority of parts specifications will at least meet the requirements of the Group IV objectives described in MIL-STD-446. The most important features of Group IV include temperatures from $-65°C$ to $+125°C$; 100,000 ft operating altitude; 1000-2000 cps vibration at 10 g; and 50-g/11-millisecond shock requirements. It must be noted that the 50-g/11-millisecond shock requirement is for the assembled equipment. For many small parts, such as resistors, capacitors, and transistors, this requirement is not meaningful because of the smaller mass of these parts and the methods used in their construction. Typical specification requirements for many of these parts may be on the order of 1000 g at 1 millisecond. For missile and space applications, nuclear radiation resistance and higher temperature capabilities must be considered as test requirements.

Many of the environmental tests are considered destructive; consequently, selection of the sample-size requirement demands that care be taken to establish an appropriate trade-off between testing costs and the required level of quality assurance. *AQL* values in the range of 4 to 6.5 per cent and *LTPD* values of 20 per cent are typical for destructive tests in most parts specifications.

PROBLEMS

1. Draw an operating characteristic curve using the hypergeometric distribution if $N = 100$ is the size of the lot; $n = 10$ is the number of samples to be tested; and 1 failure is allowed.

2. Draw an operating characteristic curve using the binomial distribution for the following conditions: $N = 1000$ is the size of the lot; $n = 5$ is the number of samples to be tested; 1 failure is allowed, and $\alpha = 0.05$, $\beta = 0.1$.

3. A military specification for a certain electronic part references a MIL-STD-105 sampling plan in the following manner: $AQL = 4$ per cent, Inspection Level I. (a) What sample size and acceptance number are required for a lot of 1000 items? and (b) for a lot of 20,000 items? (c) If, instead of Inspection Level I, Code Letter N had been used, what would the sampling requirement be for a lot of 1000 items? and (d) for a lot of 50 items? (e) What are the *LTPD* values for each of the conditions above?

4. An *LTPD* of 10 per cent is specified, with a minimum rejection number of 3. What is the significance of this requirement as applied to the sampling plan in Table 13.8?

5. Compare Table 13.8 with Tables 13.2, 13.3, and 13.4 to determine the minimum lot size that is applicable for Table 13.8.

6. The data shown below were taken by a step stress technique to determine the threshold characteristics of a certain transistor type that was subjected

to a constant acceleration (centrifuge) stress. What would be a realistic rating? (*Hint*: Plot on linear probability paper.)

| | *First Test* | | *Second Test* | |
	G Level	Cumulative Failures (per cent)	G Level	Cumulative Failures (per cent)
	20K	5	20K	2
	40K	6	30K	3
	60K	10	40K	4
	80K	20	50K	5
	100K	25	70K	10
			80K	25
			90K	30
			100K	40

Reliability Demonstration Procedures

14.1 Introduction

The objective of a reliability test program is to gain information concerning failures—specifically, the tendency of systems to fail and the effects of failure. Such information will help to control failure tendencies and their consequences. Reliability tests are thus distinguished from most other types of tests, which are primarily concerned with normal operation of equipments.

Few tests, however, can be strictly classified as either "reliability" or "non-reliability." In the broad sense, every test is intended to improve product quality, and any failure tendencies would be noted whenever observed. Any test relating to a malfunction or its consequence can be termed a *reliability test*, and there is clear overlap between reliability and engineering tests in the study of effects of failures.

Quality control and assurance tests are generally considered to be closely related to reliability tests, in that the former are essentially evaluations of parameters associated with failure probabilities.

For present purposes, a reliability test is identified by the fact that the objective is to obtain information concerning failures, particularly failure occurrence patterns. Thus, the test may be merely a data collection and analysis activity associated with some other engineering test or with the operation of a system. This suggests that the ideal reliability test program is one which makes use of all available test data.

14.2 Reliability Testing Objectives

Reliability tests can be divided into three categories: development and demonstration, qualification and acceptance, and operation.

Reliability Development and Demonstration Testing

The objectives of reliability development and demonstration tests are:

1. To determine if the design must be improved to meet the reliability requirement
2. To indicate any design changes needed
3. To verify improvements in design reliability

These test objectives are most closely associated with the development of the prototype model and are applicable to production as well as design. The nature of these tests depends on the level of complexity being considered and the type of system or subsystem being investigated (e.g., electronics, structural, propulsion, ground support).

For electronic parts, reliability development and demonstration may take the form of life tests to determine whether the part can meet its reliability requirements, and, if not, what improvements are needed. Required improvements are ascertained from a study of the failure patterns and failure mechanisms of the part. At a higher level of complexity (e.g., an electronic assembly), the same objectives apply—to determine, for example, the requirements for additional redundancy or protective devices.

With respect to structures, a developmental reliability test may have as its objective the determination of the variance in strength of welded, riveted, and bolted joints. For a propulsion subsystem, equivalent testing would determine the relative incidence of specific malfunctions, and through induced malfunctions would determine the adequacy of sensing and engine shut-down provisions.

To fulfill the objectives of reliability development and demonstration tests, the test data must be of a nature that permits insight into the failure probabilities and failure effects for a particular design. Such data form a sound basis for reliability analysis and assessment, both for the project of interest and for later related programs.

Reliability development and demonstration tests are generally directed by personnel in charge of design and development. However, their most effective application requires centralized planning and review. Through central control, related test efforts can be combined, with resulting savings in time and money.

Qualification and Acceptance Testing

The objectives of qualification and acceptance tests are:

1. To determine if a part, assembly, or end item should be accepted or rejected (either on an individual or lot basis).
2. To determine if a particular design should be considered as qualified for its intended application.

Thus, in this category of testing, the objectives differ from those of other reliability tests, particularly with respect to the accept-or-reject procedure. The acceptance of specific production items or lots is a quality-control function associated with the production phase, while design qualification is generally associated with either the design or production phase.

Reliability demonstration testing of a subsystem design may also represent the qualification testing required for inclusion of the subsystem in the next higher level of assembly.

Qualification and acceptance includes the usual quality-control function of testing and screening incoming parts. For those parts having a mean-time-to-failure requirement, quality control includes a life test to indicate compliance or noncompliance with the specified *MTBF*.

Although these tests are associated with the design and production prototype phases of a system's development, they begin early in the program with respect to the materials and components to be used in the system.

Operational Testing

The objectives of operational tests are:

1. To verify the reliability analyses performed during the project.
2. To provide data indicating necessary modifications of operational procedures and policies, as they affect reliability and maintainability.
3. To provide information to be used in later activities.

This test category is associated with the mission preparation and operation phases. Operational test conditions are not generally subject to continuing change, as is the case for developmental tests. Operational tests provide the final verification of design and allow the experimental observation of the relationship between operational performance and the individual areas of a reliability effort. These tests provide, in effect, the feedback from practice to theory.

14.3 Testing Program

A reliability test program is a dynamic procedure, changing to meet the needs of the development effort. Since much of the actual testing is conducted by various individual contractors, a central test-program management is needed to insure that the proper test specifications are applied and that test requirements are met. In addition, over-all coordination of the testing effort helps to prevent duplication and insures that there are no serious omissions.

In general, the test program should include provisions for planning and executing the tests and for analyzing the resulting data. In a large program, central data-collecting and data-analysis functions are also required.

The first step in a reliability program is the preparation of a comprehensive plan detailing the specific requirements of each category of testing for each phase of system development, and describing the testing development responsibility of each participating organization. This plan should be modified as the development program progresses.

One of the main inputs to the preparation of the test plan is a review of available data pertaining to the equipment of interest. These data may serve to satisfy certain of the test objectives and also to indicate trouble areas for objectives which require testing.

As a part, assembly, or system progresses through the various phases, it is possible to identify certain critical points in the preparation of the test plan.

During the *design concept* phase, the broad outlines of the test program are established. At this point, certain reliability development tests should be indicated. These tests can perhaps be assigned independently of design responsibilities (i.e., separate study assignments or contracts). The recommended tests in the structures area are in this category.

Requirements for reliability development and demonstration testing are formulated as detailed design responsibilities are assigned. These responsibilities should be assigned as early in the design phase as possible.

Plans for qualification testing are developed as the program proceeds toward the *production prototype* and *production* phases, so that the production model can be tested and qualified as an operational model.

14.4 Outline of Reliability Test Program

Table 14.1 is an outline of central management tasks in a reliability test program. As previously stated, some of the testing requirements will be identified as the program progresses. Certain of the individual tasks may be delegated to subcontractors or others having specific areas of responsibility.

Table **14.1** OUTLINE OF RELIABILITY TEST PROGRAMS

1. *Comprehensive Test Plan*

    1.1. Determine test requirements and objectives.

    1.2. Review existing data to see if they meet any existing requirements, thereby eliminating the need for certain tests.

    1.3 Review existing or planned functional tests and operations to determine if they can be utilized to yield reliability information.

    1.4. Determine the necessary tests to be performed.

    1.5. Allocate time, funds, and effort; develop individual test requirements; assign responsibilities.

    1.6. Review and approve test reporting procedures and forms.

    1.7. Maintain test-status information for entire program (utilizing data central facilities).

    (Each of the above is a continuing task requiring frequent updating.)

2. *Qualification and Acceptance Testing*

    2.1. Review, continuously, acceptance and qualification testing. Investigate relationships between these tests and operational reliability and recommend changes where indicated.

    2.2. Review data of acceptance and qualification tests to determine applicability to reliability test program.

3. *Reliability Development and Demonstration Testing*

    3.1. Assign general responsibilities for reliability, development, and demonstration tests.

    3.2. Determine specific tests; design and conduct tests, if necessary. Tests should include:

    (1) Electronic assembly demonstration tests

    (2) Propulsion malfunction tests

    (3) Structural integrity tests

4. *Reliability Analysis Reports*

    4.1. Develop efficient data-collection procedure.

    4.2. Prepare and disseminate reports.

    4.3. Review procedures and recommend changes, if required.

14.5 Statistical Considerations in Life Testing

Time was not considered as a reliability test condition in the discussion in the preceding chapter. This section will show how the time factor enters into reliability testing.

Kao* lists three main classifications of life tests:

*J. H. K. Kao, "The Design and Analysis of Life Testing Experiments," *Proceedings of the American Society for Quality Control*, Middle Atlantic Conference (1958).

1. *Nonvariate or stationary life tests* involve only one set of stress conditions for a specified time period. Most life tests for electronic parts and systems are in this category; they can be considered as *rating verification* procedures, and appear in procurement specifications. (Life tests may be conducted to determine failure rates under anticipated use conditions. However, these tests require either such large sample sizes or such long time periods that they are unpractical for parts or components having high reliability levels.)

2. *Univariate life tests* measure the behavior of a device when one stress is varied at a time. A *step-stress* technique can be applied to the items under consideration so that separate behavior patterns are obtained as the primary stresses (such as temperature and power dissipation) are increased.* A variation of this technique involves application of a controlled, continuously increasing stress to the devices under test. These tests are particularly useful for delineating the stress levels at which different failure modes occur.

3. *Multivariate life tests* determine how different stresses with various levels of intensity influence the items being tested. These tests are applied in product qualification or design approval, as well as in basic research and development studies. Multivariate tests utilize the techniques of statistical experimental design, examples of which include factorial, random-balance, or multiple-balance procedures.† Analysis of data from multivariate tests could become quite involved, requiring the use of somewhat complex statistical techniques; e.g., regression analysis, analysis of variance, etc.

As was shown in earlier chapters, mathematical models have been developed to fit a wide range of observed failure distributions, including the exponential, log-normal, gamma, and Weibull.‡ Sampling procedures have been developed for these distributions in a similar manner. However, the discussion in the next section will be based on life-test procedures developed for the exponential distribution, the type most frequently encountered in specifications for parts and systems.§

---

*G. A. Dodson, "High-Stress Ageing to Failure of Semiconductor Devices," *Proceedings of Seventh National Symposium on Reliability and Quality Control*, Philadelphia, Pennsylvania (January, 1961).

†As described in the following references:

D. K. Lloyd and M. Lipow, *Reliability: Management, Methods and Mathematics* (Englewood Cliffs, N.J.: Prentice-Hall, Inc., 1962).

M. Zelen, "Factorial Experiments in Life Testing", *Technometrics* (August, 1959).

‡Practical applications of these methods are discussed in *Reliability of Semiconductor Devices*, W. H. von Alven and G. J. Blakemore, ARINC Research Corporation, Publication No. 144-6-270 (December 22, 1961).

§Life test plans have also been developed which are based on nonparametric assumptions independent of the underlying failure distribution; see *Nonparametric Sampling Plans*, S. Gupta, Stanford University and Bell Telephone Laboratories (1962).

14.6 Exponential Distribution

The tests described in this section are based on the assumption that failures are distributed exponentially as a function of time. The probability density function associated with this distribution is

$$f(t) = \frac{1}{\theta}\, e^{-t/\theta} \tag{14.1}$$

$$= \lambda e^{-\lambda t}$$

where

$t =$ time to failure,

$\theta =$ mean time between failures ($MTBF$), and

$\lambda =$ constant failure rate $\left(\text{equal to } \dfrac{1}{\theta}\right)$.

Appropriate test specifications are those for which the acceptable reliability level (θ_0) and unacceptable reliability level (θ_1) are, or can be, converted to values of exponential mean life or constant failure rate.* Two general types of life tests are considered: fixed-time (or fixed number of failures) and sequential.

Replacement Tests with Fixed Time or with Preassigned Number of Failures

Acceptance tests can be conducted to a fixed time or to the occurrence of a preassigned number of failures. These tests are usually more economical than those in which all items are tested to failure, because the maximum amount of testing is known beforehand and the expected waiting time is lower. Table 14.2 illustrates the reduction in time afforded by a nonreplacement test where r failures are required before a decision can be made. If $n > r$ items are on test, a decision can be made before all test items fail. The values in Table 14.2 are the expected saving in time for a test with a preassigned number of failures, relative to a test with r items wherein all items are tested to failure.

Replacement tests terminated after a specified number of test hours will now be considered. For a given test specification—θ_0, θ_1, α, β—a sample size and critical number of failures can be computed. If T_0 is the test termination time (the maximum number of test hours *each* item will accumulate) and r_0 is the

---

*B. Epstein, *Statistical Techniques in Life Testing*, Technical Report No. 3 (ASTIA No. AD-21145) Wayne State University (October, 1958).

Table 14.2 EXPECTED SAVING IN TIME
(PER CENT) WITH TRUNCATED TEST

r/n	1	3	10	20
1	100	33	10	5.0
3		100	18	8.7
10			100	23.0

For example, if 3 failures are required in nonreplacement
testing of 10 items, the waiting time before a decision is reduced (on the average) to 18 per cent of the waiting time for
a nontruncated test.

critical number of failures, the decision rule is as follows:

If r_0 failures occur before T_0 test hours, reject the lot.

If T_0 test hours are accumulated before r_0 failures occur, accept the lot.

Replacement tests terminating after a preassigned number of failures lead
to approximately the same amount of testing for plans with identical O.C.
curves. Nonreplacement tests for either type will, on the average, require
more test hours but fewer test items.

The O.C. curve for the plan described above can be obtained from the
Poisson formula

$$L(\theta) = \sum_{x=0}^{r_0-1} \frac{\exp\left[-nT_0/\theta\right]\left(\frac{nT_0}{\theta}\right)^x}{x!} \tag{14.2}$$

where

$L(\theta)$ = the probability of accepting items with a mean life of θ,

n = the number of items put on test,

r_0 = the critical (reject) number of failures,

T_0 = the test termination time.

The quantity r_0 is determined so that

$$L(\theta_0) \geq 1 - \alpha, \, L(\theta_1) \leq \beta.$$

Example 1. An acceptance plan is to be devised such that a lot of
devices with an average mean life of 100 hr is to have a 90 per cent
chance of being accepted, and a lot of devices with an average mean life
of 20 hr is to have a 10 per cent chance of being accepted. The sampling
that provides this discrimination consists of 10 devices, each to be
tested 11 hr. The allowable number of failures for this plan would be
derived as follows:

The given values are

$$\theta_0 = 100 \text{ hr}, \quad \alpha = (1.0 - 0.9) = 0.1; n = 10$$
$$\theta_1 = 20 \text{ hr}, \quad \beta = 0.10, T_0 = 11 \text{ hr}.$$

Compute

$$\frac{nT_0}{\theta_0} = 1.1 \quad \text{and} \quad \frac{nT_0}{\theta_1} = 5.5,$$

and determine from Figure 13.3 (which gives the summation of terms of Poisson's Exponential Binomial Limit) the allowable number of failures which satisfies the conditions $L(\theta_0) \geq 1 - \alpha$ and $L(\theta_1) \leq \beta$. In this case the value of c would be 2. To find other points on the O.C. curve, other values of θ are assumed and $L(\theta)$ is read from the table. For an assumed

$$\theta_2 = 50 \text{ hr}, \quad \frac{nT_0}{\theta_2} = 2.2 \quad \text{and} \quad L(\theta) = 0.623.$$

Or, from Equation (14.2),

$$L(\theta) = \sum_{x=0}^{2} \frac{e^{-2.2}(2.2)^x}{x!} = 0.623.$$

For a replacement test, $T_0^* = nT_0$ represents the *total* number of test hours and, therefore, can be used in lieu of T_0 to index a set of sampling plans.

To compare the amount of testing for various plans, the following statistics can be used:

$E_\theta(r)$ = expected number of failures, given mean life θ;

$E_\theta(T^*)$ = expected total test time, given mean life θ;

$E_\theta(WT)$ = expected waiting time, given mean life θ.

For replacement tests with n items, the following relationships hold:

$$E_\theta(T^*) = \theta \, E_\theta(r) \tag{14.3}$$

$$E_\theta(WT) = \frac{\theta}{n} E_\theta(r) = \frac{E_\theta(T^*)}{n}. \tag{14.4}$$

Therefore, the computing of $E_\theta(r)$ enables one to determine the other two statistics for a given plan. The formula for the expected number of failures if the mean life is θ has been shown by Epstein† to be

$$E_\theta(r) = m \sum_{x=0}^{r_0-2} \frac{e^{-m}m^x}{x!} + r_0 \left[1 - \sum_{x=0}^{r_0-1} \frac{e^{-m}m^x}{x!} \right] \tag{14.5}$$

where m is the Poisson mean nT_0/θ.

†B. Epstein, *op cit*, page 513.

Values of $E_{\theta_0}(r)$ and $E_{\theta_1}(r)$ are given for various common test plans in Table 14.3.

Example 2. A lot with an acceptable mean life of 25,000 hr should be accepted 95 per cent of the time, and a lot with an unacceptable mean life of 5,000 hr should be rejected 90 per cent of the time.

Statistical conditions:

$$\alpha = 0.05 \qquad \theta_0 = 25{,}000 \text{ hr} \qquad k = 5.$$
$$\beta = 0.10 \qquad \theta_1 = 5{,}000 \text{ hr}$$

To determine a test plan to be terminated at a preassigned time ($T_0 = 1{,}000$ hr) with replacement of failed items, the steps are as follows:

1. In Table 14.3 under "Truncated Tests," for $k = 5$, $\alpha = 0.05$, and $\beta = 0.10$, obtain the values

$$r_0 = 4, \quad \text{and} \quad \frac{\chi^2_{1-\alpha,2r_0}}{2} = 1.37.$$

2. Solution of the equation at the bottom of Table 14.3 yields $n = 34$. The procedure to implement this plan is to place 34 items on test for 1,000 hr, replacing those items that fail. Terminate and accept the lot if less than four failures have occurred by 1,000 hr. Terminate the test and reject the lot if four or more failures occur before 1,000 hr.

Table 14.4 is useful for the solution of time-truncated sampling problems.

Example 3. Assume that a maximum of 75 equipment operating hours can be allocated for a reliability demonstration test. Under the conditions stated in Table 14.4, determine θ_0 and θ_1 if no more than 2 failures are allowed. Ans. $\theta_0 = 68.2$ hr; $\theta_1 = 14.1$ hr. (What is the significance of these results?)

Nonreplacement Tests (Fixed Time)

Life tests of electronic parts are based, for the most part, on the non-parametric sampling procedures of MIL-STD-105 (see Chapter 13, p. 482). However, *LTPD* (or λ) plans developed by Sobel and Tischendorf* are receiving increasing application in newer specifications for electronic parts (Table 13.8).

Example 4. What sample size and acceptance number will assure a failure rate (λ) of 5 per cent per 1000 hr ($\beta = 0.1$), and a maximum *AQL* of 1 per cent per 1000 hr ($\alpha = 0.05$)? Answer (from Table 13.8): $n = 132$, $a = 3$.

*MIL-S-19500C, *Semiconductor Devices, General Specification for* (January 19, 1962).

Table 14.3 TEST PARAMETERS AND EXPECTED NUMBER OF FAILURES FOR VARIOUS TRUNCATED AND SEQUENTIAL REPLACEMENT LIFE TESTS

$k = \dfrac{\theta_0}{\theta_1}$	α	β	Rejection Number, r_0	$\dfrac{\chi^2_{1-\alpha,2r_0}}{2}$	Truncated Tests—replacement Expected Number of Failures $E_{\theta_0}(r)$	Expected Number of Failures $E_{\theta_1}(r)$	Truncation Number, r^*	Expected Number of Failures $E_{\theta_0}(r)$	
1.5	0.05	0.05	67	54.13	54.0	66.8	201	28.0	36.7
	0.05	0.10	55	43.40	40.5	54.6	165	21.1	32.9
	0.05	0.25	35	25.87	24.0	34.0	105	12.0	23.5
	0.10	0.05	52	43.00	37.6	51.8	156	25.1	27.6
	0.10	0.10	41	33.04	32.8	40.7	123	18.6	24.4
	0.10	0.25	25	18.84	18.7	24.2	75	10.1	16.5
	0.25	0.05	32	28.02	27.3	31.9	96	18.0	15.7
	0.25	0.10	23	19.61	19.0	22.7	69	12.6	13.2
	0.25	0.25	12	9.52	9.1	11.4	36	5.8	7.6
2	0.05	0.05	23	15.72	15.6	22.9	69	8.6	13.7
	0.05	0.10	19	12.44	12.4	18.8	57	6.5	12.3
	0.05	0.25	13	7.69	7.6	12.4	39	3.7	8.8
	0.10	0.05	18	12.82	12.7	17.9	54	7.7	10.3
	0.10	0.10	15	10.30	10.2	14.8	45	5.7	9.1
	0.10	0.25	9	5.43	5.3	8.5	27	3.1	6.2
	0.25	0.05	11	8.62	8.2	10.9	33	5.5	5.9
	0.25	0.10	8	5.96	5.6	7.8	24	3.9	4.9
	0.25	0.25	5	3.37	3.2	4.7	15	1.8	2.8
3	0.05	0.05	10	5.43	5.4	9.9	30	2.9	6.1
	0.05	0.10	8	3.98	3.9	7.8	24	2.2	5.5
	0.05	0.25	6	2.61	2.6	5.6	18	1.3	3.9
	0.10	0.05	8	4.66	4.6	7.9	24	2.6	4.6
	0.10	0.10	6	3.15	3.1	5.9	18	2.0	4.1
	0.10	0.25	4	1.74	1.7	3.6	12	1.1	2.8
	0.25	0.05	5	3.37	3.2	5.0	15	1.9	2.6
	0.25	0.10	4	2.54	2.4	3.9	12	1.3	2.2
	0.25	0.25	2	0.96	0.86	1.7	6	0.61	1.3
5	0.05	0.05	5	1.97	1.9	5.0	15	1.1	3.3
	0.05	0.10	4	1.37	1.4	3.9	12	0.83	2.9
	0.05	0.25	3	0.82	0.81	2.7	9	0.47	2.1
	0.10	0.05	4	1.74	1.7	4.0	12	0.99	2.5
	0.10	0.10	3	1.10	1.1	2.9	9	0.73	2.2
	0.10	0.25	3	1.10	1.1	2.9	9	0.40	1.5
	0.25	0.05	2	0.96	0.86	1.9	6	0.71	1.4
	0.25	0.10	2	0.96	0.86	1.9	6	0.50	1.2
	0.25	0.25	1	0.29	0.26	0.8	3	0.23	0.68

Note: If either n, T_0, or T^* is specified, the other two test parameters can be determined from the relationship

$$\frac{\chi^2_{1-\alpha,2r_0}}{2} = \frac{nT_0}{\theta_0} = \frac{T^*}{\theta_0}.$$

For expected total accumulated test hours: $E_{\theta_0}(T^*) = \theta_0 E_{\theta_0}(r)$; $E_{\theta_1}(T^*) = \theta_1 E_{\theta_1}(r)$.

For expected waiting time for n items on test: $E_{\theta_0}''(WT) = \dfrac{1}{n} E_{\theta_0}(T^*)$; $E_{\theta_1}(WT) = \dfrac{1}{n} E_{\theta_1}(T^*)$.

Table 14.4 A Time-Truncated
Sampling Plan
(c = Maximum Allowable Failures;
t = Preallocated Test Time, Hours)

$\alpha = 0.1$	$\beta = 0.1$	
c	t/θ_1	θ_0/θ_1
0	2.30	21.90
1	3.89	7.34
2	5.32	4.84
3	6.68	3.84
4	7.99	3.29
5	9.27	2.94
6	10.53	2.70
7	11.77	2.53
8	12.99	2.39
9	14.21	2.28
10	15.41	2.20

Source: "Limitations of Plans Designed to Demonstrate
Minimum Life with High Confidence," by A. C. Gorski and
B. Epstein, *Proceedings, Ninth National Symposium on Reli-
ability and Quality Control*, San Francisco, California
(January, 1963). Reproduced by permission.

Sequential Life Tests: Testing with Replacement*

Sequential life tests based on an exponential failure distribution are
commonly used for complex equipments to minimize the amount of testing.
For a given θ_0, θ_1, α, and β, the following values are computed:

$$h_0 = \frac{-\ln\dfrac{\beta}{1-\alpha}}{\dfrac{1}{\theta_1}-\dfrac{1}{\theta_0}}, h_1 = \frac{\ln\dfrac{1-\beta}{\alpha}}{\dfrac{1}{\theta_1}-\dfrac{1}{\theta_0}}, S = \frac{\ln\left(\dfrac{\theta_0}{\theta_1}\right)}{\dfrac{1}{\theta_1}-\dfrac{1}{\theta_0}}. \tag{14.6}$$

If n items are put on test, continuous decisions can be made as follows:

After t test hr are accumulated with r observed failures,

reject if $nt \le (-h_1+rs)$ $\qquad\qquad$ (14.7)

*B. Epstein and M. Sobel, "Sequential Life Tests in the Exponential Case," *Annals of
Mathematical Statistics* (March, 1955). It is recommended that the student obtain a copy
of Handbook H108, *Sampling Procedures and Tables for Life and Reliability Testing*.
(Washington, D.C.: Government Printing Office) (price 50 cents).

accept if $nt \geq (h_0 + rs)$ (14.8)

continue testing if $(-h_1 + rs) < nt < (h_0 + rs)$.   (14.9)

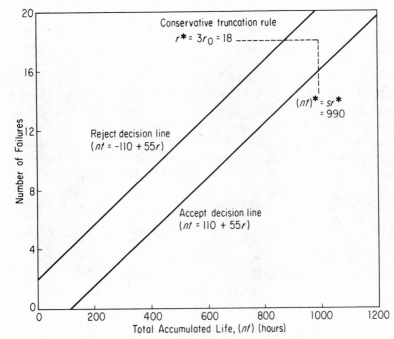

Fig. 14.1 Sequential replacement life test ($\theta_0 = 100$, $\theta_1 = 33$, $\alpha = \beta = 0.01$).

This decision criterion is shown graphically in Figure 14.1 for a specific test plan. The expected number of failures before a decision is reached is

$$E_{\theta_0}(r) = \frac{(1-\alpha)\ \ln\ \dfrac{\beta}{1-\alpha} + \alpha\ \ln\ \dfrac{1-\beta}{\alpha}}{\ln\ k - (k-1)} \tag{14.10}$$

$$E_{\theta_1}(r) = \frac{\beta\ \ln\ \dfrac{\beta}{1-\alpha} + (1-\beta)\ \ln\ \dfrac{1-\beta}{\alpha}}{\ln\ k - \dfrac{k-1}{k}} \tag{14.11}$$

where

$$k = \frac{\theta_0}{\theta_1}.$$

EXPECTED NUMBER OF FAILURES AND EXPECTED WAITING TIME

The relationships between $E_\theta(r)$, $E_\theta(T^*)$, and $E_\theta(WT)$ are the same as those for the truncated nonsequential replacement test given by Equations 14.3 and 14.4.

OPERATING CHARACTERISTIC CURVE

The O.C. curve of the sequential procedure shows the probability that the test will terminate with the acceptance of the hypothesis $H_0(\theta = \theta_0)$ for all values of the parameter θ (mean life); i.e., for any value of θ, the probability of making a correct decision can be immediately obtained once the O.C. curve has been determined. The O.C. curve thus describes what the sequential test procedure accomplishes. If this curve is denoted by $L(\theta)$—the probability of accepting H_0 when θ is the true mean life of the items being tested—then,

$$L(\theta) = \frac{A^h - 1}{A^h - B^h} \tag{14.12}$$

where

$$\theta = \frac{\left(\dfrac{\theta_0}{\theta_1}\right)^h - 1}{h\left(\dfrac{1}{\theta_1} - \dfrac{1}{\theta_0}\right)} \tag{14.13}$$

$$A = \frac{\beta}{1 - \alpha}, \quad B = \frac{1 - \beta}{\alpha}$$

and θ is determined by assigning values to h (for all real values ranging from $-\infty$ to $+\infty$) and solving for θ and $L(\theta)$. Five particular points on the O.C. curve are as shown in Table 14.5. From these points, a rough sketch can be made of the O.C. curve, and the additional points needed for more accurate detail can be ascertained.

Table 14.5 FIVE POINTS ON THE
O.C. CURVE

h	θ	$L(\theta)$
$-\infty$	0	0
-1	θ_1	β
0	S	$h_1/(h_0 + h_1)$
$+1$	θ_0	$1 - \alpha$
$+\infty$	$+\infty$	1

Graphically, the O.C. curve takes roughly the shape shown in Figure 14.2, for the case where $\alpha = 0.05$, $\beta = 0.10$.

TRUNCATION RULES

If θ is between θ_0 and θ_1, the expected number of failures is greater than both $E_{\theta_0}(r)$ and $E_{\theta_1}(r)$. In order to avoid having to test a substantial number of items, a truncation rule can be established for the number of failures or the total time accumulated, or a combination of the two. This procedure leads to a pair of decision lines which at some point begin to converge.

The following conservative truncation rules (based on the number of failures)† will have a negligible effect on the α and β errors:

1. The maximum number of failures is set at three times the number of failures required for an equivalent nonsequential test. Hence r^*, the truncation number, equals $3r_0$.

2. The maximum total accumulated test hours, nt^*, is set at sr^*, where s is the slope of the decision lines.

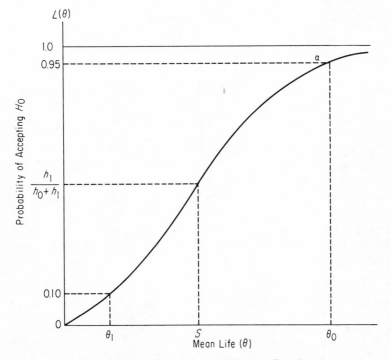

Fig. 14.2 O.C. curve for sequential sampling plan.

†These rules are used in *Sampling Procedures and Tables for Life and Reliability Testing*, Publication H-108 (Washington, D.C.: Government Printing Office).

Although the truncation number of failures may be large, it does prevent the possibility of a seemingly never-ending test. Also, the probability of r^* failures is usually very small. As an example, if $\alpha = 0.05$, $\beta = 0.05$, and $k = 3$, the rejection number for a nonsequential test is 10. The truncated sequential test would therefore be terminated at a maximum of $3(10) = 30$ failures. However, the probability is only 0.10 for the following cases: (1) requiring more than 7 failures when $\theta = \theta_0$; (2) requiring more than 12 failures when $\theta = \theta_1$; (3) requiring more than 26 failures in the worst possible case, i.e., for a value of θ between θ_0 and θ_1.

Although it is true that the rules above will provide negligible α and β errors, recent work by Burnett† using a Monte Carlo technique has shown that for the two AGREE sequential tests, the truncation time and number of failures can be reduced by approximately one-half of the conservative rule while maintaining α and β at the desired level or better.‡ The exact risks for the truncation points for the AGREE plans have been determined as follows.§

AGREE Task Group 2: $\alpha = 0.101$; $\beta = 0.117$
AGREE Task Group 3: $\alpha = 0.115$; $\beta = 0.126$.

SELECTION OF α, β AND k

The decision to accept or reject, based on test results, is closely associated with the choice of α, β, and k for the design of the sequential test plan. The values of α and β should be a compromise between the consumer and the producer with respect to the risk the consumer is willing to take in accepting what is marginal or unacceptably low, and the risk the producer is willing to take in having rejected a quality that is acceptably high. Often, economics limit available test time, and risks higher than desired may be necessary (typically, $\alpha = 0.05$ and $\beta = 0.10$ for acceptance tests associated with electronic parts; and $\alpha = 0.10$ and $\beta = 0.10$ for electronic systems).

Another variable is the discrimination ratio, k, between θ_0 and θ_1. (θ_0 is the *MTBF* associated with α; θ_1 is the *MTBF* associated with β.) The relationship between α, β, and k is such that as the risks and the value of k increase, less

†T. L. Burnett, "Truncation of Sequential Life Tests," *Proceedings, Eighth National Symposium on Reliability and Quality Control*, Washington, D.C. (January, 1962).

‡This procedure is based on the Epstein-Sobel approximations for the accept and reject lines. An alternate method, developed by L. A. Aroian, gives somewhat different results [L. A. Aroian, "Exact Truncated Sequential Tests for the Exponential Density Function," *Proceedings, Ninth National Symposium for Reliability and Quality Control* (January, 1963)].

§See "The Exact Analysis of Sequential Life Tests with Particular Application to AGREE Plans," by B. Epstein, A. Patterson and C. Qualls, *Proceedings, Aerospace Reliability and Maintainability Conference*, Washington, D.C., (May 6-8, 1963).

test time and fewer failures are required for making a decision. In summary, an analysis of the effects of α, β, and k should always be made before deciding on any one sequential plan. With the help of such an analysis, together with a thorough knowledge of the objectives of the test and the amount of resources available, including time and money to run the test, the technique of sequential analysis can be an invaluable and economical tool for decision making.

Example 5. A sequential test plan is required for the following conditions:

$$\theta_0 = 600 \text{ hr} \qquad \alpha = 0.1$$
$$\theta_1 = 400 \text{ hr} \qquad \beta = 0.1.$$

Procedure. Calculate h_0, h_1, and S (Equation 14.6) and plot the rejection and acceptance barriers (Equations 14.7 and 14.8):

$$TA = 2{,}636.4 + 486.564r,$$

and

$$TR = -2{,}636.4 + 486.564r.$$

Five points for the O.C. curve are obtained from Table 14.5:

θ	$L(\theta)$
0	0
$\theta_1 = 400$	0.1
$S = 487$	0.5
$\theta_0 = 600$	0.9
∞	1.0.

The expected waiting time to decision is obtained from Equations 14.3 and 14.4:

θ	$E_\theta(T^*)$
0	0
$\theta_1 = 400$	9,750
$S = 487$	14,292
$\theta_0 = 600$	11,154
∞	2,634.

14.7 Administrative Considerations for Demonstration Tests

AGREE Recommendations

The AGREE Report, *Reliability of Military Electronic Equipment,* gave considerable impetus to reliability testing of systems. The report of Task Group 2, AGREE, concerned the testing of equipments and systems in design and development, while Task Group 3 presented a procedure for system testing in the procurement phase. Both groups recommended sequential tests as the most favorable manner of reaching a decision. Understandably, the two tests differed in the number of equipment hours required. The preproduction test is shorter on the average and the production version requires more equipment time, as well as larger samples in accordance with the production lot size.

The two groups took a somewhat different approach with respect to the environment imposed upon the system or equipment under test. Task Group 2 considered the problems associated with prototype equipments to be sufficiently severe without imposing high environmental stress levels during the reliability test. It was also considered rather impractical to furnish elaborate environmental facilities for the short length of time required in a life test of this type. Task Group 2 recommended on-off and temperature cycling to the maximum called for in the equipment specifications. Task Group 3 took a more aggressive view of the need for simulating the usage environment in acceptance tests of a production item. Group 3 suggested four levels of testing: light (L), medium (M), high (H), and extreme (X), as described in the AGREE report and outlined in Table 14.6.

Table 14.6 SUMMARY OF AGREE ENVIRONMENTAL TESTS

Test	Temperature	Vibration	On-Off Cycling	Input Voltage
L	25°C ± 5°C	None	3 hr "on" plus time to stabilize at specified temperature	Nominal
M	40°C ± 5°C	25 ± 5 cps 1/32″ Ampl.	Same as "L"	Max. specified
H	−54°C to +55°C	Same as "M"	Same as "L"	Same as "M"
X	−65°C to +71°C	Same as "M"	Same as "L"	Same as "M"

Of principal concern in the present discussion are sequential test procedures; the two outlined in the AGREE Report are those most commonly used. Normally, the appropriate AGREE test is selected as the first approach

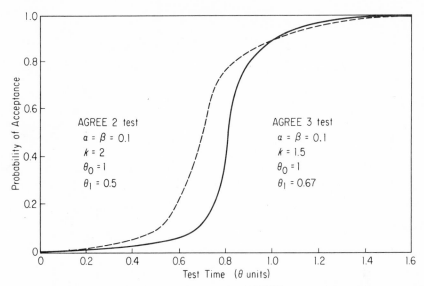

Fig. 14.3 Approximate O.C. curves for AGREE sequential test plans.

to testing any military system in design or production. Variations on these test procedures then follow as the peculiar circumstances of time, money, and equipment availability dictate.

Test procedure for *preproduction* equipments (Task Group 2) is essentially defined by $\alpha = \beta = 0.10$, and discrimination ratio $k = 2.0$. The operating characteristic curve for these conditions (Figure 14.3) is modified somewhat by early truncation of the test, which not only limits the maximum test time but also changes to some extent the probabilities, α and β. As indicated in the AGREE Report (p. 89), this test is truncated at a point in time equal to 10.3 times the value of θ_0, therein referred to as the *contract mean time between failure*. Also, as shown in the report, a decision to reject cannot be made until a minimum of 8 failures has occurred. Truncation of the test at a maximum of 15 failures, providing truncation by time has not yet taken place, is also built into the test, at which point the decision to reject is made.

The report recommended accumulation of a minimum of $3\theta_0$ test hours before acceptance, provided the test is failure free. This recommendation was based on engineering judgment "to give some assurance that equipment with unduly short life will be rejected."

The test procedure recommended by Task Group 3 for *production* equipments is also based on an α and β error probability equal to 0.10; however, the discrimination ratio (k) in this case equals 1.5. The approximate operating characteristic curve, included in Figure 14.3, is also modified by early truncation. Truncation in this case takes place at 33 times the value of θ_0,

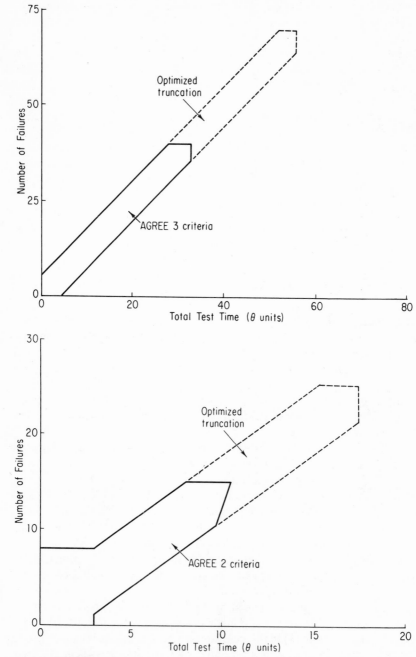

Fig. 14.4 Graphical representations of sequential test plans recommended by AGREE task groups 2 and 3 (with optimized truncation recommendations).

the contract-specified *MTBF*. This modification, as noted previously, would result in some changes in the values of α and β. Also, as indicated in the AGREE Report (pp. 136-137), 6 failures is the minimum number required before rejection of the test. A maximum of 41 failures is permitted, at which point the test is truncated and a reject decision made, provided the test has not been truncated by time. This test cannot be accepted until a minimum of 4.4-times-θ_0 test hours have been accumulated, provided no failures occur.

The early points of truncation provided in both of the test procedures above have resulted in comments from statisticians that the AGREE recommendations for truncation may introduce errors in α and β. As indicated in the discussion of "Truncation Rules," on pages 521-522, recent work employing a Monte Carlo technique indicates that the AGREE truncation points may introduce errors of the order of 15 per cent to 30 per cent in α and β, and that the optimum truncation conditions are approximately 50 per cent more than the AGREE recommended points. Figure 14.4 illustrates the truncation points for the sequential plans recommended by both task groups.

Obviously, the advantage of early truncation, as set forth in the AGREE Report, is that the test will be terminated sooner with smaller lead time requirements and lower over-all costs. Early truncation, therefore, does make a test feasible where it might otherwise impose an unbearable burden on military schedules and expenditures. Some testing is usually preferable to none, and confidence in the results will always be less than ideal simply due to time and money limitations.

Procedure for Developing Reliability Demonstration Tests

Following is a step-by-step procedure for developing reliability demonstration tests for electronic equipments.

Step 1. *Functional Description of Equipment.*

An indispensable foundation for successful testing is the establishment of a comprehensive description of the operational mission, environmental conditions (operational, storage, transportation, etc.) mission time, service life, operating personnel, and maintenance problems.

Step 2. *Selection of Test Environment.*

The test environments should be related to the anticipated use environments and are usually specified in the equipment contract. (See Table 14.6 for typical examples.)

Occasionally, "accelerated" tests have been specified for uncovering design weaknesses early in the program, when modifications can be accomplished most efficiently. They have also been suggested for reducing the required test time to prove a given reliability requirement (assuming that a reasonable acceleration factor can be determined).

For some applications the higher AGREE stress levels can be considered as accelerated conditions, and the establishing of the relationship between (1) the *MTBF* determined under accelerated conditions, and (2) the expected field reliability, has often been considered desirable. However, accelerated tests present several problems if correlation between the accelerated condition and the anticipated use condition is required. Some of these problems are

1. Failure modes in the accelerated environment must be the same as those occurring when the equipment is performing its intended function.*

2. The accelerated environments must be carefully controlled and measured, since small errors in measurement may cause large errors in the acceleration factor.

3. Failure rates for different components can, among themselves, vary considerably in the accelerated environment (see Chapter 9).

Step 3. *Test Criteria for Operational Parameters.*

To provide means for decision making and to prevent arguments between the customer and producer, criteria generated *before the start of the test* provide the basis for demonstrating whether or not the intended function has been satisfied. The criteria establish (1) the parameters required to assure proper equipment operation, (2) quantitative limits on these parameters, (3) a schedule of preventative maintenance operations, and (4) those malfunctions which would seriously reduce the effectiveness of the equipment.

Aside from the establishment of criteria of success and failure, consideration must be given to establishing those items which are to be excluded from the test.

Step 4. *Selecting the Test Plan.*

Time and cost generally prohibit an exact determination of *MTBF* for every system coming off the production line. A few systems are tested periodically, and some calculated risks are taken in accepting or rejecting a system (or a production lot) on the strength of test results. Sequential test plans are frequently used because the expected waiting time to decision is significantly less for sequential plans than for life tests terminated at a preassigned time (each method having similar O.C. curves).

At times even sequential plans prove to be too expensive, and it is necessary

---

*An example of the practical difficulties is provided by a large-scale study on communications receivers in which operation at high line voltages caused rapid failures of audio-output tubes, while operation at low line voltages caused failures of RF oscillators.

to resort to shortened fixed-time tests that provide some control over primary failure modes, but only limited assurance that the required *MTBF* is achieved.

Step 5. *Determination of Test Duration.*

The time available for reliability testing ranges from the earliest date the equipment will be available for test to the date a decision regarding the acceptability of the equipment is required. Calendar time for conducting the test can be reduced by increasing the number of equipments which are tested simultaneously; however, the availability of facilities, test personnel, and equipments to be tested must be considered.

Step 6. *Establishment of Monitoring Methods and Data Log.*

Consideration must be given to the establishment of appropriate test measurement methods (automatic or manual), and to the development of an appropriate data log in which critical test parameters as selected in Step 3 are recorded, along with data indicating time of failure. Specific data requirements are often included in equipment specifications.

Step 7. *Data Analysis.*

The reliability test itself does not indicate the true *MTBF* of the system under test. Methods given in Chapter 5 show how a reliability estimate can be made from the data developed from the life test, and, as shown in Figure 5.16, Chapter 5, a reasonable number of failures must be observed (say 10, or preferably 30 or more) so that a realistic reliability estimate can be made.

Reliability growth data should be developed, a simple matter of comparing reliability estimates of successive lots.

Step 8. *Failure Analysis.*

An indispensable portion of the reliability demonstration program is a complete analysis of each failure to determine its source. Early in the program, mechanical design weaknesses, part application or circuit design problems, and assembly or workmanship problems predominate. Good quality control procedures are mandatory at this point. If a "reject" decision is reached as a consequence of the test program, the equipment specification or contract will generally require a complete failure analysis, as well as a determination of the type of rework or design modifications that may be required. The "reject" decision may also play an important part in fee redetermination, as required in some of the newer incentive contracts.

PROBLEMS

1. A total of 168 hr of test time is available for a time-truncated reliability demonstration test. $\theta_0 = 54$ hr; $\theta_1 = 18$ hr; $\alpha = \beta = 0.1$. What is the maximum number of allowable failures?

2. It is desired to prove a maximum failure rate of 0.001 per cent per 1000 hr at the 90 per cent level of confidence for a production lot of metalfilm resistors. How many test samples are required if 2 failures are allowed in a nonreplacement fixed-time test?

3. Why is it that, for a given value of α and β, fewer allowable failures are required as the ratio of θ_0 to θ_1 increases for either fixed-time or sequential life tests?

4. Given $\theta_0 = 100$ hr, $\theta_1 = 33$ hr, $\alpha = \beta = 0.05$. (a) Plot the ACCEPT and REJECT decision lines and the O.C. curve and determine the expected waiting time if two equipments are tested simultaneously.   (b) Same as (a), except $\alpha = 0.05$, $\beta = 0.1$. (c) Same as (a), except $\alpha = 0.10$, $\beta = 0.05$.

5. What difficulty arises in applying the AGREE Task Group 3 test plan to a system with an *MTBF* requirement $\theta_0 = 5000$ hr? What would be a practical solution to this problem?

6. Comment on the use of the AGREE truncation points.

7. A reliability demonstration test has a requirement $\theta_0 = 100$ hr, $\theta_1 = 33$ hr, $\alpha = \beta = 0.1$. What is the true *MTBF* of any system that passes this test, and how can it be determined?

Effectiveness and Cost

15.1 Introduction

As indicated in Chapter 1, a fundamental objective in the building of a system is that it be capable of performing its intended function at the lowest total cost. Costs must be related to system effectiveness for such purposes as (1) comparing competitive systems, (2) determining the number of systems that are required, and (3) minimizing support costs.

Relationships between cost and effectiveness can be expressed by curves like that in Figure 15.1, in which one axis has an effectiveness measure, the other a dollar sign (more often, a million-dollar sign).

The points in Figure 15.1 represent different systems, or different ways to build a system, or to go about building it, or to operate and support it. There are many points above the curve. These represent "bad" ways of going about the job—evidently the same effectiveness can be achieved for smaller amounts of money. This, however, was not known until somebody looked at the alternatives; so the nonoptimal points form a necessary part of any cost-vs.-effectiveness curve. Unfortunately, these points are usually suppressed, leaving only the curve of Figure 15.1.

These curves are, for all but the very highest level of military analysis, a terminal output. At the highest level there might be two, representing two types of systems which contribute toward the same strategic objectives— e.g., missiles and civil defense activities. It is then possible to combine the

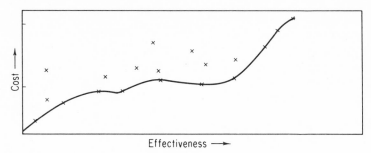

Fig. 15.1 Cost-effectiveness points and curve.

information from the two curves to find an optimal mixture. However, for contractors it generally suffices to produce one such curve for the system which they are proposing to build, or are building.

One dimension has been omitted from Figure 15.1 and that is time. The costs shown are so-called *integrated costs*, meaning total costs over the lifetime of the system. The time scale is nevertheless very important. We want to know not only what we will get for our money, but also when we will get it. Figure 15.2 presents some of the timing considerations by means of a three-dimensional drawing. The curve, representing the points of Figure 15.1, stays for a while in the time-cost plane, because time and money must be expended to produce anything; then it begins to rise from the plane until the peak effectiveness of the particular system to which the curve applies is reached. The figure does not extend the curve beyond this point, but it would maintain the same effectiveness at a roughly constant rate of expenditure for a while, after which, with obsolescence, the effectiveness would start to drop.

Again, the curve of Figure 15.2 represents, for a system, the total cost—everything that has been spent to research, develop, engineer, produce, install, operate, and support the particular system. It is evident that the actions and

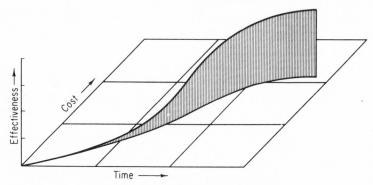

Fig. 15.2 A cost-time effectiveness curve.

materials needed for each of these phases are sufficiently different so that we need to estimate the cost of each phase separately. And from the discussion above it is clear that we must estimate the expenditure on each phase as a function of time.

The major concern of this chapter is to show how reliability and maintainability enter cost estimates. They enter chiefly in two ways:

1. by determining the number of weapons that must be acquired to achieve a given effectiveness level

2. by determining the support cost of the system (information on support cost is of value not only as a basis for deciding between competitive systems, but also for the purposes of minimizing support costs in field operations, optimizing maintenance policies, and assessing the value of proposed field modifications before they are installed on a fleet-wide basis).

The discussion of these points will be preceded by a brief section on ways of estimating the cost of other phases of system cost.

15.2   Methods of Cost Analysis

General Description

The method followed in any cost prediction is straightforward enough, but apt to be very laborious. Furthermore, the data on which any prediction must be based are difficult to collect, and the gross estimates which we are then forced to employ must be treated with a good deal of reserve.

To make any cost prediction at all, it is necessary to (1) break the expenditures down into rather small categories, (2) collect as much past experience on expenditures in each category as possible, and (3) predict from this information how much is likely to be spent in each category for the project being costed. Thereafter, all the categories must again be put together to obtain the system cost as a function of time.

The significance of this discussion is that the research for, and the development, testing, acquisition, and operation of, a new weapon system constitute a vast undertaking. It requires the best effort of many people, both in industry and in government. All of the groups of people concerned perform tasks; some of these tasks will result in hardware, some in documents. Each group performs its task as a profession, and has in the past performed roughly similar tasks for different systems, and perhaps under different circumstances. However, the major project is in essence conducted by assembling competent groups of people, experienced in their work. The smaller the groups, and the smaller the tasks into which the major project is divided, the more the output from the individual task will look like something that has been done before, and therefore the better a prediction based on past experience is apt to be.

In Section 15.3, such a process will be described for a particular expenditure heading. This section will be concluded with a listing of the major expenditure headings that must be used. The portion of the bibliography relating to Section 15.2 will list some of the fundamental references in cost analysis.

Major Categories

The major categories into which costs must be divided are as follows:

A. *Research and Development Costs*
    1. System development
    2. System test and evaluation
    3. Other system costs

B. *Investment Costs*
    1. Installations
    2. Equipment
        (a) Primary mission
        (b) Specialized
        (c) Other
    3. Stocks
        (a) Initial stock levels
        (b) Equipment spares and spare parts (initial)
    4. Initial training
    5. Miscellaneous
        (a) Initial transportation
        (b) Initial travel
        (c) Intermediate and support major command

C. *Operating Costs*
    1. Equipment and installations replacement
        (a) Primary mission equipment
        (b) Specialized equipment
        (c) Other equipment
        (d) Installations
    2. Maintenance
        (a) Primary mission equipment
        (b) Specialized equipment

　　　(c) Other equipment
　　　(d) Installations
　3. Pay and allowances*
　4. Training
　5. Fuels, lubricants, and propellants
　　　(a) Primary mission equipment
　　　(b) Other
　6. Services and miscellaneous
　　　(a) Transportation
　　　(b) Travel
　　　(c) Other (including maintenance of organization equipment)

Amortization and Depreciation

In commercial cost estimates, it is common practice to add another step to the cost-estimation procedure described in the preceding section—distributing the initial costs (Categories I and II) over the useful life of the equipment. This process is called *amortization* for Category I and *depreciation* for Category II. In military cost estimates, amortization and depreciation are not practiced because:

1. Government directives call for yearly estimates of actual expenditures; this information aids in the preparation of the annual budget.

2. In making lower-level predictions—e.g., whether or not to replace a system—the Category I and II expenditures for the old system have already been made. Therefore, a cost comparison should be made between the Category III expense of the old system and the Category I, II, and III expenses of the new.

For commercial equipments, however, the amortization and depreciation costs are a very important consideration, since the real test of usefulness of such equipments is the net revenue they produce. The investment costs can usually be covered, out of borrowed money or capital; however, the proposed equipment competes with other uses of such funds and must therefore produce either income to offset the income that the other uses would produce or make a reduction in expenditure to offset the investment made on it.

Since amortization and depreciation deal with net revenue, certain factors enter into the process:

1. The expected useful life of the equipment

2. The way in which the equipment can be evaluated for tax purposes

*For operation and maintenance.

Commercial amortization and its resultant effect on cost estimates are therefore intimately connected with the tax situation. The necessary computations are left to the tax accountant, who can use the estimates in the categories of the preceding section along with knowledge of the expected useful life of the equipment.

15.3   Costing the Support System

General Methodology

As an example of what cost coefficients are, how they can be measured, and how they can be used, we shall discuss the costing of the support function of a weapon system.

The support function comprises maintenance, service, and supply. It will be seen that the function splits naturally into additive terms for the different echelons, which are very simply connected by the values of parameters representing costs accrued and time spent at higher echelons.

To arrive at cost, we note that the principal commodities for which money is spent are

1. time
2. men
3. matériel.

So at each stage in the life of the equipment, we must obtain data on the amount of matériel, manpower, and time that the stage consumes. Each of these quantities must then be converted into money, and the total obtained to get an estimate of the cost of the stage. Many estimates, from many observations, will then furnish a statistical basis for predicting the future cost of similar stages.

Reliability and maintainability enter this process in a perfectly natural way:

Reliability, or rather the lack of it, triggers the support system, and hence determines how often a particular equipment will pass through different stages and consume the corresponding amount of funds.

Maintainability determines the amount of direct manpower needed to effect repairs at each stage. Hence the maintainability of the equipment triggers the manpower consumption for each stage. It also accounts for part of the time consumed at the stage. However, the principal component of the time consumed is in the delays in the system (shelf time, paper-shuffling time, transportation time, etc.).

A Simplified Example of a Support Function

To see this methodology in operation, let us take a very simple example:

A group of aircraft is maintained by a dual maintenance organization, consisting of flight-line (first echelon) maintenance, and shop (second echelon) maintenance.

If a complaint against an aircraft system is received, flight-line personnel try to verify the complaint; if they verify it, they

(a) perform some maintenance on the aircraft (this may consist of changing a black box), and

(b) in a certain percentage of the cases, generate some shop maintenance.

If they do not verify the complaint, they have, of course, spent some time in the investigation; however, that complaint is disposed of.

There are two supply systems, one for bits and pieces and one for reserve black boxes.

This support organization is represented in Figure 15.3.

To cost out the support of this black box during the month, the procedure is as follows:

1. Each complaint is verified, or not verified, at a manpower expenditure of t_v.

2. A fraction v, of the complaints are verified, and of these a fraction, f, are resolved by repairs on the flight line (taking t_f hr each), and a fraction $(1-f)$ are corrected by black-box changes, which take t_c hr per change.

   Let N equal the number of complaints per month. Hence the total direct man-hour expenditure, T_{DF}, on the flight line is

$$NT_{DF} = N[t_v + vft_f + v(1-f)\,t_c] \qquad (15.1)$$

   where N is the number of complaints per month.

3. To this time, the overhead time for the flight-line organization must be added, and then the cost of this time must be computed. Let us denote this by

$$NC(T_f) = \text{total cost of flight-line manpower (loaded) per month.} \quad (15.2)$$

   where $C(T)$ is the loaded flight-line maintenance manpower cost.

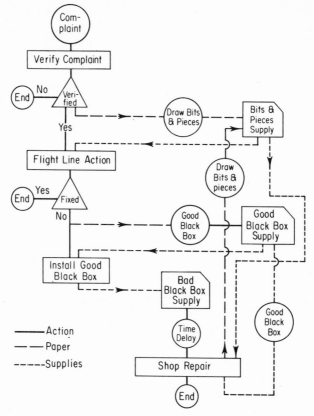

Fig. 15.3 Maintenance and supply in a two-echelon system.

4. The materials used are

   (a) vf (bits and pieces for flight-line fix), and
   (b) $v(1-f)$ (black boxes).

   To cost these we need to know the complete costs of black boxes and of bits and pieces at the flight line. These are costs which depend on the cost of the supply system. The costing of bits and pieces will be considered first.

5. Cost of bits and pieces for flight-
   line maintenance to supply $= C_f$

   Direct manpower to get them $= S_l$

   Cost of this manpower, with overhead $= C(S_l)$

   Total cost of bits and pieces on
   flight line $= C_f + C(S_l).$ (15.3)

   The result obtained in Equation 15.3 is then substituted into (a) of (4).

6. Cost of black boxes to supply.* This is where the time delay, ΔT, is charged. Evidently if a black box could be fixed at once, there would be no need to have a supply on hand. The supply, then, furnishes a buffer against the time delay, and as a first approximation we need an average of $\dfrac{\Delta T}{v(1-f)N}$ on hand to cover the black boxes needed during the delay. If the average life of one of these boxes on the base is L months, then the average cost of this buffer-supply is

$$\text{Cost of black box at maintenance} = \frac{1}{L} \cdot \frac{\Delta T}{v(1-f)N} \cdot C_c = C_B \quad (15.4)$$

where C_c is the cost of black box to supply, $C(S_c)$. To this must be added the manpower to obtain a black box. The total cost of a black box on the flight line is given by

$$\text{Cost of black box on flight line} = C(S_c) + \frac{1}{L} \cdot \frac{\Delta T}{v(1-f)N} C_c. \quad (15.5)$$

This result is then substituted into (b) of (4) above.

7. The cost of flight-line maintenance is then

$$\begin{aligned}
\text{Cost of flight-line maintenance} = \; & NC(T_F) + Nvf[C_f + C(S_l)] \\
& + Nv(1-f)\,[c(S_c) \\
& + \frac{1}{L}\frac{\Delta T}{v(1-f)N}\,C_c].
\end{aligned} \quad (15.6)$$

This cost can be split into a maintenance-manpower cost,

$NC(T_F) =$ maintenance-manpower cost, and a maintenance-materials cost,   (15.7)

$$\begin{aligned}
C\,(\text{materials}) = \; & Nvf\,[C_f + C(S_l)] \\
& + Nv(1-f)\,C(S_c) \\
& + \frac{\Delta T}{L}\,C_c.
\end{aligned} \quad (15.8)$$

The maintenance-materials cost can again be split into a supply-materials cost,

$$C\,(\text{supply materials}) = Nvf\,C_f + \frac{\Delta T}{L}\,C_c \quad (15.9)$$

---

*The method here used to charge the time delay is essentially suited to a "one time look" at the system. If the system is observed over a period of time, and black boxes are bought for supply, a better method can be used. See, for example Ref. 9.

and a supply-manpower cost

$$C(S_F) = N_v[f\,C(S_l) + (1-f)\,C(S_c)].\qquad(15.10)$$

8. The cost of the shop maintenance is computed in the same way as above, and turns out to be

Cost of shop maintenance $= Nv(1-f)\,[C(T_s) + C(S_s) + C_s]\quad(15.11)$

where

$C(T_s) =$ cost of the (loaded) maintenance manpower in the shop,

$C(S_s) + C_s =$ cost of materials at the shop level,

$C_s =$ cost of these materials at supply, and

$C(S_s) =$ cost of the (loaded) supply manpower needed to get the materials for the shop.

If we put equations 15.6 and 15.11 together we obtain the support cost for the system during the month:

$$\begin{aligned}\text{Support cost} = N\{&C(T_F) + vf(C_f + C(S_l)) \\ &+ v(1-f)\,[C(S_c) + \frac{\Delta T}{Lv(1-f)N}\,C_c \\ &+ C(T_s) + C(S_s) + C_s\}.\end{aligned}\qquad(15.12)$$

This equation will give us a cost allocation, if we have numbers for the symbols on the right. If we have estimates or predictions for these quantities, then we can use Equation 15.12 to predict support costs.

The next section will deal with the task of obtaining and predicting these numbers.

Data Inputs

The quantities which are used in the derivation of Equation 15.12 are explained below:

1. the number of complaints per month $\qquad\qquad N$
2. the fraction verified $\qquad\qquad v$
3. the manpower needed to verify $\qquad\qquad t_v$
4. the fraction fixed on the line $\qquad\qquad f$
5. the manpower needed to fix the complaint on the line $\qquad\qquad t_f$
6. the cost of the bits and pieces needed on the line $\qquad\qquad C_f$
7. the manpower needed to get the bits and pieces on the line $\qquad\qquad S_l$

8. the manpower needed to replace the black box $\qquad$ t_c

9. the cost of the replacement black box $\qquad$ C_c

10. the manpower needed to get the black box $\qquad$ S_c

11. the time delay before the bad box is fixed and ready for issue $\qquad$ ΔT

12. the time needed to fix the black box in the shop $\qquad$ T_s

13. the cost of the bits and pieces needed in the shop $\qquad$ C_s

14. the manpower needed to get the bits and pieces in the shop $\qquad$ S_s

15. the cost of loaded manpower for various categories of direct labor $\qquad$ $C(T_j)$

$$j = v, f, c, s,$$ $\qquad$ $C(S_k)$
$$k = l, c, s$$

16. the average lifetime of a black box $\qquad$ L

17. the cost of various categories of loaded manpower $\qquad$ $C(S), C(T)$

We must now indicate what these quantities represent in terms of a weapon system, and how estimates of their size may be obtained.

1. *The quantity N.*

The quantity N is composed of usage rate, reliability, and the number of black boxes in use. If there are n black boxes being used, each on the average t hr a month, and if the complaint-reliability, i.e., the mean time between complaints ($MTBC$) on the black box, is λ_c complaints per hour of use, then

$$N = \frac{nt}{\lambda_c}. \qquad (15.13)$$

The $MTBC$ can be estimated from field failure data, such as have been compiled by ARINC Research in Reference 10.

The usage rate, t, must be estimated from deployment plans, as must the number, n, of boxes in use. On a projected weapon system all these factors will be available. Hence N can be estimated.

2. *The fraction of verified complaints.*

The fraction of verified complaints, v, is another output of the observation of field failure data. If λ_f is the mean time between verified complaints, then,

$$V = \frac{\lambda_c}{\lambda_f}. \qquad (15.14)$$

3. *The fraction of verified complaints which can be fixed on the flight line.*

The fraction of verified complaints which can be fixed on the flight line, f, is estimated essentially from two pieces of information:

(a) A maintenance plan, which defines the repairs that will be made on the flight line. (Adjustments, for instance, will often be made there.)

(b) A reliability prediction in greater detail than those needed for λ_c and λ_f, namely, a breakdown of λ_f into those cases which will be fixed on the flight line, and those which will have to go to the shop. Such predictions can be obtained from a more detailed knowledge of the equipment; or, alternately, statistical values of the fraction f for similar equipments now in service can be used as estimates.

4. *The cost of maintenance manpower.*

Direct maintenance manpower is obtainable from maintainability predictions, as discussed previously in Chapter 10. Suppose the direct maintenance time required, and the corresponding hourly pay, required on the average to accomplish the maintenance actions are, respectively,

$$t_i \quad \text{and} \quad C_i,$$

where the subscript "i" refers to the skill class. Then t_i can be obtained from maintainability predictions, such as are given in Reference 12, and C_i can be obtained from lists of pay classes, together with an estimate of the useful life of maintenance men in grade. Here a suitable definition of *useful life* might be the percentage of the time in grade during which the man is actually assigned to maintenance duties.

Besides direct labor, there is in any organization a great deal of overhead labor. Much of this is concerned with scheduling and supervision, some with management, and a good deal goes to leave, training, and nonmaintenance duties of the men themselves.

In general, the loaded time will be a linear function of the direct labor time. If T_i is the loaded time (direct *and* overhead) spent in labor class i, then

$$T_i = a_i + b_i\, t_i. \tag{15.15}$$

The cost of labor is then given by

$$C(T) = \sum_i T_i\, C_i = \sum a_i\, C_i + \sum b_i\, t_i\, C_i. \tag{15.16}$$

The constants involved in Equation 15.16 are the overhead coefficients, a_i and b_i. Rough estimates of these can be made from tables of organizations and from estimated work-loads.

A good approximation to Equation 15.16 can be obtained in the form

$$C(T) = C_a + C_b \cdot t, \tag{15.17}$$

where t is the total active repair time in all labor classes. For the one case analyzed—based on seven months' worth of data for all maintenance actions on aircraft—the coefficients were roughly as in Table 15.1.

It must be stressed that these numbers, although correct for the period and place from which they were derived, have not been checked for general applicability. They are, therefore, examples, not cost factors.

Table **15.1** Manpower Cost Coefficients

Location	C_a	C_b
Flight line	$10	$2.6/hr
Shop	$9	$1.9/hr

5. *The materials costs at supply.*

The materials costs at supply, C_f, C_c, C_s, are the parameters which connect the different echelons. To estimate them we need, again, two kinds of information: (a) the average amount and kinds of materials needed to perform repairs; and (b) the cost of these materials to supply. These categories are discussed below.

(a) Estimates of the kinds and quantities of materials can be obtained either from a statistical analysis of the behavior of similar equipments in present use or from a detailed reliability analysis, based on actual schematics.

(b) The cost of these materials at the supply echelon in question will consist of

1. The cost at the higher supply echelon which supplies the one in question
2. The cost of the labor needed at the higher echelon and at the echelon in question to move the materials
3. the cost of transportation

The materials cost at one echelon then contains implicitly the accrued supply costs at all higher echelons, and thus the whole system support cost will be accounted for.

6. *The cost of supply manpower.*

The labor cost at supply, $C(S_l)$, $C(S_c)$, $C(S_s)$, must be obtained by an analysis similar to that described for the maintenance manpower. If a detailed analysis is not available, probably the best estimate obtainable is to assume that every action, i.e., every requisition and every issue, takes, on the average, about as much labor as every other. Then if the total payroll of the supply organization is divided by the number of pieces of paper generated, an estimate of the labor cost of requisitions and issue is obtained that should serve well enough.

7. *The time delay; the average life of a black box.*

The cost of time delay [involving ΔT, L, and C_c—the time delay, the average life of a black box in the (partial) system, and the cost of the black box to supply, respectively] has been given as

$$\frac{\Delta T}{L} C_c.$$

Estimation of C_c has been discussed in 5, above. Time delays must be estimated by observations on similar support organizations. The average life can be estimated from condemnation rates, return rates to higher echelons of maintenance, and the total number in circulation.

If on an average a black boxes a month are condemned, and b are returned to higher echelons of maintenance for repair, then if there are n boxes in circulation

$$L = \frac{n}{a + b}. \tag{15.18}$$

Again, a and b can be estimated from a detailed reliability analysis and a maintenance plan, or from statistical values for similar equipment. The number n is determined by the verified failure rate and by logistic policy.

Data Requirements

The basic reliability and maintainability data compiled to support the requirements of the mathematical model should meet the following requirements.

1. *The data should be available in sufficient quantity to provide significant sample sizes of the various system characteristics and parameters being studied.*

The confidence in test results increases with the quantity of observations. Accordingly, every effort must be made to acquire sufficient data from actual surveillance of the hardware in an operational environment. (If, however, adequate data of this type are not available, it is not necessary to resort to mere estimates. Several proven techniques are available for predicting the reliability and maintainability of hardware items. These techniques are discussed in Chapters 9 and 10.)

2. *The data should reflect current system conditions.*

Timely collection of input data is required if the mathematical model is to depict current conditions in the hardware it simulates. Many diverse factors affect the cost of operating and maintaining a modern weapon system. Unfortunately, at least from a model development standpoint, almost all of these factors are dynamic in nature. Reliability, for instance, may change as the result of (a) changes in the skill level of technicians, (b) engineering changes in the hardware, or (c) transition of a new system from a "debugging" phase to maturity. Maintainability may be changed by (a) revision of spare-components allocations, (b) the obtaining of more (or better) test equipment, and (c) use of more (or more highly skilled) maintenance personnel.

3. *The data should be accurate.*

The importance of using the most accurate data available cannot be over-emphasized. The stringent requirement for accurate data is generated by the intrinsic nature of the mathematical model. Many compound summing operations (or multiplications) will be accomplished during the exercising of the model; thus, any inaccuracies in the data will also be compounded during these mathematical manipulations.

4. *The data should be representative of the operational situation of interest.*

Equipment reliability and maintainability are known to be affected by the operational and maintenance environment. Until such time as the direction and extent of the various influencing factors upon system reliability and maintainability are more explicitly defined, it will be desirable to collect data from the specific operational situation that is being simulated in the mathematical model.

General Comments

GENERALITY OF THE PROCEDURE

The procedure presented in Section 15.3 will form a framework for modeling any support (combined maintenance and supply) system.

The example was given in terms of black box repairs. However, inspections, overhaul, periodic replacements, and servicing will fit into the same general scheme. For instance, a periodic replacement may be regarded as a verified complaint with a nonprobabilistic mean time between occurrences. Some of these have shop follow-ons, others may not. For these replacements we then have

$$\lambda_f \text{ is a constant set by policy;}$$

$$v = 1, f = 1 \text{ or } 0, \text{ depending on policy.}$$

CONNECTIONS BETWEEN ECHELONS

The connections between echelons are made through parameters which govern the flow of materials and the accrued cost:

f and L determine the demand placed by lower echelons on higher echelons;

C_f, C_c, C_s contain the cost accrued at higher echelons, as they enter in the cost to the lower echelons.

SIMPLIFICATIONS

The chief simplification that has been employed in this discussion is the use of averages for probabilistic variables. This has been done both in the allocation of manpower costs, and in the charging of the time delay ΔT and the bits and pieces used for various repairs.

It is only in the latter two instances—the costing of time and the supply inventories—that the use of averages introduces a significant error. The error will consist of underestimating the cost.

In actuality, supply will carry stocks which are much larger than the average monthly usage rate. They must do this for several reasons:

(a) the replenishment time—the time to replace stocks used—may be longer than a month;

(b) the usage itself will be a probabilistic variable, and stocks on hand must protect against a chance bunching of demand for one item in any replenishment period.

Much mathematical theory* exists to account for these effects, which will not be treated here. The net effect is that the cost for supply is subjected to inflationary factors which account for the safety stocks. For instance, C_f, which is the cost of supplies used at supply, will be replaced by

$$C(\Delta T_1 + \Delta T_2; \sigma), \tag{15.19}$$

where

$\Delta T_1 = $ the part of the replenishment period due to the local supply activity,

$\Delta T_2 = $ the part of the replenishment period due to higher echelons, and

$\sigma = $ a measure of the variability of usage.

Similarly the charge for the time delay ΔT, instead of being

$$\frac{\Delta T}{L} C_c, \tag{15.20}$$

will be in fact some fraction of the more complicated function

$$\mu(\Delta T + \Delta T_1 + \Delta T_2; \sigma)C_c, \tag{15.21}$$

which represents the stocks carried as a function of the total delay time and demand variability. Which part of these stocks is assigned to the particular ΔT is, to a certain extent, the decision of management.

*See Reference 13.

COST OF INSUFFICIENCIES IN SUPPORT SYSTEM

The discussion above deals with the cost of creating and operating a support system. Also related to the support system is the cost of the time during which the weapon system is unavailable—while it is being worked on by the support system. This "unavailability cost" will be briefly discussed in this section.

Essentially, the unavailability cost consists of

1. A planned portion, which includes the additional weapon systems purchased to cover both the unavailability and ineffectiveness of the primary weapon system

2. An unplanned portion, which results from the failure of the support system to live up to its specified performance

For example, if six extra missiles are purchased as back-ups for the six that are likely to be undergoing maintenance at any one time, then it may happen that at some instant eight are undergoing maintenance; hence, the weapon system is two missiles short. This effect is extremely hard to translate into dollars and cents; however, it is a *reduction in effectiveness at no change in cost*. Therefore, such a situation enters into the effectiveness side of a cost-versus-effectiveness curve.

However, the six extra missiles can easily be costed, and their cost related to the anticipated time delays in the support system. This procedure should be followed whenever a cost estimate is made of a system still going into inventory. The necessity for costing out these time delays is another reason for carefully noting the delay times in the support system. (In a commercial application the equivalent factor which must be considered is the loss of revenue resulting from system down time.)

15.4 Summary

The technique developed in Section 15.3 will furnish a cost model for any support system. The particular form this model will take in any instance depends upon the plan of support for the specific system. Such a plan of support forms part of the system description which is necessary for both cost and effectiveness predictions. Other elements of the description for a military system will be (1) the mission, with expected enemy counteraction, (2) the method of operation, and (3) a description of the physical makeup of the system and its major support elements (bases, airfields, ground support equipment, etc.).

The cost model for commercial applications should include consideration of loss of revenue due to system down time as a factor influenced by reliability and maintainability.

The description can be given either explicitly or by analogy, and will generally be a mixture. For instance, some of the hardware may be available in function-diagram form; some of the blocks may have been worked out in great detail, while the support plan and system are assumed to be "standard" for the agency which will operate the system. If this is the only information available, then it is the information that must be used to develop both cost and effectiveness predictions. The accuracy of the results will be no better than the quality of the input information; it is the duty of the analyst to make sure that it is no worse.

REFERENCES

References for Section 15.2

1. D. Novick, *New Tools for Planners and Programmers.* The RAND Corporation, P-2222, February 14, 1961.

2. A. Enthoven and H. Rowen, *Defense Planning and Organization.* The RAND Corporation, P-1640, March 17, 1959, revised July 28, 1959.

3. C. J. Hitch and R. N. McKean, *The Economics of Defense in the Nuclear Age.* The RAND Corporation, Report R-346, Cambridge, Mass.: Harvard University Press, March, 1960. See especially Part II.

4. H. Kahn and I. Mann, *Techniques of Systems Analysis.* The RAND Corporation, Research Memorandum RM-1829-1, rev., June, 1957.

5. R. N. McKean, *Efficiency in Government Through Systems Analysis.* New York: John Wiley & Sons, Inc., 1958.

6. D. Novick, *System and Total Force Cost Analysis.* The RAND Corporation, Research Memorandum RM-2695, April 15, 1961.

7. Department of Defense Instruction, *Reporting of Research, Development and Engineering Program Information.* DDR&E No. 3200.6, June 7, 1962.

8. J. W. Noah, *Identifying and Estimating R & D Costs.* RAND Publication No. RM-3067-PR (Abridged), April 11, 1962.

References for Section 15.3

9. D. E. Van Tijn, *A Method for Determining the Cost of Failures.* ARINC Research Corporation, Publication No. 206-2-298, May 21, 1962.

10. H. Balaban and A. Drummond, *Prediction of Field Reliability for Airborne Electronic Systems.* ARINC Research Corporation, Publication No. 203-1-344, December 13, 1962.

11. D. E. Van Tijn, *Second Special Report on Cost Studies.* ARINC Research Corporation, Publication No. 206-1-277, January, 1962.

12. A. Drummond, G. T. Harrison and H. R. Leuba, *Maintainability Prediction-Theoretical Basis and Practical Approach.* ARINC Research Corporation, Publication No. 207-1-275, February 15, 1962.

13. K. J. Arrow, S. Karlin and H. Scarf, *Studies in the Mathematical Theory of Inventory Production.* Stanford, Cal.: Stanford University Press, 1958.

PROBLEMS

1. Suppose the loaded manpower costs are computed from equations like Equation 15.15 for all categories of direct labor, and the coefficients are given by the following table:

Labor Category	C_a	C_b
t_v, t_f, t_c	$10.00	$2.60/hr
T_s	9.00	1.90/hr
S_l, S_c, S_s	12.00	2.20/hr

What is the cost of

(a) 1/2 hr of t_v?   (e) 1 hr of S_l?

(b) 1 hr of t_f?   (f) 1 hr of S_c?

(c) 1/2 hr of t_c?   (g) 2 hr of S_s?

(d) 2 hr of T_s?

2. Given the following values, what is the average monthly support cost of the box?

(1) $N = 20$	(7) $S_l = 1$ hr	(13) $C_s = \$50$
(2) $v = 0.6$	(8) $t_c = 1/2$ hr	(14) $S_s = 2$ hr
(3) $t_v = 1/2$ hr	(9) $C_c = \$1000$	(15) computed from
(4) $f = 0.1$	(10) $S_c = 2$ hr	Problem 1
(5) $t_f = 2$ hr	(11) $\Delta T = 1/2$ month	(16) $L = 6$ months
(6) $C_f = \$10$	(12) $T_s = 2$ hr	

Use Equation 15.4 to compute the cost of a time delay.

3. (a) Suppose the catalogue cost of the black box is $1000, as in Problem 2, but a charge of $500 is accumulated on the box before it gets down to the base, so that the cost to the base is $1500. What will then be the average monthly maintenance cost?

(b) If, in addition, the average repair time of the box on the base is increased to three weeks, what will be the total cost?

4. Suppose all the direct flight maintenance times are cut in half. What will be the monthly cost (for the black box as it is in Problem 2)?

5. Why does the answer of Problem 4 differ so little from that of Problem 2? Hint: Look at the computation in Problem 1.

Reliability Engineering,

Programming, and Management

16.1 Introduction

Compared with other accepted engineering disciplines, reliability engineering developed rather late, although the mathematical and statistical methodology required in reliability research had been available in published form for some time. During the decade starting in 1950, two types of reliability research began to be emphasized. On the one hand was the data-collection or fact-finding type, in which the problem was primarily the accumulation of failure-rate information through testing programs. The second type of research was concerned with the development of techniques by which the methods of mathematics and engineering could be applied in the solution of reliability problems. Both types of research are continuing as important elements in reliability engineering, but they have been joined by another branch of research in which interest is continually increasing—the development of methods for reliability management.

This chapter presents certain management principles which apply to the reliability engineering programs being conducted by industry and government today. For the sake of clarity, these principles are stated as if the issues involved could be decided on a black-and-white basis. This, of course, is not the case. Every reliability group will have different problems that must be solved through compromise, often dictated by considerations of cost. The one principle on which there can be no compromise is that reliability groups must have as their ultimate aims the improvement of product quality and the reduction of failure rate.

16.2   Underlying Causes of Unreliability

The management problem presented by reliability engineering is usually defined in negative terms, with emphasis on the item that fails, the mode of failure, the human errors involved, and the control actions that must be taken to prevent recurrence of this type of failure. Although this description is sound and can be used effectively, there is another, more affirmative method of defining the problem: What are the management methods that promote an atmosphere in which maximum product quality can be achieved? By emphasizing the prevention rather than the correction of trouble, the latter approach points to the constructive rather than the remedial responsibility of management.

It is necessary to begin with a study of the underlying causes of unreliability, but the principles set forth are such that they can be applied in the prevention of new modes of failure as well as in the correction of old modes.

Insofar as management is concerned, the causes of unreliability should be classified not by mode or mechanism of failure, but rather in terms of the characteristics of raw materials, the extent of technical "know how," and the effectiveness of management control over fabrication and testing processes. Pressure for new products with better performance has accelerated the speed with which new materials have been used. In the case of certain missiles, some of the new materials are subjected to such extreme stresses that only by operational testing of prototype models can the physical and chemical reactions of the materials be determined. Furthermore, variation in the characteristics of materials has resulted in many infrequently occurring new modes of failure. Microminiaturization has increased the criticality of variations in materials and fabrication processes. The rapidity with which the state of the art has been pushed has increased the amount of knowledge required for prediction and control of infrequently occurring modes of failure. This shortage of knowledge exists with respect to both failure data and theoretical aspects of the physics and chemistry of performance.

Compressed time schedules in research and development phases have promoted the use of novel designs based on "cut and try" methods without benefit of knowledge gained from use of basic transfer functions. While design and production of new equipment have been speeded up, the problems of reliability engineers have been increased. Indeed, sometimes these problems have been the prime motivation for the organization of a company reliability group.

The success of the reliability engineering program depends, of course, on the maintenance of certain management disciplines, such as design-stress analysis, qualification testing, quality-control procedures, and acceptance testing. An important cause of unreliability is the breakdown of these disciplines. But it is significant that, in some fields of engineering, management disciplines have been enforced so well that an extremely high level of reliability has been achieved. The classic example is the structural field, where almost 100 per cent reliability is common. Engineering stress analysis is capable of providing the bridge builder or the building contractor with designs which "guarantee" a failure-free product. Quotes are used around the word "guarantee" because there have been cases in which bridges fell and buildings collapsed, but such failures are so rare that people never hesitate to drive over a new bridge or to be the first tenant to occupy a new office building or home.

The difference between the failure mechanisms of structures and communication equipments, missiles, or satellites is worthy of some consideration. The characteristics of structures that make high reliability an achievable objective are the single mode of failure and the relatively complete understanding of the physics involved in failure. Structural failures result from a load of sufficient size to break the structure; and the strength and stress relationships of the materials are sufficiently well known to permit accurate computation of adequate safety margins. When this theoretical work is done, it is a relatively simple matter for the construction and inspection engineers to monitor the fabrication and thereby insure proper compliance with design.

By contrast, communications equipments and missiles have many modes of failure; the physics and chemistry of the reactions are not well understood; the stress analysis is difficult to perform and cannot be done as accurately, if it is done at all; and conformance to design in the process of fabrication is difficult to monitor and to test. The very existence of these problems tends to weaken management discipline by creating ways for error and oversight to occur. This, then, is the problem of management: to keep such errors and oversights to a minimum.

16.3   Establishment of Top Management Reliability Policy

An important characteristic of the operations of the reliability group must be a positive approach as expressed in its charter. The group will be extremely ineffective if it views itself only as a policing agency to see that errors are avoided. The reliability group must do this to be sure, but it must do many other things as well. The group must be dynamic, attempting to achieve reliability by a sequence of positive actions that usually start at the top and are carried on down through every level of the company.

Since the operations of the reliability group involve all departments, from design to production, it is essential that the reliability philosophy of top management be made known throughout the company. Each department needs guidance as to how it should work with the reliability engineers, and such guidance is provided by the charter of the reliability group. This charter, then, presents the management reliability policy for company-wide information and guidance. It must outline the top management views on the following six aspects of the reliability group and its operation:

1. Purpose and objectives

2. Responsibilities

3. Organizational relationships

4. Operational methods

5. Personnel

6. Top management audit, or the methods for reporting reliability status to top management.

These areas are considered individually below, and a policy statement is suggested for each area.

Purpose and Objectives of Reliability Group

The charter must contain a statement of the purpose and objectives of the reliability group, justifying its establishment by specific reference to the required output of the group in terms of the underlying causes of unreliability and the positive steps to be taken in avoiding failure occurrence. Such a statement might be as follows:

"The company shall establish and maintain an aggressive reliability program which shall be company-wide in scope. All departments of the company are therefore required to support and cooperate with the reliability group to insure the achievement of its objectives as stated herein. The purpose of this program is to attain maximum system effectiveness and product quality by providing organized knowledge and by enforcing necessary disciplines for the production, control, measurement, and reporting of experience on frequently and infrequently occurring modes of failure. The knowledge referred to above includes parts, components, and vendors' qualifications, as well as state of the art in design practices."

The statement of purpose above is only a model which must be tailored to fit the requirements peculiar to each application. It is important to note that the purpose is extremely broad. The reliability group has an important role to play in the development and manufacture of a high-quality product. Depending on the situation, it can function as (1) a library for basic data, (2) a policing agency to enforce proper practices, (3) a consultant on advances in design, (4) an educational agency to instruct in reliability principles, (5) a preacher or salesman to "sell" the reliability philosophy, and (6) an operating department in parts and components, vendor qualification, and in-plant control engineering. It is important to stress the research function of the reliability group. It must contribute methods research to support desired company growth, a necessary adjunct to company effort if it is to maintain its competitive status.

Responsibilities of Reliability Group

In considering the responsibility assignment for reliability, the question may be asked, "Who is primarily responsible for reliability?" or "Who gets the blame if the product is unreliable?" A quick response to these questions might be, "The head of the reliability department." However, when it is recalled that a reliability program must be company-wide, the general manager must take the ultimate responsibility. Thus, some of the problems of management are above the power of a department or program head unless this head has company-wide authority. In this latter case, he then becomes an arm of the general manager. In the final analysis, responsibility must lie in the area of the general manager's office whenever the activity is inter-departmental.

Instead of asking about primary responsibility for reliability, it might be better to determine which of the various reliability activities are the responsibility of company personnel at levels below the office of the general manager. Thus:

1. Who is responsible for reliability research and training?
2. Who is responsible for reliability creation?
3. Who is responsible for reliability assurance?
4. Who is responsible for project reliability management?
5. Who is responsible for reliability considerations for equipments already in use?

The general reliability policy statement should include the answers to the five questions above. Model policy statements are as follows:

(a) *Reliability Research and Training.* A reliability director, who reports directly to the general manager, will be responsible for collection and development of knowledge, practices, skills, and methods required for reliability creation and assurance and for making these available as needed on a company-wide basis. Dissemination of such information shall be accomplished through reports, consultation, seminars, and various training programs.

(b) *Reliability Creation.* The design departments have the responsibility for building reliability into their designs. Similarly the purchasing department will be responsible for procurement of parts and components which meet prescribed reliability requirements, and the production department will be responsible for building reliability into production hardware.

(c) *Reliability Assurance.* Reliability assurance will be provided by design and production reliability-assurance groups within the engineering and quality departments, respectively. (As an alternative, these groups can be set up under the direct control of the reliability group director, with all reports going to him.)

(d) *Reliability Project Management.* A reliability project manager will be appointed for each project to insure the execution and effectiveness of the project reliability program and to cooperate with other project programs in interchanging information and promoting company uniformity in reliability. It is permissible for one manager to have responsibility for more than one project.

(e) *Use Reliability.* The reliability director shall also be responsible for that activity which relates to evaluation of equipments in operating service. The purpose of this activity shall be not only to determine degree of customer satisfaction but also to provide feedback for future design effort.

Organizational Relationships

The foregoing typical policy statements imply a choice of two methods of organization. All reliability research and assurance groups can be collected together in one line organization under a reliability director or they can be separated and placed in other departments. Advantages can be presented for both approaches, and industry furnishes numerous examples of very successful reliability programs, some using centralized organization and some using a separation of the group into various departments. The final decision in this matter should be made only after due consideration of individual company problems and the activities of the reliability group. These problems and activities are discussed in Section 16.4.

Operational Methods

To be dynamic, a reliability program must comprise clearly defined and vigorously executed activities that are tailored to program needs and cover all levels of company operation. This requirement is met by having management control reliability practices through a series of documents as proposed in the following general policy statement:

"Continuing management control over reliability practices will be established and maintained by publishing and implementing the provisions of the following sequence of documents:

(a) *Reliability Policies Manual.* The reliability director will issue and revise as required a reliability policies manual which will

1. identify specific activities essential to achieving reliability objectives,

2. provide policy guidance for each such activity, and

3. define departmental responsibilities in terms of these activities.

(b) *Reliability Procedures Guides.* The reliability director will identify all the procedures which must be prepared for each department to enable it to implement the provisions of the policies manual; the director will also provide a guide for each such procedure.

(c) *Department Procedures and Instructions.* Each functional department will issue and maintain department procedures as required to fulfill the provisions of the policies manual and any further operating instructions as required."

Personnel

The implementation of a good reliability program involves so many specialized disciplines that personnel are usually expert in only one or two fields, not in all of them. In order to assure an effective distribution of skills, serious attention must be given to the specific requirements of particular reliability programs. Naturally, guarantees of status and long-term career potential will help to attract the best talent.

The reliability policy statement on personnel can be quite simple, as indicated by the following typical example:

"The reliability director is responsible for determining the specialized skills required for implementing the reliability program through the described activities; for establishing job categories and preparing job descriptions; and, with the cooperation of the personnel department, for procuring the personnel with the required skills."

Audit

The reliability director is responsible for continuous audit and supervision of the over-all company reliability program, including any programs that are limited to the project level, and for monitoring of the reliability programs of vendors who supply parts and components. This responsibility is covered by a policy statement of the following form:

"The reliability director is responsible for monitoring the over-all reliability program of the company and the reliability programs of vendors who supply parts and components, including all necessary incoming inspection tests designed to verify attainment of acceptable reliability levels."

16.4   Defining and Planning Reliability Activities

In Section 16.4, pages 557-572, a group of 26 reliability activities that form the substance of a good reliability program are discussed in relation to the 5 major areas of responsibility set down in Section 16.3—namely, reliability research and training, reliability creation, reliability assurance, reliability project management, and use reliability.

Naturally, not all the activities discussed here are applicable to all programs, but they do provide some sort of basis on which to select activities for a newly organized reliability group, and they offer a check list in terms of which existing reliability programs can be evaluated.

Reliability Research and Training

Those reliability activities which pertain to the general areas of research and training are discussed below.

DATA AND METHODS RESEARCH

This activity includes performance of laboratory tests for the purpose of generating data on strength, strength deviation, and failure rates for materials and parts. It also includes performance of analytical studies to develop either technical methods for prediction, measurement, or control of infrequently occurring modes of failure, or management methods for controlling any type of reliability activity.*

*In filing information on methods research, it is useful to include terminology and lists of definitions, as these are an important reflection of the methods followed by the using organization.

Although it is extremely difficult to describe the assignment of responsibility for various details of a research program, there seem to be two areas in which responsibility is most likely to be assigned to certain departments or certain people—namely, the areas of stress analysis and new methods.

(a) *Stress Analysis*

During feasibility studies and early design, the reliability group may be asked to assist in analysis of internal and environmental stresses. In some organizations, this type of research is performed by an operations research group or by design engineers. Insofar as stress analysis requires a broad knowledge of research programs of other agencies, the reliability group or the operations research group might be in a better position to handle the assignment than the design-engineering group.

(b) *New Methods*

In the later phases of a program, the reliability group has a broader and more obvious research responsibility, due in part to its previously noted potential for research in statistical methods. There is also a possibility of significant research in other areas, such as materials, fabrication methods, theoretical physics and chemistry, various branches of engineering, and management methods.

EXPERIENCE-RETENTION GUIDES AND CHECK LISTS

This activity includes preparation of company manuals (for instance, reliability policy manuals or preferred-part manuals), guides, and check lists,* and selection of external documents giving statements of design principles, management philosophies, and other forms of conceptual guidance. An example of such a document would be a specification containing paragraphs that must be classified as guidance material rather than as measurable contractual requirements.

Consultation between reliability-creation personnel and part-application or other specialists should be included under this activity.

Educational activities are intended to do three things: (1) to sell the idea of reliability work, (2) to assist personnel in the use of reliability reports, and (3) to teach the use of reliability techniques. It is reasonable to say that these three objectives are listed in order of a time priority: first sell, then apply, and finally teach the skills. The time priority does not imply an importance priority.

---

*Guides in general provide narrative descriptions of knowledge that should be part of the education of design and manufacturing engineering personnel. A good example is a parts application guide that describes the general characteristics of electronic parts and the most common types of misapplication.

Check lists are brief itemized lists of points that should be remembered by a reliability specialist when reviewing a design or production document.

To accomplish the objectives cited above, the educational activity must include all levels of training and motivation, from placement of posters in the factory to research seminars conducted for experienced reliability specialists. Each reliability program plan should include a specific schedule of lectures for supervisory personnel, lectures for working-level personnel, and either on-the-job training or lectures for reliability-specialist trainees.

In smaller companies, in-plant activities may well be limited, and employee training may be carried out by cooperation with other companies or by using consultants for formal courses and seminars. Even some larger companies have found the latter approach quite acceptable. The basic goal and responsibility are the same, whether the training activity is carried out by company personnel or by outside consultants.

Selected case histories of failure analysis and recurrence-prevention action provide perhaps the best possible means of effective training and motivation. Consequently, brief descriptions of these case histories should be filed under this activity. The reliability information file should also include references to training or indoctrination methods in industry or to specific university or technical-society training courses. These references should include moving pictures, slogans, publicity campaigns, or any other form of motivation.

Reliability Creation

It is a well-known fact that inherent reliability is established by the basic design. Hence, reliability must be "built in" or created during the design phase. The specific activities essential to reliability creation are discussed in the following ten subsections.

STUDIES TO ALLOCATE, PREDICT, AND IMPROVE RELIABILITY

The allocation, prediction, and improvement activity includes most of the effort of equipment designers. Reliability discipline requires designers to start off with an allocation of the amount of unreliability "allowed" in subsystems and components. This allocation must be based on past knowledge of the approximate complexity of each "black box" in the system block diagram and the corresponding part count, average part-stress levels, and generic failure rates. The allocation activity also includes all subsequent reliability predictions. As the design develops, actual part counts, individual part stresses, and failure rates are used in refining the original predictions.

The reliability-improvement aspect of this activity requires each designer and supplier of equipments or components to identify the modes of failure of his product and to propose action to reduce the frequency-of-failure probability as affected by design, manufacturing, or inspection procedures. Whenever a reliability prediction shows that the system goal cannot be achieved with parts reflecting the existing state of the art, it is necessary to prepare a reliability-improvement plan for the most frequently occurring parts or the parts with the highest failure rate.

PART AND CIRCUIT SELECTION AND STRESSING

Selection and stressing of parts and circuits cover all the detailed design work performed by circuit designers and packaging engineers—including the preparation of parts lists for every item in every circuit, together with a record of the principal circuit stresses and the corresponding manufacturer's rating, and the calculation of the operating temperature and mechanical stresses on every individual part. This amount of detail is essential for reliability prediction in accordance with specification MIL-STD-441.

TESTING TO FAILURE AND ADJUSTMENT OF SAFETY MARGINS

This testing activity includes all design and development tests performed in order to determine the actual strength distribution or random failure rate of a particular part required for a particular purpose.

Whenever a single mode of failure is of primary importance and it is practical to measure the mean strength and strength deviation of the part, such testing is essential for reliability prediction as well as for incorporation of adequate safety margins.

SPECIFICATIONS, TEST PLANS, CRITERIA, AND REPORTS

This activity encompasses all types of documentation* that are necessary to implement production of the design. It includes procurement specifications, process specifications, receiving inspection criteria, production alignment criteria, and final-acceptance and reliability-demonstration test plans.

To be effective, the results of a reliability analysis performed at the design stage must be suitably documented and distributed to using departments of the company. Here the emphasis is on the words "suitably documented" and "using": it is essential that the author of a report keep before him the needs of the particular reader who will be making use of the material, and he must always strive for a high level of technical quality. The reliability co-

---

*Each project reliability program plan should include an index to the document that will be required for execution of the project. The reliability information file should include examples of particularly effective types of design documentation. Examples of company directives that require a formal "design inventory" should also be referenced.

ordinator can promote good communication by setting high standards of sensitivity to the needs of the using departments, and by assuring that no report which does not meet the required standards is permitted to leave the reliability group.

The processing of unsatisfactory-condition reports by the reliability group is a major aspect of the reporting activity in terms of time and money, and should do much more than merely publicize failure experience. If proper authority is vested in the reliability coordinator, failures will be followed up and remedial action by the appropriate department will be insured. This action may imply (a) redesign, if excessive stress or parts application is the cause of trouble; (b) improved parts, if quality is below that currently obtainable; (c) improved fabrication, if poor workmanship is involved; or (d) modified packaging, if in-transit damage has been observed. In the usual situation, the reliability group is responsible for locating the trouble and calling it to the attention of the department involved, but is not responsible for developing a "fix," although reliability engineers generally recommend correction methods whenever they can. It is customary, however, for the reliability group to monitor the process of obtaining a remedy and to have authority to force remedial action when required. The other department involved must, of course, have a right of appeal to top management if it believes the reliability group is in error. The reliability group should assure uniformity of reporting trouble and describing the remedy by issuing standard reporting forms.

RECURRENCE CONTROL

Many reliability problems are essentially repetitive in nature. Hence, it is beneficial to employ methods which are specifically designed to minimize the number of recurrences. This is accomplished through control at the design activity, at the inspection activity, or at the production activity.

(a) *Recurrence Control at Design Activity*

Recurrence control at the design activity includes the studies required to decide upon a design change, negotiation of the most efficient procedures for effecting the change, and implementation of the decision to make the change in order to remove the cause of a recurring failure. The actual documentation of the change and review of the document are covered in the preceding section.

(b) *Recurrence Control at Inspection Activity*

Recurrence control at the inspection activity usually involves the addition of a requirement to an inspection procedure so as to decrease the probability of escape of the particular type of manufacturing discrepancy that caused the failure. In those cases where the inspection documents are already adequate, the action will take the form of further training and motivation of the personnel who failed to follow the document.

The reliability information file should give brief descriptions of case histories where imaginative engineers have conceived radically new methods of inspection to prevent potential failures from being accepted. All the methods of nondestructive testing, such as X-ray and eddy-current testing, should be included in these case histories.

(c) *Recurrence Control at Production Activity*

Recurrence control at the production activity covers any modification that may be introduced into manufacturing, assembly, or production-testing techniques to prevent recurrence of failure. In general, such control will consist of additions to, or modifications of, production documents. In those cases where the documents are already adequate, the action will consist of further training and motivation of the personnel who failed to follow the documents.

The reliability information file should contain case histories where changes in manufacturing methods or industrial engineering concepts have been applied successfully to prevent recurrence of failures.

DATA CENTER AND FAILURE-RATE SURVEILLANCE

This activity includes the operation of either a company data center or a project data center. The most fundamental type of data that is processed for reliability purposes is failure-rate data. Other important types include mean values, mean deviations, discrepancy or rejection rates, failure-isolation time, operating time, and down time.

PRODUCTION AND INSPECTION DOCUMENTATION

This activity includes preparation of all types of documentation subsequent to design documentation. It includes manufacturing flow charts, assembly instructions, detailed process instructions, inspection plans, and inspection instructions.* It does not include plans for the maintenance of surveillance over production by quality-assurance reliability specialists.

Each reliability program plan should include a manufacturing flow chart on which the document number controlling each step in manufacturing is listed. Such charts should be accompanied by an index to the documents.

Even though acceptance tests are part of the quality-assurance function and are not essential to actual manufacturing of the product, the documents are included under this activity because there is an urgent need to provide

---

*Even when certain documents are classified as "company confidential," the customer has the right to know that they exist and that they are capable of ensuring reliability in the product.

the customer with a clear picture of the manufacturing process and to provide the manufacturing supervisor with a complete picture of what will be going on.

SUPPLIER SELECTION AND CONTROL

This activity includes complete plans for controlling vendors and sub-contractors. Such control is a mandatory part of most reliability specifications.

Some of the specific aspects of this activity are vendor plant surveys, vendor rating systems, and requirements that suppliers prepare written reliability program plans.

Control of suppliers requires them to pursue all or at least some of the activities listed in Section 16.4.

MANUFACTURING SUPERVISORY CONTROLS

This activity includes any procedure for bringing the superior experience and skill of manufacturing supervisors to bear on reliability achievement. It includes verbal and visual controls exerted by lead men and the preparation of control charts by manufacturing personnel to help lead men evaluate the quality of their own work.

Each reliability program plan should include a clear statement of the company's position in regard to the responsibilities of manufacturing supervisors, and the distinction between these responsibilities and those of quality-assurance inspectors.

PROCESS MONITORING AND ASSEMBLY TESTS

This activity includes the performance of production tests other than acceptance tests. All tests that are performed because the manufacturing supervisor wishes to avoid rejection at a later acceptance-test stage are included. Process-monitoring tests are included when they are used to adjust processes rather than to accept or reject production lots.

Reliability Assurance

Once reliability has been designed into a product, it is necessary that any subsequent degradation be eliminated, or at least minimized. This is accomplished by the review, testing, and inspection activities discussed below.

A review of company and supplier designs should be made after a design is completed, but there is some difference of opinion as to whether this should be done by reliability specialists or by design-review groups. One view is that design is properly the sole responsibility of the design engineer and that reliability engineers should stay out of this area. At the other extreme, some reliability groups are given veto power over design engineers by the requirement that all designs be initialed by the reliability coordinator before they can be processed. An intermediate approach requires coordination of the two groups, with the reliability group serving in an advisory or consulting capacity with varying degrees of authority to present objections to higher organizational levels for resolution of differences. In this intermediate arrangement, the reliability group may be responsible for advising or dictating selection of parts and suppliers.

We might argue that the best general rule is to place the reliability group in such an organizational position that it can appeal directly to top management without being under control of the design group. Even though this is a good general rule, we can find exceptional cases in which the reliability group is very effective even though it is at a relatively low organizational level and even though it is under the control of the chief design engineer. Success is more dependent on personalities and capabilities than on the organization of a company, but the organization can set the scene for easier operation by assuring that certain difficulties are avoided.

Whatever organization is used, it is very important that the reliability specialist not compete with the designer. The former's ability to perform review is based on organized knowledge of what has happened. Under no circumstances should he recommend an alternative design, although he may provide information to the effect that experience with an alternative design has resulted in a lower failure rate.

For effective design review, it is essential that reviewers be provided with comprehensive check lists based on analysis of failures that have occurred in similar equipments.

This activity covers the practice of having a formal design-review meeting. Each reliability program plan should include a definition of how such meetings will be conducted, when they will be scheduled, and who will attend. In most cases it will be necessary for each company to prepare a report describing the manner in which design-review meetings are conducted.

The reliability information file should include references to published descriptions of design-review procedures used by other companies.

FUNCTIONAL AND ENVIRONMENTAL PROOF TESTS

This activity includes all types of qualification testing as well as any type of functional or environmental testing performed to prove that a design meets the specified requirements. The activity does not include research tests performed to provide data, design-development tests used merely to guide the designer in making his own decisions, or parts tests performed in order to develop data for parts application manuals.

Each contract should indicate clearly whether a demonstration of failure rate is a required part of qualification testing. In general, it is not practical to provide the large samples and extreme test times required to demonstrate failure rates for parts or simple equipments. Consequently, contracts may limit the extent of qualification testing performed as proof that the design meets the functional requirements, and they may permit a delay in demonstration of failure rate by testing until production quantities are available or even until the required number of service operating hours has accumulated. This decision is justified because analytical and prediction methods may provide adequate demonstration of design reliability.

PRODUCTION AND INSPECTION DOCUMENT REVIEW

This activity includes review of purchasing, manufacturing, and inspection documents by reliability specialists.

Each reliability program plan should include an index to production documents together with a section indicating responsibility for writing and reviewing.

IN-PROCESS INSPECTION AND DOCUMENT COMPLIANCE

This activity includes surveillance by quality-assurance reliability specialists to determine if production documents are actually being followed. Each progam plan should clearly identify who has this responsibility. Also, the manufacturing flow charts should identify those steps in manufacturing where in-process inspection must be performed, because at a later stage the opportunity to inspect may have been lost by further assembly.

Vendor surveys performed before a product is in production are part of the activity involved in vendor selection. However, subsequent visits to a vendor's plant will include surveillance of document compliance by quality-assurance personnel.

"Setup checks" are part of this activity because, as the setup must comply with the controlling document, the checking of the setup is a check of document compliance.

All tests that the customer requires be performed by the quality division as an independent evaluation of manufacturing—including receiving inspection, in-plant assembly tests, and final acceptance tests on equipment—are covered by this activity.

As final acceptance testing of production equipment provides the normal opportunity for demonstration of reliability in terms of mean time between failures, reliability demonstration tests normally are also part of this activity. Acceptance testing of production lots also makes use of control-chart methods, and parts screening tests, the latter being used to remove potential premature failure items from lots that have been accepted.

FAILURE REPORTING AND ACTION ASSIGNMENT

Ordinarily, a fairly detailed company report is required to describe the complicated steps involved in failure detection, reporting, action assignment, recurrence prevention, and experience retention. This activity includes the decisions assigning responsibility for corrective action to the inspection, manufacturing, or design department but it does not include the corrective action itself.

The reliability information file should contain copies of the failure reporting requirements that are contained in basic customer documents.

Administration of Reliability Projects

The administration of projects involves such activities as the preparation of contracts and specifications, the working out of procedures and time schedules for projects, and the auditing of projects to see that contractual commitments are met. These activities are described in more detail in the following subsections.

ESTABLISHMENT OF CONTRACTUAL COMMITMENTS

As one of the first and most important steps in the preparation of a contract, an evaluation of the existing state of the art should be made in order to guarantee the setting of realistic goals. Both parties to the contract should participate in this evaluation so as to prevent misunderstanding at a later date after irrevocable commitments have been made.

The writing of the contract involves certain routine tasks, such as the preparation of "boiler plate" paragraphs, in addition to the more onerous task of preparing the reliability specification that will form the body of the contract. A few points that can make the difference between an effective and an ineffective contract warrant emphasis.

For example, in some cases the methods of proof of reliability are not detailed in the contract or reliability specification but are left to subsequent negotiation. This is dangerous unless the procuring agency appreciates the trade-off between cost and proof of reliability, and unless the contract is "cost-plus-fixed-fee" or has renegotiation clauses with sufficient flexibility to eliminate endless arguments over costs of testing. It must be recognized that proof of reliability does not improve reliability. The procuring agency must decide how much the proof is worth to them. They should have a firm rejection right for failure to pass the proof test. Even with a rejection privilege, it is necessary to recognize that the vendor will place a dollar value on his risk, and the product will be priced accordingly.

With respect to reliability, the optimum situation for contracting occurs when award is made to a contractor who has maintained an effective reliability program for some time. Past performance is extremely important in selecting a vendor and awarding a contract. Whenever a contract is awarded to a company which does not set up a new reliability program, the procuring agency must assume the responsibility for careful and continuous surveillance, even to the extent of conducting training courses in the vendor's plant. It may even be desirable in such a case to have the contract spell out in detail the basic activities of the vendor's reliability effort.

Incentive and penalty clauses are other means that have been found effective in assuring performance in reliability contracts.

PROJECT PROGRAMMING, TASKS, AND LIAISON

This activity covers preparation of procedures that apply specifically to a given project. They should identify general procedural documents which are suitable for direct application to the project, as well as types of personnel who will be involved in the execution of the project. Individual names may be included to the extent that the customer requires.

The time-phasing aspect of the activity requires the preparation of program plans with specific recognizable steps which are scheduled for accomplishment on stated dates. The liaison aspect includes establishment of agreed forms of communication with the customer or with suppliers.

TASK AUDITS, REVIEWS, AND STATUS REPORTS

The auditing activity includes any method for determining compliance with a reliability program plan and for submission of periodic reliability reports to the customer. It covers the review of attained levels of reliability as well as the data submitted in proof that these levels have been achieved.

The reliability director can never be sure of the degree of success of all company reliability activities unless he has an adequate numerical measure of progress or growth in reliability. In April 1958, the Office of the Director of Guided Missiles published a report entitled, *Proposed Reliability Monitoring Program for Use in the Design Development and Production of Guided-Missile Weapons Systems*, prepared by C. A. Beyer, Chairman of the Ad Hoc Committee for Guided Missile Reliability.* This report is a significant contribution to the literature on reliability theory, for it describes very clearly the basic requirements for a good reliability monitoring program. Although it is oriented toward guided missiles, the presentation is equally applicable with minor changes to the monitoring of reliability in any equipment type.

The results of a monitoring program are usually displayed as a plot of attained reliability versus time. For this purpose, two kinds of time are needed—calendar time and project state, which is really the project time schedule with slippages removed. In other words, the plot can be presented in either or both of two forms:

1. Attained reliability vs. calendar time
2. Attained reliability vs. level of progress with respect to a time base which may or may not match the planned schedule.

Management needs to know whether or not the attained levels of reliability meet time schedules and also whether they meet acceptable levels at each stage in the cycle from engineering design to production.

An idealized reliability growth cycle is shown in Figure 16.1. This curve, obtained from the guided missile report previously cited, was prepared to show eight suggested monitoring points, identified by the development stage at each point. The variation in the length of the abscissa intervals between monitoring points is indicative of the variations in time required to complete the different stages in the development cycle.

Even though the growth cycle is described as "idealized," it is dangerous to assume that the development of such a curve is an unattainable goal for any properly planned and vigorously executed reliability program. Many reliability studies now completed show reliability predictions and measurements at *all* stages in the development cycle. To be sure, points on the growth curve vary with respect to confidence intervals according to the stage of development, the type of equipment, the budget for testing, and many other factors. This variation in confidence does not imply any justification for failure to exert every effort to perform the monitoring function.

*Later issued as U.S. Air Force Specification Bulletin No. 506.

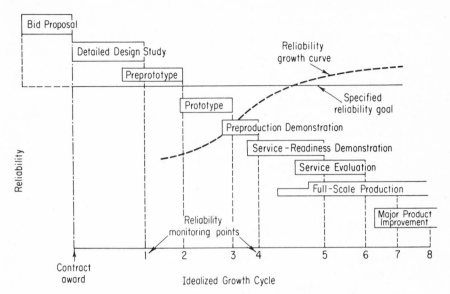

Fig. 16.1 Reliability monitoring program.

The ODGM report suggests that the first monitoring point fall after the detailed design study. This is consistent with the view that reliability monitoring is not of major concern in a proposal-feasibility analysis. However, from a reliability standpoint, as distinguished from monitoring, it is wise to begin to think about reliability in the earliest stages.

The emphasis on monitoring reliability should not be interpreted as an implication that this is the major function of the reliability group. It is probably more proper to say that monitoring is a necessary but minor function of the reliability group. The real purpose of this group is to assist in improving product quality in relation to available budget, and measurement of progress by a reliability growth curve is really only an indication of the level of accomplishment; it is not the accomplishment itself.

PERT AS A MANAGEMENT TOOL

Within recent years, several formal techniques for assisting program managers have been developed. While the details of these systems have varied, they have all been intended to serve the same basic purpose: to summarize a wealth of information in a manner which permits management to determine readily the existence of problem areas or forecast future problem areas. In addition, it is desired that, in so doing, these management tools provide an indication of adequacy of program progress.

Of the various management methods which have been developed, probably the most significant is that which is known as "PERT" (Program Evaluation and Review Technique). Originally developed by the Navy Special Projects Office, it has since received the endorsement of both the Department of Defense and the National Aeronautics and Space Aministration. Its application is not confined necessarily to reliability activities, but its increasing use warrants brief consideration here.

The basic PERT system is not a device for measuring reliability of a product. Rather, it is a means for coordinating the many disciplines involved in the reliability technology.

The essential terms associated with the PERT system are

Event—An unambiguous point in time in the life of a project.

Activity—Technological operation which consumes time, money, and manpower. Each activity is characterized by a specific initial event and terminal event.

Network—A visual presentation of events and activities which depicts interdependencies.

An example of a PERT network is shown as Figure 16.2 Note that events are joined together by activities in a manner which illustrates constraints. The method utilized in developing any PERT network is to combine various technologies, such as design, testing, drafting, and reliability, into a single system. However, it is possible to select any one area of endeavor, such as reliability, and confine the network to the events and activities associated with this area. Such a network would be of considerable value to a reliability manager in administering his program.

Possibly the most difficult task associated with a PERT program is the initial selection of significant program events. Reliability program specifications are usually general in nature. Hence, it becomes necessary to translate these generalities into carefully defined occurrences which can be associated with specifically defined hardware items. In selecting events which are to be

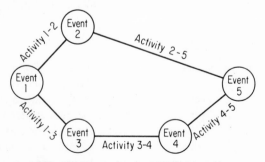

Fig. 16.2 Example of a PERT network.

controlled, there are certain factors which must be considered. The more significant of these include the following:

(a) Events must represent instantaneous points in time.

(b) They must be identifiable as the responsibility of a particular working group within the organization.

(c) The activity joining two distinct events must be capable of definition.

(d) The time, cost, and manpower expenditure for activities joining two events must be capable of estimation.

Once a network has been developed, there are a variety of subsequent procedures which could be followed. These include:

(a) Estimating time required to complete each activity.

(b) Estimating cost to accomplish each activity.

(c) Estimating resources (in terms of materials or manpower to accomplish each activity).

In some variations of the basic PERT system, multiple estimates are made. For example, instead of formulating a single estimate of time to accomplish a specific activity, it is possible to supplement an "expected" time with an optimistic and a pessimistic estimate. In so doing, statistical processes are employed to determine the probability of accomplishing an event by a given time.

The most important single advantage of a management system such as PERT is considered to be that it forces a contractor to define the elements of his reliability program. In many complex reliability efforts this could represent the most difficult task of all. Many other advantages are apparent, the primary ones including the following:

(a) It can be used in establishing schedules and cost budgets.

(b) It assists in forecasting total costs and time.

(c) It can be used in illustrating time and cost status.

(d) It assists in the analysis of manpower requirements.

(e) It can be used to determine effects of simulated program changes.

(f) It provides a basis for time/cost trade-offs.

(g) It assists in the making of vital management decisions.

This brief discussion of the PERT system is not intended to provide the reader with a detailed understanding of the system. There are several publications—such as *PERT Summary Report, Phase* 1 *and Phase* 2, published by Special Projects Office, Bureau of Weapons, and *DOD and NASA*

Guide, PERT Cost, published jointly by the Department of Defense and the National Aeronautics and Space Administration—which adequately fulfill this need. However, it is intended to suggest that the basic PERT system does define a basic approach which can be amplified to full advantage in administering a program of reliability activities.

Use Reliability Assistance

Once his product has been accepted by his customer, a contractor will often assume that reliability responsibility ceases to exist. While in a legal sense this may be true, the contractor should maintain activities which provide the customer with use assistance. Also, he should be vitally concerned not only with customer satisfaction, but also with the feedback of data for future design effort.

OPERATION AND MAINTENANCE INSTRUCTIONS

One activity in this area includes the writing of operating and maintenance manuals and the presentation of instruction courses to user personnel. Maintenance instructions should include techniques for marginal checking or failure anticipation and a discussion of stockpile surveillance practices.

This activity may occur either before or after a product has been accepted by the customer.

USER TESTS, DATA ANALYSIS, AND EVALUATION

Each reliability program report should define clearly the development phase at which responsibility for the equipment is transferred from the contractor to the user and the extent of the contractor's participation in field operations thereafter.

The contractor often assists with proving-ground or operational tests performed by the user for the purpose of evaluating reliability, as well as with the collection and analysis of failure data incidental to normal use of the equipment.

16.5   Reliability in the Company Organization Chart

The discussion of the objectives and activities of the reliability group has laid the groundwork for establishment of the group in the proper place on the company organization chart. In this context, "proper place" means the best place as a general rule, but it does not mean the only possible place. Industrial surveys have shown wide variation in the placement of the reliability group within the company. They also have shown that a given position

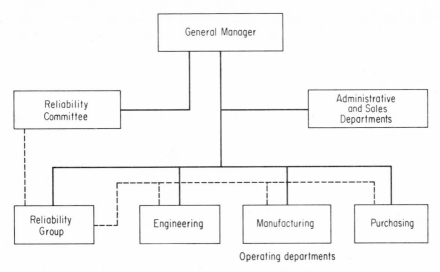

Fig. 16.3 Reliability in the organization chart.

is effective in one case and ineffective in another. This variation does not disprove the thesis that there is a best arrangement, one that creates an environment in which it is easier to obtain success even with such handicaps as opposition by certain departments.

An ideal plan might take the form shown in the chart in Figure 16.3. In this chart, direct responsibility is denoted by solid lines, and functional authority in reliability activities is denoted by dotted lines. For simplicity, no attempt is made to show all departments or the delegation of general managerial responsibilities through offices of the various managers. Also omitted are lines indicating relationships between the departments and the reliability committee—a point which will be discussed in describing the rationale of the plan.

It is immediately apparent that the reliability group is placed in parallel with the engineering, manufacturing, and purchasing departments. Some of the reasons for this parallel arrangement have already been given,* but it is well to restate the fundamental reason here: The reliability group is particularly concerned with one aspect of equipment quality or system effectiveness—namely, the pattern or frequency of failure. If an improvement in reliability can be made only at the expense of a characteristic within the purview of another department—for example, increased cost, size, or weight, decreased performance, or complicated maintenance—the reliability group must be in a position equal to that of the other department involved, and the responsibility for making the trade-off decision must be placed at a level above that

*See Section 16.4, page 564.

of the competing departments. In this chart, the line of authority leads directly to the general manager. It is, of course, entirely consistent with this argument for the general manager to delegate the trade-off responsibility to an appropriate manager.

In some companies, the reliability group functions very successfully in an advisory capacity, at a level below that of the departments. However, there are certain disadvantages in this arrangement. For example, the head of a design engineering group might be in a very embarrassing position in supporting reliability rather than performance; he would be in an even more embarrassing position if he supported reliability in a disagreement with a production engineering group that was on the same organizational level as the design group itself. Indeed, the design engineer in charge of the department would be very hard pressed to assume a company-wide philosophy in support of reliability in any discussion; his interest and responsibilities would have to be biased in favor of his own department.

However, it should be emphasized again that the reliability coordinator does not necessarily have to report to top management for the reliability effort to succeed. Cooperation on a company-wide basis can make almost any system work. A good organization is protection against some of the problems that arise as a result of personnel conflict, but it is not a guarantee of success; capable people are a necessity.

The reliability committee, shown above the reliability group on the chart, is usually composed of a representative from each department, including the reliability coordinator and some representative from the general manager's office. This committee is responsible for a number of functions. It sets company reliability policy, which really means that it determines the responsibilities of the reliability group and can prescribe principles of operation of the group. It also monitors the reliability activities by reviewing the periodic malfunction reports and the problems involved in obtaining "fixes." A major responsibility of the committee is to serve as an "appeals court" for settlement of conflicts which cannot be resolved by mutual agreement between the reliability group and other departments. In brief, the reliability committee authorizes and monitors the activities of the reliability group and adjudicates disagreements involving that group.

As indicated by the dotted lines, the reliability group is responsible for surveillance over all matters relating to reliability in all operating departments. This means that it studies reliability problems and calls them to the attention of the appropriate department. In turn, the department involved is required, usually within a specified time, to examine the situation and report its evaluation and the appropriate "fix" to the reliability group. The reliability group will then follow up to assure compliance. There is always a chance that fundamental disagreement will preclude such smooth operation. As discussed previously, such disagreements are resolved by the reliability committee. In most cases the reliability coordinator and the representative of the involved department are both disqualified from voting.

For the many reasons previously stated, we are accepting here the thesis that the company should have a centralized reliability group. However, in large companies it is sometimes decided to let each department have its own reliability group in addition to the company group, the point being that such a departmental group can avoid a majority of the problems which might otherwise arise. This departmental organization is based on sound reasoning, provided, of course, that each department head is sold on the idea. It may work even if the department head is opposed, but the chances of success are severely reduced. In the face of serious opposition, the department reliability group must assume a selling job as one of its first tasks, just as in the case of the company reliability group. It is to be hoped that the selling might be somewhat easier for the departmental group, since it can work more closely with the particular department than can the company group. Naturally, the company group must contribute its support in every way possible.

16.6   The Future of Reliability Engineering

The history of equipment designed in the recent past gives forceful emphasis to the vital role of reliability in present-day and future systems. However, the rapidity of change which characterizes equipment development programs has severely taxed the ability of reliability groups to meet the heavy demands placed upon them. Management can help to ease this situation by utilizing knowledge of current trends in planning for the future and by accepting the need to be ready for new challenges when they arise. More specifically, management must be prepared to make long-term reliability commitments involving not only continuation of past functions or even development of new methods, but often assumption of completely new functions by reliability groups.

Reliability groups in turn must recognize the expanded scope of their activities and must define their new problems accurately. For example, due to the fact that increased complexity and high reliability requirements have greatly increased unit cost, it is necessary to develop new methods of evaluating high-cost items in cases where there is understandable hesitation to provide large quantities for testing—perhaps destructive testing. These methods may depend on new techniques for synthesizing results of component tests, repeated use of items in nondestructive tests, more complete research into the physics and chemistry of operation (which means better understanding of failure mechanisms), or perhaps new statistical methods (in addition to simple failure frequency) for using reliability growth rate to predict ultimate levels of reliability. With high-production items, field observation feedback is a ready source of a large volume of useful data. With low-production items, field experience is limited, and, if the items are costly, every possible new method must be examined. Methods research is critical here, and it constitutes one of the most important challenges to reliability groups in the future.

In addition to reliability, the other factors influencing system effectiveness must be analyzed in order to optimize effectiveness through appropriate trade-off relationships. At present the emphasis is on such factors as maintainability, repairability, intrinsic availability, and operational readiness. The reliability engineer must view reliability in the wider perspective of all the attributes of system effectiveness, and management must support programs in which this broad view is taken.

PROBLEMS

1. What distinct phases exist in the life cycle of any equipment?

2. What are the significant milestones (or check points) that should be established in developing R & D and production programs for a new system?

3. Some military agencies now require a "Dependability Plan" to be established for each new system at the time its performance characteristics are being developed. Such a plan outlines the basic requirements for reliability, maintainability and, in some cases, serviceability, as well as the methods by which requirements are to be achieved and demonstrated. Develop, for a hypothetical equipment, a "Dependability Plan" that includes reliability and maintainability requirements, significant milestones, and significant elements that should be included in the contractor's dependability assurance program.

4. What action might management take if desired goals of system reliability, maintainability, and performance (although feasible) cannot be achieved due to budget limitations?

Government Documents
Establishing and Supporting
Reliability and Maintainability Requirements

APPENDIX A

GOVERNMENT DOCUMENTS ESTABLISHING AND SUPPORTING RELIABILITY REQUIREMENTS

*MIL-STD---(DOD) RELIABILITY PROGRAM MANAGEMENT (IN PROCESS)

WS-3250 (WEPS) RELIABILITY PROGRAM REQUIREMENTS

NPC-250-1 (NASA) RELIABILITY PROGRAM REQUIREMENTS

AR-705-25 (Army) RELIABILITY PROGRAM FOR MATERIEL AND EQUIPMENT

RELIABILITY PROGRAM SPECIFICATIONS:

MIL-R-27542(II)(USAF) AEROSPACE SYSTEMS AND SUBSYSTEMS (REV. "A" PROPOSED)

MIL-R-22732(II)(SHIPS) Shipboard and Ground Electronic Equipment

RELIABILITY IN DESIGN, DEVELOPMENT AND PRODUCTION OF EQUIPMENTS AND SUBSYSTEM SPECIFICATIONS

MIL-R-26674(USAF) Production - Ground Electronic Equipment

MIL-R-27070(USAF) Development - Ground Electronic Equipment

MIL-R-27173(USAF) Ground Checkout Equipment

MIL-R-26484A (II) (USAF) DEVELOPMENT - SYSTEMS AND SUBSYSTEMS

MIL-R-26667A(II)(USAF) PRODUCTION ELECTRONIC EQUIPMENT

MIL-R-22256(USAF) Design - Equipment And Systems

MIL-R-19610(WEPS) PRODUCTION - ELECTRONIC EQUIPMENT (INACTIVE)

*MIL-STD-781(DOD) RELIABILITY TEST PROCEDURES

MIL-R-22973(WEPS) INDEX DETERMINATION FOR AVIONIC EQUIPMENT

MIL-R-26667A(II)(USAF) DEMONSTRATION REQUIREMENTS

MIL-R-5521(IEL) PRODUCTION ELECTRONIC EQUIPMENT

MIL-R-23094(WEPS) ASSURANCE FOR PROD. AVIONICS EQUIPMENT

MIL-M-9933(II)(USAF) QUICK REACTION CAPABILITY ELECTRONIC EQUIPMENT

SPEC. BULLETIN 506 (USAF) MONITORING

*MIL-STD-721A DOD DEFINITIONS

MIL-STD-441 (DOD) Military Electronic Equipment

WR-41 (WEPS) RELIABILITY EVALUATION

*MIL-STD-756A(DOD) PROCEDURE FOR RELIABILITY PREDICTION

OTHER RELIABILITY DOCUMENTS

SUPPORT DOCUMENTS

MAINTAINABILITY
MIL-M-26512B (USAF)
MIL-M-23313 (SHIPS)
MIL-M-45765
WS-3099+(WEPS)
SCL-4301B 1516)

TRAINING
MIL-T-4860C(USAF)
MIL-T-26046(I)(DOD)
MIL-T-26137B(USAF)
MIL-T-27382(USAF)

REPORTS
MIL-R-1800.B(I)(WEPS)
MIL-R-1839A(2)(WEPS)

TEST REPORTS
MIL-T-9107(2)(II)(USAF)

DATA
MIL-D-9308(2)(USAF)
MIL-D-9812(II)(USAF)
MIL-D-26239A(USAF)
MIL-D-70327(2)(DOD)

TEST EQUIPMENT
MIL-STD-4515B
MIL-T-945A(2)(DOD)
MIL-T-8306A(II)(WEPS)
MIL-T-21200(IASG)

INSTALLATION
MIL-I-8700(IASG)
MIL-E-25366A(USAF)

VIBRATION
MIL-STD-167

WIRING
MIL-W-73A(GH)(DOD)
MIL-W-5088B(IASG)
MIL-W-5400F (ASG)
MIL-W-8660(USAF)
PD-E-53(IABMA)

SAMPLING
MIL-STD-105C
MIL-STD-414
DOD-HDBK-106
DOD-HDBK-108

ENCLOSURES
MIL-STD-108(D)(II)
MIL-C-172C-(DI)(DOD)
MIL-E-2036C(4)(NAVY)

EQUIPMENT TYPES
MIL-STD-243

DESIGN
MIL-STD-4598(II)
MIL-E-4158C(USAF)
MIL-E-5400F (ASG)
MIL-E-8189B(IASG)
MIL-W-9411A(2)(USAF)
MIL-E-16400 (SHIPS)
MIL-E-19600A(WEPS)
ANA BLTN. 444
ADII4274 (ASTIA)
AD48556 (ASTIA)
AD48907 (ASTIA)

PROVISIONING
MIL-B-5005A(2)(DOD)
MIL-E-17362D (SHIPS)
DOD INST. 415I.7
MCP 71-650
MCP 71-673
PP-SIG-SE-1A
WR-1
WR-2

TEST METHODS
MIL-STD-202(B)
MIL-STD-446A
MIL-T-4807A(USAF)
MIL-E-4970A(USAF)
MIL-E-5272C(I)(ASG)
MIL-T-5422E(2)(IASG)
MIL-T-18303(WEPS)

HUMAN FACTORS
MIL-H-22174(WEPS)
MIL-H-26707(USAF)
BSD 61-99(USAF)

QUALITY CONTROL
MIL-Q-9858(2)(DOD)
NPC 200-2 (NASA)
NPC 200-3 (NASA)
DCAS EX. 62-10 (USAF)
DOD-HDBK - III
NAV-1-1034 APP "A"
USAF SPEC. BLTN. 523
ASD-TR-61-363

ENVIRONMENTAL FACTORS
MIL-STD-210A(I)
MIL-T-152811(DOD)
USAF SPEC. BLTN. 106A
USAF SPEC. BLTN. 115(I)
USAF SPEC. BLTN. 523
ASD-TR-61-363

INTERFERENCE
MIL-E-605(C)(USAF)
MIL-I-6181D(2)(USAF)
MIL-I-11690(A3)(SHIPS)
MIL-I-26600(2)(USAF)
PD-C-186(ABMA)

PRESERVATION AND PACKAGING
MIL-P-116(DI)(DOD)
MIL-P-9024B(USAF)
USAF Spec. Bltn. 56E
MCP 71-163

REFERENCE DOCUMENTS
AD 148868(ASTIA) RADC Reliability Notebook TR-58-111
PB-121839 Navy Reliability Design Handbook
NAV-AER-16-1-519 Handbook Preferred Circuits
NAVSHIPS 900-193 Rel. Stress Analysis For Electronic Equipment
NAVSHIPS 93820 Handbook For Prediction Shipboard and Shore Electronic Equipment Reliability
NEL - Suggestions For Designers of Electronic Equipment
AGREE REPORT - Reliability of Military Electronic Equipment
RELIABILITY ENGINEERING NOTES - Diamond Ordnance Fuze Laboratory
NAVSHIPS 900-16 Reliability Analysis Data For Systems
MIL-HDBK-217(DOD) Reliability Stress and Failure Rate Data for Electronic Equipment
PSMR-1 Parts Specification Management For Reliability
IDEP 1 and IDEP 2
USAF BLTN. - 519. Bibliography Reliability Documents (USAF)

AN ASTERISK (*) DENOTES A DOD TRI-SERVICES DOCUMENT

†ADDED BY AR INC RESEARCH

Prepared by Contract Technical Requirements, Martin Marietta Corporation (June 1, 1963). Reprinted with permission.

GOVERNMENT DOCUMENTS ESTABLISHING MAINTAINABILITY REQUIREMENTS

MAINTAINABILITY PROGRAM SPECIFICATIONS

AR-705-26 (Army)
Maintainability Program
for Materiel and Equipment
16 April 1963

WR-30 (WEPS)
Weapon Readiness
Achievement Program.
1 May 1963

MIL-M-26512B (USAF)
Maintainability Requirements
For Aerospace Systems And
Equipment.   23 March 1962

MIL-M-23313 (SHIPS)
Maintainability Require-
ments For Shipboard And
Shore Electronic Equip-
ment And Systems.
12 June 1962

MIL-G---(PROPOSED) (CE)
Maintainability-Generator
Sets, Engine Driven, A/C,
15K. W -150K. W.
(H. L. Yoh Company)
May 1962
(Not released)

XAV---(PROPOSED) (WEPS)
System Effectiveness/
Maintainability -
Avionics Systems.
8 August 1962
(Not released)

EXH.---(PROPOSED)(USA OMC)
Maintainability
Requirements.
1 June 1962
(Drafted From MIL-M-26512)
(Not Released)

OTHER MAINTAINABILITY SPECIFICATIONS

MIL-M-99330 (USAF)
Maintainability And
Reliability Program-Quick
Reaction Capability
Electronic Equipment.
20 June 1962

SCL-4301B (SIG-CORP)
Maintainability
Design.   29 March 1962

MIL-M-26512B APP. "B"
Maintainability Verification
Of Predictions For Systems
And Equipment.
23 March 1962

MIL-STD-778
Definitions for
Maintainability
Engineering
(Proposed)

MIL-M-26512B APP. "A"
Maintainability Test And
Demonstration Requirements
For Systems and Equipment.
23 March 1962

WS-3099-1 (WEPS)
Maintainability
General Specification
24 October 1962

MIL-STD-829 (USAF)
Terms and Definitions
for Maintainability
12 September 1962

DESIGN GUIDES

USA OMC - Maintainability Design Factors.
NAVSHIPS 94324 - Maintainability Design Criteria Handbook. (SHIPS)
NAVTRADEVCEN 330-1 Series - Design For Maintainability
ORD P 20-134 - Maintenance Engineering Guide For Ordnance Design
ASD-TR-61-424 - Integrated System Design For Maintainability.
WADC-TR-56-218 - Electronic Equipment Design Guide
ASD-TR-61-381 - Guide To Design For Maintainability Mechanical Equipment
AFSCM 80-1 - Handbook Instructions - Aircraft Designers.
AFSCM 80-3 - Handbook Instructions - Personnel Subsystem Designers
AFSCM 80-5 - Handbook Instructions - Ground Equipment Designers
AFSCM 80-6 - Handbook Instructions - G. S. E. Designers.
AFSCM 80-8 - Handbook Instructions - Missile Designers.

REFERENCES

AFR 66-29A - Maintainability - Weapon, Support, Command And Control Systems.
AFSCR 80-9 - Maintainability Policy For R & D.
AFM 66-1 - Organizational and Field Maintenance.
AMCM 66-2 - Maintenance Engineering Methods And Management.
AMCM 66-5 - Maintainability In Air Force Equipment
RADC-TN-60-5 - Methods of Maintainability Measurements And Predictions.
AD-148868 - RADC Reliability Notebook.
BuWeps Instr. - 4423.2 - Mtns. Eng. Review Team (MERT) Policy and Procedures.
BuWeps Instr. - 4710.1A - Progressive Aircraft Rework Program (PAR)
BuWeps Instr. - 4700.2 - Naval Aircraft Maintenance Program Manual.
EIA - Maintainability Bulletins.

† ADDED BY ARINC RESEARCH

Prepared by Contract Technical Requirements. Martin Marietta Corporation
(March 25, 1963). Reprinted with permission.

Answers to Problems

Chapter 2

1. (a) 1000

(b) 720.

2.

	0 *Defectives*	1 *Defective*	2 *Defectives*
With replacement	512	384	96
Without replacement	336	336	48

3.

	0 *Defectives*	1 *Defective*	2 *Defectives*
With replacement	0.516	0.387	0.096
Without replacement	0.467	0.467	0.066

4. 1/10.

5. $R_A = 0.9025$; $R_B = 0.9375$; $R_C = 0.896$.

6. Minimum number: (a) 4; (b) 6.

7. (a) $(0.95)^8 = 0.664$

(b) $\sum_{k=6}^{8} \binom{8}{k} (0.95)^k (0.05)^{8-k} = 0.994$

(c) $\left[\sum_{k=3}^{4} \binom{4}{k} (0.95)^k (0.05)^{4-k} \right]^2 = 0.972$.

Chapter 3

1.   *Set* 1   *Set* 2
    (c) $\bar{x} \approx 98.5$ $\bar{x} \approx 94.79$
    (d) $s \approx 29.5$ $s \approx 93.5$.

2. $P[L > S] = 0.005\%$.

3. (a) $\sigma = 312.5$ micromhos

    (b) $P\left[1600 > \dfrac{x_1 + x_2}{2} > 2000\right] = 0.07$

    (c) $P\left[1600 > \dfrac{x_1 + \ldots + x_{10}}{10} > 2000\right] = 0.00006$.

4. $\psi(t) = e^{m(e^t - 1)}$, mean and variance both equal to m.

5. $\theta = \dfrac{1}{n} \sum ti$.

Chapter 5

1. $\hat{\mu} = 763\,\text{hr}$; $\hat{\sigma} = 409\,\text{hr}$
    95 per cent confidence interval $= (639, 887)$.

2. $\hat{\mu} = 1.15$; $\hat{\sigma}(\ln t) = 0.442$
    95 per cent confidence interval $= (0.89, 1.48)$.

3. (a) $\theta = 185\,\text{hr}$
      95 per cent confidence interval $= (129, 278)$
    (b) $\beta \cong 0.8$; $\alpha \cong 65$.

Chapter 6

1. (a) $\theta_s\text{med} = 21.2\,\text{hr}$ (b) $\hat{R}_M(8) = 0.67$
       $\theta_s\text{lower} = 7.2\,\text{hr}$ $\hat{R}_L(8) = 0.34$
       $\theta_s\text{upper} = 54.5\,\text{hr}$ $\hat{R}_U(8) = 0.86$.

2. (a) $\theta_s = 80\,\text{hr}$ (c) $\theta_A = 333\,\text{hr}$
    (b) $\theta_s = 48\,\text{hr}$ $\theta_B = 133\,\text{hr}$
                                $\theta_C = 500\,\text{hr}$.

3. $\theta_{B'} = 2\theta B = 266\,\text{hr}$
    $\theta_{S'} = 68.5\,\text{hr}$.

4. $\theta_A = 29\,\text{hr}$ $\hat{R}_A(8) = 0.932$
    $\theta_B = 154\,\text{hr}$ $\hat{R}_B(8) = 0.951$
    $\theta_C = 475\,\text{hr}$ $\hat{R}_C(8) = 0.983$.

Chapter 7

1. Series arrangement, 0.4; parallel-series, 0.64; series-parallel, 0.72; switched parallel-series, 0.55; switched series-parallel, 0.55; triple redundancy on A, 0.496; triple redundancy on B, 0.70.

2. $P_S = [1-(1-P_aP_b)^3]\,[1-Q_cQ_d]P_E$.

3. (a) 0.90

   (b) 0.91

   (c) 0.9744.

4. R_a/dollar $= 0.9/\$1000$

   R_b/dollar $= 0.91/\$1000$

   R_c/dollar $= 0.9744/\$800$.

Chapter 8

1. (a) When $q_s/q_o = 0.05$ and $q_o \geq 0.143$; use 3 parallel elements.

   (b) When $q_s = 0.1$, $q_o/q_s = 0.5$; use 2 series elements.

2. (a) $q_o' = 0.953$; use 9 parallel elements and 1 series element.

   (b) $q_o = 0.33$; use 1 parallel element and 2 series elements.

3. (a) $R_A = 0.85$; $R_B = 0.72$

   (b) $R_A = 0.31$; $R_B = 0.31$

   (c) $R_A = 0.098$; $R_B = 0.135$.

Chapter 9

1. (a) 170,000 hr at 25°C; 58,000 hr at 85°C.

   (b) $\hat{R}$ (1 yr) $= 0.95$ (at 25°C); $\hat{R}$ (1 yr) $= 0.86$ (at 85°C).

2. (a) $\hat{\theta} = 80,000$ hr
   (b) $\hat{\theta}(lcl) = 69,000$ hr } for normal operation.
   (a) $\hat{\theta} = 53,000$ hr
   (b) $\hat{\theta}(lcl) = 45,700$ hr } for heater voltage $= 6.9$V.

3. $\hat{R}$ (without redundancy) $= 0.921167$
   $\hat{R}$ (with redundancy) $= 0.995186$.

4. Approximately 0.000035 (35 out of 10^6 8-hr flights).

5. The valves may fail open or closed. Mechanical interference, contamination, or freezing due to moisture accumulation can disrupt normal operation in any mode. Failing open, the valve continues to discharge fuel after engine start-up. Failing closed, the valve fails to provide proper cool-down flow prior to start-up and fails to bleed pump during start-up. Failing open, system failure modes are 1, 2, or 4; failing closed, system failure modes are 2 or 4 (see Table 9.16). Failing open, the system failure mechanism is pump cavitation, especially if the interstage

valve is affected. Operation at a high mixture ratio will result in tube wall burnout. In any case, fuel loss through bleed vents will result. Failing closed, the system failure mechanism is severe pump cavitation resulting in pump destruction of cool-down, and bleed is not accomplished. Partial fuel pump cavitation will result in engine operation at a high mixture ratio which may damage tube walls.

6. (a) Equation I: $\hat{\theta}$ = 1550 hr; lcl = 710 hr; ucl = 3385 hr.

(b) Equation II: $\hat{\theta}$ = 1662 hr; lcl = 752 hr; ucl = 3673 hr.

(c) Equation III: $\hat{\theta}$ = 1268 hr; lcl = 342 hr; ucl = 4708 hr.

(d) Equation IV: $\hat{\theta}$ = 1152 hr; lcl = 284 hr; ucl = 4682 hr.

Chapter 10

1. Median = 2 hr.   95% confidence limits = 1.66 hr. and 2.35 hr.

2. (c) MSI = 124.8

4. 29

5. (a) 186 minutes

(b) 11,471 minutes

(c) 57.4 minutes

(d) 228 minutes

Chapter 11

1. (a) Median repair time = 1.48 hr
   Mean repair time = 2.38 hr
   P_{Ai} = 0.95

(b) P_A = 0.909

(c) P_{OR} = 0.912

(d) P_R = 0.884

(e) P_{SE} = 0.806.

2. (a) 0.6885

(b) 0.6059

(c) 0.5756

Chapter 12

1. (a) 7.3V max.; 4.9V min.

(b) 6.3V max.; 5.6V min.

3. (a) $q'_{fo} = (q_{fo})^2 (1-q_{cs})$

(b) $q'_{fs} = q_{cs}[2q_{fs} - (q_{fs})^2]$

where

q_{fo} = probability that a fuse will blow spuriously
q_{cs} = probability that the load will short
q_{fs} = probability that a fuse will fail to blow on an overcurrent.

Problem assumes resistor has a reliability of unity. Note that total probability of failure is the sum of (a) and (b).

Chapter 13

3. (a) $n = 50, c = 4$

  (b) $n = 300, c = 20$

  (c) $n = 300, c = 20$

  (d) $n = 300, c = 20$

  (e) 16 per cent for (a); 9 per cent for (b), (c) and (d).

5. $N > 200$.

Chapter 14

1. $c = 5$. (Rejection number is 6.)

2. $n = 532,300$.

4. (a) $N+(a) = 145 + 54.6\,r$ (c) $N+(a) = 108.6 + 54.6\,r$
    $N+(r) = -145 + 54.6\,r$ $N+(r) = -111.5 + 54.6\,r$
    $E_{\theta_0}(WT) = 143.5$ hr $E_{\theta_0}(WT) = 134$ hr
    $E_{\theta_1}(WT) = 99.8$ hr $E_{\theta_1}(WT) = 80.5$ hr.

  (b) $N+(a) = 110.9 + 54.6\,r$
    $N+(r) = -144.4 + 54.6\,r$
    $E_{\theta_0}(WT) = 108$ hr
    $E_{\theta_1}(WT) = 89.3$ hr

Chapter 15

1. (a) \$11.30   (e) \$14.20

  (b) \$12.60   (f) \$14.20

  (c) \$11.30   (g) \$16.40.

  (d) \$12.80

2. (a) Monthly cost of flight-line manpower   \$363.20

  (b) Monthly cost of black box replacement   \$154.20

  (c) Monthly cost of flight-line maintenance   \$629.00

  (d) Monthly cost of shop maintenance   \$855.36

  (e) Monthly cost of total black box maintenance \$1,484.36.

3. (a) \$1,561.46

  (b) \$1,686.11.

4. (a) Monthly cost of flight-line manpower   \$340.00

  (b) Final answer   \$1,461.16.

Index of Authors

Index of Subjects

Active element group (AEG), 187–89
Addition rule, 202
Additive rule, 202
Adjustment factors in prediction:
　for stress, 311, 318, 320, 327–28
　for tolerances and use conditions, 292, 293
Advisory Group on Reliability of Electronic Equipment (AGREE), 192, 522, 524–27
　allocation method, 192–95
　environmental tests, 524–27
American Institute for Research, 375–77
Application problems, 436
AQL, 471
ARINC Research Corporation, 157, 189, 195, 242, 300, 309–11, 378, 380, 433, 541
Assumptions:
　of constant failure rate, 284
　validity of, 168
Autonetics, 309–10
Availability (operational readiness), 7, 398, 401–3, 406–7

Bayes' Theorem, 53
Bertrand's "Box Paradox," 53
Biological growths, 451–53
Block:
　definition of, 198
　diagram, 200, 245–47, 279–82, 286–90, 325

Block (*Cont.*):
　failure rates, 292–93
Boeing Company, 309–10
BTL, 309–10

Censored observation, 143, 149
Chi-square goodness-of-fit test, 170–71
Collins Radio, 309–10
Combinations, 23–24, 40–45
Component and part types:
　alternators, 309
　amplifier, 192
　attenuators, 312
　batteries, 309
　capacitors, 188, 275, 309, 310, 438, 486
　chokes, 188
　coils, 309
　connectors, 309, 312
　crystals, 309, 312
　diodes, 188, 192, 216, 300, 309, 313, 439, 440
　filters, 314
　fuses, 309
　generators, 309, 315
　gyros, 187, 309
　heaters, 309, 314
　inductors, 188, 309, 314
　jacks, 188
　klystron, 313
　lamps, 309
　magnetron, 121, 124, 125, 129, 311